34

University of Lowell
Alumni/Lydon Library

AF797804

FIRE AND FLAME RETARDANT POLYMERS

FIRE AND FLAME RETARDANT POLYMERS

Recent Developments

Albert Yehaskel

NOYES DATA CORPORATION
Park Ridge, New Jersey, U.S.A.
1979

TH 1074.5 .Y43
Yehaskel, Albert.
Fire and flame retardant
polymers :

Copyright © 1979 by Noyes Data Corporation
No part of this book may be reproduced in any form
without permission in writing from the Publisher.
Library of Congress Catalog Card Number: 78-70742
ISBN: 0-8155-0733-X
Printed in the United States

Published in the United States of America by
Noyes Data Corporation
Noyes Building, Park Ridge, New Jersey 07656

FOREWORD

The detailed, descriptive information in this book is based on U.S. patents, issued since July 1976, that deal with fire and flame retardant polymers.

This book serves a double purpose in that it supplies detailed technical information and can be used as a guide to the U.S. patent literature in this field. By indicating all the information that is significant, and eliminating legal jargon and juristic phraseology, this book presents an advanced, technically oriented review of fire and flame retardant polymers, their manufacture and use.

The U.S. patent literature is the largest and most comprehensive collection of technical information in the world. There is more practical, commercial, timely process information assembled here than is available from any other source. The technical information obtained from a patent is extremely reliable and comprehensive; sufficient information must be included to avoid rejection for "insufficient disclosure." These patents include practically all of those issued on the subject in the United States during the period under review; there has been no bias in the selection of patents for inclusion.

The patent literature covers a substantial amount of information not available in the journal literature. The patent literature is a prime source of basic commercially useful information. This information is overlooked by those who rely primarily on the periodical journal literature. It is realized that there is a lag between a patent application on a new process development and the granting of a patent, but it is felt that this may roughly parallel or even anticipate the lag in putting that development into commercial practice.

Many of these patents are being utilized commercially. Whether used or not, they offer opportunities for technological transfer. Also, a major purpose of this book is to describe the number of technical possibilities available, which may open up profitable areas of research and development. The information contained in this book will allow you to establish a sound background before launching into research in this field.

Advanced composition and production methods developed by Noyes Data are employed to bring these durably bound books to you in a minimum of time. Special techniques are used to close the gap between "manuscript" and "completed book." Industrial technology is progressing so rapidly that time-honored, conventional typesetting, binding and shipping methods are no longer suitable. We have by-passed the delays in the conventional book publishing cycle and provide the user with an effective and convenient means of reviewing up-to-date information in depth.

The Table of Contents is organized in such a way as to serve as a subject index. Other indexes by company, inventor and patent number help in providing easy access to the information contained in this book.

15 Reasons Why the U.S. Patent Office Literature Is Important to You –

1. The U.S. patent literature is the largest and most comprehensive collection of technical information in the world. There is more practical commercial process information assembled here than is available from any other source.

2. The technical information obtained from the patent literature is extremely comprehensive; sufficient information must be included to avoid rejection for "insufficient disclosure."

3. The patent literature is a prime source of basic commercially utilizable information. This information is overlooked by those who rely primarily on the periodical journal literature.

4. An important feature of the patent literature is that it can serve to avoid duplication of research and development.

5. Patents, unlike periodical literature, are bound by definition to contain new information, data and ideas.

6. It can serve as a source of new ideas in a different but related field, and may be outside the patent protection offered the original invention.

7. Since claims are narrowly defined, much valuable information is included that may be outside the legal protection afforded by the claims.

8. Patents discuss the difficulties associated with previous research, development or production techniques, and offer a specific method of overcoming problems. This gives clues to current process information that has not been published in periodicals or books.

9. Can aid in process design by providing a selection of alternate techniques. A powerful research and engineering tool.

10. Obtain licenses — many U.S. chemical patents have not been developed commercially.

11. Patents provide an excellent starting point for the next investigator.

12. Frequently, innovations derived from research are first disclosed in the patent literature, prior to coverage in the periodical literature.

13. Patents offer a most valuable method of keeping abreast of latest technologies, serving an individual's own "current awareness" program.

14. Copies of U.S. patents are easily obtained from the U.S. Patent Office at 50¢ a copy.

15. It is a creative source of ideas for those with imagination.

CONTENTS AND SUBJECT INDEX

POLYURETHANES

PHOSPHORUS-CONTAINING ADDITIVES

Triesters of Pentavalent Phosphorus

J.A. Albright and T.C. Wilkinson; U.S. Patent 4,083,825; April 11, 1978; assigned to Velsicol Chemical Corporation describe triesters of pentavalent phosphorus acid having the formula:

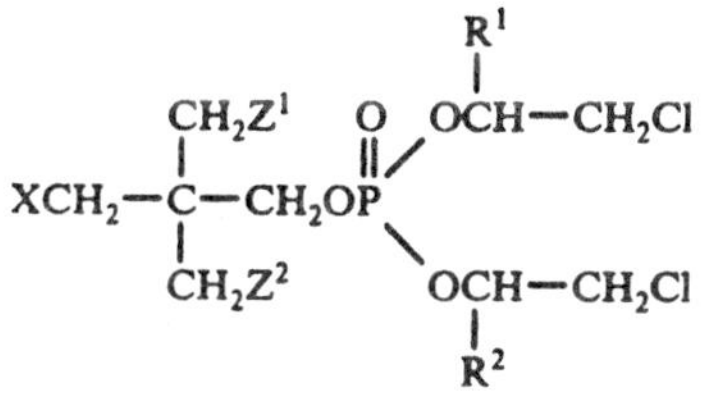

where X is chlorine, bromine, or hydrogen, Z^1 and Z^2 are independently selected from bromine, chlorine, or hydrogen; R^1 is methyl, chloromethyl, bromomethyl or hydrogen; and R^2 is methyl, chloromethyl or hydrogen, except that R^2 is chloromethyl when R^1 is bromomethyl. In addition, a nonscorching plastic composition containing polyurethanes and the triesters of pentavalent phosphorus acid is described.

The polyurethanes used include rigid foams, semirigid foams, flexible foams, rubbers and adhesives. The following examples illustrate the process.

Example 1: To a suspension containing 200 ml of methylene chloride and 145.6 grams of neopentyl glycol was added, dropwise, 192.4 grams of phosphorus trichloride while the reaction mixture was maintained at 10°C. After addition of the phosphorus trichloride was completed the reaction mixture was heated to 35°C and maintained at this temperature for one-half hour. Methylene chloride was then removed from the reaction mixture under vacuum. The temperature of the reaction mixture was then lowered to below 20°C and chlorine gas was introduced until the reaction mixture was yellow in color. The excess chlorine was then removed and a titanium tetrachloride catalyst was introduced. The reaction mixture was then heated to 80°C and 259 g of epichlorohydrin was added dropwise. The reaction mixture was maintained at 80°C for 1 hour after addition of epichlorohydrin was complete. The resulting product was washed with aqueous ammonia followed by two washings with deionized water. The product was dried under vacuum and filtered hot to yield 542 g of a clear liquid which was substantially bis(dichloropropyl) 2,2-dimethyl-3-chloropropyl phosphate.

Example 2: To a suspension containing 62.4 g of neopentyl glycol in 200 ml of methylene chloride was added 82.5 g of phosphorus trichloride. The reaction mixture was kept below 10°C during the addition of the phosphorus trichloride. The reaction mixture was then heated to 40°C and kept at this temperature for one-half hour. The mixture was then brought to below 20°C and 96 g of bromine was added while cooling the mixture to keep the temperature below 20°C. The methylene chloride solvent was then removed under vacuum and 0.8 g of titanium chloride catalyst was introduced and the mixture was heated to 60°C.

One hundred eleven grams of epichlorohydrin was added to the mixture. This resulted in an exothermic reaction which raised the temperature to between 80° and 90°C. The reaction mixture was held at 80°C for one-half hour after the addition of the epichlorohydrin. The reaction mixture was then cooled and the product was washed with aqueous ammonia followed by two washings with deionized water. Thereafter the material was dried and filtered. The resultant yellow liquid weighed 286 g and was substantially 3-bromo-2,2-dimethylpropyl 1-bromo-3-chloro-2-propyl 1,3-dichloro-2-propyl phosphate.

Example 3: To a solution of 153.5 g of phosphoryl chloride and 2.0 g of titanium tetrachloride, heated to 50°C, was added 325 g of tribromoneopentyl alcohol. The resulting reaction mixture was heated for 4 hours at 80°C. One hundred eighty-five grams of epichlorohydrin was added to the reaction mixture and the resulting reaction system was kept at 90°C for 2 hours. At the end of this period the product was washed with deionized water, followed by an aqueous ammonia wash, and again washed with deionized water. The washing was followed by drying and filtering. The end product was 430 g of substantially bis-(1,3-dichloro-2-propyl) tribromoneopentyl phosphate.

Example 4: One hundred seventy-three grams of dichloroneopentyl glycol was added, at room temperature, to 137 grams of phosphorus trichloride. The resultant reaction mixture was heated to 90°C and maintained there for 4 hours. The mixture was then cooled and 150 ml of methylene chloride was added. The resulting solution was kept below 20°C and chlorine gas was added below the surface of the liquid. The methylene chloride was then removed under vacuum.

Titanium tetrachloride (0.1 g) was introduced into the reaction mixture and the mixture was heated to 80°C for 4 hours. The resulting product was washed with deionized water, aqueous ammonia, and thereafter dried and filtered. The product was substantially bis-(1-chloro-2-propyl) trichloroneopentyl phosphate, a medium viscosity liquid weighing 226 g.

Example 5: Four hundred twelve grams of phosphorus trichloride was cooled in ice to 10°C and 786 g of dibromoneopentyl glycol was added. An endothermic reaction developed with the evolution of hydrogen chloride. After 2 hours the ice bath was removed and the product was heated to 50°C for 1 hour, then aspirated for 1 hour at this temperature. The product was then cooled and 500 ml of methylene chloride was added. One hundred ninety-seven grams of chlorine was then added, keeping the resultant exothermic reaction below 30°C. Upon complete addition of chlorine, the methylene chloride was removed under vacuum and a viscous clear liquid resulted. This intermediate was then used in Example 6 to form bis-(2-chloroethyl) 2,2-bis-bromomethyl-3-chloropropyl phosphate.

Example 6: To 397 g of the viscous clear liquid of Example 5 was added 3 g of titanium tetrachloride. This mixture was heated to 65°C and 88 g of ethylene oxide was added. The resulting reaction was exothermic and the addition took 2 hours. The reaction mixture was aspirated for one-half hour at 60°C and washed with water followed by washing with aqueous ammonia and two further water washes.

Celite brand of diatomaceous earth filter aid and Celkate brand of magnesium silicate filter aid were added and the product was dried under vacuum. After filtering 428 g of a colorless liquid resulted. This liquid was substantially bis-(2-chloroethyl) 2,2-bis-bromomethyl-3-chloropropyl phosphate.

The compounds of the above general formula can be incorporated with the polyurethanes at any processing stage in order to prepare the compositons. In general, this is undertaken prior to fabrication either by physical blending or during the process of forming polyurethanes per se.

Example 7: Separate flexible polyurethane foam compositions (designated 1 through 9 in the following table) were prepared via the following procedure: Formulation No. 1 was the control and did not contain a phosphate compound.

Approximately 500 g polyol (Voranol 3010, a glycerine based propylene oxide polyol of about 3,000 MW and a hydroxyl number of about 46 to 47), 5.0 g surfactant [L 5720 (Union Carbide), a silicone based material) 35.0 g Freon fluorocarbon blowing agent and 45 g of each flame retardant compound indicated in the table were mixed together.

293.0 g of toluene diisocyanate was added to the polyol mix and allowed to settle to the bottom. 2.25 g of amine catalyst Niax A-1 was mixed with 15.0 g of water and the resultant mixture was added to the polyol mix and allowed to float on a portion of the surface. 1.5 g of stannous octoate was added to the polyol mix and the mixture was stirred for 6 seconds at 1,300 rpm.

The mixture was quickly discharged into a gallon paper container. The foam was then allowed to rise. After the foam had completely risen, it was allowed to set and was then postcured at 121°F for 30 minutes.

Portions of the samples of each formulation (No. 1 through 9 in the table) prepared according to the above described procedure were then subjected to a standard flammability test and to a color determination test. The results of these tests are shown in the table on the following page. The scorch and flammability of the flexible urethane foams were measured by the Gardner Colorimeter and the California Vertical Burn Test.

Formulation No.	Compound	Gardner Colorimeter Test				California Vertical Burn Test: Maximum Char Length			California Vertical Burn Test: Average Char Length		
		L	a_L	b_L	ΔE	Initial	Aged 2 wks 150° F.	Aged 2 wks 250° F.	Initial	Aged 2 wks 150° F.	Aged 2 wks 250° F.
1	Control	75.2	−0.5	+0.3	—	12.0	12.0	12.0	12.0	12.0	12.0
Process											
2	Example No. 6	72.3	−1.3	+2.1	3.5	5.8	5.4	4.5	5.0	5.4	3.6
3	Example No. 1	69.1	−0.7	+0.4	6.1	8.0	6.9	8.1	6.8	6.4	7.1
4	Example No. 3	71.2	−0.3	−0.5	4.2	6.4	5.9	5.5	5.7	5.3	5.0
5	Example No. 4	71.1	−0.3	+0.0	4.1	7.8	8.9	8.4	6.6	7.7	6.9
6	Example No. 2	71.1	−1.0	+2.8	4.9	5.7	5.8	4.3	5.3	5.0	3.7
Comparative											
7	Bis-(bromopropyl) chloroethyl phosphate	64.4	−0.5	+0.3	10.9	5.4	5.9	—	4.7	5.4	—
8	3-bromo-2,2-dimethylpropyl 1-bromo-2-propyl 1-chloro-2-propyl phosphate	63.3	−0.5	+0.5	11.9	6.5	6.7	5.0	5.5	5.8	4.4
9	Tris-(chloropropyl) phosphate	69.9	−0.3	−0.4	5.4	6.8	7.2	12.0	6.2	6.6	12.0

As is evident from the table, the polyurethane compositions of this process, i.e., Formulations 2 through 6 in the table, are both flame retardant and nondiscolored. The control Formulation 1 in the table, while not discolored, was not flame retardant. The tris(chloropropyl) phosphate containing polyurethane composition (Formulation 9 in the table), while having a relatively low discoloration was not fire retardant after aging for 2 weeks at 250°F and gave the same results on the Burn Test as the control.

The bis-(bromopropyl) chloroethyl phosphate (Formulation 7 in the table) and 3-bromo-2,2-dimethylpropyl 1-bromo-2-propyl 1-chloro-2-propyl phosphate (Formulation 8 in the table) polyurethane compositions, while being flame retardant, were relatively highly discolored. Thus all three of these polyurethane compositions (No. 7 through 9 in the table) failed to meet the criteria necessary for flame retardant nonscorching polyurethane compositions.

The term scorch, includes discoloration and/or loss of physical properties of the polyurethane composition. This phenomenon is generally observed in the center of the polyurethane composition.

Vinyl Phosphonate Esters

K.S. Shim; U.S. Patent 3,991,019; November 9, 1976; assigned to Stauffer Chemical Company describes a process for forming a flame retardant article which comprises the addition to a normally flammable substrate of an effective amount of at least one compound having the formula:

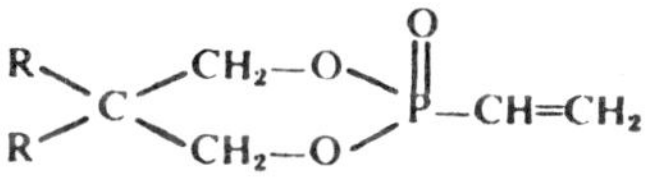

where R is selected from the group consisting of the straight and branched C_1-C_{12} alkyl groups, the phenyl group, and bromo and chloro substituted C_1-C_{12} alkyl groups, and the bromo and chloro substituted phenyl groups, to confer flame retardancy on the substrate. These vinyl phosphonate esters are primarily intended for use as flame retardants, particularly for polyurethane foams, thermoplastic and thermoset resins and cellulosic fibers.

Example 1: The compound shown below is prepared as follows.

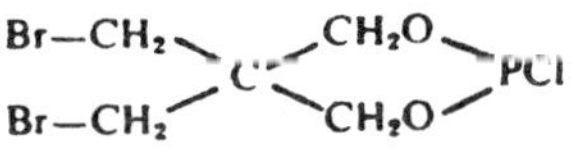

In a 2-liter flask containing 458 g (1.75 mols) of dibromoneopentyl glycol and dry benzene (1,000 ml) was added dropwise phosphorus trichloride, 460 g (3.34 mols), at 60°C. Upon completion of addition, solvent and excess phosphorus trichloride was stripped off to yield 540 g (95%) of clear liquid product, boiling point: 120°C at 0.05 mm Hg, analysis: P = 9.4% (theory 9.5%); Br = 53.8% (theory 49.2%); Cl = 9.9% (theory 10.8%).

Example 2: Formation of Dibromoneopentyl Glycol Chloroethyl Phosphite – In a 500 ml flask was placed the compound from Example 1 (133 g, 0.41 mol) and dry hexane (150 ml). While stirring vigorously ethylene oxide was added through a sintered glass gas dispersing tube. The reaction temperature was maintained at 40°C using an ice bath since it was quite exothermic. Ethylene oxide was added until the reaction temperature began to drop, yield = 97%.

Example 3: Formation of the Chloroethyl Phosphonate Ester of Dibromoneopentyl Glycol – In a 250 ml flask was placed 200 g of the compound from Example 2 and 0.1 g iodomethane. The resulting mixture was heated to 192°C and held at that temperature for 30 minutes to give 198 g of clear liquid, analysis: P = 9.6% (theory 8.9%); Br = 37.9% (theory 43.3%); Cl = 8.6% (theory 9.6%).

Example 4: Formation of the Vinyl Phosphonate Ester of Dibromoneopentyl Glycol – Into a 250 ml flask was placed the compound from Example 3 (92 g, 0.25 mol), o-dichlorobenzene (110 ml) and sodium acetate (30 g, 0.35 mol). The resulting mixture was heated to 110° to 115°C and kept at that temperature over a period of 5 hours, after which the insoluble salts were separated through a filter and the organic layer was washed with water, dried over magnesium sulfate and stripped of solvent to give a yellowish oil, analysis: P = 8.4% (theory 8.6%); Br = 42.8% (theory 44.6%).

Example 5: The product from Example 4 was used in formulating a polyurethane foam utilizing the following reagents:

Reagent	Amount, g
Polyol (Voronol 3000)	100
Example 4 product	15
Silicone L-520 surfactant	1.5
Water	3.6
Amine catalyst (A-1)	0.12
Dabco 33 LV	0.35
Blowing agent (Cl_3CF)	3.0
DMF	3.0
T-10 (stannous octoate)	0.60
Toluene diisocyanate	45.8

The rise time of the foam was 120 seconds at ambient temperature. The foam that was produced had a density of 2.03 lb/ft and had an average air flow of 4.3 ft^3/min.

When three samples of the foam were tested for flame retardancy using ASTM-1692 the samples self-extinguished between 80 and 95 seconds after 60 to 70 millimeters of foam had burned. The average burn rate was 0.72 mm/sec for all three samples.

A number of compounds having the following structural formula:

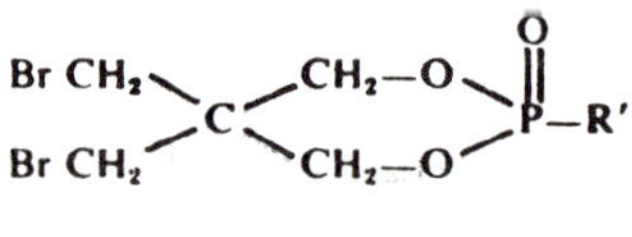

where R' is defined in the table on the following page were incorporated in the polyurethane foam formulation set forth above and were tested for flame retardancy. The table sets forth the burn rate:

R′	Average Burn Rate (mm/sec)
$OCH_2CH{=}CH_2$	1.30
$N(CH_3)_2$	0.73
OCH_2CH_2Cl	0.70

Substituted Phosphate Esters

W. Batorewicz; U.S. Patent 4,054,543; October 18, 1977; assigned to Uniroyal, Inc. describes a flame retardant composition comprising a polyurethane polymer and a flame retarding effective amount of phosphate ester of the formula

$$RO[(A)_m(B)_n]H$$

integrally part of the polymer. In the formula A has the structure

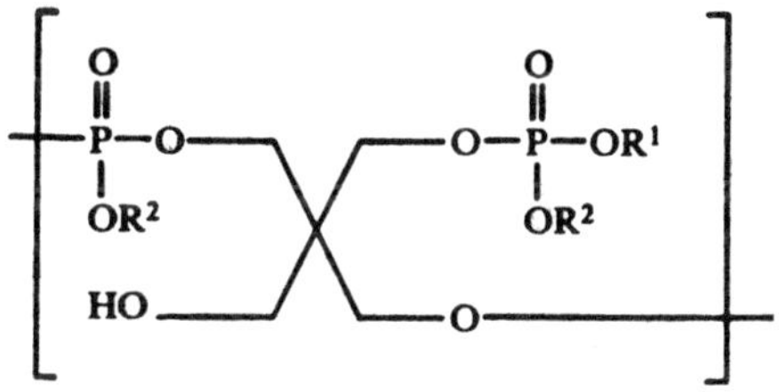

and B has the structure

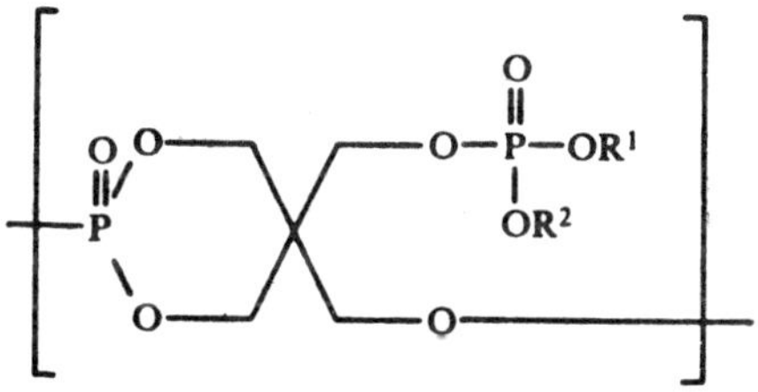

where R, R^1 and R^2 are the same or different and are linear or branched primary radicals selected from alkyl having 1 to 6 carbon atoms, and chloroalkyl, bromoalkyl and hydroxyalkyl having 2 to 6 carbon atoms; m and n are integers from 1 to about 3 and the sum of m plus n is from 2 to about 4.

The flame retardants are useful for rigid polyurethane foams. Polycyclic phosphorohalidites are used for the preparation of the phosphate esters. These phosphorohalidites precursors have the general formula $(C_5H_8O_4P_2)X_uY_v$, where the substituents X and Y may either be halogen or alkoxy radicals.

These precursors may also be represented by the structural formulas on the following page.

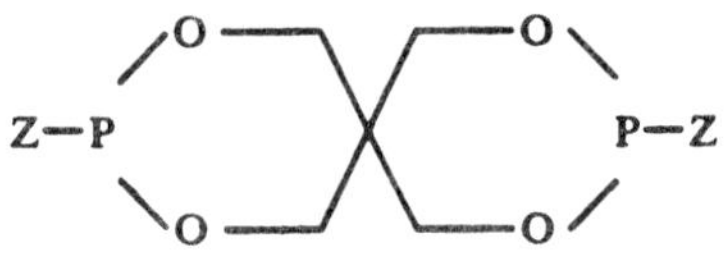

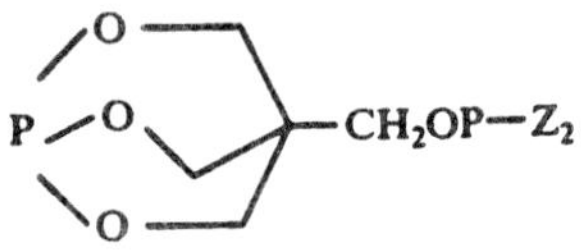

These are isomers and Z has the meanings of X or Y as defined above. Polycyclic phosphorohalidites may be prepared from phosphorus trihalides such as PCl_3 or PBr_3 and pentaerythritol according to the method described by H.J. Lucas et al, *JACS,* Vol. 72, p. 491-497 (1950).

Example 1: 3,9-Dichloro-2,4,8,10-Tetraoxa-3,9-Diphosphaspiro[5.5] Undecane – This polycyclic phosphorochloridite precursor is prepared in the following manner. To a 2-liter, 3-necked round bottom flask equipped with a condenser, a stirrer, a thermometer, a dropping funnel and a nitrogen inlet tube, is added PCl_3 (280.0 g, 2.04 mols) dropwise to a stirred suspension of pentaerythritol (136.0, 1.0 mol) at 35° to 50°C by means of a water bath.

During the addition the system is kept under a gentle nitrogen sweep to facilitate the removal of the HCl formed. When the PCl_3 addition is completed, the temperature of the reaction mixture is slowly raised to reflux over about a one-hour period. The mixture is then kept under reflux, using a vigorous nitrogen sweep, until a homogeneous solution is obtained, i.e., about one to two hours. Trace amounts of white precipitate may separate from the reaction solution upon cooling to room temperature. This small amount of precipitate, consisting mostly of pentaerythritol, is removed by filtration.

The filtrate contains the desired product, 3,9-dichloro-2,4,8,10-tetraoxa-3,9-diphosphaspiro[5.5] undecane, and is essentially free of by-products. The phosphorochloridite precursor thus obtained is used in the subsequent reactions without further purification.

Example 2: Methyl Phosphate Ester – This example represents the general procedure for the preparation of the compounds of this process by oxidation of a phosphorochloridite with chlorine in presence of an alcohol.

Methanol (240 g, 7.5 mols) is stirred and cooled to 5° to 10°C by means of an acetone-ice bath in a reaction vessel fitted with a thermometer, an addition funnel, and a gas inlet tube connected to a chlorine gas cylinder. Chlorine (2 mols) and the phosphorochloridite (1 mol) of Example 1 are added simultaneously to the methanol maintaining a slight chlorine excess, as evidenced by the greenish color of the reaction medium.

During the addition, the reaction mixture is maintained at about 25° to 35°C by means of the acetone-ice bath as well as by controlling the rates of addition of reactants. When the addition of phosphorochloridite is completed, the resulting greenish solution is distilled under 10 to 20 mm Hg pressure at 30° to 50°C in order to remove most of the HCl formed during the reaction. The residual acidic solution is then treated with propylene oxide at 40° to 60°C until neutral to moist litmus paper. Low boiling materials are then further removed from the reaction solution by distillation under 10 to 20 mm Hg and subsequently 0.5 mm Hg at about 100°C. The residue is the methyl phosphate ester, a clear fluid oily material. Analysis: P, 12.5%; Cl, 0.8%; Hydroxyl No. 203. In Examples 3 through 5 inclusive, the spirocyclic compound described in and prepared essentially according to Example 1 is used exclusively.

Example 3: Methyl Phosphate Ester – The procedure of Example 2 is repeated except that 15 mols of methanol are placed in the reaction vessel, the overall methanol to phosphorochloridite mol ratio being 15. Analysis: P, 16.4%; Cl, 0.5%; Hydroxyl No. 287.

Example 4: Ethyl Phosphate Ester – The procedure of Example 2 is repeated except that 15 mols of ethanol are used in place of the methanol, the mol ratio of ethanol to phosphorochloridite being 15. The resultant product is a clear fluid oil. Analysis: P, 14.7%; Cl, 0.4%; Hydroxyl No. 190.

Example 5: Ethyl Phosphate Ester – The procedure of Example 4 is repeated except that 8 mols of ethanol are used, the overall molar ratio of alcohol to phosphorochloridite being 8. The product is a clear fluid oil. Analysis: P, 14.6%; Cl, 0.7%; Hydroxyl No. 130.

Example 6: n-Butyl Phosphate Ester – The procedure of Example 2 is repeated except that methanol is replaced by n-butanol, and the molar ratio of n-butanol to the phosphorochloridite employed is 7.0. The product is a white waxy solid. Analysis: P, 13.6%; Hydroxyl No. 105.

Example 7: 2-Chloroethyl Phosphate Ester – The procedure of Example 2 is repeated except that methanol is replaced by 2-chloroethanol, and the ratio of 2-chloroethanol to the phosphorochloridite employed is 4.5. The product obtained is a clear oil. Analysis: P, 13.8%; Cl, 19.8%; Hydroxyl No. 91.

Example 8: Rigid Polyurethane Foam – This example illustrates the utility as well as the efficacy of the compounds of this process in rigid polyurethane foam. The foams are prepared using a conventional one-shot process employing the recipes listed in Table 1 (Formulation). For this purpose, all the ingredients except the polyisocyanate are thoroughly mixed in a beaker and thereafter the polyisocyanate is added while rapidly agitating the mixture with an air-driven stirrer. The liquid contents of the beaker are then poured into a mold where within a short period of time a foam develops. The foam is cured overnight at room temperature, and the flame resistance is determined by the Oxygen Index test (ASTM D-2863).

Experiment A does not include a flame retardant, and Experiment E contains a well-known flame retardant used commercially in rigid polyurethane foams. Experiments A and E are outside the scope of this process. In Table 1 the flame resistance data of the foams are summarized.

The results indicate that in all examples where the compounds of the process are employed as the flame retardant, the oxygen indices achieved are higher than those of the blank (A) and of the comparison experiment (E). The improvement amounts to over 30% (Experiments C and D) when compared to the commercial compound at substantially the same concentration. Even when only one-half of the compound of this process is incorporated in the polyurethane an improvement is achieved (B) over the comparison foam (E).

TABLE 1: COMPARISON OF RESULTS OF EXAMPLE 8

Experiment	A	B	C	D	E
Formulation parts by weight					
Polyol(1)	100	87.6	75	74.2	74.6
Surfactant(2)	2	2	2	2	2
Triethylenediamine(3)	1	—	—	1	—
Dimethylethanolamine(3)	—	2	2	—	2
Trichlorofluoromethane	30	30	30	45	30
methylenebis(phenyl isocyanate) (crude) (5)	136	121	127	115	126.5
Phosphate of Example 5	—	12.4	25	—	—
Phosphate of Example 7	—	—	—	25.8	—
Commercial Flame Retardant(6)	—	—	—	—	25.4
Results					
Oxygen Index	20.6	23.6	24.5	24.9	23.5
% Improvement(8)	—	3.4	34.	48.	—

(1) Propylene oxide adduct of N-aminoethylpiperazine; molecular weight ~350; Hydroxyl No. ~530.
(2) Silicone compound, DC-193 of Dow Corning Corp.
(3) Catalyst
(4) Blowing agent
(5) Product of Rubicon Chemicals, Inc.
(6) O-O-diethyl N,N-bis(hydroxyethyl)aminomethyl phosphonate.
(7) ASTM D-2863
(8) Calculated according to equation

$$\% \text{ Improvement} = [OI_I - OI_E/OI_E - OI_0] \times 100$$

wherein

OI_I = Oxygen index of composition of process
OI_E = Oxygen index of Example E (comparison)
OI_0 = Oxygen index of composition without flame retardant (A)

Salts of Benzenephosphonic Acid and Melamine

A process described by *E. Simon; U.S. Patent 4,061,605; December 6, 1977* relates to the identification of salts of benzenephosphonic acid and melamine that are useful as additives for reducing the flame propagation of normally flammable organo-polymers.

The physical and chemical properties of the benzenephosphonic acid-melamine addition products may be modified by varying the mol ratios of the reactants. The salts of benzenephosphonic acid and melamine may be functionally inert or reactive when incorporated as additives, depending on their composition and the polymer system involved. Thus, applied to polyurethane foams, it is desirable that the additive be characterized as inert during the foaming reaction, whereas applied to monomeric-type epoxy systems, they may advantageously have the dual function of entering into the oxirane curing mechanism as well as imparting flame retardency; in either case, the resultant products are markedly improved in their resistance to flame initiation and flame propagation.

The fire retardants comprise aqueous reaction products between benzenephosphonic acid and melamine within the mol % range of 33⅓ to 40 mol % of benzenephosphonic acid.

The fire retardants are conceptualized as addition products formed between the amino group(s) of melamine and the acidic hydroxyl group(s) of benzenephosphonic acid.

The procedure used in the preparation of the addition products include the steps of predissolving the benzenephosphonic acid in hot (up to boiling) distilled water, adding the melamine powder slowly with continuous agitation to the hot solution, maintaining heating and stirring for approximately 15 to 30 minutes, cooling, filtering and drying the precipitate to constant weight at approximately 120°C. Moisture removal may be accomplished at ambient or reduced pressures or by spray drying. Recrystallization is readily accompanied, if desired, by dissolving the compound in a sufficient volume of hot (preferably boiling) distilled water, cooling, filtering, removing the bulk water from the crystals, and drying to constant weight at approximately 120°C.

Example 1: Benzenephosphonic Acid-Melamine Addition Product Salts as Flame Retardants for Polyurethane Foamed Plastics — The polyurethane foam used was CPR's (division of the Upjohn Company) Isocyanate 2018, a two-component flexible polyurethane foam having a free rise density of approximately 3 pcf. The benzenephosphonic acid-melamine salt additives were used in mol ratios of 2:3, 1:1, 1:2, and 3:2 respectively. The weight percent of the salt additives used was up to 25 wt % of the total, or approximately 5.3 to 33.3 parts of additive per 100 parts of foam mix.

Flame retardancy was evidenced for all of the cases noted, using a Fisher type of burner as the igniting source, with best results for the mol ratios of benzenephosphonic acid to melamine of 2:3 and 1:2 respectively at a preferred weight percentage loading of at least 10%. These results are in marked contrast to the behavior of the cured foam without additive, which when tested in the same manner showed rapid flaming and vigorous flame propagation.

For the mol ratio of benzenephosphonic acid to melamine greater than 50%, the acidity of the salt, used as a flame retardant additive in polyurethane foaming systems, would have to be compensated by the addition of alkaline neutralizing agents, such as a tertiary amine.

The applicability of the benzenephosphonic acid-melamine addition products as flame retardants was further demonstrated by the relative ineffectiveness of a 25 wt % loading of equal parts by weight of borax, sodium metaphosphate, sodium metasilicate, and aluminum sulfate (either to inhibit flame initiation or flame propagation) when incorporated in the flexible polyurethane foam system.

Example 2: Stability to Moisture Leaching of Benzenephosphonic Acid-Melamine Salt — A benzenephosphonic acid-melamine having a mol ratio of 2:3 respectively was used as an additive to CPR's foaming system of Example 1, at 25 wt % of the total weight. A sample of the cured foam, with all faces freshly cut, weighing 2.25 g was totally immersed in 60 ml of distilled water for 26.5 hours at ambient temperature. After drying to constant weight at 250°F, the foam showed no impairment in resisting flame initiation or flame propagation when soaked in the flame of a Fisher-type burner.

Example 3: Test of Benzenephosphonic Acid and Melamine as Neat Additives to a Polyurethane Foaming System – Benzenephosphonic acid and melamine were added without prior reaction to the polyurethane foam of Example 1, in a mol ratio of 2:3 respectively and a combined weight percentage of 25 of the total weight of the components. This resulted in almost complete collapse of the foam, inadequately cured, in contrast to results in Examples 1 and 2 in which there was no adverse effect on the foaming reaction for the same weight percentage and mol ratio of the additive when prereacted.

Example 4: Benzenephosphonic Acid-Melamine Addition Product Salt as Flame Retardant for Rigid Polyurethane Foamed Plastics – This is an extension of Example 1, in which a rigid polyurethane foam is substituted for the flexible foam. At 25 wt % (of the total), the salt of 2 mols benzenephosphonic acid to 3 mols melamine changed the cured foam (CPR's Isocyanate 380) from one easily ignited with vigorous flame propagation, to one that was self-quenching.

HALOGEN-CONTAINING ADDITIVES

Diesters of Tetrabromophthalic Anhydride

J.P. Baldino and J. Feltzin; U.S. Patent 3,989,653; November 2, 1976; assigned to ICI United States Inc. describe diesters comprising the reaction product of tetrabromophthalic anhydride and a polyhydroxy compound containing at least 3 hydroxyl groups. The diesters are prepared, in one step, by reacting tetrabromophthalic anhydride with the polyhydroxy compound at a temperature of from about 130° to 190°C in the presence of an alkaline material. The diesters are especially useful in the preparation of fire retardant, self-extinguishing, polyurethane foams.

Any polyhydric alcohol which contains at least 3 hydroxyl groups and which does not decompose at the reaction temperatures employed may be utilized in preparing the diesters. Representative polyhydric alcohols include both aliphatic and aromatic compounds such as, for example, trimethylolpropane; pentaerythritol, glycerol; prodendro sorbitol, i.e., the reaction product of propylene oxide and sorbitol; arabitol; xylitol; adonitol; mannitol; dulcitol; sorbitol; 1,2,4-benzenetriol; phloroglucinol; 1,2,3,5-benzenetetrol; 1,2,4,5-benzenetetrol and benzenehexol. Of these, it is especially preferred to employ glycerol resulting in the preparation of the diglyceryl ester of tetrabromophthalic anhydride.

It has been found to be essential to include in the reaction mixture an alkaline material in an amount sufficient to neutralize any residual acid present in the tetrabromophthalic anhydride. Any material an aqueous solution of which has a pH greater than 8 may be employed. Representative materials include, for example, alkali metal and alkaline earth metal hydroxides; carbonates and bicarbonates, such as sodium hydroxide, sodium carbonate, sodium bicarbonate, potassium hydroxide, and calcium carbonate; ammonium hydroxide; and salts such as sodium acetate.

The foams produced may be used as void fillers, board stock, sprayed coatings, and in any other application where polyurethane foams are conventionally employed and where improved fire retardancy and self-extinguishing properties are required.

In order to more conveniently describe the components of the foam-forming compositions utilized in the examples, the components are identified by their trademarks or a representative symbol.

Papi is polymethylene polyphenylisocyanate.

TBPDG refers to the diglyceryl ester of tetrabromophthalic anhydride.

VC-611 refers to an organic phosphorus compound (Mobil Chemical Company).

Fyrol 6 refers to diethyl N,N-bis(2-hydroxyethyl)aminomethyl phosphonate.

L-5410 refers to a silicone surfactant (Union Carbide Corporation).

Dabco R8020 refers to a catalyst.

Freon R-11B refers to trichlorofluoromethane.

Polyol A refers to a material having a hydroxyl number of from 310 to 320 prepared by condensing propylene oxide and partially anhydrized sorbitol wherein 9.9 mols of propylene oxide are utilized per mol of partially anhydrized sorbitol.

In the foam-forming compositions in the examples, the amounts are expressed in parts by weight. In evaluating the polyurethane foams, the following standard tests were employed. The oxygen index of the foams was measured in accordance with ASTM D-2863-70. The weight retention of the foams was determined in accordance with the Butler Chimney Test described in *J. Cellular Plastics,* Vol. 3, 497 (1967). The fire retardancy is expressed in terms of the percent of original weight retained by a sample and a higher value indicates a more fire retardant foam.

Example 1: Into a 500 ml three-necked round bottom flask equipped with a stirrer, nitrogen inlet tube, and condenser, there were added 104 g of glycerin and 1.0 g of sodium acetate. The mixture was heated to 130°C and 464 g of tetrabromophthalic anhydride were then added. The temperature of the reaction mixture was allowed to rise slowly to 190°C and the reaction mixture was maintained at that temperature until the acid number reached less than 1. The final product was an amber solid having an acid number of 0.6, a percent bromine of 50, and a hydroxyl number of from 345 to 360.

Example 2: A polyol blend was prepared comprising: 50 g of the diester prepared in Example 1, 3.7 g of Polyol A, and 6.3 g of VC 611. A polyurethane foam-forming composition was prepared comprising: 46 g of the polyol blend described above, 44 g of Papi, 0.5 g of L-5410, 1.0 g of Dabco R8020, 8.0 g of Freon R-11B, and 0.5 g of water. A polyurethane foam prepared from the above composition had an oxygen index value of 26.5, a percent weight retention of 87-88, and a density of 2.3 pcf.

Example 3: A foam-forming composition was prepared comprising:

Papi	47.0
TBPDG	20.2
10-Prodendro sorbitol	17.7
VC-611	2.5
L-5410	0.5
Dabco R8020	1.5
Water	0.5
Freon R-11B	10.0

Foam prepared from this composition had the following properties:

Oxygen Index	26.6
Weight retention, (%)	85.5

Substituted Bis(2,4,4,4-Tetrachlorobutyl) Acetals

J.S. Babiec, Jr. and R.J. Turley; U.S. Patent 3,968,170; July 6, 1976; assigned to Olin Corporation describe a group of halogenated acetals. These acetals, which are derived from the reaction of 2,4,4,4-tetrachlorobutanol with an aldehyde, are useful as additives for reducing the burning rate of polyurethane foam.

These acetals may be represented by the following formula.

$$X_aR[CH(OCH_2CHClCH_2CCl_3)_2]_n$$

where X is chlorine, bromine, or a mixture; a is an integer from 0 to 4; R is hydrogen, alkyl, alkene, aryl, alkaryl, or aralkyl; and n is 1 to 2.

The following list is illustrative of the aldehydes which may be used in preparing the acetals: formaldehyde, acrylaldehyde, chloroacetaldehyde, chloral, benzaldehyde, glyoxal, etc.

In preparation of polyurethane foam, the acetals represented by the formula above are incorporated as additives in the polyurethane forming reaction mixture before foaming. Conveniently, they are first mixed in with the polyol reactant, and the other ingredients used in making the foam then added to this mixture. Any suitable proportion of these additives may be used which is effective in reducing the combustibility of the foam without otherwise interfering with the foaming reaction or materially altering the properties of the resulting foam. For example, the acetals may be used in proportions of 15 to 35 parts per every 100 parts by weight of total polyol which is employed in preparing the polyurethane foam.

Increased effectiveness of the halogenated acetal additive is achieved by incorporating in the foam-forming formulation, prior to foaming, antimony oxide or antimony oxyhalide as a coadditive. By combining both additives in the foam, a synergistic combustion retarding effect is attained.

As such the acetals of this process are of utility in making polyurethane foam for use in those applications where such a cellular material is desired or required which exhibits a certain measure of reduction in combustibility or burning rate. These applications include, for example, the manufacture of furniture cushions, carpet underlayments, and mattresses.

Example 1: Bis(2,4,4,4-Tetrachlorobutyl) 3-Chloropropional – In a reaction vessel equipped with a thermometer and a mechanical agitator, 106 g of 2,4,4,4-tetrachlorobutanol were dissolved in 75 ml of benzene. The solution was then saturated with anhydrous hydrochloric acid at 0°C and in the presence of 30 g of anhydrous magnesium sulfate. An ice bath was used to control the temperature. Acrolein was thereafter added to the mixture dropwise and at such a rate as to maintain the temperature inside the reaction vessel below 2°C. After a total of 14 g of acrolein were added, the mixture was agitated for an additional hour while being maintained within a temperature range of 0° to 2°C. It was then allowed to stand at room temperature for three days.

The benzene was then removed by distillation under partial vacuum. Any residual water present was removed by adding a fresh quantity of benzene and stripping this off with the residual water at reduced pressure. The remaining reaction product mixture was dissolved in benzene followed by filtration to remove undissolved solids. After stripping off the benzene, the bis(2,4,4,4-tetrachlorobutyl) 3-chloropropional product was recovered from the reaction mixture by the extraction with a solvent mixture of 360 ml methanol and 240 ml water. The lower acetal phase was separated from the methanol-water system and stripped of volatiles by heating at 45° to 50°C and 2 mm Hg as a light brown liquid.

The structure of bis(2,4,4,4-tetrachlorobutyl) 3-chloropropional was confirmed by nuclear magnetic resonance as follows, based on the empirical formula $C_{11}H_{15}Cl_9O_2$. Calculated: 26.51% carbon, 3.03% hydrogen, and 64.03% chlorine. Found: 27.16% carbon, 2.91% hydrogen, and 63.68% chlorine.

Example 2: Polyurethane Foam – A flexible polyurethane foam was prepared from the following ingredients in the indicated proportions:

Ingredients	Proportions
Oxypropylated glycerin, MW 3,000	100 g
Water	4 ml
Triethylene diamine catalyst*	0.4 ml
Stannous octoate catalyst	0.3 ml
Surfactant**	1.5 ml
Toluene diisocyanate (80/20 mixture of 2,4-2,6-isomers)	50 g

*This material, purchased under the trademark Dabco 33LV, contains about 33% triethylene diamine, the balance being substantially dipropylene glycol.

**This surfactant, purchased under the trademark DC-190, is a block copolymer of polydimethylsiloxane and a polyester resin.

To the above mixture, 25 g of bis(2,4,4,4-tetrachlorobutyl) 3-chloropropional were added. Procedurally, this was blended in with the oxypropylated glycerin before adding the other ingredients of the foam forming reaction mixture. The total mixture was then poured into an open-top, cardboard box and allowed to expand into a flexible foam block which was oven cured at 100°C for 20 minutes.

The combustibility of the foam was investigated using a standardized method described in ASTM D-1692-68 for testing the flammability of plastic sheeting and cellular plastics.

Following the procedure described in the ASTM method, a specimen, measuring 6 inches long, 2.0 inches wide and 0.5 inch thick, was cut from the foam block prepared above. The specimen was ignited on one end. As a result, it burned but was not completely consumed by the flame. Rather, the combustion ceased after a while and before the sample was completely consumed. The extent of the burn was measured and found to be limited to an average of 4.5 inches of the total length of the sample. Also the average burn rate was 3.5 inches per minute.

For purposes of comparison, another foam was prepared exactly as described above except for the exclusion of the bis(2,4,4,4-tetrachlorobutyl) 3-chloropropional from the foam forming reaction mixture. When a specimen of this foam was subjected to the same combustibility test described above, it exhibited no resistance to burning. Thus it burned through and was completely consumed by the flame. The average burn rate was 6.3 inches per minute.

The above example and comparison demonstrate the effectiveness of the acetals of the process in reducing the combustibility and burning rate of polyurethane foam.

Example 3: Polyurethane Foam – The identical procedure of Example 2 was followed except that there were also included in the foam forming reaction mixture 10 g of antimony trioxide. When the combustibility of the resulting foam was checked, the test specimen exhibited further reduction in combustibility over that of Example 2. Thus the average extent of burning here was 2.2 inches and the average burn rate was 2.8 inches per minute. This example demonstrates the synergistic reduction in combustibility which results from using a coadditive mixture of an acetal of the process and antimony trioxide.

PHOSPHORUS- AND HALOGEN-CONTAINING ADDITIVES

Bis(2,4,6-Tribromophenyl) Phosphonates

D.R. Breckenridge; U.S. Patent 4,083,818; April 11, 1978; assigned to Ethyl Corporation describes compounds such as methylbis(2,4,6-tribromophenyl) phosphonate and phenylbis(2,4,6-tribromophenyl) phosphonate useful as flame retardants for organic polymers and resins such as polyurethane foams.

The compounds have the formula:

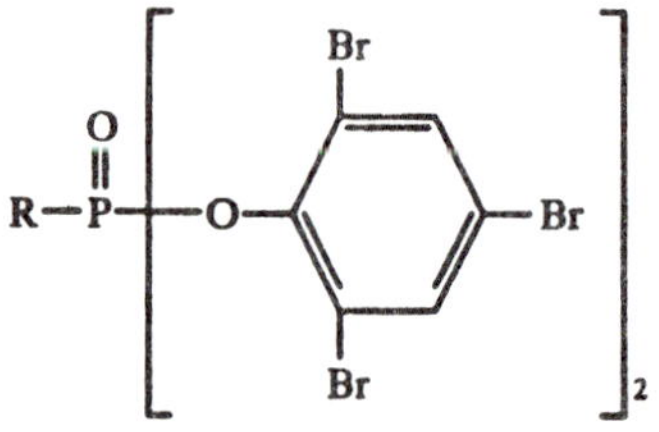

wherein R is a hydrocarbyl radical, preferably alkyl or aryl and having up to about 8 carbon atoms. These compounds can be made by reacting the corresponding $RPOCl_2$ compound and brominated phenol, preferably in the presence of a halogen halide acceptor such as pyridine or other tertiary amine.

The above phosphonates are used in polyurethane foam produced by reaction of an organic polyisocyanate and a polymer selected from the class consisting of hydroxyl terminated polyethers and hydroxyl terminated polyesters and having incorporated a flame retardant amount of a compound of the formula shown above. These foams are produced by incorporating the desired amount of one or more compounds in the formula in the mixture to be formed.

Example 1: To a solution of 2,4,6-tribromophenol (100 g, 0.30 mol) in toluene (600 ml) was added 22.4 g of 50% NaOH (11.2 g, 0.28 mol). After removing the water by azeotropic distillation, the mixture was cooled and methyl phosphonic dichloride (16.0 g, 0.13 mol) made up to 52 ml in toluene was added at 37° to 48°C in 11 minutes (exotherm). After refluxing 1 hour and 20 minutes, the mixture was cooled, washed with water, caustic and again with water, and the organic layer dried over $MgSO_4$ and filtered.

After stripping, there was obtained 44 g of a near-white solid (46.8% of theoretical product weight). The product is bis(2,4,6-tribromophenyl) methyl phosphonate. Similar results are obtained when 2,3,5-tribromophenol and 2,3,4-tribromophenol are used in place of 2,4,6-tribromophenol, and when KOH is used in place of NaOH.

Example 2: To a solution of 2,4,6-tribromophenol (71.0 g, 0.215 mol) in toluene (200 ml) and pyridine, (40 g, 0.51 mol) is added a solution of methyl phosphonic dichloride (12.0 g, 0.098 mol) in toluene (29 g). The addition took 20 minutes at 22.5° to 35°C (exotherm).

After refluxing at 112°C for 2 hours and 50 minutes, the mixture was cooled, washed with water, caustic (10%), and then water, and the organic layer dried over $CaCl_2$. After filtering and vacuum stripping, 31 g of bis(2,4,6-tribromophenyl) methyl phosphonate was obtained.

Analysis (in percent) —Found: C, 21.9; H, 1.30; Br, 64.3; Cl, 0.0. Calculated: C, 21.6; H, 0.98; Br, 66.5; Cl, 0.0.

Similar results are obtained when 2,3,5-tribromophenol and 2,3-4-tribromophenol are utilized in the process of the above example. Similar results are obtained when from 2 to 5 mols of the bromophenol are employed for each mol of methyl phosphonic dichloride. Similar results are obtained when from 1 to 100 mols toluene are used per each mol of phenol and when equivalent volumes of benzene and petroleum ether are substituted for the toluene. Similar results are obtained when the reaction is conducted from 1 to 10 hours utilizing a temperature of from 50° to 150°C. Similar results are obtained when the pyridine is replaced with methyl pyridines or trimethyl amine.

Following the procedure of the above example, ethyl phosphonic dichloride reacts with the phenol to yield bis(2,4,6-tribromophenol)ethyl phosphonate (1). Likewise, bis(2,4,6-tribromophenyl)butyl phosphonate (2), bis(2,4,6-tribromophenyl)-octyl phosphonate (3), bis(2,3,5-tribromophenyl)benzyl phosphonate (4), and bis-

(2,3,4-tribromophenyl)-2-ethyl phenyl phosphonate (5) are prepared by reacting the corresponding brominated phenol with the corresponding phosphonic dichloride.

Example 3: To each of five separate batches of 100 g of polyoxypropylene polyol having an average molecular weight of 2,920 and a hydroxyl number of 56, were added 1.0 g silicone surfactant, 0.3 g tertiary amine blowing catalyst, and 4.0 g distilled water. To two of the above dispersions were added 10.0 g and 15.0 g bis(2,4,6-tribromophenyl) methyl phosphonate. To two other dispersions were added 10.0 g and 15.0 bis(2,4,6-tribromophenyl) phenyl phosphonate. No flame retardants were added to the fifth dispersion.

Each of the above dispersions were mixed thoroughly for 10 seconds. Next, 0.2 g of stannous octoate catalyst was added and mixed for 5 seconds. Next, 50.0 g of an 80:20 blend of the 2,4- and 2,6-isomers of toluene diisocyanate were added and stirred rapidly for 10 seconds or until creaming was noted. The resulting admixtures were poured into 8 inch x 8 inch x 4 inch boxes and allowed 3 to 5 minutes for the foaming reaction to take place. Then the foams were placed in an oven and cured for 30 minutes at about 120°C.

The foams were then removed from the oven, aged for seven days at ambient temperature, and cut into test specimens for flammability testing in accordance with ASTM-D-1692-59T. The foam recipes and test results are shown below.

Materials	Foam A	Foam B	Foam C	Foam D	Foam E
Polyol GP-3030	100	100	100	100	100
Surfactant 192	1	1	1	1	1
DABCO 33-L	0.3	0.3	0.3	0.3	0.3
Stannous octoate	0.2	0.2	0.2	0.2	0.2
Distilled water	4	4	4	4	4
Toluene diisocyanate, 80:20	50	50	50	50	50
Bis(tribromophenyl)-methyl-phosphonate	—	10	15	—	—
Bis(tribromophenyl)-phenyl phosphonate	—	—	—	10	15
Foam density, pcf	1.61	1.90	1.85	1.85	1.9

Foam	ASTM D-1692-59T Results Foam A	Foam B	Foam C	Foam D	Foam E
Self extinguishing time, Seconds	—	44	—	—	—
Distance Burned, Inches	5.0	2.5	5.0	5.0	5.0
Rate of Burning, Inches per minute	7.0	3.4	3.0	4.6	4.1

In the tables above, the amounts of materials given are in grams. Similar results are obtained when from 2 to 40 parts per hundred of compounds 1 through 5 above are substituted for the phosphonates shown in the above examples.

Bis(2,3-Dibromopropyl) Chloroalkyl Phosphates

J.A. Albright; U.S. Patent 4,083,826; April 11, 1978; assigned to Velsicol Chemical Corporation describes a compound of the formula

$$(BrCH_2CHBrCH_2O)_2\overset{\overset{\displaystyle O}{\|}}{P}O{-}R{-}Cl$$

where R is a straight or branched chain alkyl group containing from 2 to 12 carbon atoms and a polymeric composition comprising a polyurethane and a flame retarding amount of the desired compound.

The following is a partial listing of compounds of the formula: bis(2,3-dibromopropyl)-2-chloropropyl phosphate, bis(2,3-dibromopropyl)-3-chloropropyl phosphate, bis(2,3-dibromopropyl)-6-chlorohexyl phosphate and bis(2,3-dibromopropyl)-2-chloroethyl phosphate.

The compounds of the formula are useful flame retardants in polymeric compositions selected from the group consisting of polyurethane, including flexible and rigid foams and elastomers, polyester, both saturated and unsaturated polyester, styrenated polyester and cellulosic polyester blends, e.g., 35/65, 50/50, and 65/35 cotton/polyester blends.

Example 1: Preparation of Bis(2,3-Dibromopropyl)-2-Chloroethyl Phosphate – Into a three necked five liter flask equipped with mechanical stirrer, thermometer, condenser and dropping funnel was charged 924 g of phosphoryl chloride and 10 g titanium tetrachloride. The solution was heated to 65°C and 2,616 g of 2,3-dibromopropanol was added dropwise over a four hour period. Upon complete addition, the solution was heated to 80°C for four hours. After cooling to 60°C, gaseous ethylene oxide (290 g) was added over a three hour period. A slightly exothermic reaction was observed. The resulting product was washed with an equal volume of water at 50°C, followed by an aqueous ammonia wash and two additional water washes.

The volatiles were removed by steam stripping. The product was treated with Celite and Celkate, dried and filtered. The resulting 2,709 g of product was analyzed and the following results obtained: viscosity = 1300/28°C; Acid No. = 0.1 mg KOH/g sample; % Br = 55.61 (theory = 57.2); % Cl = 6.66 (theory = 5.4); % water = 0.05.

In a similar manner, other compounds within the scope of the formula, e.g., bis(2,3-dibromopropyl) 2-chloropropyl phosphate, bis(2,3-dibromopropyl) 3-chloro propyl phosphate, bis(2,3-dibromopropyl) 6-chlorohexyl phosphate, bis(2,3-dibromopropyl) 1-chloro-2-propyl phosphate and bis(2,3-dibromopropyl) 12-chlorododecane phosphate can be prepared when the corresponding oxide is employed.

Example 2: An 80/20 mixture of toluene 2,4- and 2,6-diisocyanate, Type I, toluene diisocyanate (TDI) was placed in a first tank of a Martin Sweets Modern Module No. 3A urethane foam equipment modified for six components. Bis-(2,3-dibromopropyl)-2-chloroethyl phosphate (1 kg) was mixed with 10 kg of Pluracol GP 3030 polyol in a second tank. (Pluracol GP is a polypropylene glycol having a molecular weight of approximately 3,000 and a hydroxyl number of approximately 56). Stannous octoate catalyst was placed in the third tank. Into a fourth tank was placed a silicone surfactant. A water/Dabco 33LV mixture (3/0.29 to 0.46) was added to a fifth tank, (Dabco 33LV, Houndry Process & Chemical Company in a 33% solution of triethylene diamine in dipropylene glycol.) All of the above components were simultaneously mixed using a size 3 pin type mixer at 3,000 rpm in the following ratio:

Component	Parts per Hundred Parts Polyol (php)
Flame retardant	10
Polyol	100
TDI	39.0
Stannous octoate	0.21-0.3
Silicone surfactant	1.0
Water	3.0
Triethylene diamine/dipropylene glycol solution, 33%	0.29-0.46

The mixture was dropped during the mixing procedure into a 14 inch x 6 inch Adstrom cardboard box. After the flexible foam came to full height it was postcured in a forced air oven at 210° to 220°F for 30 minutes.

After allowing the foam to sit for at least 7 days, the foam was then subjected to several tests and the data obtained therefrom are reported in Table 1. The same procedure was used to make other flexible foams having different flame retardant load levels. These foams were also subjected to the same tests and the data obtained therefrom are also reported in Table 1.

TABLE 1: COMBUSTIBILITY OF FLEXIBLE URETHANE FOAM

	Flame Retardant Load Level (php)				
	0	2.5	5	10	20
Oxygen index, %*	19	23	23.5	24	27.5
MVSS302**					
Initial	***3.9 (10)†	0 (0)	0 (0)	0 (0)	0 (0)
Aged††	3.46 (4.8)	0 (0)	0 (0)	0 (0)	0 (0)
Aged†††	3.66 (10)	–	–	0	0

*ASTM D-3863-74.
**Motor Vehicle Safety Standard 302, Department of Transportation.
***Numbers without parentheses are in/min burn rate based on burn past the timing line.
†Numbers in parentheses are inches of burn distance from the timing line.
††Aged as per ESBM25P3-A, Ford Motor Corp.
†††Aged as per FBML TM 32-10, General Motors Corp.

Example 3: A rigid foam was prepared using the following basic formulation:

Component	Parts per Hundred Parts Polyol (php)
Polyol (Thanol R-350-X)	100
Silicone glycol surfactant (Dow Corning 193)	2

(continued)

Component	Parts per Hundred Parts Polyol (php)
Trichlorofluoromethane (Freon 11B)	35
Polyisocyanate (Mondur MRS)	135

The polyol, surfactant, and fluorocarbon blowing agent were combined in a master batch based on 1,000 g of polyol to minimize loss of flowing agent. The following procedure was used to prepare the foam:

The polyisocyanate was weighed into a tared, 10 ounce, paper cup (allowances being made for holdup) and the cup set aside while the remaining ingredients were weighed out and mixed.

The polyol master batch was weighed out, in the proper amount to give 100 g of polyol, in a one quart, untreated, paper cup.

The 30 g of the flame retardant of Example 1 were then weighed into the same one quart cup.

The contents of the one quart cup were mixed at 1,000 rpm for 5 seconds.

The polyisocyanate was then added and stirring at 1,000 rpm continued for 10 seconds.

The mix was poured into a 5-pound, untreated, paper tub and allowed to rise.

After the foam was tack-free, and substantially cured, it was set aside for at least 7 days prior to subjecting the foam to an Oxygen Index Test, ASTM D-2863-74. The results of the test are reported in Table 2.

The same procedure was used to make another foam at the same load level and also containing a different flame retardant additive. This foam was also subjected to the same Oxygen Index Test (O.I.) as the above foam and the data are also reported in Table 2.

TABLE 2

Flame Retardant	Load Level, php	O.I.,%
Control	0	21.0
Example 1	30	26.5
Tris(2,3-dibromopropyl)phosphate	30	26.0

Bis(Haloalkyl)Nitrilo(Trimethylene Phosphonate)

J.A. Albright; U.S. Patent 4,054,544; October 18, 1977; assigned to Velsicol Chemical Corporation describes compounds of the formula

$$N[CH_2\overset{O}{\overset{\|}{P}}(OR)_2]_3$$

where each R is independently selected from the group comprising halogenated alkyl radicals having from 2 to 6 carbon atoms and from 1 to 4 halogen substituents.

The following compounds are illustrative of the preferred flame retardants bis(2,3-dibromopropyl)nitrilo(trimethylene phosphonate), bis(2,3-dichloropropyl)-nitrilo(trimethylene phosphonate), bis(3-bromopropyl)nitrilo(trimethylene phosphonate), bis(3-chloropropyl)nitrilo(trimethylene phosphonate), and bis(2-chloroethyl)nitrilo(trimethylene phosphonate).

The compounds of this process are useful as flame retardants in polyurethane polymeric compositions comprising a polyurethane polymer and a flame retarding amount of the above described compounds. Polyurethane polymers embrace not only the rigid and flexible foams, but the solid elastomeric compositions.

Example 1: Preparation of Bis(3-Bromopropyl)Nitrilo(Trimethylene Phosphonate) – Phosphorus trichloride (137 g; 1 mol) was dissolved in 100 ml of methylene chloride and 278 g (2 mols) of a bromopropanol mixture was added dropwise over a period of 2 hours with stirring and external cooling in ice to keep the temperature below 15°C. Upon complete addition, the reaction mixture was warmed to 35°C, held at that temperature for 1 hour, and then cooled again to 10°C. Water (18 g; 1 mol) was added dropwise keeping the temperature below 15°C.

Upon complete addition, excess hydrogen chloride was removed under vacuum and the solution was then heated for 1 hour at 40°C. The methylene chloride was evaporated to yield 319 g of a colorless oil. Infrared absorption (2460 cm^{-1}) confirmed the bis(3-bromopropyl)phosphite structure.

To 66 g of a 37% aqueous formaldehyde solution (10% excess) was added 15 g of a 29% aqueous ammonium solution with stirring. A slight exothermic reaction developed. The solution was constantly stirred for one-half hour and then 250 g of the above bis(3-bromopropyl)phosphite was added dropwise over a 1 hour period. The mixture was stirred at room temperature for an additional 1 hour and then heated to 60°C and held at that temperature for 8 hours.

The organic layer was separated and washed twice with water. The resulting viscous straw-colored oil was dried at 100°C/1.5 mm of mercury for 2 hours. The infrared showed the disappearance of the POH absorption. Analysis for $C_{21}H_{42}Br_6NO_9P_3$ – Calculated: bromine, 46.8%; phosphorus, 9.08%. Found: bromine, 43.25%; phosphorus, 9.32%.

Example 2: Preparation of bis(2-chloroethyl)nitrilo(trimethylene phosphonate) was carried out in a similar fashion.

Example 3: A rigid foam was prepared using the basic formulation prepared in Example 3 of the previous patent, U.S. Patent 4,083,826. However, in the preparation, only 10 g of the flame retardant of Example 1 were weighed into the one quart cup.

After the foam was tack-free, and substantially cured, it was set aside for at least 7 days prior to subjecting the foam to an Oxygen Index Test, ASTM D-2863-74, and a Compressive Strength Test, ASTM D-1621-73. The results of the tests are reported in the table on the following page.

The same procedure was used to prepare other rigid foams at different load levels or containing a different flame retardant additive. These foams were also subjected to the same Oxygen Index Test and Compressive Strength Test as the above foam and the data are also reported in the following table.

Flame Retardant	Load Level, php[a]	OI, percent	Maximum Compressive Strength psi
Control	0	21	29.5
Example 1	10	24.0	31.1
	20	25.5	34.6
	30	27.0	35.3
Example 2	30	25.0	

[a]parts per hundred polyol.

The table clearly shows that the compounds of this process possess an excellent degree of flame retardant efficacy in polyurethane polymeric compositions and further do not detract from, and in fact help improve, the physical properties of the polymeric compositions.

Diols Containing Phosphorus and Halogen

S. Altscher and J.C. Goswami; U.S. Patent 4,086,194; April 25, 1978; assigned to Stauffer Chemical Company describe a polyurethane foam containing an effective amount for flame retardancy of the foam formed from a compound of the formula:

$$R\overset{O}{\overset{\|}{P}}\text{—}(OCH_2\overset{X}{\overset{|}{C}}H[CH_2]_nOH)_2$$

where n is an integer from 1 to 8, R is selected from the group consisting of C_1-C_{20} alkyl, aryl and halo-substituted C_1-C_{20} alkyl and aryl and X is halogen.

The above diol is formed by the reaction of halides of alkyl or aryl phosphonic acids having the formula $RP(O)X_2$, where R and X are as defined above, with an epoxy-group-containing alcohol of the formula:

$$H_2C\overset{O}{\text{—}}CH\text{—}(CH_2)_n\text{—}CH_2OH,$$

where n is an integer from 1 to 8, such as glycidol. The molar ratio of such reactants is 1:1 to 1:4. These diols are used as flame retardants in polyurethane.

Example 1: Into a three-necked 500 ml flask fitted with a stirrer, condenser, funnel and a thermometer were charged the following materials: methylene chloride, 100 g and glycidol, 80 g.

The thus charged flask was first cooled in an ice bath (to approximately 3°C) and then 90.3 g of chloromethyl phosphonic dichloride was added dropwise from the funnel keeping the reaction temperature below 40°C. The addition took approximately 3 hours. The reaction (including the addition of chloromethyl phosphonic dichloride) was carried out under a atmosphere of dry nitrogen gas. After the addition was complete, the reaction temperature was raised (by heating the flask in an oil bath) to 60°C and the reaction was continued at this temperature for an additional hour. After the reaction was over, methylene chloride solvent was removed by applying vacuum, first at room temperature, followed by heating to 60°C. The product (152 g) was a highly viscous material which could be readily dispersed in polyether polyol. The phosphorus and chlorine contents of the product are expected to be approximately 9.8% and 33%, respectively.

Example 2: This example illustrates the preparation of a flame retarding polyurethane foam from the reaction product (at 15 parts level) in conjunction with a haloalkyl ester of phosphoric acid flame retardant (5 parts level). The foam prepared from this combined 20 parts of flame retarding additives per 100 g of polyol passed the initial flammability (ASTM D-1692) as well as the dry heat aging (ASTM D-1564) tests. The formulation with this mixed additive is as follows:

	Grams
Thanol F-3002 polyol	100.0
Reaction product	15.0
Tris(dibromopropyl) phosphate	5.0
Triethylamine	0.2
L-540 silicon surfactant	0.8
H_2O	4.0
N-ethylmorpholine	1.2
A-1 catalyst	1.0
Stannous octoate (50% in dioctyl phthalate)	1.6
Toluene diisocyanate	56.0

Foam characteristics: foam rise and gelling good

Density: approximately 2 lb/ft^3

Initial flammability (ASTM D-1692): foam is self-extinguishing

Average extinguishment time (T_{SE}): 42 seconds
Average distance burnt: 2.1 inches
Burning rate: 3 inches per minute

Dry heat aging test (ASTM D-1564): foam is self-extinguishing

Average extinguishment time (T_{SE}): 43 seconds
Average distance burnt: 2.2 inches
Burning rate: 3.1 inches per minute

Low Viscosity Composition of 2,3-Dibromopropanol and Tris(2,3-Dibromopropyl) Phosphate

T.C. Wilkinson; U.S. Patent 3,997,449; December 14, 1976; assigned to Michigan Chemical Corporation describes a low viscosity flame retardant comprising a mixture of 2,3-dibromopropanol and tris(2,3-dibromopropyl) phosphates. The low viscosity flame retardants of this process have special utility in both rigid and flexible polyurethane polymers.

This mixture comprises 15%, 2,3-dibromopropanol and 85% tris(2,3-dibromopropyl) phosphate and has a viscosity of from 1,500 to 2,400 cp at a temperature of 20° to 30°C.

The low viscosity flame retardant is useful in many polymeric compositions. It finds particular utility in flexible and rigid polyurethane polymers where the viscosities of the reactants and the reactant mixtures are critical to machine preparation of the desired polymer.

Example: An 80/20 mixture of toluene 2,4- and 2,6-diisocyanate, Type I, (referred to as TDI) was placed in the first tank of a Martin Sweets Modern Module No. 3A urethane foam equipment modified for simultaneous addition of up to six components. A mixture of 15 parts 2,3-dibromopropanol and 85 parts tris(2,3-dibromopropyl) phosphate (0.5 kg) was mixed with 10 kg of polyol (Voranol 4291). Stannous octoate catalyst was placed in a third tank (T-11, a stannous octoate plus mineral oil mixture). Into a fourth tank was placed a silicone surfactant (L-540). Water/triethylene diamine (Dabco 33LV), N-ethyl morpholine (referred to as NEM) mixture was added to a fifth tank. (Dabco 33LV is a 33% solution of triethylene diamine and dipropylene glycol.)

All of the above components were pumped at a predetermined rate simultaneously into a chamber and were mixed using a size 3 pin type mixer revolving at 3,000 rpm in the following ratio:

Component	Parts by Weight
Low viscosity flame retardant mixture, 15% 2,3-dibromopropanol 85% tris(2,3-dibromopropyl) phosphate	5
Polyol	100
Silicone surfactant	1.0–1.5
NEM	0.10
Triethylene diamine	0.40–0.42
Stannous octoate catalyst	0.10–0.18
Water	3.0
TDI–103 to 105 index	

The mixture was dropped during the mixing procedure into a 14 x 14 x 6 inch Adstrom cardboard box. After the foam came to full height it was postcured in a forced air oven at 210° to 220°F for 30 minutes. After allowing the foam to sit for at least 7 days, the foam was then subject to the several tests listed in Tables 1 and 2. Different load levels were also tested.

TABLE 1: PHYSICAL PROPERTIES OF MOLDED URETHANE FOAM CONTAINING LOW VISCOSITY FLAME RETARDANT

	. . . Flame Retardant Load Level*. . . .		
	Control	2.5 php*	5.0 php
Density, pcf***,†	2.40	2.44	2.64
Air flow, scfm††,†	1.22	1.27	1.42
	1.35	1.61	1.61
Tensile, psi†	18.3	23.3	25.4
Aged†††	26.7	27.5	31.4
Elongation, %†	134	162	175
Aged†††	191	193	198
Tear, pli†,§	1.89	2.73	2.30
Ball rebound, %	33	25	28

(continued)

TABLE 1: (continued)

	. . . Flame Retardant Load Level*. . . .		
	Control	2.5 php	5.0 php
ILD, lb/50 in^2 †,§§			
25%	46.2	39.7	41
65%	111.5	102	106
25% R§§§	36.4	32.0	34
Compression set, C_d, %†			
50%	8.18	5.93	7.0
50% aged†††	18.1	9.56	10.4
90%	6.00	3.63	5.5
90% aged†††	21.0	16.6	22.7

*Flame retardant composed of 15% 2,3-dibromopropanol, 85% tris(2,3-dibromopropyl) phosphate mixture.
**php is parts per hundred polyol.
***pcf is pounds per cubic foot.
†Dry heat aged, as per ASTM D-2406.
††scfm is standard cubic feet per minute.
†††Autoclave aged as per ASTM D-2406.
§pli is pounds per linear inch.
§§ILD is indentation load deflection.
§§§R is return.

TABLE 2: COMBUSTIBILITY OF MOLDED URETHANE FOAMS CONTAINING LOW VISCOSITY FLAME RETARDANT

	. . . Flame Retardant Load Level*. . . .		
	Control	2.5 php	5.0 php
Oxygen index**	19.5	23.0	23.5
MVSS302, in/min***			
Initial	pass 3.05 (10)†	pass 0 (0)	pass 0 (0)
Aged††	pass 2.92 (10)	pass 0 (0)	pass 0 (0)
Aged†††	pass 3.19 (10)	pass 0 (0)	pass 0 (0)
Vertical burn test§			
Maximum char, inches			
Initial	complete burn	ND§§	4.7
Aged 2 wk at 150°F	complete burn	ND	5.8
at 250°F	complete burn	ND	4.5
Average char, inches			
Initial	complete burn	ND	4.4
Aged 2 wk at 150°F	complete burn	ND	5.3
at 250°F	complete burn	ND	3.2

(continued)

TABLE 2: (continued)

	. . . Flame Retardant Load Level*		
	Control	**2.5 php**	**5.0 php**
Afterburn, sec	not applicable	ND	none
Rating	fail	ND	pass

*Flame retardant composed of 15% 2,3-dibromopropanol, 85% tris(2,3-dibromopropyl) phosphate mixture.
**ASTM D-3863-74.
***MVSS302 is the Motor Vehicle Safety Standard 302, Department of Transportation.
†Numbers in bracket give distance in inches.
††Aged as per FBMS TM 10-12, General Motor Corp.
†††Aged as per ESBM25P3-A, Ford Motor Corp.
§State of California, Bureau of Home Furnishing, Department of Consumer Affairs, Standard 117.
§§ND means not determined.

Tables 1 and 2 clearly show that the low viscosity flame retardant of this process can effectively flame retard molded urethane foam without detrimentally affecting the foam's physical properties and, it should be noted, in many instances actually improving the physical properties of the foam.

OTHER ADDITIVES

Smoke Reducing Acids

A process described by *W.J. Kauffman and E.J. Quinn; U.S. Patent 4,069,173; January 17, 1978; assigned to Armstrong Cork Company* relates to fire retardant polyisocyanurate and polyurethane foams having reduced smoke levels when burned and to methods for producing such foams. The foams are produced by incorporating into the polyisocyanurate or polyurethane: (a) polycarboxylic cycloaliphatic acids, (b) polycarboxylic heterocycloaliphatic acids, (c) polycarboxylic aromatic acids and/or (d) polycarboxylic naphthalenic acids having 3 nonadjacent carboxylic acid groups or any configuration with more than 3 carboxylic acid groups up to and including acids containing one carboxylic acid group per carbon atom, except that in the case of (d) the number of COOH groups must be two less than the number of carbons in the ring system. Some anhydride groups may also be present in the composition.

The acids may be incorporated during the foaming step or after the foam is formed. The polycarboxylic acid compounds may be used at 10 to 20% by weight of the total composition.

Example 1:

Typical Foam Formulation

	Parts by Weight
Polyol–Multranol-7100 (high MW triol, Hydroxyl No. 49, equiv wt–1145)*	31.7
Acid additive	Varied
Surfactant–Tegostab B3640 (fire retardant silicone surfactant)	0.8

(continued)

	Parts by Weight
Catalysts	
Dabco LV-33 (triethylene diamine)	0.2
T-9 (stannous octoate)	0.2
N,N',N'-tris(dimethyl aminopropyl) hexahydro-s-triazine	0.8
Blowing agent–UCON or Freon (fluorocarbon BP 113F)	10.0
Isocyanate–Upjohn Isonate 143L (liquid methylene diphenyl isocyanate)*	40.0

*NCO/OH = 1.0/0.1

The above foam was prepared using a standard one-shot foaming technique followed by postcuring the foam for 10 minutes at 80°C. Foam samples were cut and tested according to the standard procedure established in ASTM D-2843-70, commonly called the Rohm & Haas XP2 Smoke Chamber Test.

Example 2: Experimental results were obtained for a series of foams containing various acids. The smoke value per gram (SV/g) was obtained by dividing the Rohm & Haas smoke result by the sample weight of the one-inch cube that was tested. The acid equivalent (AE) for the acids employed was calculated by multiplying the number of acid groups in the molecule times the weight of the acid groups in the molecule times the weight of the acid used and dividing the result by the compound molecular weight.

When the acid equivalent is in turn divided by the total weight of the foam mix, excluding the blowing agent weight since it is lost during foaming and aging, the acid equivalent per gram of foam weight (AE/g) is obtained. The smoke value per acid equivalent (SV/AE) is calculated by dividing the smoke value per gram by the acid equivalent per gram of foam weight. The smoke-reduction efficiency factor (SF) is determined by normalizing the smoke value per acid equivalent to that of fumaric acid. Using the above method, the following results were obtained.

Smoke Results

Acid Employed	Rohm & Haas SV	SV/g	AE/g	SV/AE	SF
2,3,4,5-Tetrahydrofuran tetracarboxylic acid	12	10	0.001926	5,192	2.53
Fumaric acid	32	27	0.002059	13,113	1.00
Maleic acid (couldn't foam, deactivated catalyst)	–	–	0.002059	–	–
Itaconic acid	89	66	0.001836	35,948	0.36
Adipic acid	92	86	0.001636	52,567	0.25
1,2,3-Propane tricarboxylic acid	68	63	0.002024	31,126	0.42
Citric acid	75	58	0.001865	31,099	0.42
Aconitic acid	62	41	0.002059	19,913	0.66
Nitrilotriacetic acid	40	29	0.001876	15,458	0.85
Ethylenediamine tetra-acetic acid	57	42	0.001636	25,672	0.51

As demonstrated by these results, 2,3,4,4-tetrahydrofurantetracarboxylic acid is more than 2½ times as effective as a smoke reductant in the same foam system than the commonly used fumaric or adipic acids. Smoke results were obtained using 10 g acid additive in the standard foam formulation of Example 1.

Example 3: When the standard foam formulation of Example 1 contains only 7.9 g polyol (NCO/OH = 1.0/0.025) and the corresponding foam is prepared, it forms a rigid structure and is fire retardant. A comparison of the results for 2,3,4,5-tetrahydrofurantetracarboxylic acid and fumaric acid as smoke reductants in this type formulation is given below. Smoke results were obtained using 10 g acid additive in the foam formulation.

Acid Employed	Rohm & Haas SV	SV/g	AE/g	SV/AE	SF
None	25	25	–	–	–
Fumaric acid	39	39	0.002059	18,941	1.00
Tetrahydrofuran tetracarboxylic acid	9	9	0.001926	4,674	4.05

The above results again show the superiority of tetrahydrofurantetracarboxylic acid and in this case show the use of the material in a fire retardant, rigid foam system.

Example 4: Polycarboxylic aromatic acids were used in the basic foam formulation of Example 1 in amounts as shown in the following table. Comparison with other acids is also set forth in the table. Results were obtained using 10 g acid additive in the foam formulation.

Acid Employed	Rohm & Haas SV	SV/g	AE/g	SV/AE	SF
Mellitic acid	17	11	0.00209	5,281	1.78
Trimesic acid	8	10	0.001706	5,862	1.60
Trimellitic acid	18	16	0.001706	9,379	1.00
o-Phthalic acid	81	86	0.001438	59,805	0.16
Isophthalic acid	25	16	0.001438	11,127	0.84
Terephthalic acid	43	31	0.001438	21,558	0.44
Pyromellitic acid	46	87	0.001881	46,252	0.20
1,4,5,8-Naphthalene tetra-carboxylic acid	15	13	0.001571	8,275	1.13
Aurin	82	78	0.000873	89,347	0.10
Benzoic acid (couldn't foam)					

As demonstrated by these results, the trimesic and mellitic acids are at least 1½ times more efficient smoke reductants in the system listed than the commonly used trimellitic acid. The 1,4,5,8-naphthalenic tetracarboxylic acid of the process also shows results superior to other acids.

Organosilicates as Smoke Retardants

K. Ashida, M. Ohtani and S. Ohkubo; U.S. Patent 4,072,636; February 7, 1978; assigned to Mitsubishi Chemical Industries Limited, Japan describe a process for preparing a smoke retardant polyisocyanurate foam by reacting an organic polyisocyanate in the presence of a blowing agent and an isocyanate-trimerization catalyst. This is characterized in that the reaction is conducted in the presence of an organosilicate having the formula $Si(OR)_4$ or $R_nSi(OR')_{4-n}$ where R and R' are the same or different and represent aliphatic, alicyclic and aromatic hydrocarbon groups having 1 to 12 carbon atoms or substituted derivatives and n is an integer of 1 to 3.

Typical organosilicates having the formula $Si(OR)_4$ include tetramethyl silicate, tetraethyl silicate, tetraisopropyl silicate, tetra-n-butyl silicate, tetraisobutyl silicate, etc.

Typical organosilicates having the formula $RSi(OR')_3$ include methyltriethoxysilane $CH_3Si(OC_2H_5)_3$, butyltrimethoxysilane $C_4H_9Si(OCH_3)_3$, ethyl-tri-n-butoxysilane $C_2H_5Si(OC_4H_9)_3$, methyltriamyloxysilane $CH_3Si(OC_5H_{11})_3$, phenyltriisopropoxysilane $C_6H_5Si(O\text{-iso-}C_3H_7)_3$, etc.

The amount of the organosilicate relative to the polyisocyanurate foam is 0.2 to 3 wt % as Si. For example, when tetraethyl silicate is added at a ratio of 0.2 wt %, flame resistance and elimination of the bursting property are found; however, a slight deformation is observed when the product is contacted with a flame for a long time. When the proportion is 0.5 wt % as Si (3.7 wt % as tetraethyl silicate), the degree of deformation is reduced.

When it is 0.7 to 1.5 wt % as Si (5.2 to 11.1 wt % as tetraethyl silicate), deformation is essentially not present. With regard to smoke-generation, sufficient smoke-suppression is found by the addition of the organosilicate at a content of more than 0.5 wt % as Si, where a small degree of white smoke is generated by flame contact.

The effects of the organosilicate used in the process are:

(a) excellent smoke-suppression,

(b) an elimination of the bursting property,

(c) a surface active effect (the effect of foam uniformity is achieved although it is slightly weaker than that of the commercial organosilicone surfactants for polyurethane foam, and a foam having fine cellular voids can be prepared without using such organosilicone surfactants), and

(d) the solubilizing effect (solubility of the alkali metal salt type isocyanate trimerization catalyst is increased thereby accelerating trimerization) and the like.

Example 1: 100 g of polymeric isocyanate (Isonate-580 having 140 of NCO equivalent) was mixed with 22 g of trichloromonofluoromethane to prepare A solution. On the other hand, 10 g of tetraethyl silicate (tetraethoxysilane) (0.98% as Si to total weight of the foam), 2 g of a silicone surfactant (SH-193), 1.0 g of N,N',N''-tris(dimethylaminopropyl) hexahydro-sym-triazine (referred to as HHT) and 2.0 g of 30% ethylene glycol solution of potassium acetate were admixed to prepare B solution.

The A solution and the B solution were charged in a paper cup and the mixture was vigorously stirred to form a foam having a density of 0.035 g/cm³. The characteristics of the foam were as follows. When the foam was touched to a flame, the foam had no bursting property and was not substantially deformed and expanded.

Smoke-generation:	
maximum % light obscuration	90
time for reaching maximum % light obscuration	40 seconds

(continued)

Surface flammability:	
% weight retention	84
Flame resistance:	
flame penetration time	85 minutes
After-flame time:	0 second
Friability:	
% weight loss	58

As a reference, a foam was prepared by the process except removing tetraethyl silicate from the abovementioned components. The smoke-generation of the foam was tested to give 96% as the maximum % light obscuration and 30 seconds as the time for reaching the maximum % light obscuration. When the foam was touched to a flame, deformation, bursting and expansion were found and the color of the smoke was black.

Example 2: In accordance with the process of Example 1 using the same components under the same conditions except adding 2.0 g of tetraethyl silicate (0.21% Si), a foam having a density of 0.035 g/cm^3 was prepared under a cream time of 17 seconds and a rise time of 50 seconds. The characteristics of the foam were as follows:

Smoke-generation:	
maximum % light obscuration	89
time for reaching maximum % light obscuration	31 seconds
Surface flammability:	84.5 %
Flame resistance:	120 minutes
After-flame time:	0 second
Friability:	66 %

When the foam was touched to a flame, no bursting was found, though a slight deformation was found.

Example 3: A foam was prepared using an inorganic silicon compound. In accordance with the process of Example 1 using the same compounds under the same conditions except adding 3 g of SiO_2 instead of tetraethyl silicate and changing the amount of trichloromonofluoromethane from 22 g to 20 g, a foam having a density of 0.025 g/cm^3 was prepared at 15° to 16°C under a cream time of 10 seconds and a rise time of 60 seconds.

The characteristics of the foam were as follows: As a reference, a foam having a density of 0.036 g/cm^3 was prepared in accordance with the abovementioned process except adding 5 g of tetraethyl silicate instead of SiO_2 under a cream time of 15 seconds and under a rise time of 45 seconds.

	This Example	Reference
Smoke-generation		
Maximum % light obscuration	97	90
Time for reaching maximum, sec	28	40
Surface flammability, %	72	83.8
Flame resistance, min	48	85
After-flame time, sec	0.5	0
Friability, %	95.5	58

The weight percent of Si to the total weight of the foam is 1.11% as to SiO_2 and 0.53% as to tetraethyl silicate and the Si percent of the foam using SiO_2 is higher than that of the foam using tetraethyl silicate; however, the characteristics of the foam using SiO_2 are inferior. It is clear from this example, that smoke suppression effect depends not only on Si percent, but seems to depend on the condition of Si dispersed in the foam.

PHOSPHORUS- AND HALOGEN-CONTAINING REACTANTS

Phosphonic Acid Dialkylesters

A procedure described by *H. Hübner, J. Blahak and H.-J. Meiners; U.S. Patent 4,061,697; December 6, 1977; assigned to Bayer AG, Germany* relates to flame retarding agents and processes in which they are used in the production of polyurethanes. These compounds are represented by the general formula:

(1)

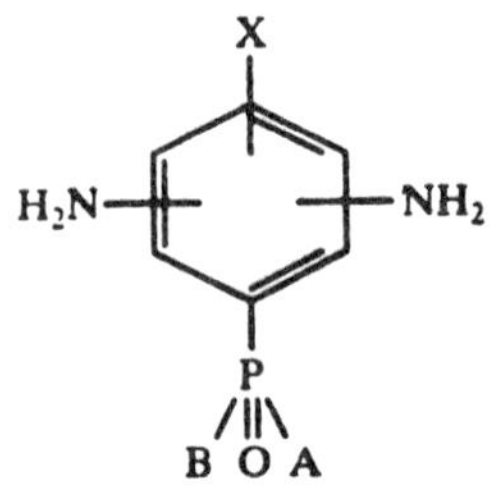

where A and B, which are identical or different, represent optionally branched C_1-C_{17} alkyl groups or –OR groups where R represents an optionally branched C_1-C_8 alkyl group, and X represents hydrogen, halogen, an optionally branched C_1-C_8 alkyl group, a C_6-C_{12} aryl or aralkyl group, NH_2, CF_3, CN, COOR', SO_3R',

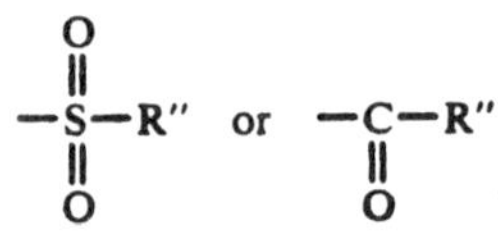

where R' represents hydrogen or an optionally branched alkyl or cycloalkyl group having 1 to 10 carbon atoms and R'' represents an optionally branched alkyl or cycloalkyl group having 1 to 10 carbon atoms.

These compounds are prepared by catalytic reduction of the corresponding polynitro precursors in the presence of Raney nickel at pressures 30 to 100 excess atmospheres, preferably 50 to 80 excess atmospheres and temperatures of 20° to 120°C, preferably 35° to 60°C, using as a solvent, the alcohol which corresponds to the radical R in the phosphinic or phosphonic acid ester.

The nitrosubstituted aromatic phosphoric acid alkyl esters are generally prepared as follows: The aromatic halogen compound is introduced into the reaction vessel, if desired as a mixture with an inert solvent and/or catalyst. The trialkylphosphite is then added in small portions while maintaining the reaction mixture at a temperature of 0° to 150°C, preferably 50° to 120°C. The total quantity of phosphite added is about 0.9 to 3 mols, preferably 1 to 2 mols and most preferably 1 to 1.5 mols per mol of halogen compound. When the calculated quantity of alkyl halide has been formed (which can be checked, for example, by means for a gas meter), the reaction is stopped (if necessary with cooling) and any excess trialkylphosphite and solvent may be removed under vacuum or alternatively the reaction mixture may be extracted with petroleum ether after removal of the solvent.

The polyurethane may be prepared from polyisocyanates, high molecular weight and/or low molecular weight polyols and other compounds containing groups which are reactive with isocyanates, characterized in that the compounds which are reactive with isocyanates include those of the general formula (1), and which are used in such an amount that the polyurethane contains at least 0.5% by weight of phosphorus. The isocyanates which may be used as starting components include aliphatic, cycloaliphatic, araliphatic, aromatic and heterocyclic polyisocyanates.

Example 1: 2,6-Diamino-4-Carbomethoxybenzenephosphonic Acid Dimethyl Ester — 45 g (0.134 mol) of 2,6-dinitro-4-carbomethoxybenzenephosphonic acid dimethyl ester are dissolved in 550 ml of methanol and hydrogenated at 35°C and 60 excess atmospheres after the addition of 10 g of Raney nickel. The catalyst is filtered off after 4 hours and the solvent is then evaporated off and the residue is freed from any remaining solvent in a high vacuum. Yield: 32 g, 82% of the theoretical, MP 150° to 151°C; identity confirmed by IR and NMR spectra.

Example 2: 2,6-Diamino-4-Carbomethoxybenzenephosphonic Acid Diethyl Ester — The following compounds are used in a manner analogous to Example 1: 80 g (0.221 mol) of 2,6-nitro-4-carbomethoxybenzenephosphonic acid diethyl ester, 1,000 ml of ethanol and 15 g of Raney nickel. Yield: 61 g, approximately 91% of the theoretical, MP 105° to 107°C. The IR and nuclear resonance spectra confirm the chemical structure of the reaction product.

Example 3: A mixture of the following components is reacted in a closed mold:

100 parts by weight of a polypropylene glycol having an OH number of 28 which has been started on trimethylolpropane and modified with ethylene oxide to result in 60% of primary hydroxyl end groups,

3.1 parts by weight of N-methylmorpholine,

0.5 parts by weight of N,N-dimethyl-ethanolamine,

1.0 parts by weight of a silicone stabilizer according to the general formula:

$R = CH_3$,
$R' =$ phenyl,
$a =$ 0 to 4

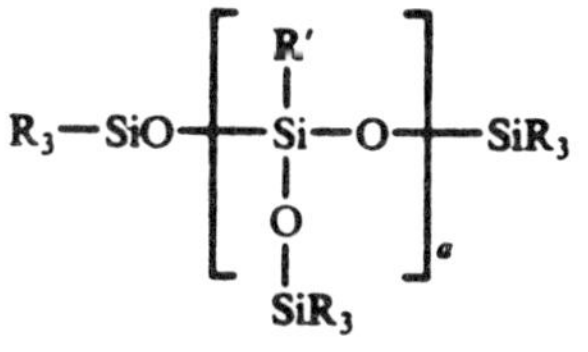

5.0 parts by weight of 2,6-diamino-4-carboisobutoxybenzenephosphonic acid diethyl ester, and

465 parts by weight of the polyisocyanate described below:

20 parts of 1,2-propylene glycol are added to a mixture of 225 parts of a mixture of 80% by weight of 2,4-tolylenediisocyanate and 20%

by weight of 2,6,tolylenediisocyanate and 274 parts of 4,4'-diphenylmethane diisocyanate at 60°C and the resulting mixture is reacted inside a metal mold for 30 minutes. The temperature is raised to 130°C after the addition of 1 part of β-phenylethyl-ethyleneamine. The trimerization reaction which occurs at this temperature is stopped after 2½ hours, when the NCO content of the reaction mixture is 26.5%, by the addition of 1 part of methyl p-toluene sulfonate.

The polyisocyanate obtained after dilution with 624 parts of an 80/20 mixture of 2,4- and 2,6-tolylene diisocyanate has an NCO content of 38.4% by weight, a viscosity of 24 cp at 25°C and a refractive index n_D^{50} = 1.5738.

A foam which has the following mechanical properties is obtained:

Gross density	(DIN 53 420)	36 kg/m³
Tensile test	(DIN 53 571)	100 KPa
Elongation at break	(DIN 53 571)	110%
Pressure test	(DIN 53 577)	4.5 KPa
Pressure deformation residue	(DIN 53 572)	7.9%
Combustion test according to ASTM-D 1692-68		
Assessment: self extinguishing		
Average burning path: 83 mm		
Average extinction time 44 seconds.		

Microcapsules Containing Tris(2,3-Dihalopropyl) Phosphate

D.N. Vincent and R. Golden; U.S. Patent 3,968,060; July 6, 1976; assigned to Champion International Corporation describe a process whereby microcapsules having an average diameter below 5 microns and containing a flame retardant material are incorporated into polymeric materials, such as polyurethane foams without affecting the structural integrity of the cell walls of the foam.

The capsules are chemically bonded to the foam, as well as physically incorporated in it by employing polyhydroxy polymer-containing capsule wall materials which react with the isocyanate used in the polyurethane. The minute capsules are incorporated into paint to give increased hiding power as well as flame resistance, and are coextruded into filaments.

The introduction of comparatively large diameter microcapsules, such as those having an average diameter of 20 microns or greater can result in the weakening of the cell structure of the foam. The cell wall or rib structure of the urethane foam has a typical size of, for example, 10 to 25 microns. Accordingly, when comparatively large diameter microcapsules are introduced into the polyurethane foamable mixture, the resulting foam contains numerous stress points in the foam at the points where the microcapsules have been introduced into the cell walls.

By employing microcapsules having an average particle diameter less than 5 microns, only a relatively small fraction of the total thickness of the cell wall or rib of the foamed structure contains the microcapsules. Thus, the stress points are obviated providing a finished article having antiflammability properties without a reduction or weakening of the physical properties of the foamed material.

Any suitable fire retardant may be employed in the microcapsules whether in liquid or solid form. Thus, the liquid fire retardant material or solution of such fire retardant material may be any one or a mixture of several classes of fire retardant chemical compounds such as phosphate esters, halogenated phosphate esters, phosphonate esters, halogenated phosphonate esters, phosphites, halogenated paraffins, halogenated aromatics and other halogenated organic compounds generally recognized as having fire retardant properties.

The microcapsular walls of the microcapsules to be employed in polyurethane foams are formed from polyhydroxy polymers, such as methylcellulose, poly(vinyl alcohol), starch, etc. By employing polyhydroxy bearing polymers in the formation of the microcapsular walls, there are hydroxyl groups available to chemically react with the polyisocyanates of the polyurethane foam forming mixture to result in a chemical bonding between the microcapsular walls and the polyurethane structure. Any suitable process for forming microcapsules may be employed including that described in U.S. Patent 3,418,656. Suitable polyhydroxy polymers include poly(vinyl alcohol), methylcellulose, starch hydroxyethylcellulose, etc.

Example 1: Thirty-two grams of a 3:1 molar adduct of toluene diisocyanate and trimethylolpropane are dissolved in 64 g of a low viscosity chlorinated aliphatic hydrocarbon (commercially available as Exchlor 4 and containing 45% chlorine) and then mixed thoroughly with 800 g of tris(2,3-dibromopropyl) phosphate (commercially available as Fyrol HB-32). The resulting oily solution is emulsified in 3,434 g of a 7% by weight of a high molecular weight, 87% hydrolyzed, poly(vinyl alcohol) solution (commercially available as Elvanol 50-42 G) containing 240 g on a dry basis, in a large Waring-type blender until the emulsion droplets have an average particle size of 3 microns.

The emulsion is then cured for 2 hours at 60°C to harden the capsule walls. The cooled product is diluted at a ratio of 1:3 with water and allowed to stand for 3 days.

The supernatant liquid is poured from the sediment, and the sediment is redispersed in 3 liters of fresh water and permitted to settle once it began. The sediment is then spray-dried at 52% by weight solids at a temperature of 65° to 93°C to give a dry, free-flowing powder containing 85% by weight volatile oils. The dry powder comprises microcapsules having an average particle diameter of 3 microns and containing the flame retardant bis(2,3-dibromopropyl) phosphate.

Example 2: Sixteen grams of the microcapsules as produced in Example 1 are combined with 300 g of a latex flat paint. The mixture is stirred at high shear to provide a flame retardant, water-based paint. When applied to a substrate, the paint provides an opaque, flame-resistant coating.

Example 3: Forty-eight grams of an adduct of toluene diisocyanate and trimethylolpropane are dissolved in 64 g of the low viscosity, chlorinated aliphatic hydrocarbon of Example 1, and the resulting solution is mixed with 80 g of tris(2,3-dichloropropyl) phosphate. The resulting solution is emulsified in 2,350 g of an 8% by weight poly(vinyl alcohol) (commercially available as Elvanol 52-42 G) employing a homomixer in order to obtain microcapsules having an average particle diameter less than 2 microns. The resulting emulsion is diluted with 300 g of water and cured for 2 hours at a temperature of 60° to 65°C. The product is then diluted on a 1:1 basis with water, and the capsule solids are separated by centrifugation. The sediment is diluted to 1 gallon and spray-dried at 11.6% solids while at a temperature of 65° to 135°C to give a dry, free-flowing powder.

Example 4: Microcapsules with an average particle diameter of <1 μ are provided by dissolving 4 g of the adduct of toluene diisocyanate and trimethylolpropane of the previous examples in 45 g of a chlorinated aliphatic hydrocarbon. The resulting solution is emulsified in 150 ml of a 7 wt % high molecular weight, 87% hydrolyzed poly(vinyl alcohol) solution containing 0.25 g of sodium lauryl sulfate (30% by weight). The emulsion is cured for 3 hr at 60° to 70°C.

Example 5: This example demonstrates the incorporation of microcapsules into regenerated cellulose fibers. Ten grams of filter paper are treated with caustic, then with carbon disulfide, and finally dissolved in dilute caustic in order to form an aqueous solution of sodium cellulose xanthate. After suitable aging or ripening of the viscose solution, 1.5 g of microcapsules made in the manner of Example 4, but containing tris(2,3-dibromopropyl) phosphate and spray-dried, are dispersed in the sodium cellulose xanthate solution.

The resulting mixture is filtered and then passed through a spinneret and extruded into a spinning solution containing 8 to 10% sulfuric acid, about 15% sodium sulfate, 1% zinc sulfate and 6% glucose. The resulting solution from the spinneret is coagulated in the bath as a filament of regenerative cellulose containing the flame retardant microcapsules. The rayon fibers are then collected and upon testing have greatly reduced flammability, while their other physical characteristics remain similar to untreated rayon.

Example 6: The following example demonstrates production of a urethane foam containing the microcapsules. Rigid, Freon-blown urethane foam is prepared by admixing 11.7 g of a polyol resin component, i.e., a polypropylene glycol, and 18.3 g of polyisocyanate. This results in a rapid initiation of an exothermic reaction and production of a rigid foam.

A second batch is prepared with the same proportions, but additionally, 3 g of the spray dried capsules of Example 5 containing 85% active flame retardant are thoroughly dispersed in the resin component before admixing the resin and the isocyanate.

The resulting foam shows no impairment of physical properties due to the presence of the microcapsules. On the other hand, the untreated foam burns vigorously when ignited, while the foam containing the flame retardant microcapsules is self-extinguishing.

Example 7: In a manner similar to Example 6, a rigid, carbon dioxide blown urethane foam is prepared by reacting 10.6 g of a polyester base resin with 9.4 g of a polyisocyanate in order to provide an exothermic reaction with rapid formation of a rigid foam. The procedure is repeated with the exception that 1 g of spray-dried microcapsules of Example 3 are thoroughly dispersed in the resin component before combining the resin and the isocyanate. The resulting foam is self-extinguishing, while the microcapsule-less foam burns vigorously.

OTHER REACTANTS

Substituted Amines in Rigid Polyurethanes

S. Mori and T. Fujita; U.S. Patent 4,088,614; May 9, 1978; assigned to Dai-Ichi Kogyo Seiyaku Co., Ltd., Japan describe a process for producing a flame resistant rigid polyurethane foam comprising reacting a mixture consisting of:

(a) a polyether polyol and a polyester polyol, each having an active hydrogen atom,

(b) a polyisocyanate, the polyisocyanate being used in an amount of 0.5 to 3 equivalents per 1 equivalent of the polyether polyol and polyester polyol,

(c) at least one blowing agent selected from water and a volatile foaming agent, the water when present being used in an amount of less than 5% by weight and the volatile foaming agent when present being used in an amount of less than 70% by weight, the weights being based on the total weight of the starting materials,

(d) an emulsifying agent, the emulsifying agent being used in an amount of less than 2% by weight based on the total weight of the starting materials, and

(e) at least one amine compound selected from general formulas (1), (2), (3), (4), (5), (6) and (7).

(1)

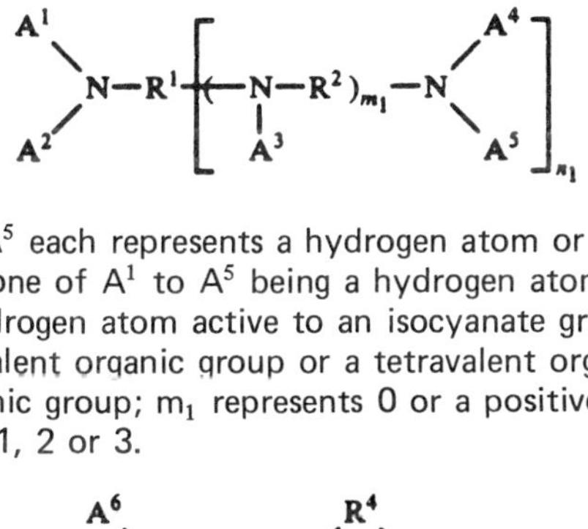

A^1, A^2, A^3, A^4 and A^5 each represents a hydrogen atom or a monovalent organic group, with at least one of A^1 to A^5 being a hydrogen atom or an organic group which contains a hydrogen atom active to an isocyanate group; R^1 is a divalent organic group, a trivalent organic group or a tetravalent organic group; R^2 represents a divalent organic group; m_1 represents 0 or a positive integer smaller than 8; and n_1 represents 1, 2 or 3.

(2)

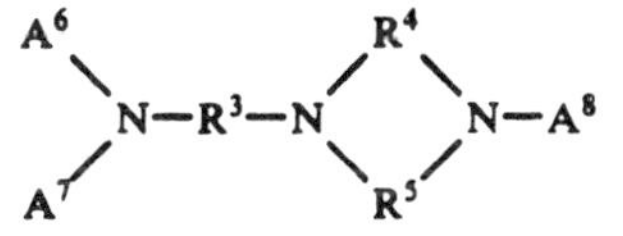

A^6, A^7 and A^8 each represents a hydrogen atom or a monovalent organic group, with at least one of A^6 to A^8 being a hydrogen atom or an organic group which contains a hydrogen atom active to an isocyanate group; R^3, R^4 and R^5 each represents a divalent organic group.

(3)

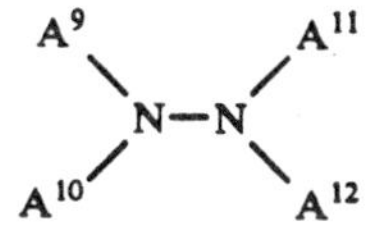

A^9, A^{10}, A^{11} and A^{12} each represents a hydrogen atom or a monovalent organic group, with at least one of A^9 to A^{12} being a hydrogen atom or an organic group which contains a hydrogen atom active to an isocyanate group.

(4)

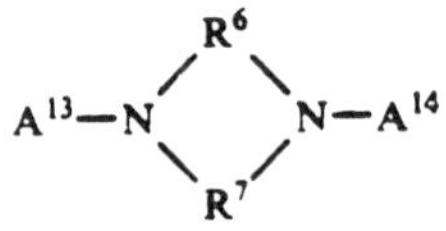

A^{13} and A^{14} each represents a hydrogen atom or a monovalent organic group, with at least one of A^{13} and A^{14} being a hydrogen atom or an organic group which contains a hydrogen atom active to a isocyanate group and R^6 and R^7 each represents a divalent organic group.

(5)

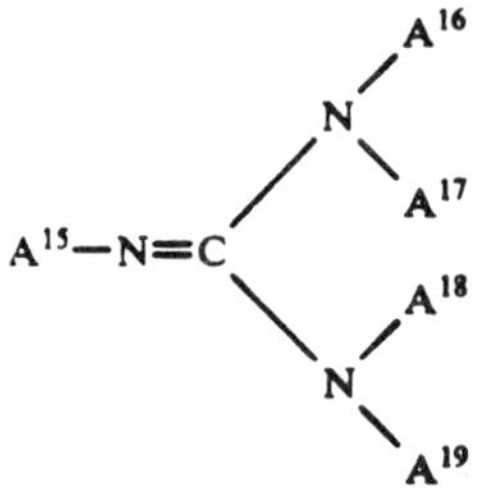

A^{15}, A^{16}, A^{17}, A^{18} and A^{19} each represents a hydrogen atom or a monovalent organic group with at least one of A^{15} to A^{19} being a hydrogen atom or an organic group which contains a hydrogen atom active to an isocyanate group.

(6)

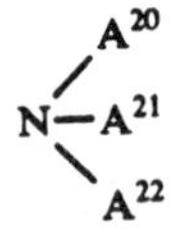

A^{20}, A^{21} and A^{22} each represents a hydrogen atom or a monovalent organic group, with at least one A^{20} to A^{22} being a hydrogen atom or an organic group which contains a hydrogen atom active to an isocyanate group.

(7)

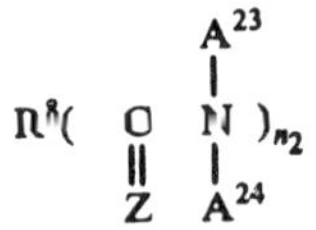

R^8 represents a hydrogen atom, a monovalent organic group, a divalent organic organic group, or a trivalent organic group. A^{23} and A^{24} each represents a hydrogen atom or a monovalent organic group, at least one of the R^8, A^{23} and A^{24} is a hydrogen atom or a group having a hydrogen atom active to an isocyanate group. Z represents an oxygen atom or a sulfur atom, and n_2 is an integer of 1 to 3.

The weight ratio of the amine compound to the polyether polyol and polyester polyol is from 7:3 to 3:7.

By blending the amine compound represented by general formula (1), (2), (3), (4), (5), (6) or (7), individually in the production of polyurethane foam, it is possible to render the polyurethane foam flame resistant, but more excellent flame resisting effects can be obtained by using a combination of two kinds of any amine compounds.

A combination of an amine compound wherein at least one member of the monovalent groups A^1 to A^{24} is a hydrogen atom and an amine compound wherein all the monovalent organic groups A^1 to A^{24} are hydroxyl-containing groups or those having hydroxyl-containing groups and organic groups having no active hydrogen atom is preferred.

The polyurethane foam obtained using the above amine compounds without using a flame retarder is self-extinguishing with a burning distance of 14 to 22 mm as tested in accordance with ASTM D 1692-68. When two kinds of amine compounds were used without adding a flame retarder, the polyurethane foams thus prepared show self-extinguishing properties with a burning distance of 0 to 4 mm using the ASTM D 1692-68 test.

A flame retarder can be used in combination with the amine compound or compounds although as is apparent from the above results, the flame resisting effects can be obtained without using such flame retarder. Suitable flame retarders which can be used in such a case are phosphorus-containing compounds. Examples of these phosphorus-containing compounds are those compounds containing phosphorus only and which do not contain any halogen, such as, for example, triphenyl phosphate, tricresyl phosphate, diethyl-N,N-bis(2-hydroxyethyl)aminomethyl phosphonate.

The amount of phosphorus-containing compound used can be less than 0.5% by weight to the weight of polyurethane foam. The polyether polyol or the polyester polyol used can be selected to achieve the desired polyurethane foam. That is, as the polyether polyol or polyester polyol for rigid polyurethane foam, polyols having a hydroxyl value of about 150 to 900 mg KOH/g can be used individually or as a mixture. The polyether polyol can be prepared by the addition polymerization reaction of an organic compound having at least two active hydrogen atoms and an alkylene oxide.

Examples of organic compounds having at least two hydrogen atoms which can be used as the raw material for producing the polyether polyol suitable for this process are an alkylene glycol such as ethylene glycol, diethylene glycol, triethylene glycol, propylene glycol, dipropylene glycol, tripropylene glycol, butane-2,3-diol, butane-1,3-diol, hexane-2,5-diol, octadecanediol, octadecenediol, cyclohexane-1,4-dimethanol, glycerine, trimethylolethane, trimethylolpropane, pentaerythritol, mannitol, sorbitol, and sucrose.

Also, suitable examples of alkylene oxides which can be used for the addition polymerization reaction are ethylene oxide, propylene oxide, butylene oxide, epichlorohydrin, and the like. Moreover, the polyester polyol used can be prepared by the polymerization reaction of a polyol having at least two active hydroxyl groups and a polybasic acid.

Examples of the polyols are, as in the case of polyether polyols, an alkylene glycol such as ethylene glycol, diethylene glycol, triethylene glycol, propylene glycol, dipropylene glycol, tripropylene glycol, butane-2,3-diol, butane-1,3-diol, hexane-2,5-diol, octadecanediol, octadecenediol, cyclohexane-1,4-dimethanol, glycerine, trimethylolethane, trimethylolpropane, pentaerythritol, mannitol, sorbitol, sucrose, etc.

Examples of suitable polybasic acids are adipic acid, succinic acid, suberic acid, sebacic acid, oxalic acid, methyladipic acid, terephthalic acid, isophthalic acid, 1,2,4-benzenetricarboxylic acid, thiodiglycolic acid, thiodipropionic acid, maleic acid, fumaric acid, citraconic acid, itaconic acid, trimellitic acid, a dimerized or trimerized fatty acid, tartaric acid, etc.

The polyisocyanate used is a material conventionally used for producing polyurethane foams and can be appropriately represented by the general formula:

$$R^{26}(NCO)_{n_3}$$

where R^{26} represents an aliphatic group, an aromatic group, a substituted aliphatic group, or a substituted aromatic group and n_3 has a mean value of 1.5 to 3.0. Typical examples of polyisocyanate are ethylene diisocyanate, ethylidene diisocyanate, propylene diisocyanate, butylene diisocyanate, etc.

In preparing polyurethane foams, a polyether polyol (and/or a polyester polyol), polyisocyanate, a foaming agent, an amine compound, and other additives, for examples, dyes, pigments, fillers, antioxidants, plasticizers and the like are uniformly blended in a short period of time followed by foaming and molding and then the foamed product is heated externally or is allowed to stand to complete the curing of the product. These polyurethane foams find applications in fields of automobiles, railway vehicles, aircrafts, etc.

Polyhydroxy Compounds and Polyether Polyols

T. Yukuta, T. Ohashi, M. Kojima, A. Suzuki and Y. Taniguchi; U.S. Patent 4,066,579; January 3, 1978; assigned to Bridgestone Tire Company Limited, Japan describe flame and smoke resistant flexible polyurethane foam produced by reacting a polyisocyanate with a mixture of

(A) polyether polyol containing secondary hydroxyl group of at least 50 mol percent of total hydroxyl content and having a number average molecular weight of 3,000 to 6,000 and

(B) a low molecular weight polyhydroxyl compound, an equivalent weight ratio of the low molecular weight polyhydroxyl compound (B) to the polyether polyol (A) being 0.5 to 2.0. If desired, the flame retardancy of the foam may be enhanced by the addition of a small amount of flame retardant.

Example 1: A mixture of 100 parts of poly(oxypropylene) triol GP-3000 (hydroxyl value: 56, molecular weight: about 3,000) as a component (A), 3.0 parts of trimethylolpropane as a component (B), which was previously melted at 60°C, the equivalent weight ratio of the component (B) to the component (A) being 0.68, 3.0 parts of water, 0.1 part of triethylenediamine, 0.3 part of stannous octoate, 3.0 parts of silicone surfactant SH-193 (a surfactant for rigid polyether urethane foam) and 5 parts of trichloromonofluoromethane

(F-11) was stirred for about 30 seconds to obtain a homogeneous solution. When the solution was stirred at high speed together with 43.6 parts (NCO index: 100) of tolylene diisocyanate TDI-80 (2,4-isomer:2,6-isomer = 80:20), the reaction mixture is emulsified to become creamy in about 13 seconds. When the creamy mixture was immediately charged into a paper mold, the mixture began to foam, and the foam rise was completed in about 170 seconds. The resulting flexible polyurethane foam had a uniform cell structure with a small proportion of closed cells, yielding completely open cell-type structure after crushing.

The burning test of the resulting polyurethane foam according to ASTM D-1692-59T showed that the burning distance of the foam was 3.8 cm and the burning time was 22 seconds and the foam had an excellent self-extinguishing property. The smoke generation coefficient C_{Amax} of the foam was 56.6, when the value of the coefficient measured according to JIS A-1321-1970 was converted to a basis of 30.0 kg/m^3 of foam density.

The physical properties of the polyurethane foam obtained in this example are shown in the following Table 1 together with the results of the above described burning and smoke tests.

TABLE 1

	Example 1
Density (kg/m^3)	25.9
Tensile strength (kg/cm^2)	0.32
Elongation (%)	80
ILD (kg/200 mmϕ)25%	7.5
Compression set (%)	
(50% deflection, 70°C, 22 hr)	9.1
Resilience (%)	25
Burning test	self-
(ASTM D-1692-59T)	extinguishing
Smoke test	
(JIS A-1321-1970), C_{Amax}	56.6

Examples 2 through 6: The procedure of Example 1 was repeated, except that the amount of component (B) of trimethylolpropane, silicone surfactant, and isocyanate was varied as shown in Table 2, to obtain flexible polyurethane foams.

The results of the burning test and smoke test, and the physical properties of the resulting flexible polyurethane foams are also shown in Table 2. As seen from Table 2, all of the resulting flexible polyurethane foams are self-extinguishing, and have a smoke generation coefficient C_{Amax} lower than that of general purpose polyurethane foams shown in the following Examples 7 and 8.

TABLE 2

	. . . .	. . . .	Example Number	. . . .	. . . .
	2	3	4	5	6
Compounding recipe (parts)					
(A) Triol GP-3000	100	100	100	100	100
(B) Trimethylolpropane	4.4*	6.6*	4.4*	4.4*	4.4*
	(1)	(1.5)	(1)	(1)	(1)

(continued)

TABLE 2: (continued)

	Example Number				
	2	3	4	5	6
Stannous octoate	0.3	0.1	0.3	0.3	0.1
Triethylenediamine	0.1	0.2	0.1	0.1	0.2
Water	3.0	3.0	3.0	3.0	3.0
Silicone surfactant (SH-193)	3.0	3.0	4.0	3.0	3.0
Trichloromonofluoromethane (F-11)	5.0	5.0	5.0	5.0	5.0
Tolylene diisocyanate (TDI-80)	46.2** (100)	50.5** (100)	46.2** (100)	43.9** (95)	48.5** (105)
Creaming time (sec)	15	12	12	13	12
Rising time (sec)	120	120	110	130	90
Burning test (ASTM D-1692-59T)	self-extinguishing				
Smoke test (JIS A-1321-1970)					
Smoke coefficient C_{Amax}	63.7	61.0	57.8	48.1	65.4
Physical properties					
Density (kg/m³)	25.5	25.4	25.0	25.1	28.1
Tensile strength (kg/cm²)	0.47	0.41	0.34	0.45	0.56
Elongation (%)	98	87	89	108	75
ILD (kg/200 mmφ)$^{25\%}$	5.9	5.1	6.6	5.4	8.6
Compression set (%) deflection, 70°C, 22 hr)	21.8	20.0	9.3	15.3	8.1
Resilience (%)	21	19	25	18	21

*The numeral in parentheses is the equivalent weight ratio of trimethylolpropane of component (B) to the hydroxyl group of triol of component (A).

**The numeral in parentheses is the isocyanate index. The isocyanate index is the percentage of the amount of polyisocyanate to the whole amount of hydroxyl compound and other compounds having active hydrogen in equivalent basis.

Examples 7 and 8: In order to clarify the feature of the process, conventional polyurethane foams were produced according to the compounding recipes shown in Table 3 in the same procedure as described in Example 1. The results of the burning test and smoke test and the physical properties of the resulting foams are also shown in Table 3.

TABLE 3

	Example Number	
	7	8
Compounding recipe (parts)		
(A) Triol GP-3000	100	100
Stannous octoate	0.26	0.26
Triethylenediamine	0.1	0.1
Water	4.2	4.2
Silicone surfactant (L-520)	1.7	—
Silicone surfactant (SH-193)	—	1.7
Tolylene diisocyanate (TDI-80)	53.5* (108)	53.6* (108)
Burning test (ASTM D-1692-59T)	flammable	flammable
Smoke test (JIS A-1321-1970)		
Smoke coefficient C_{Amax}	82.7	86.3
Physical properties		
Density (kg/m³)	27.6	29.2
Tensile strength (kg/cm²)	0.60	0.61
Elongation (%)	80	74
ILD (kg/200 mmφ)$^{25\%}$	16.0	23.2
Compression set (%) (50% deflection, 70°C, 22 hr)	2.5	3.3
Resilience (%)	39	30

*The numeral in the parentheses is the isocyanate index.

In Example 7, a flexible polyurethane foam was produced in the same manner as described in Example 1, except that silicone surfactant L-520 (a surfactant for flexible polyurethane foam) was used and a low molecular weight of polyhydroxyl compound of component (B) was not used. The resulting foam is a so-called general purpose flexible polyurethane foam. The burning test and smoke test of the resulting foam showed that the foam was flammable and had a smoke generation coefficient noticeably higher than that of the foam of the process.

In Example 8, a flexible polyurethane foam was produced in the same manner as described in Example 1, except that a low molecular weight polyhydroxyl compound of component (B) was not used. The resulting foam is flammable and has a smoke generation coefficient higher than that of the foam of the process. The polyurethane foams produced are flexible polyurethane foams having a density of 20-50 kg/m^3. They are useful for sheet cushion materials in various vehicles, aircrafts, and ships and materials of various buildings, furnitures and interiors.

Cyano-Substituted Organosiloxane-Polyoxyalkylene Polymers

B. Prokai and B. Kanner; U.S. Patent 4,058,493; November 15, 1977; assigned to Union Carbide Corporation describe a process for producing flexible polyurethane foam which comprises simultaneously reacting and foaming a reaction mixture containing:

(a) a polyester polyol reactant having an average of at least two hydroxyl groups per molecule;

(b) a polyisocyanate reactant containing at least two isocyanato groups per molecule;

(c) a blowing agent;

(d) a catalyst comprising an amine; and

(e) a foam stabilizer comprising polysiloxane polyoxyalkylene copolymers having the following average composition.

$$(R)_{3-(t+u)}\overset{(E)_u}{\underset{(Q)_t}{Si}}O[\overset{R}{\underset{R}{Si}}O]_x[\overset{R}{\underset{Q}{Si}}O]_{y-(t+w)}[\overset{R}{\underset{E}{Si}}O]_{z-(u+v)}\overset{(E)_v}{\underset{(Q)_w}{Si}}(R)_{3-(v+w)}$$

R is alkyl having from 1 to 10 carbon atoms; Q is a cyano-bearing ether group having the formula, $-(O)_qR'OR''CN$, wherein q is 0 or 1, R' is bivalent alkylene having from 3 to 8 carbon atoms and R'' is bivalent alkylene of 2 to 4 carbon atoms. E is a polyoxyalkylene block having the formula, $-(R^\circ)_p-(OC_aH_{2a})_b-OG$, wherein R° comprises a bivalent alkylene group a carbon atom of which is bonded to silicon, G comprises a monovalent hydrocarbon group having from 1 to 12 carbon atoms, p is zero or one, $-(OC_aH_{2a})_b-$ is a polyoxyalkylene chain constituted of at least 75 wt % of oxyethylene wherein b has an average value from 3 to 30.

Each of t, u, v and w is independently 0 or 1 provided each of the sums t + u and v + w is 0 or 1; each of the sums t + w and u + v is independently 0, 1 or 2; x is zero or a positive number having an average value up to about 20; y has an average value from 2 to 20; and z has an average value from 2 to 30.

The polyoxysiloxane-polyoxyalkylene polymers of this process are effective stabilizers of flexible polyester polyol-based polyurethane foam and offer particular advantage in the formation of flame retarded foam.

The flame retardants that can be employed in producing flame retarded flexible polyester foams can be chemically combined in one or more of the other materials used (e.g., in the polyol or polyisocyanate), or they can be used as discrete chemical compounds added as such to the foam formulation. The organic flame retardants usually contain phosphorus or halogen, both phosphorus and halogen, or phosphorus and nitrogen. Halogen, when present, is chlorine and/or bromine.

Flame retardants of the discrete chemical variety include: 2,2bis(bromomethyl)-1,3-propanediol; 2,3-dibromopropanol; tetrabromophthalic anhydride; brominated phthalate ester diols such as those produced from tetrabromophthalic anhydride, propylene oxide and propylene glycol; tetrabromobisphenol A; 2,4,6-tribromophenol; pentabromophenol; brominated anilines and dianilines; bis(2,3-dibromopropyl) ether of sorbitol; tetrachlorophthalic anhydride; chlorendic acid; chlorendic anhydride; diallyl chlorendate; chlorinated maleic anhydride; etc.

Of the above flame retardants, those of the discrete chemical compound variety which contain groups reactive with hydroxy or isocyanato groups can be used as reactants in producing the polyester polyols or can be reacted with organic polyisocyanates to produce modified polyols or polyisocyanates having chemically combined flame retardant groups. Such modified polyester and polyisocyanates are useful as reactants in this process. In such cases, due regard must be given to the possible effect of the functionality of the compound on the other properties (e.g., degree of flexibility) of the resulting foam.

The flexible polyester urethane foams produced can be used in the same areas as conventional flexible polyester urethane foams. For example, they can be used as textile interliners, cushioning materials for seating, for packaging of delicate objects, and as gasketing materials.

In work that is similar, *B. Prokai and B. Kanner; U.S. Patent 4,058,494; November 15, 1977; assigned to Union Carbide Corporation* describe a process for producing flexible polyurethane foam which comprises using in the reaction mixture a polyether polyol reactant in place of the polyester polyol component mentioned above.

Modified Diisocyanate Mixtures

Y. Sawachika, H. Kawakami and S. Minegishi; U.S. Patent 4,075,137; Feb. 21, 1978; assigned to Kohkoku Chemical Industry Co., Ltd., Japan describe a process for preparing flame retarding polyurethane foams from a polyhydroxyl compound, diisocyanate mixture, a blowing agent, a catalyst and any other auxiliary agent if necessary.

This comprises mixing the polyhydroxyl compound having at least two active hydrogen atoms per molecule, a molecular weight of 1,000 to 10,000 and where at least 10% by weight of the hydroxyl groups are primary hydroxyl groups, and a diisocyanate mixture containing 10 to 60% by weight of a modified diisocyanate prepared from the reaction of a diisocyanate with a compound expressed by the formula

$$(1)\qquad CH_3{-}NH{-}\overset{\overset{\displaystyle X}{\|}}{C}{-}NH{-}CH_3$$

wherein X is an oxygen or sulfur atom.

The polyhydroxyl compounds can be prepared by adding butylene, propylene or ethylene oxide to a reaction initiator having at least two hydroxyl groups, such as, glycol, glycerine, and trimethylolpropane, to make the molecular weight up to 1,000 to 10,000.

Organic diisocyanates suitable for use in the process of this method include aliphatic diisocyanates (such as hexamethylene diisocyanate), cycloaliphatic diisocyanates (such as dicyclohexylmethane diisocyanates) and aromatic diisocyanates (such as tolylene diisocyanate, diphenylmethane diisocyanate and xylene diisocyanate). The compounds suitable to use in this process to prepare the modified diisocyanate include, 1,3-dimethylurea, 1,3-diphenylurea and 1,3-diethylthiourea.

In carrying out the reaction between compound (1) and the diisocyanate, it is important that a molar ratio of 1:3.6 to 25.6 be maintained and that the reaction temperature be less than 150°C. Otherwise, trivalent and tetravalent isocyanates are produced which are not suitable.

By applying this process, flame retarding polyurethane foams having low smoke generation properties that have been otherwise impossible to produce, can successfully be prepared. These polyurethane foams exhibit self-extinguishing properties as defined by the flammability test ASTM D 1692-68 and a small smoke generation coefficient as defined in JIS A 1321.

Comparative Example: For the sake of comparison, a prior art process for preparing the flame retarding polyurethane foams will be described as follows. A composition consisting of the following components was vigorously stirred for about 7 seconds with an agitator and allowed to rise freely to a foam to obtain a mass of polyurethane foam. The characteristics of this foam were as follows: foam density, 1.2 lb/ft^3; tensile strength, 15.6 lb/in^2; tear strength, 3.4 lb/in; 25% ILD, 22 lb/50 in^2; and elongation, 180%.

These foam properties were measured by ASTM D 1564-71. The foam was self-extinguishing in the flammability test as defined by ASTM D 1692-68. Further, the smoke-generation coefficient as estimated by JIS A 1321 was 60.

	Parts by Weight
Polyol (glycerine base, OH value 55, MW 3,000)	100
Water	3.5
Stannous octoate	0.6
Triethylenediamine	0.1
N-methylmorpholine	1.0
Silicone surfactant	1.5
Tris(2-chloroethyl) phosphate	13
Tolylene diisocyanate	45

Examples of the present process are as follows.

Example 1: Preparation of Diisocyanate Mixture — 176 g of 1,3-dimethylurea was gradually added to 1,740 g of a 80/20 isomeric mixture of 2,4- and 2,6-tolylene diisocyanates at 110°C, then the temperature was elevated to 150°C under a protecting atmosphere of nitrogen and the reaction was continued for 12 hours with the temperature maintained, to obtain a diisocyanate mixture which contains 45.1% by weight of modified diisocyanate. When cooled, the mixture showed the following properties: NCO content, 35.1% and viscosity, 148 cp/25°C.

Preparation of Polyurethane Foam —

	Parts by Weight
Polyol (glycerin base, OH value 25, MW 6,000, primary OH group content 60%)	100
Water	3.5
Triethylenediamine	0.1
Triethylamine	0.6
Silicone surfactant	0.5
Diisocyanate mixture	50

A mixture consisting of the above components was vigorously stirred with an agitator for 7 seconds and allowed to rise freely to a foam to obtain a polyurethane foam. This foam had the following characteristics: foam density, 2.1 lb/ft^3; tensile strength, 18.2 lb/in^2; tear strength, 3.5 lb/in, 25% ILD, 17.6 lb/50 in^2; and elongation, 150%.

The foam was self-extinguishing as defined by ASTM D 1692-68. Data in this example and the comparison example are shown in Table 1. Further, the smoke generation coefficient as estimated by JIS A 1321 was 14, which was much smaller than those exhibited by the foams prepared by prior art processes.

Example 2: Preparation of Diisocyanate Mixture — 135 g of 1,3-dimethylthiourea was gradually added to 1,810 g of a 80/20 isomeric mixture of 2,4- and 2,6-tolylene diisocyanates, and the temperature was elevated up to 150°C under a protecting atmosphere of nitrogen, the whole mixture was kept stirred at the same temperature for 12 hours, to obtain a diisocyanate mixture which contains 30.2% by weight of modified diisocyanate. When cooled, the product mixture showed the following characteristics: NCO content, 40.0% and viscosity, 69 cp/25°C.

Preparation of Polyurethane Foams —

	Parts by Weight
Polyol (glycerine base, OH value 25, MW 6,000, primary OH group content 60%)	100
Water	3.5
Triethylenediamine	0.1
Triethylamine	0.6
Silicone surfactant	0.5
Diisocyanate mixture	43.0

A composition consisting of the above components was vigorously agitated for 7 seconds with a stirrer, and then allowed to rise freely to a foam to obtain a mass of polyurethane foam. The characteristics exhibited by the foam were as follows: foam density, 1.9 lb/ft^3; tensile strength, 13.9 lb/in^2; tear strength, 2.9 lb/in; 25% ILD, 15.4 lb/50 in^2; and elongation, 130%.

The foam was also self-extinguished as defined ASTM D 1692-68. Comparison with the Comparative Example above is shown in Table 1. The smoke generation coefficient was estimated to be 15 following JIS A 1321. This is illustrated in Figure 1.1 in comparison to Comparative Example above.

TABLE 1: FLAMMABILITY BY ASTM D 1692-68

	Foam in Example 1	Foam in Example 2	Foam in Comparative Example
Burning extent (mm)	50	60	70
Extinguishment time (sec.)	26	35	50
Burning rate (mm/min.)	115	103	84
Flammability	self-extinguishment	self-extinguishment	self-extinguishment
Remark			
Foam density:	2.1 lb/ft^3	1.9 lb/ft^3	2.1 lb/ft^3 Prepared in the presence of flame retardant containing halogen and phosphorus.

FIGURE 1.1: VARIATIONS OF SMOKE GENERATION PROPERTIES

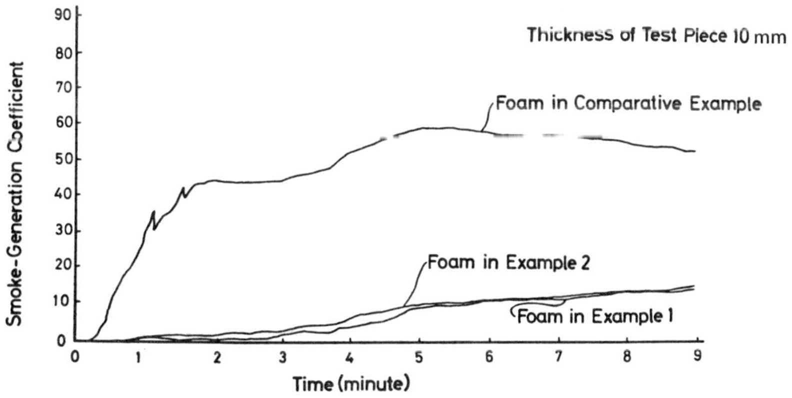

Source: U.S. Patent 4,075,137

In related work, *Y. Sawachika and H. Kawakami; U.S. Patent 4,071,484; Jan. 31, 1978; assigned to Kohkoku Chemical Industry Co., Ltd., Japan* describe several other diisocyanate mixtures used in preparing unshrinkable, flexible flame retarding polyurethane foams.

Elastomeric Polyurethane Compositions

J.C. Fletcher, J.T. Howarth, S. Sheth, K.R. Sidman and A.A. Massucco; U.S. Patent 4,069,212; January 17, 1978 describe a flame retardant spandex type of polyurethane resin suitable for forming into fibers. This is formed from a mixture comprising:

(a) A diisocyanate selected from the group consisting of arylene diisocyanates, 4,4'-diphenylmethane diisocyanate and mixtures of these.

(b) A halogen-containing difunctional polyol selected from the group consisting of dibromoneopentyl glycol, polyesters of dibromoneopentyl glycol, esters of the product obtained from the reaction of ethylene oxide and tetrabromophthalic anhydride, esters of the product obtained from the reaction of propylene oxide and tetrabromophthalic anhydride and esters of the product obtained from the reaction of ethylene oxide, propylene oxide and tetrabromophthalic anhydride, the polyol being substantially free of corresponding monofunctional and trifunctional impurities.

(c) A urea-producing compound selected from the class consisting of diamines and water.

The composition defined in (b) is first reacted with a stoichiometric excess of the composition defined in (a) to form an isocyanate terminated prepolymer, the halogen-containing polyol is present in an amount sufficient to render the prepolymer substantially nonflammable in ambient environments. The prepolymer is then extended by reacting it with the composition defined in (c).

Example 1: A halogen-containing polyester based polyurethane was prepared using dibromoneopentyl glycol. The purified brominated neopentyl glycol was reacted with toluene diisocyanates in a 1:2 molar ratio, respectively following which the isocyanate-terminated product was reacted with a polybutyl adipate having a molecular weight of approximately 1,200. The resulting polymer possessed the following properties:

Bromine, wt %	10-11
Oxygen index	24-25
Ultimate tensile strength, psi	7,000
Ultimate elongation, %	500
Elasticity	Good
Recovery	Good
Color	Water white
Soft segment, wt %	60.6
Toluene diisocyanate, wt %	12.1
Brominated neopentyl glycol, wt %	18.3

It is noted that the oxygen index of 25 for the polyurethane containing the brominated neopentyl glycol compares with an oxygen index of 19 to 20 for a normal spandex type of polyurethane such as Lycra. Such an oxygen index, i.e., 25, would be sufficient to offer protection in normal air environments, conventional spandex type polyurethanes on the other hand being flammable in normal air environments.

Example 2: The adipic acid ester of a diol obtained by reacting brominated phthalic anhydride and ethylene and propylene oxides was prepared. The resulting adipate had an acid number of 6.83 and a hydroxyl number of 41.4. The hydroxy-terminated adipate was then reacted with 4,4'-diphenylmethane diisocyanate in a molar ratio of 1:2, respectively to produce an isocyanate-terminated polyadipate prepolymer. This prepolymer was then extended with hydrazine into a spandex type polyurethane. The resulting material was a strong, hard, flexible and clear material having an oxygen index of 30.

Example 3: The procedure of Example 2 was followed except azelaic acid was substituted for the adipic acid. The material produced was strong, hard, flexible and clear, had greater elasticity than the adipate prepared as per Example 2 and had an oxygen index of 28.

Examples 1 through 3 demonstrate the preparation of spandex type polyurethanes having incorporated in them, as part of the polymer backbone, a halogenated polyol. As observed from the results of the examples, the polyurethanes produced have oxygen indexes of 25 or greater meaning that fibers and ultimately fabrics woven or knit from them could be used under atmospheric conditions without fear of the materials being flammable. This is to be compared with a conventional type spandex polyurethane which has an oxygen index of 19 to 20 and which in normal atmosphere is flammable.

It should be noted that by incorporating certain of the fire retardant additives noted above into the polyurethanes prepared in Examples 1 through 3, the fire retardant properties of the materials can be enhanced even more.

Example 4: A series of elastomeric formulations comprised of a commercially available spandex type polyurethane (Lycra) and various fire retardant additives were prepared. Table 1 shows the compositions of the formulations and the results of the oxygen and flammability tests carried out on the formulations.

TABLE 1

	Sample Number											
	1	2	3	4	5	6	7	8	9	10	11	12
Spandex (Lycra), %	10	10	10	10	20	20	20	20	40	40	40	40
Hexabromobenzene, %	45	55	65	75	40	48	56	64	30	36	42	48
Tris(bromochloropropyl) phosphate, %	45	35	25	15	40	32	24	16	30	24	18	12
Oxygen index	40-50	45-50	50-55	55-60	35-40	40-45	–	40-45	35	35-40	30	30

(continued)

TABLE 1: (continued)

	Sample Number 1	2	3	4	5	6	7	8	9	10	11	12
Flammability tests, rate of propagation, in/sec												
Burning 35% O_2, 10 psi	–	SE*	–	–	SE	SE	SE	–	–	0.71	–	–
Burning 100% O_2, 6.2 psi	–	1.66	–	–	1.0	1.25	1.42	–	–	1.66	–	–
Tensile strength, psi	373	335	277	310	1,157	918	560	727	1,750	1,950	1,657	1,726
Elongation, %	600	563	500	125	625	638	600	488	713	650	563	575
Control, Spandex alone												
Oxygen index	19											
Tensile strength, psi	7,100											
Elongation, %	700											

*Self-extinguishing.

Example 5: A series of formulations of various fluoroelastomeric resins in combination with certain flame retardant additives and curing agents was prepared and extruded into fibers, cured at 400°F for 16 hours and their physical and fire retardant properties evaluated. The compositions of the formulations are shown in Table 2 whereas the results of the physical and fire retardant property tests are shown in Table 3.

TABLE 2

	Formulation, parts by weight						
Ingredients	**Control**	**A**	**B**	**C**	**D**	**E**	**F**
Viton A	—	100	—	—	—	—	—
Viton B	100	—	80	80	100	100	100
Viton LM	—	—	20	20	—	—	—
Decabromobiphenyl	100	100	100	100	100	100	100
Zinc Oxide	—	—	—	—	—	10	—
Lead Oxide	15	15	15	15	15	—	15
Dyphos (dibasic lead phosphate)	—	—	—	—	—	10	—
Tricresylphosphate	—	10	10	10	10	10	10
Antimony Oxychloride	—	—	—	—	—	—	20
Diak #3	4	4	4	4	—	4	4
Diak #4	—	—	—	—	4	—	—

TABLE 3

Fiber Formulation	Weight Loss on Cure (%)	Tensile Strength (psi)	Elongation (%)	Oxygen Index
Control	2.7	1400	250	85
A	6.9	1200	410	85
B	6.3	1200	450	85
C	5.0	1180	375	85
D	1.9	1500	450	85
E	7.3	1490	400	85
F	6.4	1240	350	85

Example 6: A formulation comprised of 100 parts of Viton B, 50 parts of decabromodiphenyl, 15 parts of lead oxide, 5 parts of tricresyl phosphate and 4 parts of N,N-dicinnamylidene-1,6-hexane diamine was blended on an unheated two-roll mill. The formulation was then granulated to particles approximately 1/16 inch in diameter and fed to an extruder where it was converted into 12 continuous filaments having diameters of approximately 13 mils. The extruder was run at 50% of capacity, producing approximately one-third pound of fiber per hour (equivalent to 2,000 ft/hr of fiber).

As the fibers exited the die, they were directed individually through guides to vessels where they were collected in coiled form. The fibers were removed from the vessels and uncoiled in Pyrex trays into an open, loose pattern. During the uncoiling procedure the fibers were sprinkled liberally with talc to prevent fiber-to-fiber adhesion. The trays were then placed in air-circulating ovens at 400°F for 16 hours to cure the fibers. Following curing, the fibers were wound on spools. The fibers thus produced had physical and fire retardant properties similar to those shown in Table 3.

Example 7: This example demonstrates the production of fabrics incorporating the fire retardant elastomeric fibers of this process with and without nonelastic fibrous materials such as fiberglass, certain types of nylons, particularly those containing isophthaloyl chloride, polybenzimidazoles, etc. Fibers produced from fluoroelastomeric compositions (Viton B) such as those set forth in Table 2, were made into jersey knits from 800 denier fiber. The fibers exhibited excellent two-way stretch.

In another case an 800 denier fiber produced from a fire retardant fluoroelastomeric composition similar to those of Table 2 and containing Viton B and a 200 denier continuous filament Nomex were made into a warp knit, 36-gauge power net. The resulting fabric exhibited two-way stretch.

A 120-needle jersey consisting of Beta Fiberglas 150 and a fluoroelastomeric composition similar to those shown in Table 2 and containing Viton as a fluoroelastomer was produced. The Viton fiber, elongated 50 to 70%, was laid in a weft direction. The fabric exhibited stretch in the weft direction only.

A power knit fabric was also produced from a thermally stabilized polybenzimidazole fiber and a flame retardant fluoroelastomeric fiber produced from compositions similar to those shown in Table 2.

The fabrics produced from the nonelastic fibers discussed above and the flame retardant elastomeric fibers disclosed herein can be used in many applications, particularly in space exploration. For example, the fabrics can be used to produce instrument caps, sleep-restraint gear, space suits, flight suits, fabrics for aircraft, etc. These compositions find application in any environment where it is desired to use an elastomeric flame resistant material having good mechanical and recovery properties. Thus for example the compositions can be used to produce open and closed cell foam coatings, tubing, hoses, gaskets and seals.

Use in Flame Retardant Mattresses

A process described by *T.P. Dougan; U.S. Patent 4,092,752; June 6, 1978;* as-

signed to The Upjohn Company relates to cushioning materials and is concerned with a box spring and mattress combination.

The matress consists of a core of flame retarded flexible polyurethane foam which is the product of reaction, under foam producing conditions, of:

(a) a polyisocyanate mixture which contains from 5 to 95% by weight of toluene diisocyanate and from 95 to 5% by weight of polymethylene polyphenyl polyisocyanate containing from 40 to 70% of methylenebis(phenyl isocyanate), the remainder of the polymethylene polyphenyl polyisocyanates having a functionality higher than 2.0;

(b) a polyether polyol having an equivalent weight from 500 to 2,500 and a functionality from 2.0 to 4.0;

(c) from 2 to 20% by weight, based on weight of final foam, of antimony oxide;

(d) from 1 to 15% by weight, based on weight of final foam, of a polyhalogenated member selected from the class consisting of polyhalogenated aliphatic diols and polyhalogenated aromatic compounds;

(e) from 0 to 15% by weight, based on weight of final foam, of alumina trihydrate; and

(f) in the case when the amount of alumina trihydrate is 0, a chlorinated hydrocarbon polymer in an amount such that the resultant foam contains from 2 to 8% by weight of chlorine.

The core of flame retarded polyurethane foam is enclosed in a covering of flame retardant ticking, and the box spring has a noncombustible frame.

The combination of box spring and mattress provided is characterized by the ability to pass very stringent tests for evidence of hazards created by exposure to fire.

Illustratively, in a test which involved igniting a pile of newspapers under the box spring and mattress and allowing the resulting fire to burn without control, a box spring and mattress combination fabricated in accordance with this process resisted ignition throughout the whole period for which the paper was burning.

POLYESTERS

ADDITIVES

Polybrominated Aromatic Compounds

V. Mischutin; U.S. Patent 3,974,310; August 10, 1976; assigned to White Chemical Corporation describes a process for flameproofing disperse-dyeable synthetic materials. This entails a process similar to dyeing but using colorless pigments with flame retardant properties. The colorless pigments are highly brominated aromatic and cyclic aliphatic compounds, which are extremely heat stable, high melting solids, which in addition are easily pulverized by milling to any desired particle size, ranging from 1 to 2 microns in average diameter.

Once the particles of the brominated aromatic compounds have been reduced to the optimum size, they are dispersed in water, at any concentration, by slowly adding them to water under strong agitation with a high-speed, high-shear mixer. In some cases a strong surface active agent must be added to effect uniform and thorough wetting of the flame retardant particles. After addition of the solid has been completed, the resulting dispersion can be stabilized by adding small quantities of a protective colloid such as carboxymethylcellulose, ethoxy- or methoxycellulose of selected molecular weight to achieve an optimum viscosity to prevent the fine particles from settling. The resulting milky, viscous paste remains stable for an indefinite length of time, if stored under normal storing conditions.

The dispersion obtained as described above can be applied to disperse-dyeable synthetic materials by exhaustion or padding techniques, normally employed to dye these materials. The dispersion will be referred to as "FR Dispersion." The FR Dispersion can be applied by three different techniques: (1) batch process, at atmospheric pressure in presence of dye carrier with or without a dye levelling agent, commonly known as atmospheric dyeing process; (2) batch process, under pressure, with or without a carrier and/or levelling agent, commonly referred to as pressure dyeing process; and (3) continuous process, at atmospheric pressure without the use of carriers and/or levelling agents, commonly called the thermosol process.

The following aromatic compounds can be used as flame retarding materials: polybrominated biphenyl, e.g., hexabromobiphenyl; polybrominated biphenyl oxide, e.g., decabromobiphenyl oxide; and polybrominated benzene,e.g., hexabromobenzene.

The process is applicable to textile materials, in staple, tow, yarn, woven, nonwoven, circular knitted, tricot knitted, raschel knitted, double knitted, crimped, texturized, flocked, tufted, etc., form, to continuous films, sheets, etc., used for packaging, decorations, insulation, etc., and any other material dyeable by the pad-dry-thermosol process.

The substrate to be treated may be in the form of staple, tow, filaments, yarn, thread, continuous fiber, woven, knitted and nonwoven fabrics, carpet, sheets, films and the like. The quantities to be used will vary depending on the construction of the material, its flammability and the flame retardant specification to be met. The following examples illustrate the process.

Example 1: 70 parts of finely pulverized hexabromobenzene having an average particle size of 1.5 microns were slowly added to 30 parts of water, under rapid agitation, containing 0.25% of an ethoxylated nonylphenol type wetting agent. After addition was completed the agitation was continued for an extra 15 minutes. To this dispersion 10 parts of a 4% hydroxyethylcellulose solution in water having a viscosity of approximately 300,000 cp were added and the resulting dispersion was agitated for an additional 15 minutes. This dispersion was designated I.

A piece of spun woven polyester fabric weighing 10 g fastened to a rod was immersed in 150 ml of water at room temperature with a pH of 6.0 adjusted with monosodium phosphate contained in a beaker. This beaker was placed in an oil bath. The rod was now attached to a mechanism which imparted an up and down motion at a rate of 20 strokes per minute. The oil bath was now heated by means of an electrical resistance until the temperature of the water reached 50°C. Heating was now interruped and 2.5 g of self-emulsifiable biphenyl predissolved in 2.5 g of water at 50°C were added as a swelling agent.

This solution was maintained at 50°C for 10 minutes to allow the biphenyl to penetrate the fabric. At this point 2 g of dispersion I and 1 g of dispersion Blue 1 (Color Index No. 42025) were added. This solution was now heated, at a rate of 0.5°C per minute, to the boil The bath was maintained at the boil for 2 hours. At this point the beaker was removed from the oil bath and placed in a cooling bath. The rod with the fabric was removed from the beaker and rinsed using running water at room temperature. The fabric was removed from the rod and dried in an oven at 105°C.

The fabric was now folded and sewn with a non-flame-retarded cotton thread to form a seam. The flammability along the seam was now tested using method DOC FF3-71 giving a void area of 1.5 inches initially and 2.0 inches after 50 launderings and tumble dryings. A similar fabric non-flame-retarded was consumed entirely when tested by this method.

Example 2: 70 parts of finely pulverized tetrabromo-o-chlorotoluene having an average particle size of less than 2 microns was dispersed in water in the same manner as dispersion I. This dispersion was designated II.

A piece weighing 10 g of a tufted polypropylene carpet with a jute backing was treated similarly as in Example 1, but using 1.0 g of dispersion II. After dyeing cycle was completed the carpet was dried at 80°C until dry. The dyed carpet exhibited approximately 10% less color than a sample dyed similarly but without the flame retardant.

The flammability of the treated carpet was tested using flammability standard for carpets and rugs DOC FF1-70, giving a burn 1 inch in diameter, whereas the control carpet was totally consumed when tested by this procedure.

Example 3: 70 parts of hexabromobiphenyl having an average particle size of 2 microns and a melting point between 136° and 150°C was dispersed in 30 parts of water in the same fashion as dispersion I. This suspension was designated dispersion III. 20 parts of dispersion III were dissolved in 80 parts of water and 0.5 part of an alkylaryl sulfonate wetting agent were added to this solution.

A piece of 100% polyester double-knit fabric was immersed in this solution and the excess removed by squeezing it between two rollers. The fabric picked up 110% of its weight of the solution and was now air dried in an oven at 105°C. The particles of hexabromobiphenyl were noticeable on the surface and would become loose on shaking the fabric. The treated material was now baked at 200°C for 90 seconds in a forced air oven. The baking had the effect of melting the flame retardant and driving it into the core of the fiber. No dusting off was observed after the baking operation. The treated fabric was judged flame retardant when tested according to standard for the flammability of children's sleepwear DOC FF3-71.

Emulsions of Tris-Haloaliphatic Phosphates

R.E. Smith; U.S. Patent 4,076,630; February 28, 1978; assigned to Velsicol Chemical Corporation describes an aqueous tris-haloaliphatic phosphate emulsion where the emulsifier for the tris-haloaliphatic phosphate is a water-soluble mono- or bis-haloalkyl ester of phosphoric acid or a salt. These emulsions are used as durable flame retardant finishes for textile materials. Two or more of the emulsifying agents may be used together, such as two different salts of a bis-halopropyl ester of phosphoric acid or a salt of a bis-halopropyl ester of phosphoric acid and a salt of a mono-halopropyl ester of phosphoric acid.

The flame retarding tris-haloaliphatic esters of phosphoric acid are especially difficult to formulate in aqueous compositions. These esters are generally not water-soluble and do not form an effective aqueous solution. Therefore, it is necessary to utilize an emulsion. Unfortunately, these compounds are also difficult to formulate into aqueous emulsions which are stable enough to be useful even when a wide variety of emulsifying agents has been tried. However, an emulsion-forming concentrate can be produced by mixing the flame retardant ester in the water-soluble emulsifiers described above. The concentrate is stable and is suitable in forming a stable aqueous emulsion.

The fire retardant aqueous emulsions are used for treating flammable materials and fabrics such as cellulosic, manmade and natural, exemplified by cotton, rayon and sisal; polyester; polyester/cotton; nylon, acrylic; and di- and triacetate. The concentrate comprises (1) a flame retardant tris-haloaliphatic phosphate, and (2) an emulsifying agent selected from a water-soluble haloalkyl mono- or bis-ester of phosphoric acid, a water-soluble salt of the acid, or mixtures thereof. The concentrate may contain only one tris-haloaliphatic phosphate or a mixture of two, or even more, tris-haloaliphatic phosphates. Thus, in effect, the concentrate contains at least one tris-haloaliphatic phosphate flame retardant. Likewise, the emulsifier component of the concentrate may consist of only a single water-soluble haloalkyl mono- or bis-ester of phosphoric acid; a mixture of two or more haloalkyl mono- or bis-esters of phosphoric acid; only one water-soluble salt of mono- or bis-haloalkyl ester of phosphoric acid; a mixture of two or more of the salts; or a mixture of one or more of the salts with one or more of the acids.

The concentrate is formulated by mixing together the fire retardant phosphate esters with the emulsifying agent. The concentrate should contain at least 1 part by weight of emulsifier for every 10 parts by weight of flame retardant. This ratio of 1:10 is necessary in order to form a stable aqueous emulsion of the concentrate. If less emulsifying agent is present, the resultant emulsion is rather unstable. The upper limit of the concentration of emulsifier in the concentrate should be about 10 parts by weight of the emulsifying agent for about 1 part by weight of the flame retardant. If the amount of flame retardant falls below this ratio, the efficacy of the resultant finish for textiles as a flame retardant generally becomes unsatisfactory. The following examples illustrate the process.

Example 1: A 67% ammonium salt of bis(2,3-dibromopropyl) phosphoric acid was prepared as follows: 400 g of bis(2,3-dibromopropyl) phosphoric acid were dissolved in 100 g of water; to this solution were added 100 g of 28% NH_4OH. The resultant liquid was light yellow in color and had a pH of about 7.

Example 2: To 70 g of the ammonium salt of bis(2,3-dibromopropyl) phosphoric acid of Example 1 were added 30 g of tris(2,3-dibromopropyl) phosphate and a clear emulsion concentrate was produced. This resultant emulsion contained about 23% water, 30% tris(2,3-dibromopropyl) phosphate, and 48% ammonium bis(2,3-dibromopropyl) phosphate.

Example 3: 33.3 g of the emulsion concentrate from Example 2 were added to 966.7 g of water with stirring until a stable milky emulsion was formed which contained 1% of tris(2,3-dibromopropyl) phosphate. Two thousand grams of additional water were added to this emulsion. Into this 3,000 g bath was introduced a piece of 100% spun polyester knitted fabric weighing 100 g. At this point the fabric to liquor ratio was 1:30 and the emulsion contained 10 g of tris(2,3-dibromopropyl) phosphate or 10% based on the weight of the fabric. The temperature of the batch containing the fabric was raised to 205° to 212°F and the bath was agitated.

As the temperature was increased the bath began to decrease in cloudiness, indicating an absorption of the tris(2,3-dibromopropyl) by the polyester fiber. The bath was held at 200° to 212°F, with agitation, for a period of about 2 hours. No foaming was observed. The fabric was removed from the bath, hand-rinsed and tumble dried. Bromine counts were measured by the x-ray fluorescence spectrometric method using a General Electric XRD-5 spectrometer on the treated fabric and an untreated fabric. The bromine count of the untreated fabric was 49 and that of the treated fabric was 11,651. This confirmed the visual observation that the tris(2,3-dibromopropyl) phosphate had been absorbed by the polyester fiber.

The treated fabric is usually cured by heating for a short period of time, usually from about 1 to 5 minutes, at temperatures of about 180° to about 210°C. The curing fixes the flame retardant in the fabric so that the fabric retains its flame retardancy even after repeated washings.

The method described in this example, known as Atmospheric Exhaustion, is only one of the many methods by which the emulsions may be applied to fabrics to impart flame retardancy. Other techniques may also be used. Among the more common ones known in the art are the so-called Pressure Exhaustion method and the Pad Dry Cure method.

The Pressure Exhaustion method is generally similar to the Atmospheric Exhaustion method described above except that the bath containing the fabric is heated to higher temperatures, in the range of about 140°C, and is kept under superatmospheric pressures. With this method no curing step is needed to fix the flame retardant, i.e., the tris-haloaliphatic phosphate, to the fabric as the higher temperatures existent in the bath cause the phosphate to be fixed to the fabric.

These Exhaustion methods are very efficient in that substantially all, on the order of about 90% and even higher, of the tris-haloaliphatic phosphate is absorbed from the bath onto the fabric. The fact that the use of the emulsions of the process does not result in foaming is of particular importance in the Exhaustion methods, especially the Pressure Exhaustion method, of application of the flame retardant. The use of heretofore known emulsions of tris-haloaliphatic phosphates had resulted in foaming, due to the presence of various organic emulsifiers and surfactants necessary to form an emulsion of the water-insoluble tris-haloaliphatic phosphate upon heating and/or agitation of the bath, thereby presenting certain problems which require solution. The solution has been to add an antifoaming agent to the emulsion. However, the antifoaming agents have generally tended to somewhat adversely affect the application and retention of the flame retardant onto the fabric and interfere with the performance of the flame retardant.

In the Exhaustion method the bath contains a relatively low amount of the concentrate of the process, generally from about 0.1 to about 25%, with from about 0.2 to about 15% being preferred, and with from about 0.2 to about 10% being more preferred. The flame retardant properties of a fabric treated with an aqueous flame retardant finish of the process and in accordance with the above-described Atmospheric Exhaustion method is exemplified below.

Example 4: Thirty pounds of an emulsion concentrate was prepared by mixing 18 lb of 80% aqueous ammonium bis(2,3-dibromopropyl) phosphate, that is, 14.4 lb of solid ammonium bis(2,3-dibromopropyl) phosphate and 3.6 lb of water, with 12 lb of tris(2,3-dibromopropyl) phosphate. This 30 lb of emulsion concentrate was added to 500 lb of cold water and stirred to form a milky emulsion. To this emulsion is added an additional 4,500 lb of water and the resultant composition is stirred.

Into this aqueous flame retardant finish were placed 240 lb of 100% polyester circular knitted polyester fabrics, which had previously been scoured for 15 minutes at 140°F, and rinsed twice. The bath was heated to 190°F at about 1° to 2°F/min and kept at 190°F for about 1 hour. The fabric was then removed from the bath, rinsed, dried and heat set. The treated fabric was put through 50 laundering cycles.

The flame retardancy of the fabric and the durability of the flame retardant treatment were evaluated using the procedures established by the "Standard for the Flammability of Children's Sleepwear," U.S. Department of Commerce FF 3-71.

The Char Length, in inches, of the treated fabric of this example was 2.4 inches.

Flame Retardant Fiber Blend

A.J. Rosenthal, A.S. Forschirm, and B.P. Barnes; U.S. Patent 4,035,542; July 12, 1977; assigned to Celanese Corporation describe a process whereby an assemblage of fibers is provided which exhibits highly satisfactory flame retardant characteristics in spite of the fact that discrete fibers are included within the mixture which would normally burn when exposed to flame. Included in intimate physical mixture with the fibers which would normally undergo combustion are discrete additive fibers consisting primarily of a chlorinated and/or brominated aromatic polymer having the inherent ability to render the mixture as a whole nonburning when subjected to the flame.

Particularly preferred additive fibers are formed primarily of an aromatic polyester formed from the reaction of tetrabromobisphenol A, isophthalic acid, and terephthalic acid. Also, a minor concentration of an oxide of antimony (e.g., antimony trioxide or antimony pentoxide) is intimately dispersed throughout the additive fibers.

Representative additive fibers used consist primarily of a chlorinated and/or brominated aromatic polyester of the recurring structural formula:

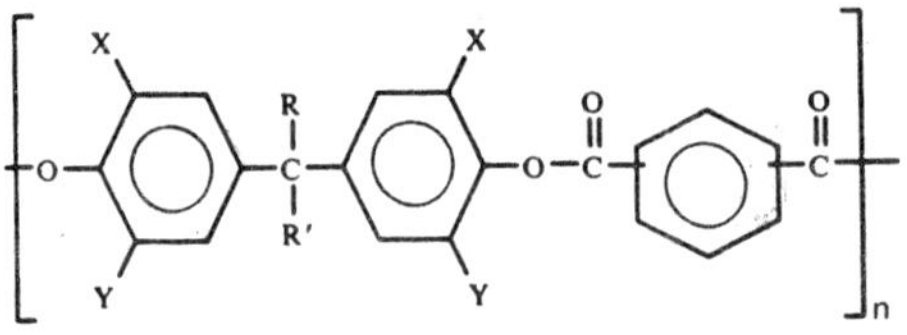

where X is chlorine or bromine, Y is hydrogen, chlorine or bromine, R and R' may be the same or different and represent lower alkyl groups (e.g., 1 to 5 carbon atoms) or hydrogen or together constitute a cyclic hydrocarbon group, and n = at least 25, e.g., about 40 to 400. The end group of the polymer illustrated in the formula commonly is –OH or –COOH.

The process is suited for use in conjunction with burnable polyester synthetic fibers such as polyethylene terephthalate, polypropylene, and natural fibers such as cotton and wool. Common polyester/cotton blends, e.g., polyethylene terephthalate/cotton blends, containing about 35 to 90% by weight polyester and 10 to 65% by weight cotton can be effectively rendered flame resistant. Additionally, polyamide/cotton or polyamide/cotton/polyester blends can be effectively rendered flame retardant.

The burnable fiber component may be provided in any one of a variety of physical configurations, e.g., fluff, slivers, yarns, tows, rovings, fibrils, filaments, etc., and may consist of staple or continuous fibers. Any discontinuous fibers selected commonly have an aspect ratio of at least 100. The fiber mixture may be utilized in a wide variety of applications, e.g., thermal insulation, carpets, textiles, apparel, etc. The following example illustrates the process.

Example: A synthetic polymer containing chlorine chemically bound to an aromatic ring is formed by reacting with stirring 190.9 parts by weight of tetrachlorobisphenol A in about 2,800 parts by weight methylene chloride solvent with 75.6 parts by weight isophthaloyl chloride and 32.4 parts by weight terephthaloyl chloride in the presence of 116 parts by weight of triethylamine acid acceptor at 40°C. The reaction is carried out with agitation for 3 hours. When the reaction is complete triethylamine is extracted with a 3% hydrochloric acid solution and the reaction mixture is washed with water until a pH of 6 is achieved.

The resulting chlorinated polyester is recovered by precipitation with methanol and possesses the appearance of a white fibrous flake and the structural formula illustrated above where X and Y are chlorine groups, R and R' are methyl groups, and n = about 80. The chlorinated aromatic polyester has a chlorine content of about 27 wt %, a melting point above 338°C, and exhibits an inherent viscosity of about 0.8 deciliter per gram determined at a concentration of 0.1 wt % in a solvent which is a mixture of 10 parts by weight phenol and 7 parts by weight trichlorophenol.

One hundred parts by weight of the chlorinated aromatic polyester are dissolved in 300 parts by weight of methylene chloride spinning solvent. The solution is filtered and 2 parts by weight of antimony trioxide are introduced. The resulting solution containing dispersed antimony trioxide is deaerated, and then at a solution temperature of about 70°C is extruded through a chrome-plated stainless steel spinneret having 20 circular holes of 42 microns diameter each. The as-spun filamentary material is passed into an air chamber provided at 70°C which flows concurrently and where the filamentary material is completely solidified and subsequently is wound up at a rate of 200 meters per minute. The filamentary material next is hot drawn at a draw ratio of about 4:1 by contact with a 12-inch hot shoe provided at about 315°C.

The drawn filamentary material is crimped by passage through a steam stuffer box and is cut into 1½-inch lengths having an antimony trioxide content of about 2 wt % based upon the weight of the chlorinated aromatic polyester. Fifty-six parts by weight of the chlorinated aromatic polyester fibers are carded with 44 parts by weight of drawn and crimped polyethylene terephthalate fibers having a length of about 1½ inches to form a uniform physical mixture of the fibers.

When a portion of the blend is knitted into a circular hoseleg and subjected to a methane diffusion flame for three seconds as described in the mounting and ignition description of the standard Children's Sleepwear Test, i.e., DOC FF 3-71, it is found that the specimen does not undergo burning. More specifically, it is found that no burning occurs after removal of the methane diffusion flame and there is no melt drip. The char length averages 1 inch. Also, when the blend is subjected to an ordinary match flame for three seconds it is nonburning. The presence of the chlorinated aromatic polyester renders the entire blend nonburning.

In control tests where solely the polyethylene terephthalate fibers apart from the blend are subjected to the same methane diffusion and match flames in air, they frequently burn vigorously in each instance until totally consumed and drip molten droplets which sometimes continue to burn after falling to the floor of the test chamber.

Linear Polymeric Metal Phosphinates

C.E. Kramer; U.S. Patent 4,078,016; March 7, 1978; assigned to Celanese Corporation describes a process whereby a composition having flame retardant properties is provided which comprises a blend of a synthetic aromatic polymer containing chlorine and bromine with a linear metal phosphinate polymer. The aromatic polymer used is a halogenated polyester as described in the previous patent (U.S. Patent 4,035,542).

The metal phosphinate polymers may be illustrated by the following formulas:

(I) $$[M(OPRR'O)_m]_n$$

(II) $$[M'(B)_z(OPRR'O)_m]_n$$

(III) $$[M''(H_2O)_a(OH)_c(OPRR'O)_q \cdot pH_2O]_n$$

where M is a member selected from the group consisting of Co, Mn, Cr, Ni, Zn, Zr, Ti, and M' is from the group consisting of Ti and Zr; R and R' may be the same or different and represent alkyl or aryl radicals; B is selected from the group consisting of oxo, alkoxy, and aryloxy, z equals 1 when B equals oxo, 2 when B equals aldoxy and aryloxy and 0 when m equals 3; M" is Cr(III); a + p = 1; q may vary from 1 to 3; m may vary from 1 to 3; c may vary from 0 to 3; and n is at least 10 (e.g., from about 10 to about 300).

These metal phosphinates may be made by metathetic reactions, either by fusion of the reactants or in solvents. Samples prepared from solution have as high degrees of polymerization as those prepared without solvents provided the stoichiometry is the same. Examples of the kinds of metathesis that have been employed may be illustrated by the following equations:

$$Zn(C_2H_3O_2)_2 + 2(C_8H_{17})_2P(O)OH \xrightarrow{C_2H_5OH} 1/x\langle Zn[OP(C_8H_{17})_2O]_2\rangle_x + 2CH_3COOH$$

$$CoCl_2 + 2NaOP(C_6H_5)_2O \xrightarrow{H_2O} 1/x\langle Co[OP(C_6H_5)_2O]_2\rangle_x + 2NaCl$$

$$Cr(C_5H_7O_2)_3 + 2(C_6H_5)_2P(O)OH \longrightarrow 1/x\langle Cr(C_5H_7O_2)[OP(C_6H_5)_2O]_2\rangle_n + 2C_5H_8O$$

$$Ti(O_2C_6H_4)(OC_3H_7)_2 + 2CH_3(C_6H_5)P(O)OH \xrightarrow{C_6H_6} 1/x\langle Ti(O_2C_6H_4)[OP(CO)_3(C_6H_5)O]_2\rangle_x + 2C_3H_7OH$$

More specifically, the metal phosphinate polymers of formula I may be prepared by reacting a phosphinic acid with a bivalent four coordinate metal acetylacetonate acetate, with or without a basic catalyst such as pyridine as described in U.S. Patents 3,681,265 and 3,255,125.

Polymer blends comprising blends of any of the above-described halogenated aromatic polymers with one or more of the linear metal phosphinate polymeric additives may be utilized in any of the coating, impregnating, molding, and fiber spinning applications known to those skilled in the art where it is desirable to provide fire retardancy to the resulting end product. For example, these compositions may be used for preparing such diverse items as camera films, blow molded bottles, extruded and blown films, extruded and shaped articles such as handles, tubes, sheets, rods, fibers, particularly with polyester fibers and textiles based on these fibers. For instance, the fabrics, textiles, wool coverings, hospital cubical draperies, children's sleepwear, flight suits, slippers, upholstery, thread, apparel, etc. may be formed from the same.

Example 1: Preparation of Brominated Aromatic Polymer Substrates – A halogenated aromatic polyester is prepared by condensing tetrabromobisphenol A with a 60:40 isophthalic:terephthalic acid chloride mixture. More specifically, 201.7 parts by weight tetrabromobisphenol A, 46.05 parts by weight of isophthaloyl chloride and 30.8 parts by weight of terephthaloyl chloride are reacted to form a brominated aromatic polyester in the presence of about 2,600 parts by weight methylene chloride solvent and 82 parts by weight of triethylamine acid acceptor.

The contents of the reaction zone are heated at about 40°C with agitation for 3 hours. When the reaction is complete, triethylamine is extracted with a 3% hydrochloric acid solution and the reaction mixture is washed with water until a pH of 6 is achieved. The resulting brominated polyester which serves as the substance which is blended with metal phosphinate polymeric substituents in the following examples is recovered by precipitation with methanol. The brominated aromatic polyester has the appearance of a white, fibrous flake. The brominated aromatic polyester has a bromine content of about 48 wt %, a melting point of about 265°C and exhibits an inherent viscosity of about 0.75 deciliter per gram determined at a concentration of 0.1 wt % in a solvent which is a mixture of 10 parts by weight of phenol and 7 parts by weight trichlorophenol.

Preparation of Control – The polymer thus prepared is cast to form a film and ground to 60 mesh on a Wiley mill with a 50/50 polyethylene terephthalate/cotton mixture at levels of 20, 30, and 40%. The mixture so formed is pressed into pellets and thin cross sections are cut. These samples serve as a control for LOI determinations subsequently performed on the polymer/additive blends of the following examples.

Preparation of Ti Phosphinate Polymer – A titanium phosphinate of the recurring structural formula $[TiO(OP(C_6H_5)_2O)_2]_n$ where n is at least 10 is prepared according to the procedure in U.S. Patent 3,457,195. Thus, a solution of 17.19 grams (0.091 mol) of freshly distilled titanium chloride ($TiCl_4$) in 150 ml of benzene is added to a stirred solution of 44.783 g (0.182 mol) of ethyl diphenyl phosphinate in 200 ml of benzene. The yellow adduct is filtered off, washed with benzene and dried overnight in vacuo. This compound is heated with agitation to 160° to 175°C at which temperature evolution of C_2H_5Cl proceeds quite readily. Heating is continued in vacuo for about 20 hours at 180°C to complete the reaction.

The intermediate polymer thus formed $[TiCl_2(OP(C_6H_5)_2O)_2]$ is suspended in boiling water until no further liberation of chloride ions can be detected (10 hours).

Preparation of Blend Test Samples – The halogenated aromatic polyester prepared above by condensing tetrabromobisphenol A with a mixture of isophthalic and terephthalic acid is blended with the titanium phosphinate to form two different mixtures having brominated aromatic polyester/additive ratios of 90:10 and 80:20, respectively. Each mixture is cast to a film and ground to 60 mesh on a Wiley mill with a 50/50 PET/cotton mixture at levels of 20, 30, and 40%. Half-gram samples are pressed into pellets and thin cross sections are cut therefrom for LOI determinations. The resulting LOI values obtained are set out in the table below.

Example 2: Chromium hydroxyaquo biphenyl phosphinate of the recurring structural formula $[Cr(H_2O)_a(OH)_c(OP(C_6H_5)_2O)_2 \cdot pH_2O]_n$ where n is at least 10, and a + p is 1, and c may vary from 0 to 3 is prepared by adding a solution of $NaOP(C_6H_5)_2O$ (4.80 g, 0.0200 mol) and K_2CO_3 (0.69 g, 0.0050 mol) in 50 ml of water with stirring to a solution of $CrCl_3 \cdot 6H_2O$ (2.66 g, 0.100 mol) in 50 ml of THF (tetrahydrofuran). The reaction mixture is brought to boiling and the THF allowed to evaporate.

An oily product separates and solidifies upon evaporation of the THF. After most of the THF has been removed, additional water (about 50 ml) is added, and the suspension is kept boiling until the precipitate can be easily ground with a spatula. The powdered solid is then collected on a filter, and thoroughly washed with water. Solutions of the compounds had intrinsic viscosities of about 0.04 dl/g which increased to about 0.4 dl/g on standing at 55°C for a few days in a $CHCl_3$ solution.

The phosphinate polymer so prepared is blended and tested in accordance with the procedures outlined in Example 1. The resulting LOI values are set out in the table below.

Example No.	% Additive Blended with Control	LOI of Control/Additive Blended with 50/50 - PET/Cotton		
		20%	30%	40%
(1) Control		18.2	20.8	23.1
(1) $[TiO(\phi_2PO_2)_2]_n$	10	18.7	21.5	24.2
(2) $[Cr(OH)(O_2)(\phi_2PO_2)_2]_n$	10	19.5	23.1	26.7

Inverse Interfacial Polymerization Technique for Preparing Aromatic Polyesters

M.M. Besso and L. Saferstein; U.S. Patent 4,066,623; January 3, 1978; assigned to Celanese Corporation describe a process for preparing certain halogenated aromatic polyesters by an inverse interfacial polymerization technique. Furthermore, this process offers a method for substantially reducing the amount of low molecular weight fraction which accompanies the formation of these polyesters when prepared by standard interfacial polymerization techniques by (1) inverting the order of addition of one phase to the other, i.e., by adding the aqueous phase to the organic phase, (2) controlling the rate of addition of the aqueous phase to the organic phase, and (3) adding the catalyst to the organic phase prior to the addition of the aqueous phase.

The halogenated aromatic polyesters described herein have been used to produce a number of inherently nonburning fibrous materials which offer the public a great degree of fire safety, particularly when fibrous articles are required for use in fire-control environments, e.g., children's sleepwear, suits for fire fighters, hospital furnishings, uniforms for military and civilian flight personnel.

Example 1: Preparation of Organic Phase – 14.2 g (0.07 mol) of a 60:40 mixture of isophthaloyl chloride and terephthaloyl chloride are charged into a one-liter reaction vessel with an overhead stirrer. Three hundred and twenty milliliters (0.427 g) of methylene chloride are added to the same reaction vessel and stirred until the acid chlorides are dissolved. Subsequent to the dissolution of the acid chlorides 0.235 g of triethylbenzylammonium chloride (phase transfer catalyst) are added to the reaction vessel.

Preparation of Aqueous Phase – 38.0747 g (0.07 mol) of 4,4'-isopropylidene-2,2',6,6'-tetrabromodiphenol (i.e., tetrabromobisphenol A) are added to and dissolved in a vessel containing 318 ml of water and 5.88 g (0.14 mol) of sodium hydroxide.

Reverse Interfacial Polymerization – The aqueous solution is then placed in a dropping funnel and added to the organic solution under conditions of agitation (i.e., stirring) at the rate of 3.3 mol percent per minute based on the total number of mols of the bisphenol originally utilized to prepare the aqueous phase. This rate may also be expressed as a total addition time of 30 minutes (i.e., at a uniform rate of addition) which corresponds to 33.3% of a total polymerization time of 90 minutes.

Upon completion of the addition of the aqueous solution the reaction mixture is stirred for an additional hour. The mixture is then allowed to separate into two layers. The very viscous lower methylene chloride layer contains the polymer and is separate from the upper water layer. The organic layer is washed with distilled water until the water washings give a negative chloride ion test with silver nitrate. The methylene chloride solution is then added to acetone in an air-driven blender to precipitate the polymer.

The polymer is dried at 85°C in a vacuum oven for 16 hours. The aromatic polyester prepared via this inverse addition technique has an intrinsic viscosity of 1.31 dl/g measured in chloroform, a kinematic viscosity of 158 cs, and most importantly, a gel permeation chromatogram of this polymer shows a bimodal molecular weight distribution with the lower molecular weight fraction comprising only 3% of the entire weight of the polyester.

Example 2: The same procedure as described in Example 1 is repeated with the exceptions that 4,4'-isopropylidene-2,2',6,6'-tetrachlorodiphenol (i.e., tetrachlorobisphenol A) is utilized to prepare the aqueous phase instead of tetrabromobisphenol A and that a 70:30 mixture of isophthaloyl and terephthaloyl chloride, respectively, is utilized in preparing the organic solution instead of the 60:40 ratio.

The results which are substantially the same as Example 1, show a bimodal molecular weight distribution having a low molecular weight fraction comprising about 2% of the entire weight of the polymer.

Comparative Example: The same solutions are prepared as described in Example 1 with the exception that the phase transfer catalyst is added to the aqueous solution. The order of addition is then reversed and the organic phase is added dropwise to the aqueous phase in the same manner as described in Example 1. The intrinsic viscosity and kinematic viscosity of the resulting polymer are similar to the values cited in the example above, and the gel permeation chromatogram again shows a bimodal molecular weight distribution; however, the low molecular weight fraction comprises 15 to 20 wt % of the entire polymer.

Poly(2,4,6-Tribromostyrene)

A process described by *L. Buxbaum and F. Breitenfellner; U.S. Patent 3,975,354; August 17, 1976; assigned to Ciba-Geigy Corporation* relates to glass fiber-reinforced flame resistant thermoplastic polyester molding compositions which contain a saturated polyester and, optionally, synergistic substances, and are characterized in that 3 to 30 wt % of poly(2,4,6-tribromostyrene), relative to the total composition, are admixed.

It has been found that poly(2,4,6-tribromostyrene) can be mixed with a polyester and the molding composition thus produced can be used for the manufacture of moldings without decomposition phenomena leading to discoloration or embrittlement of the material at the requisite mixing and/or processing temperatures. The polyester molding composition should contain 6 to 20 wt % of poly(2,4,6-tribromostyrene) and preferably its degree of polymerization is 100 to 6,000, especially 500 to 4,000.

Suitable saturated polyesters in the molding compositions are the homopolyesters and copolyesters which are obtained in accordance with known processes from polyester-forming monomers, for example, dicarboxylic acids, diols and hydroxycarboxylic acids. The viscosity of these polyesters depends on the end use and on economic aspects. In general, the saturated polyesters have a relative viscosity of 1.2 to 4.0, preferably 1.3 to 3.5.

In particular, the molding compositions are suitable for use as "engineering plastic" materials which can be used for the manufacture of moldings such as gear-wheels, containers for chemicals or foodstuffs, machine components and appliance components, sheets and also semifinished goods which can be shaped by machining.

Example 1: Manufacture of Polybutylene Terephthalate – Three thousand grams of dimethyl terephthalate, 2,800 g of 1,4-butanediol and 1.78 g of titanium tetraisopropylate are introduced into a 10-liter reactor equipped with a stirrer, nitrogen inlet and separating column. 98.5% of the theoretical amount of methanol are distilled off over the course of 4.5 hours while stirring and passing nitrogen through the reactor; at the same time the temperature of the reaction mixture rises to 200°C.

The mixture is then transferred into a second reactor, a vacuum of 0.5 mm Hg is applied over the course of 1.5 hours and at the same time the reaction temperature is raised to 250°C. After a further 3 hours the reaction is terminated, the melt is extruded through dies to form ribbons and the ribbons are chilled in water at 18°C and comminuted to form colorless cylindrical granules of size 2 x 3 mm, the material having a relative viscosity of 2.13.

Example 2: Manufacture of the Molding Compositions – 56 wt % of polybutylene terephthalate manufactured according to Example 1, which has been dried for 10 hours at 100°C in vacuo, are mixed with 14 wt % of poly(2,4,6-tribromostyrene) and 30 wt % of glass fibers in a tumbler mixer and the mixer is then granulated by means of an extruder, at temperatures of 240°C. The granules are dried at 100°C for 10 hours and then injection-molded by means of an injection molding machine to form test specimens for the measurements. The temperature of the intake zone is 220°C, the barrel temperature is 240°C, the temperature of the output zone is 260°C and the mold temperature is 80°C. The cycle time for the production of the moldings is 45 seconds. The characteristic data are listed in the table below. The poly(2,4,6-tribromostyrene) used has a density of 2.3 g/cm³ and a bromine content of 69%.

Example 3: Test specimens of the following composition are manufactured analogously to Example 2: 53.3 wt % of polybutylene terephthalate, 13.3 wt % of poly(2,4,6-tribromostyrene), 4.8 wt % of antimony trioxide and 28.6 wt % of glass fibers. The characteristics are listed in the table below.

Example 4: (Comparison Example) – Test specimens consisting of 70 wt % of polybutylene terephthalate and 30 wt % of glass fibers are manufactured under the conditions of Example 2. The characteristic data are listed in the table below.

Example 5: (Comparison Example) – A mixture of 59.5 wt % of polybutylene terephthalate, 7.0 wt % of decabromodiphenyl, 3.5 wt % of antimony trioxide and 30.0 wt % of glass fibers is converted to test specimens analogously to Example 2. The characteristic data are listed in the table below.

Mechanical Properties of Glass Fiber-Reinforced PBT/Poly(2,4,6-Tribromostyrene)

Ex. No.	Density (g/cm^3)	Tensile Strength* (kg/cm^2) DIN 53,455	Flexural Strength* (kp/cm^2) ASTM D 790	Flexural Modulus* (kp/cm^2) DIN 53,457	Impact Strength* (cm-kp/cm^2) DIN 53,453	Notched Impact Strength* (cm-kp/cm^2) DIN 53,453	Heat Distortion Point ISO/R 75 (°C) Method A	Inflammability ASTM D 635 (cm/min)
1	–	1,220	1,980	85,600**	32.4	11.5	–	self-extinguishing, specimen thickness, 4 mm
2	1.6998	1,160	1,830	92,300**	–	–	–	noninflammable, specimen thickness, 4 mm
3	1.53	1,330	2,020	85,000	40	8	205	combustibility, 1.2 cm per min; specimen thickness, 4 mm
4	1.66	1,200	2,000	90,000	24	6	205	noninflammable, specimen thickness, ½ inch

*At 23°C.
**Flexural modulus according to ASTM D 790.

Synergism of Iron Compounds with Polyester Resins

A process described by *E. Dorfman, W.T. Schwartz, Jr., and R.R. Hindersinn; U.S. Patent 4,013,815; March 22, 1977; assigned to Hooker Chemicals and Plastics Corporation* is directed at fire retardant polymer compositions which comprise an α,β-ethylenically unsaturated polyester of a polycarboxylic compound and a polyhydric alcohol which contains a halogen in a proportion of greater than about 4 wt %, and an effective fire retardant proportion up to about 5 wt % of iron or an iron compound, where the proportions are based on the weight of the unsaturated polyester and a copolymerizable unsaturated monomer.

The halogen-containing polyesters of this process are those where the halogen is provided by a Diels-Alder adduct of a hexahalocyclopentadiene and a polycarboxylic compound or polyhydric alcohol containing aliphatic carbon-to-carbon unsaturation. The preferred adducts are chlorendic acid or chlorendic anhydride.

Various iron compounds can be employed in the compositions depending on the end results desired. Thus, where it is desired to retain the translucent character of the unsaturated polyester colorless iron salts can be employed, or iron salts that permit the transmission of light when incorporated in the polyester resin system can be used. The various iron oxides and other iron salts, including various organic iron compounds can be employed.

The unsaturated polyester resin containing the iron compound is copolymerized with an ethylenically unsaturated monomer, optionally in the presence of a reinforcing agent or filler, to provide the final polymerized product.

A foamed-type article may be made by incorporating a small percentage of a foaming agent such as sodium bicarbonate into the solution of unsaturated polyester dissolved in monoolefin and thereafter effecting the copolymerization in the presence of catalyst and heat to produce the foamed article. Formulations which are useful for making molding embodying the compositions of this process may be made by mixing into the unsaturated linear polyester and olefinic crosslinking agent mixture, an inert filler such as chopped fiber glass rovings, macerated fabric, asbestos fibers, mica, etc., which serve as fibrous reinforcing media and incorporating a small percentage of a mold lubricant, catalyst and/or promoter. Auxiliary fire retardant additives such as hydrated alumina can be used to add the fire retardance provided by the halogen and iron.

The following examples illustrate the process. All temperatures are in degrees centigrade and all parts are by weight, unless otherwise indicated.

Example 1: (Resin I) – An unpolymerized liquid unsaturated polyester resin was prepared by esterifying about 157 parts of ethylene glycol and 38 parts of diethylene glycol with about 668 parts of 1,4,5,6,7,7-hexachlorobicyclo-(2.2.1)-5-heptene-2,3-dicarboxylic acid and about 136 parts of fumaric acid. About 30 parts of styrene and about 70 parts of product produced by the esterification reaction were mixed together until complete solution was obtained to give a clear, substantially colorless solution of liquid polyester resin having a viscosity of about 30 poises at 25°C on a Gardner bubble viscometer and having a chlorine content of about 30 wt % of the total.

Example 2: (Resin II) – An unpolymerized liquid unsaturated polyester resin was prepared by esterifying about 76 parts of propylene glycol with about 128 parts of 1,4,5,6,7,7-hexachlorobicyclo(2.2.1)-5-heptene-2,3-dicarboxylic acid and about 76 parts of fumaric acid. About 28 parts of styrene and about 55 parts of product produced by the esterification reaction were mixed together until complete solution was obtained to give a clear, substantially colorless solution of liquid polyester resin having a chlorine content of about 19 wt % of the total.

Example 3: (Resin III) – A unpolymerized liquid unsaturated polyester resin was prepared by esterifying about 397 parts of ethylene glycol with about 232 parts of fumaric acid and about 1,258 parts of tetrachlorophthalic anhydride to provide a polyester of acid number 28 mg KOH/g. About 720 parts of sytrene were mixed with 1,674 parts of the esterification product until complete solution was obtained to give a clear substantially colorless solution of liquid polyester having a chlorine content of about 26%.

Example 4: (Resin IV) – An unpolymerized liquid unsaturated polyester resin was prepared by esterifying about 19 parts of diethylene glycol and about 140 parts of ethylene glycol with 118 parts of maleic anhydride, 57 parts of phthalic anhydride, 375 parts of tetrabromophthalic anhydride and 3 parts of sodium acetate. About 271 parts of styrene and 0.1 part of hydroquinone were mixed with polyester until complete solution was obtained to give a clear, substantially colorless solution of liquid polyester having a viscosity of about 4 poises at 25°C on a Gardner bubble viscometer and having a bromine content of about 26%.

Example 5: (Resin V) – A mixture of 194.5 parts (0.5 mol) of chlorendic acid, 98.1 parts (1.0 mol) of maleic anhydride, 217.9 parts (1.0 mol) of 1,2-dibromopropanol and 24.8 parts (0.40 mol) of ethylene glycol and 36.8 parts (0.40 mol) glycerol were heated, under a steady nitrogen sparge, to and maintained at about 165°C, while collecting the water formed as a distillate. After about 12 hours, the acid number of the reaction mixture had decreased to 46. A small quantity, about 0.10 part, of tolylhydroquinone was added to the mixture and the reaction was continued for an additional four hours at which time the acid number was about 34.

The resultant polyester product was analyzed and found to contain 26.0% bromine, 20.0% chlorine and the average molecular weight of the polymer chains was 1,150. An amount of styrene equivalent to about one-third the weight of the polyester mass was added to the hot mass and the resultant solution was cooled to ambient temperature. The viscosity (Brookfield) of the solution was 1,620 cp at 25°C.

Examples 6 through 10: Tables 1 through 5 below show the results of fire retardance tests by the ASTM D-757-65 test method on castings of the resin prepared as described in Examples 1 through 5 (Resins I through V, respectively). The castings were prepared by mixing 100 parts by weight of resin with iron compound in the amount indicated in the tables, 0.2 part of 12% cobalt octoate, and other additives as are indicated in the tables.

After at least 10 minutes' stirring to assure uniform dispersion of the immiscible material the catalyst, methyl ethyl ketone peroxide (60% in dimethyl phthalate), 1 part was added, stirred for 1 to 2 minutes, and then the mixture was allowed to cure at 25°C for 16 hours and at 65°C for 8 hours. The castings were then removed from the mold, cut into specimen bars for testing by ASTM D-757-65 test method, for smoke by ASTM D-283-70, and for heat deflection temperature (identified in the tables as "HDT"), by ASTM D-648-72 test method. Data on light absorption due to smoke obscuration was acquired by a computer which printed out the light absorption and percent smoke at one second intervals.

The "Smoke Density Rating" of this modification of the ASTM test is the percent smoke at 60 seconds. Samples which were used for the ASTM D-648-73 test measured 0.5" x 0.5" x 0.125".

TABLE 1: FIRE RETARDANT SYNERGISM OF IRON COMPOUNDS WITH RESIN I

Casting No.	Compound	Weight Percent	HDT °C	ASTM D-757-65 Burn Rate (in/min)	ASTM D-757-65 Flameout Time (secs)	Smoke Density Rating, %	Char, %
A	none		115	0.20	160	60	18
B	Antimony oxide	5	82	0.08	120	48	33
C	Antimony oxide	1	86	0.09	140	52	31
D	Iron oxide, yellow	1	79	0.04	0	37	43
E	Iron oxide, yellow	5	85	0.05	0		
F	Iron oxide, yellow	10	76	0.05	0		
G	Iron oxide, red	0.5	100	0.13	105		
H	Iron oxide, red	0.75	102	0.11	90		
I	Iron oxide, red	1	94	0.08	70		
J	Iron oxide, red	1.25	101	0.09	80		
K	Iron oxide, red	1.5	102	0.09	87		
L	Iron oxide, red	5	86	0.09	40		
M	Iron oxide, red	10	86	0.08	25		
N	Iron oxide, black	0.75	81	0.06	62		
O	Iron oxide, black	1	81	0.06	0	32	47
P	Iron oxide, black	1.25	83	0.05	27		
Q	Iron oxide, black	2.0	83	0.06	0		
R	Iron oxide, black	3	83	0.06	0		
S	Ferric acetate	1	106	0.08	122		
T	Ferric formate	1	107	0.09	145		
U	Ferrous tartrate	1.5	106	0.12	165		
V	Ferric sulfate	1	81	0.09	125		
W	Ferric tungstate	3	81	0.07	165		
X	Ferric tungstate	4	82	0.06	0		
Y	Ferric ferrocyanide	0.5	79	0.13	180		
Z	Ferric ferrocyanide	2	80	0.10	140		
AA	Ferric sulfate	0.5	83	0.11	165		
AB	Ferric sulfate	1.5	86	0.08	130		
AC	Ferric molybdate	3	82	0.08	102		
AD	Ferric borate	2	82	0.11	115		
AE	Ferrous stannate	1	84	0.07	35		
AF	Ferrous stannate	2	84	0.08	0	40	
AG	Ferrous stannate	4	82	0.06	0	38	
AH	Ferrous tungstate	1	82	0.16	130		
AI	Ferrous chloride	1	70	0.11	130		
AJ	Ferrous sulfate	1	83	0.10	145		
AK	Ferrous ammonium sulfate	1	83	0.12	145		
AL	Ferric chlorendate	2	116	0.12	70		
AM	Ferric hexachloronorbornene carboxylate	5	96	0.11	100		
AN	Ferric acrylate	1	82	0.18	100	46	
AO	Ferric acrylate	2	82	0.12	115	45	
AP	Ferric acetylacetonate	1	105	0.10			

TABLE 2: FIRE RETARDANT SYNERGISM OF IRON COMPOUNDS WITH RESIN II

Casting No.	Compound	Weight Percent	HDT °C	ASTM D-757-65 Burn Rate (in/min)	ASTM D-757-65 Flameout Time(secs)	Smoke Density Rating, %	Char, %
BA	none		91	0.43	180	44	19
BB	Antimony oxide	1	93	0.17	170	44	27
BC	Antimony oxide	2	94	0.16	165	45	28
BD	Antimony oxide	3	94	0.10	180	39	27
BE	Iron oxide, red	1	141	0.14	140	26	26

TABLE 3: FIRE RETARDANT SYNERGISM OF IRON COMPOUNDS WITH RESIN III

Casting No.	Compound	Weight Percent	HDT °C	ASTM D-757-65 Burn Rate (in/min)	ASTM D-757-65 Flameout Time(secs)	Smoke Density Rating, %
CA	none		78	0.55	180	76
CB	Antimony oxide	1	76	0.20	145	75
CC	Antimony oxide	2	77	0.17	135	77
CD	Antimony oxide	3	76	0.17	155	75
CE	Iron oxide, yellow	1	76	0.40	153	69
CF	Ferric sulfate	1	76	0.39	180	71
CG	Iron oxide, black	1	76	0.34	167	76

TABLE 4: FIRE RETARDANT SYNERGISM OF IRON COMPOUNDS WITH RESIN IV

Casting No.	Compound	Weight Percent	HDT °C	ASTM D-757-65 Burn Rate (in/min)	ASTM D-757-65 Flameout Time(secs)	Smoke Density Rating, %
DA	none		74	0.17	0	70
DB	Ferric sulfate	1	74	0.13	0	63
DC	Antimony oxide	1	75	0.16	0	66
DD*	none		83	0.29	157	71
DE*	Iron oxide, red	0.5	81	0.17	0	74

*Castings DD and DE were run with different sample of Resin IV than Castings DA, DB, DC.

TABLE 5: FIRE RETARDANT SYNERGISM OF IRON COMPOUNDS WITH RESIN V

Casting No.	Compound	Weight Percent	HDT °C	ASTM D-757-65 Burn Rate (in/min)	ASTM D-757-65 Flameout Time(secs)	Smoke Density Rating, %
EA	none		86	0.10	0	52
EB	Iron oxide, yellow	1	90	0.13	0	53
EC	Iron oxide, yellow	2	92	0.13	0	48
ED	Iron oxide, yellow	3	84	0.13	0	49
EE	Antimony oxide	1	85	0.07	0	54
EF	Antimony oxide	2	87	0.08	0	52
EG	Antimony oxide	3	85	0.07	0	53

In the testing of synergism with Resin V, it was seen in Table 5 that the D-757 test was insufficient in differentiating synergistic effects of iron and antimony compounds. The samples that contained levels of 1% or more of fire retardant additives did not ignite, so the "burning rate" which was variable and poorly reproducible, was really a "charring rate."

Resin V materials were therefore evaluated by another method than the ASTM D-757 test. Samples of material were heated at 10°C per minute while the weight loss of the samples was observed continuously by the thermogravimetric analysis technique. Results are shown in Table 6. At low rates of heating to 600°C, it was observed that there was little change in weight loss after the samples had reached 400°C. It was observed that Resin V lost weight more rapidly than Resin I which is consistent with the lower Time to Flameout of Resin V.

TABLE 6: SYNERGISM IN WEIGHT LOSS ON HEATING (10°C/min) OF RESINS I AND V

Resin	Additive	Weight Percent	Char Residue % 400° C	Temperature, ° C 30% Weight Loss
V	None	—	18	308
V	Iron oxide	1	47	264
V	Antimony oxide	5	16	313
I	None	—	36	327

With 1% iron oxide added to Resin V, it was observed that weight loss occurred much more rapidly and that a 30% weight loss had already occurred at 264°C as compared to a comparable weight loss at 308°C for Resin V without additive or with antimony oxide. The addition of iron oxide also catalyzed the formation of a large char residue (47%) which made less material available for gaseous fuel and less material available for smoke formation. From these data it is concluded that under flaming conditions where Resin V is exposed to conditions of extremely rapid temperature rise, neither the addition of iron compounds nor antimony compounds will improve the fire retardant and low smoke properties much; but in oven-like conditions where the temperature rise to the flaming condition is slower, the Resin V compositions that contain a synergistic amount of iron compound will give improved fire retardance and low smoke properties.

Polymeric Phosphonate Resins

K. Moedritzer, U.S. Patent 4,087,403; May 2, 1978; assigned to Monsanto Company describes a process for the production of polymeric phosphinate resins and combinations of such polymeric phosphinates with organic polymers such as polyesters, e.g., polyethylene terephthalate. These polymeric resins have the following general molecular structure as condensed products

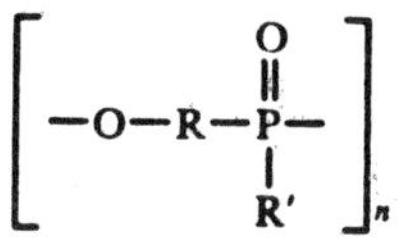

where R is a phenylene or naphthylene group in which the O substituent is nonvicinal to the P=O, R' is an alkyl group of 1 to 15 carbon atoms or an aryl group of 6 to 15 carbon atoms, and n is from 5 to 50 units. The successive repeating units provide a P-O-C aromatic linkage which is desirable for fire retardant properties, and thermal stability. The provision of oxygen atoms in the backbone of the polymeric molecular structure as P-O-C (aromatic) also imparts resistance to hydrolysis, since the oxygen is less accessible to hydrolyzing agents such as acids, bases, water and atmospheric humidity.

The use of an O substitute which is nonvicinal refers to the employment of meta and para positions on the R ring as a phenyl relative to the position of the phosphorus. With naphthyl as R, the O substituent is also nonvicinal, e.g., the 1.3, 1.4, 1.5, 2.4, and 1.6 positions, for the two substituents. The substituent groups are thus located on carbon atoms separated by at least 1 carbon atom. The R and R' groups may also have chlorine or bromine substituents.

The physical or chemical combination of the polymeric phosphinates with the organic polymers makes it possible to obtain flame retardant products, for example, fibers suitable for carpeting and clothing. The resin may be used as an additive applied to the organic polymer in a molten state before spinning. However, the polyphosphinate resin may also be applied from a solution (e.g., in an aliphatic alcohol such as methanol, ethanol or isopropanol, or a ketone such as methyl ethyl ketone) directly to sheets or other shaped forms, including fibers or fabrics of the organic polymer.

The present condensation products are useful in combination with organic polymers generally to reduce combustibility. The normally flammable organic polymers which are rendered fire retardant may be natural or synthetic but are preferably a solid synthetic polymer, more preferably, a nylon or ester-type polymer. Examples of the polymer are cotton, wool, silk, paper, natural rubber, and paint, and also the high molecular weight homopolymers and copolymers of amides, e.g., nylon 66 and nylon 6.

Example 1: Preparation of the Starting Monomer – A charge of p-bromoanisol is employed in the amount of 84.7 g (0.45 mol). The above charge is placed together with 9.0 g of nickel bromide as a catalyst in a 250 ml round-bottom flask fitted with an ice-cooled condensation head, a dropping funnel, and a thermometer. The other reactant is diethyl phenylphosphonite $C_6H_5P(OC_2H_5)_2$ employed in the proportion of 86.58 g (0.44 mol).

The latter compound is added dropwise to the reaction mixture which is maintained at 175°C. The mixture slowly darkens during this addition which is conducted over a period of about 2 hours. The reaction vessel is then allowed to cool and the reaction mixture purified. A total of 48 g of ethyl bromide is collected, indicating the formation of about a 100% yield of ethyl p-methoxyphenylphenylphosphinate. The latter is hydrolyzed (by refluxing with concentrated HCl) to p-methoxyphenylphenylphosphinic acid and then demethylated by treatment with boron tribromide to give the desired p-hydroxyphenylphenylphosphinic acid, the starting material of Example 2.

Example 2: The reactor vessel is charged with 2.5 g of purified thionyl chloride (21.0 mmol), together with 5 ml of dry benzene and 4.684 g (20 mmol) of p-$HOC_6H_4P(C_6H_5)OOH$ (p-hydroxyphenylphenylphosphinic acid). The vessel is sealed and shaken at room temperature for 8 hours. The vessel is then charged with 5 additional grams of thionyl chloride and further shaking continued until the solids have all gone into solution. After 24 hours of agitation, the vessel is opened and the solution evaporated free of solvent at room temperature over a vacuum. The reaction mixture is a viscous liquid which upon heating to 150°C at atmospheric pressure becomes liquid. Heating the mixture results in evolution of hydrogen chloride. The dehydrochlorination proceeds without a catalyst, although the addition of 0.1 wt % of pyridine hydrochloride accelerates the dehydrochlorination. The reaction mixture is heated to a temperature of 250°C for 2 hours.

The reaction mixture upon further heating under oil pump vacuum (about 0.1 mm Hg) and the use of a dry ice trap to collect volatile components results in the collection of unreacted materials in the cold trap, and the production of a residual product of the above formula in the reaction vessel. The product is soluble in methanol, methyl ethyl ketone and is precipitated from such solutions by the addition of water. The melting point of the product, condensed p-hydroxyphenylphenylphosphinate, is about 225°C, and the molecular weight is equivalent to about 10 monomeric units, e.g., a molecular weight of about 2,050. The end groups are H on the oxygen and a chlorine at the phosphorus. With the meta isomer as the starting compound, the meta-substituted polymer is obtained.

Example 3: A para-substitured polymeric composition of Example 2 is used as an additive chemical in an amount 10 wt % of the total composition to impart fire retardant properties to polyethylene terephthalate. The latter is milled with the additive and then formed into sheets 0.07 inch thick for testing purposes.

Utilizing the method of testing fire retardant properties, that is, ASTM D-2863-70, entitled "Standard Method of Test for Flammability of Plastics Using the Oxygen Index Method," the oxygen index value is 24.0, in comparison to unmodified polyethylene terephthalate which has a value of 19.0. The meta-isomer of the polymer has similar properties.

Polymeric Phosphorus Esters

A process described by *K. Moedritzer; U.S. Patent 4,073,829; February 14, 1978; assigned to Monsanto Company* relates to flame retardant compositions which are phosphorus-containing esters, and which are particularly adapted to prevent crystallization and exuding or oiling out when these compositions are employed in organic polymers.

These esters have a total of at least three different alcohol and phosphorus-containing moieties: (A) an aliphatic alcoholic moiety of an alcoholic reactant of 5 to 20 carbon atoms, and having from 2 to 10 hydroxyl functional groups, with the proviso that there is no hydrogen atom on the carbons which are beta to the hydroxyl oxygen; (B) a moiety of a phosphorus-containing acid halide, ester or amide reactant of 0 to 20 carbon atoms, and having from 1 to 4 functional groups selected from the group consisting of halogen, ester, (–OR) and amide ($-NR_2$) groups where R is an aliphatic radical of 1 to 6 carbon atoms; and (C) a different alcoholic moiety as defined in (A) or a different phosphorus-containing moiety as defined in (B), the ester having a degree of aggregation of 2 to 100.

The table below shows typical alcohol and phosphorus moieties from representative reactants. The esterifiable groups described above are bonded at the valences shown for the respective compounds.

Aliphatic Alcohol Component (A)

$-OCH_2-C(CH_3)_2-CH_2O$ — 2-functional (e.g. neopentyl glycol moiety)

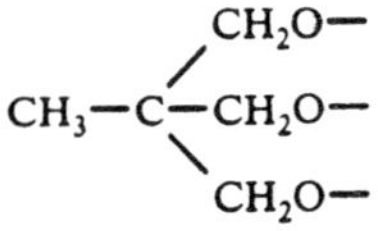

3-functional (e.g. trimethylolethane moiety)

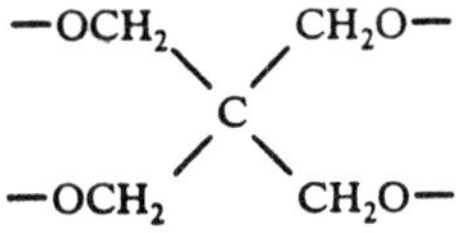

4-functional (e.g. pentaerythritol moiety)

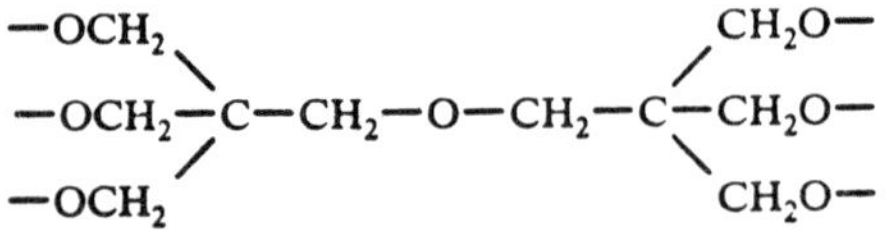

(6-functional) (e.g. dipentaerythritol moiety; and higher oligomeric forms such as polypentaerythritol, also containing the tri and tetra forms).

Phosphorus Component (B)

$R_2P(O)-$ — mono-functional (e.g. dimethylphosphinyl moiety)

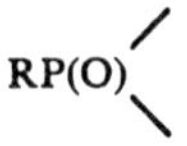

2-functional (e.g. chloromethylphosphonyl moiety)

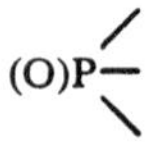

3-functional (e.g. phosphoryl moiety)

4-functional (e.g. bis(phosphonyl) methane moiety)

R_2P- — mono-functional (e.g. diphenylphosphino moiety)

$RP=$ — 2-functional (e.g. methylphosphinidene moiety)

$P\equiv$ — 3-functional (e.g. phosphinidyne moiety)

Each of (A), (B) and (C) are present at from 1 mol percent to 99 mol percent, with the proviso that the number of alcoholic groups on the alcoholic reactants is equal to the number of the functional groups on the phosphorus-containing reactants.

A representative polymeric structure is shown on the following page, based upon the reaction of $CH_3C(CH_2OH)_3$, $C_6H_5POCl_2$ and $OPCl_3$.

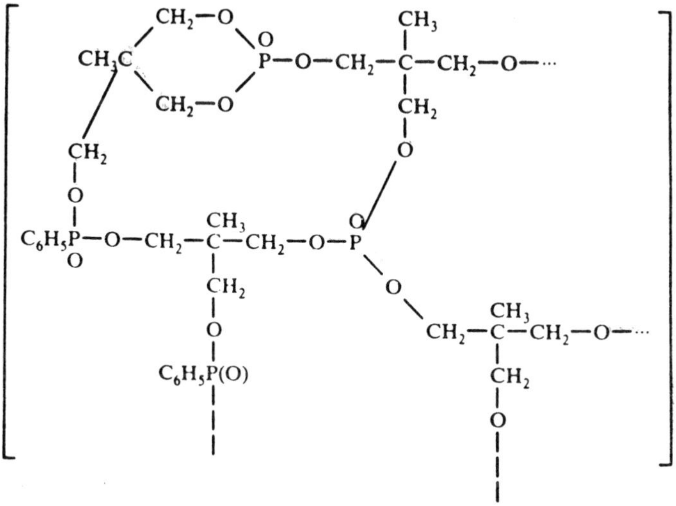

The process for the preparation of the ternary esters is conducted in the temperature range of 50° to 250°C. Control of the proportions of the (A), (B) and (C) components permits controlling the physical properties of the product. A solvent is not essential, but a nonreactive solvent such as benzene, toluene, octane, or decaline may be employed. The use of a catalyst is also optional; if desired, magnesium chloride or an organic base, e.g., piperidine, pyridine, piperazine, or picoline accelerate the rate of reaction.

The products useful as flame retardants may be applied as a solution to shaped or formed polymeric objects such as films, fibers, extruded and foamed products utilizing a solvent system such as alcohol, e.g., ethanol, or acid amides such as dimethylformamide. The esters can also be mixed with the molten organic polymer or a solution of such polymers in a solvent, before extrusion from spinnerettes to form fibers.

The polymeric esters are useful in combination with organic polymers or resins generally to reduce combustibility. The normally flammable organic polymers which are rendered flame retardant in accordance with the process may be natural or synthetic but are preferably a solid synthetic polymer, more preferably a nylon or ester-type polymer. Examples of the polymer are cotton, wool, silk, paper, natural rubber, and paint, and also high molecular weight homopolymers and copolymers of amides, e.g., nylon 66 and nylon 6.

Example 1: The use of dipentaerythritol as the alcohol component is shown in this example, while the phosphorus components are phosphorus oxychloride and chloromethylphosphonic dichloride. The chemical reaction vessel is charged with 25.9 g of dipentaerythritol (0.1 mol). This reactant is slurried into 250 ml of toluene and heated to reflux while stirring under a nitrogen atmosphere. The phosphorus components, 0.1 mol of $POCl_3$ and 0.15 mol $ClCH_2POCl_2$, in the respective amount of 15.4 g and 25.1 g are added slowly to the reaction mixture in the presence of 0.5 g of pyridine as a catalyst. The reaction temperature is about 80°C for a period of 24 hours, after which the precipitated polymeric ester is removed by filtration and washed with toluene, followed by drying at 100°C under vacuum. The ester does not support combustion, is insoluble in water, chloroform, dimethylformamide and toluene. The ester has the formula

$$\{[P(O)][(ClCH_2P(O)]_{1.5}[OCH_2)_3CH_2OCH_2(CH_2O)_3]\}_n$$

where n is about 20. The polymer does not support combustion and improves the flame retardancy of organic polymers. The use of this product as a flame retardant improves the flame retardancy of polyethylene terephthalate.

Example 2: The production of a three-component phosphorus-containing ester from two different alcoholic components, namely pentaerythritol and neopentylglycol and using phosphorus oxychloride as the phosphorus component is carried out in accordance with the equation shown below.

$$\tfrac{3}{4}\,C(CH_2OH)_4 + \tfrac{1}{2}(CH_3)_2C(CH_2OH)_2 + 1\tfrac{1}{3}\,POCl_3 \xrightarrow{-4HCl}$$

MW	136.15	104.15	153.4	153.4
g	20.42	10.42	40.9	40.9
moles	0.15	0.1	0.266	0.266

$$1/n\{[C(CH_2O)_4]_{0.75}[(CH_3)_2C(CH_2O)_2]_{0.5}[P(O)]_{1.33}\}_n$$

The two alcoholic components are charged to a reaction vessel together with 250 ml of toluene and 0.5 g of pyridine. The reaction mixture is brought to reflux temperature at about 110°C and the phosphorus oxychloride diluted to 250 ml with toluene is slowly added. After a 3-hour reaction period, the supernatant liquid is checked as being free of chlorine and hydrogen chloride evolution is completed. The product is removed from the reactor and dried in a desiccator giving a product which is composed of colorless hard crystals softening at 60°C. The use of the above product in the flame retardancy test gives an Oxygen Index of 25 with polyethylene terephthalate as the organic polymer (a control test without additive is about 19 O.I.)

Example 3: The use of two different phosphorus-containing components is shown in the example which proceeds according to the equation below:

$$C(CH_2OH)_4 + 1/6\,POCl_3 + 1\tfrac{3}{4}\,CH_3POCl_2 \xrightarrow{-4HCl}$$

MW	136.15	153.4	1.33
moles	0.24	0.04	0.42
g	32.68	6.13	55.86

$$1/n\{[C(CH_2O)_4][P(O)]_{0.17}[CH_3P(O)]_{1.75}\}_n$$

The pentaerythritol is charged into a reaction kettle together with 250 ml of toluene and 0.5 g of piperidine. After the mixture is heated to refluxing temperature at about 110°C the two phosphorus components are slowly added in a solution of 250 ml of toluene. The heating is continued until HCl evolution diminshes and the supernatant liquid shows only a trace of chlorine. The crystalline product is removed by filtration and washed with toluene and subsequently dried under vacuum. The product has an initial melting point of 85°C.

The above phosphorus ester when employed in a flame retardancy test gives an Oxygen Index of 23 when employed to the extent of 10 wt % of the total composition in polyethylene terephthalate.

Example 4: The use of a halogen-substituted alcohol material as one of the components is shown in this example which employs the equation shown below.

$$[(HOCH_2)_3CCH_2]_2O + 3[(HOCH_2)_2C(CH_2Cl)_2] + 6[ClCH_2POCl_2] \xrightarrow{-12HCl}$$

MW	254.3	173	167.4
moles	0.05	0.15	0.3
g	12.7	26.0	50.22

$$1/n\{[(OCH_2)_3CCH_2OCH_2(CH_2O)_3][(ClCH_2)_2C(CH_2O)_2]_3[ClCH_{12}P(O)]_6\}_n$$

The two alcohols are charged into an esterification kettle with 250 ml of toluene as a solvent and 0.5 g of pyridine. The reaction mixture is heated to 110°C and slow addition started of the phosphorus acidic component dissolved in 250 ml of toluene over a period of about 3 hours. The reaction is continued until hydrogen chloride evolution ceases, after which the solid product is removed. When mixed with polyethylene terephthalate as a 10% additive by weight, the Oxygen Index is 24 to 24.5.

Decabromo Diphenyl Ether

T. Aoyama, Y. Kimata, H. Okasaka, H. Kodama and N. Yoda; U.S. Patent 3,971,752; July 27, 1976; assigned to Toray Industries, Inc., Japan describe a flame retarding thermoplastic polyester composition reinforced by glass fiber which comprises 100 parts by weight of a glass fiber reinforced thermoplastic polyester, 2 to 9 parts by weight of at least one flame retarding agent selected from the group consisting of decabromodiphenyl ether and decabromodiphenyl, and 12 to 30 parts by weight of antimony trioxide, where the ratio by weight of bromine atom to antimony atom is within the range of 0.15 to 0.29.

The thermoplastic polyester may be a linear aromatic polyester which is constituted by at least one dicarboxylic acid component selected from the group consisting of terephthalic acid, 2,6-naphthalene dicarboxylic acid and 1,2-bis-(4-carboxyphenoxy) ethane, and at least one diol component selected from the aliphatic diols having a carbon number of 2 to 6.

These polyesters may be modified by replacing at most 20 mol percent of the main dicarboxylic acid or diol components with other copolymerizable dicarboxylic acids or diols, respectively. The glass fiber may be commercially available, and either short fiber type or long fiber type may be used in the process. Such polyesters have been used as engineering plastics, for example, as raw materials for mechanical parts of various machines, electrical equipment and parts of motor cars.

Example: Polybutylene terephthalate having an intrinsic viscosity of 0.80 (o-chlorophenol solvent at 25°C) was mixed in a ratio as shown in Table 1 with decabromobiphenyl ether (DBE), antimony trioxide and 6 mm length of chopped strand glass fiber, and extruded at 250°C into pellets using a 65 mmϕ extruder.

These pellets were dried at 130°C for 5 hours in a vacuum oven and subsequently molded into test pieces using an in-line screw-type injection molding machine maintained at 250°C on the cylinder and nozzle and 40°C on the mold.

The molded specimens were tested for tensile properties, flexural properties, impact strength, heat distortion temperature and flammability by ASTM D-638, D-790, D-256, D-648 and UL-94 methods, respectively. Furthermore, the long-term test of durability for thermal degradation was carried out at 170°C in air. The results are summarized in Table 1 and Table 2.

As shown in the tables, when Br/Sb ratio by weight is changed keeping the total content of DBE and antimony trioxide at 25 parts by weight, Br/Sb ratio of more than 0.25 does not give a remarkable improvement in respect to flame resistance and mechanical properties. On the contrary, the best durability for thermal degradation in long-term test was obtained at Br/Sb ratio of 0.25.

TABLE 1

	Experiment Number					
	EX-501	EX-502	EX-503	EX-504	EX-505	EX-506
Blend composition						
PBT, weight parts	70	70	70	70	70	70
Glass fiber, weight parts	30	30	30	30	30	30
DBE, weight parts	2	5	8	13	17	21
Sb_2O_3, weight parts	23	20	17	12	8	4
Br/Sb, weight ratio	0.09	0.25	0.47	1.08	2.12	5.23
Mechanical properties						
Tensile strength, kg/cm^2	1,260	1,280	1,280	1,300	1,270	1,280
Tensile elongation, %	3.0	3.2	3.0	2.9	3.0	3.0
Flexural strength, kg/cm^2	1,810	1,790	1,850	1,830	1,850	1,820
Flexural coefficient, $kg/cm^2 \times 10^4$	8.20	8.15	7.95	8.10	8.05	8.10
Izod impact strength						
Notched (½" t), kg-cm/cm	5.9	6.0	5.8	6.1	6.2	6.1
Unnotched (⅛" t), $kg\text{-}cm/cm^2$	54.1	55.3	57.3	56.2	56.8	57.2
Heat distortion temperature at 18.6 kg/cm^2 fiber stress, °C	206	204	202	201	200	201
Flammability						
⅛" t	94VE-2	94VE-0	94VE-0	94VE-0	94VE-0	94VE-0
1/16" t	94HB	94VE-0	94VE-0	94VE-0	94VE-0	94VE-0
1/32" t	94HB	94VE-0	94VE-0	94VE-0	94VE-0	94VE-0

TABLE 2

Heat Aging at 170°C (days)	Unnotched Izod Impact Strength Retention, %			
	Experiment Number			
	EX-502	EX-503	EX-505	EX-506
	Br/Sb Ratio			
	0.25	0.47	2.12	5.23
0	100	100	100	100
10	87	52	40	42
20	82	49	40	37
30	80	48	39	32
40	79	46	38	31
50	79	42	38	30
60	78	40	32	27

Zirconium-Containing Additive for Molding Composition

M. Morikawa, T. Aoyama and M. Yada; U.S. Patent 4,070,332; January 24, 1978; assigned to Toray Industries, Inc., Japan describe a process whereby thermoplastic polyester molding resins are prepared by blending about 5 to 25 parts by weight of a halogenated aromatic compound and about 3 to 15 parts by weight of a zirconium compound with about 100 parts by weight of a high molecular weight linear polyester. Additionally, antimony trioxide, to improve the flame retarding properties of the polyester resin, and reinforcing fibers such as glass fibers or asbestos to enhance the nondripping properties of the product during burning, are combined.

Examples of thermoplastic polyesters utilized are polyethylene terephthalate, polyhexamethylene terephthalate, polyethylene-2,6-naphthalate; polybutylene-2,6-naphthalate; polyhexamethylene-2,6-naphthalate, and poly-1,4-cyclohexylene dimethylene terephthalate. The aromatic compounds are selected from the following groups:

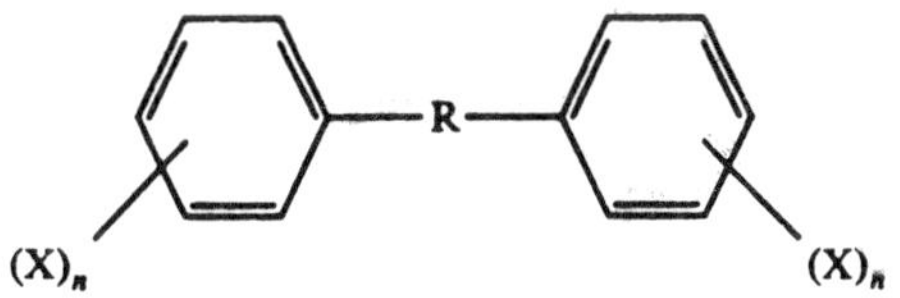

where R represents oxygen, sulfur dioxide, carbon dioxide, methylene, or phosphonates; X represents hydrogen, chlorine, or bromine with at least one chlorine or bromine; and n is 1 to 5. For example, hexabromobiphenyl ether, hexabromobiphenylsulfone, or decabromobiphenyl ether may be used. In the process a zirconium compound, preferably a powdered zirconium dioxide, is employed. Other zirconium compounds which contain a significant zirconium oxide content may be used. These include calcined products obtained by calcination of zirconium hydroxide, ammonium zirconate, titanium zirconate, calcium zirconate, and/or zinc or other metallic zirconates.

Antimony trioxide is preferably used as an auxiliary flame retarding agent in combination with the zirconium compound. The antimony compound is provided in an amount of about 1 to 20 parts by weight per 100 parts polyester by weight. Referring to fibrous reinforcement, it has been found that improved nondripping and shape-retaining properties are obtained when the fibrous reinforcement is present in an amount of about 5 to 50 parts by weight per 100 parts by weight of thermoplastic polyester. The fibrous reinforcing agent may be selected from the group consisting of glass fibers, asbestos, fibers, and gypsum fibers, silicate fibers and whiskers.

Thermoplastic polyesters, such as polyethylene terephthalate and polybutylene terephthalate, are highly desirable for use as engineering plastics and as raw materials for mechanical parts for various machines, electrical equipment and parts for automobiles, for example. Because of their excellent heat resistance and their high durability under thermal degradation, these synthetic resins are highly suitable for such uses.

Example 1: Polybutylene terephthalate (PBT) having an intrinsic viscosity of 1.26 (in o-chlorophenol at 25°C) was mixed in a ratio as shown in Table 1 with decabromobiphenyl ether (DBE), antimony trioxide, zirconium oxide and 2 mm in length of chopped strand glass fiber. The blend was extruded at 250°C on a cylinder and nozzle, and at 40°C in a mold. The molded specimens were tested for tensile properties and flammability according to ASTM specification D-638 and UL 94 methods, respectively. The results, summarized in Table 1, show that zirconium oxide when added to the flame retardant composition of glass reinforced PBT, prevents dripping of flaming particles, and that excellent mechanical properties are retained.

TABLE 1

	Run Number					
	1	2	3	4	5	6
Composition, parts by weight						
PBT	100	100	100	100	100	100
Glass fiber	40	40	40	45	50	40
DBE	12	14	5	12	12	12
Antimony trioxide	5	3	12	3	0	5
Zirconium dioxide	3	8	6	8	10	0
Flammability*						
1/16"	V-0(ND)	V-0(ND)	V-0(ND)	V-0(ND)	V-1(ND)	V-0(D)
1/8"	V-0(ND)	V-0(ND)	V-0(ND)	V-0(ND)	V-0(ND)	V-0(D)
Mechanical properties						
Tensile strength, kg/cm^2	1,867	1,865	1,980	2,059	2,230	1,780
Tensile elongation, %	3.2	4.1	3.0	2.8	3.1	3.2

*V-0(ND) is fulfilled rating V-0 of UL standards without any dripping and V-0(D) is fulfilled rating of V-0 of UL standards with dripping.

Example 2: One hundred parts of PBT having an intrinsic viscosity of 1.30 (in o-chlorophenol at 25°C) were blended with 40 parts of glass fiber, 20 parts of tetrabromobisphenol A and 5 parts of an additive as shown in Table 2. Following the procedure of Example 1, the molded specimens for UL-94 flammability tests were molded. The results as shown in Table 2 demonstrate that the addition of zirconium oxide to glass-reinforced flame retardant PBT is particularly effective to prevent the dripping of flaming particles during burning.

TABLE 2

Run No.	Additive	Flammability (1/16" thick sample)
7*	None	V-2
8	Zirconium dioxide	V-0(ND)
9*	Zinc oxide	V-0(D)
10*	Talc	V-2
11*	Titanium oxide	V-2

*Control example.

Aqueous Emulsions of Halogenated Phosphoric Esters

M. Tomita, M. Iimura and A. Ueno; U.S. Patent 4,054,720; October 18, 1977; assigned to Meisei Chemical Works, Ltd., Japan describe a flame retardant composition for polyester consisting of an aqueous emulsion of a halogenated phosphoric ester. The halogenated phosphoric ester used for the flame retardant composition may be selected from the group consisting of tris(2,3-dibromopropyl) phosphate, tris(2,3-dichloropropyl) phosphate, tris(2-chloroethyl) phosphate and tris(bromochloropropyl) phosphate.

This phosphoric ester is emulsified with use of a nonionic surfactant emulsifier which may be a compound obtained by esterification of a polyethylene glycol ether or a polypropylene glycol ether of a higher alcohol or an alkylphenol with a phosphoric esterifying agent. The phosphoric esterifying agent may be a member selected from the group consisting of phosphorus pentoxide, phosphorus oxychloride and phosphorus trichloride.

One of the most preferred compounds for the emulsifier is a monophosphoric ester obtained by esterification of an addition product of 5 to 15 mols of ethylene oxide to a saturated or unsaturated alcohol having 8 to 20 carbon atoms with phosphorus pentoxide. The amount of the nonionic surfactant emulsifier should be within the range of 5 to 20 wt % with respect to the amount of the halogenated phosphoric ester. The flame retardant compositions have a good workability and find their greatest usefulness in reducing and retarding the flammability of polyester articles such as polyester film or fibrous material.

Example: An addition product of 1 mol of oleyl alcohol and 5 mols of ethylene oxide was esterified with phosphorus pentoxide to obtain phosphoric ester containing monoester compound at a content of 80%. Four and one-half parts of the above phosphoric ester and 30 parts of perchloroethylene were added to 90 parts of tris(2,3-dibromopropyl) phosphate and mixed uniformly at 40°C. Then, 33 parts of water were added to the resultant mixture with continuous stirring by homomixer to obtain 157.5 parts of homogeneous milk-white emulsion. The obtained emulsion was very stable after standing for a long time. Then, 26 parts of the above emulsion were diluted with water to 100 parts to form a flame retarding treatment bath. Polyester taffeta was soaked in a bath of the resultant treatment liquid and squeezed to wet pick-up 65%. Then, this treated polyester taffeta was dried at 120°C for 5 minutes.

For controls, various treatment baths were prepared by emulsification of tris-(2,3-dibromopropyl) phosphate with different amounts of the above phosphoric ester and with other conventional surfactants. Polyester taffeta was treated with each treatment bath in the same manner as the above. Each treated cloth was measured for flammability by the Oxygen Index Method (JIL S-K-7201-1972). The results are shown in the following table.

Properties of Treatment Bath

	. . . . . . Composition of Treatment Bath	. . . . . .					
Test No.	Concentration of TDPP in a Bath (%)	Emulsifier	Amount of Emulsifier (based on TDPP)* (%)	Concentration of Perchloroethylene in a Bath (%)	Stability of Emulsion in a Bath	Wet Pickup (%)	O.I.**
1	15	–	–	85	–	65	31.5
2	15	phosphoric ester***	5	5	good	65	32.0

(continued)

	Composition of Treatment Bath						
Test No.	Concentration of TDPP in a Bath (%)	Emulsifier	Amount of Emulsifier (based on TDPP)* (%)	Concentration of Perchloroethylene in a Bath (%)	Stability of Emulsion in a Bath	Wet Pickup (%)	O.I.**
3	15	phosphoric ester***	7.5	5	good	65	32.5
4	15	phosphoric ester***	10	5	good	65	32.5
5	15	phosphoric ester***	15	5	good	65	32.0
6	15	phosphoric ester***	20	5	good	65	31.5
7	15	polyoxyethylene stearyl ether	5	5	bad	65	27.0
8	15	polyoxyethylene stearyl ether	10	5	bad	65	26.5
9	15	polyoxyethylene stearyl ether	20	5	good	65	25.0
10	15	polyoxyethylene sorbitan monooleate	5	5	bad	65	26.5
11	15	polyoxyethylene sorbitan monooleate	10	5	fairly good	65	25.0
12	15	polyoxyethylene sorbitan monooleate	20	5	good	65	23.0
13	–	untreated cloth	–	–	–	–	23.0

*TDPP is tris(2,3-dibromopropyl) phosphate.

**Oxygen Index.

***Described in the example.

As is clear from the above results, the treatment baths containing treating agents (Tests No. 2-6) showed an excellent flame retarding effect and stability during the process, in comparison with the baths containing the other surfactants (Tests No. 7-12). And, the cloths treated with treating agents of the process showed the same or more excellent properties as that of the cloth treated with perchloroethylene solution of tris(2,3-dibromopropyl) phosphate (Test No. 1).

Polyphenyl Phosphonates

P. Couchoud; U.S. Patent 4,061,615; December 6, 1977; assigned to Rhone-Poulenc-Textile, France describes a process relating to shaped articles of polyesters which have good resistance to flame, which are characterized by the fact that they are formed of a polymethylene copolyester containing 60 to 95 mol percent of terephthalate units and 5 to 40 mol percent of isophthalate units and that they contain 2 to 40 wt % of a phosphorus and/or halogen compound.

The phosphorus and/or halogen compound which can be employed may be 1,3-bis(diphenylphosphinoxide) 2,2-di(bromomethyl)propane, dibromoneopentyl glycol polyphenylphosphonate and, in particular hydroquinone polyphenylphosphonate and monobromohydroquinone polyphenylphosphonate.

The molecular weight of the polyphenylphosphonates which can be used is not critical, but it is generally preferred to employ polyphenylphosphonates of a specific viscosity of at least 0.10, measured on the basis of a 1% solution in dimethylformamide. Thus, the dibromoneopentyl glycol polyphenylphosphonate may have 3 or 4 recurrent units while the last two polyphenylphosphonates may have a degree of polymerization of 20 to 25.

In accordance with this process, the fireproofing agent can be mixed with the mass of molten polymer, for instance, in an extruder before its extrusion in the form of filaments, strands or films. The fireproofing agent may also be applied, in the form of solution, in a solvent such as methanol, acetone, trichloroethylene, etc., by impregnating the finished or semifinished textile articles such as threads, fibers, strands, fabrics, knitted goods, clothing, etc., with the solution; the impregnation may be followed by a heat treatment either dry or with steam. However, in this latter case the resistance of the fireproofing agent to laundering is not as good.

In these examples, the height-destroyed and number of ignited drop tests are carried out in accordance with Test AATCC 34-69; the limiting oxygen index (LOI) is determined in accordance with ASTM standard B 2863-70 relative to plastic test pieces, modified and adapted to textile samples of a size of 5 x 16 cm mounted on a rectangular frame; the intrinsic viscosity is measured on the basis of a solution of 1 wt % by volume in ortho-chlorophenol at 25°C; and the viscosity index of the polymer is determined on the basis of this same solution by means of the following formula:

$$IV = \frac{\text{specific viscosity}}{\text{concentration}} \times 1{,}000$$

in which the concentration is expressed in grams of polymer per 100 cc of solution.

Examples 1 through 3: 12%, with reference to the polymer, of powdered monobromohydroquinone polyphenylphosphonate is introduced by mixing into the following polymers:

Control: Dimethylene polyterephthalate of an intrinsic viscosity of 0.65;
Example 1: Dimethylene terephthalate/isophthalate copolymer in mol proportions of 92:8, of a viscosity index of 810, melt viscosity at 290°C of 2,450 poises and softening point of 244°C;
Example 2: Dimethylene terephthalate/isophthalate copolymer in mol proportions of 88:12, a viscosity index of 825, melt viscosity at 290°C of 2,320 poises and softening point of 231.6°C;
Example 3: Dimethylene terephthalate/isophthalate copolymer in mol proportions of 80:20, viscosity index of 820, melt viscosity at 290°C of 2,100 poises and softening point of 216°C.

These different polymers are extruded through a spinnerette having 23 orifices of 0.34 mm diameter and the yarns are wound up at 500 m/min. They are then stretched over a finger at 80°C and a plate at 70°C in the case of the control and finger at 70°C and plate at 80°C in the case of Examples 1 through 3 and at rates of 4.5 for the control, 4.3 for Examples 1 and 2 and 3.2 for Example 3.

The characteristics of the yarns thus obtained are tabulated below. The fire properties were examined on knitted material after desizing, after 20 washings, and after dry cleaning with perchloroethylene at 35°C for one hour.

Example No.	Control	1	2	3
Size in dtex	62	84	82	88
Tenacity in g/tex	30	36	38.8	27.3
Elongation %	20	18	14	17.2

(continued)

Example No.	Control	1	2	3
Weight/m² of the knitted material examined g	150 doubled	200 doubled	200 doubled	450
After desizing:				
LOI	29.2	31.5	32.5	36.1
Height destroyed in cm	10	5	6	7.5
No. of drops ignited	0	0	0	0
After 20 washings:				
LOI	29.2	28.2	26.3	38
Height destroyed in cm	25	7	6	8
No. of drops ignited	0	0	0	0
After dry cleaning:				
LOI	26.5	32	33	41
Height destroyed in cm	10	10	6	5
No. of drops ignited	0	0	0	0

Phosphorus-Containing Agents

A process described by *W.F.H. Borman; U.S. Patent 4,033,927; July 5, 1977; assigned to General Electric Company* provides flame retardant thermoplastic compositions, useful, e.g., for injection molding, compression molding, transfer molding, extrusion and film formation. These compositions comprise:

(1) a normally flammable high molecular weight linear polyester resin, and

(2) a phosphorus-containing flame retardant additive in a minor proportion based on the composition but in an amount at least sufficient to render the polyester resin flame retardant.

The additive useful in this process comprises a family of chemical compounds of the general formula

where X is –O– or a carbon-to-phosphorus bond and, (a) when X is –O–, R is a brominated aryl group and Y and G are brominated oxyaryl groups; and (b) when X is a carbon-to-phosphorus bond, R is aryl; Y is aryl, oxyalkyl, oxycycloalkyl or oxyaryl; and G is aryl, oxyaryl or a single or repeating unit of the formula

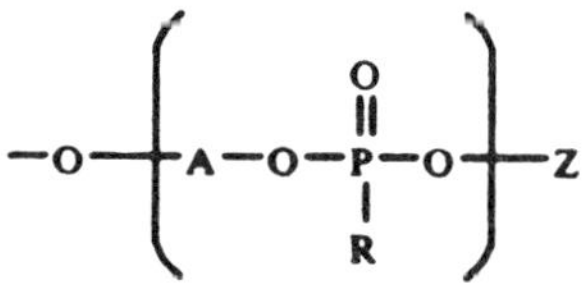

where R is aryl, A is a divalent alkylene, cycloalkylene, arylene or polyarylene radical, which may be interrupted by alkylidene, oxygen-, nitrogen-, sulfur-, and phosphorus-containing groups; and Z is a terminal group which may be, e.g., hydrogen, alkyl, cycloalkyl or aryl; and, when taken together, G and Y are divalent dioxyalkylene, dioxycycloalkylene or dioxyarylene groups.

Particularly efficient phosphorus-containing agents are brominated arylphosphates, polyalkylenearyl phosphonates and triarylphosphine oxides. The high molecular weight, normally flammable, linear polyesters used in the compositions are polymeric glycol esters of terephthalic acid and isophthalic acids, and if desired, minor amounts (up to 15 wt %) of difunctional aliphatic acids. Illustratively, high molecular weight polyesters will have an intrinsic viscosity of at least about 0.4 dl/g as measured in o-chlorophenol, a 60/40 phenol-tetrachloroethane mixture or a similar solvent at 25° to 30°C. The following procedure illustrates the preparation of a representative flame retardant additive used in the composition of this process.

Tris(4-Bromophenyl) Phosphate: 4-Bromophenol, 43 g, is dissolved in a solution of 12 g of NaOH in 100 ml of water. To this solution in a nitrogen-purged high-speed stirring reactor is added 0.2 g of methyl triphenyl phosphonium bromide. The solution is stirred rapidly while 7.7 g of $POCl_3$ dissolved in 100 ml of methylene chloride is added in a thin stream over a 2-minute period. The reaction mixture is stirred rapidly for an additional 3 minutes and then more slowly for 10 minutes. After separating the phases, the organic layer is washed with dilute caustic, then with water until neutral, and dried over $MgSO_4$. The solvent is removed in vacuo and the resulting crude product is recrystallized once from n-heptane. The product weighs 20.4 g (72.5% yield); white crystals, melting point, 108° to 110°C.

In an analogous manner replacement of 4-bromophenol with stoichiometric quantities of 2-bromophenol; 3-bromophenol; 2,4-dibromophenol; 2,6-dibromophenol; 3,4-dibromophenol; 3,5-dibromophenol; 2,4,6-tribromophenol; and 1,3,6-tribromo-2-naphthol provides the corresponding bromo-aryl phosphates.

Example 1: The following formulations are prepared, and melt-cast or injection-molded into bars and the Oxygen Indices are determined:

Example	1	1A*
Ingredients (parts by weight)		
poly(1,4-butylene terephthalate)	100	100
tris(4-bromophenyl) phosphate	10	10
% Phosphorus in composition	0.5	0.5
% Bromine in compositon	3.9	3.9
Combustibility		
Oxygen Index, %	30.0	27.0

*1A - Specimens are injection molded, instead of melt-cast.

It is thus demonstrated that tris(4-bromophenyl) phosphate at a concentration of only 10 pph (parts per hundred), and without additional synergist, imparts a high oxygen index of 27 to 30% to normally flammable poly(1,4-butylene terephthalate). This high efficiency as a flame retardant agent is realized at a phosphorus content of only 0.5% and a bromine content of 3.9%, based on the weight of the composition.

Tris(4-bromophenyl) phosphate is an effective flame retardant agent even at the 5 pph level. The injection molded bars (Example 1A) in addition to having a high Oxygen Index, are self-extinguishing with a SE-O rating in a simulated Underwriters Laboratories UL 94 test. The intrinsic viscosity of the composition of Example 1 after processing is 0.70 dl/g, indicating no degradation by the additive.

Because of their excellent physical, mechanical, chemical, electrical and thermal properties and their enhanced flame resistance, the flame retardant polyester compositions have many and varied uses. The formulations may be used alone or mixed with other polymers and may contain various fillers, such as wood flour, diatomaceous earth, carbon black, glass fibers, etc., as well as pigments and dyes, stabilizers, plasticizers, and the like.

Halogen- or Phosphorus-Containing Additives for Foamed Compositions

A process described by *T.J. Dolce; U.S. Patent 4,002,581; January 11, 1977; assigned to General Electric Company* relates to foamed thermoplastic compositions which include a polyester resin, such as poly(1,4-butylene terephthalate); an ionically crosslinked copolymer of an α-olefin and an α,β-ethylenically unsaturated mono- or dicarboxylic acid and a polytetrafluoroethylene resin.

Especially useful are branched high-melt viscosity poly(1,4-butylene terephthalate) resins, which include a small amount of a branching component containing at least three ester-forming components. The branching component may be one which provides branching in the acid unit portion of the polyester, or in the glycol unit portion, or it may be a hybrid. Illustrative of such branching components are tri- or tetracarboxylic acids, such as trimesic acid. The relative amount of branching component is always kept at a minor proportion, e.g., from 0.01 to 3 mol percent based on the terephthalate units.

The ionic copolymers comprise a polymer of an α-olefin having the general formula $RCH{=}CH_2$ where R is a radical selected from the group consisting of hydrogen and alkyl radicals having from 1 to 8 carbon atoms, the olefin content of the polymer being at least 50 mol percent based on the polymer. The α,β-ethylenically unsaturated monocarboxylic acid used has from 3 to 8 carbon atoms, the acid monomer content of the polymer being from 0.2 to 25 mol percent based on the polymer. Furthermore, the polymer contains a metal ion selected from the group consisting of ions of alkali metals and alkaline earth metals uniformly distributed through the polymer in sufficient quantity to neutralize at least 10% of the carboxylic acid monomer.

The preferred alkali metal ion is Na^+ although K^+ and Li^+ ions may be employed. Preferred polymers will employ ethylene as the α-olefin and acrylic or methacrylic acid as the α,β-ethylenically unsaturated carboxylic acid monomer.

The compositions will comprise from about 95 to about 99.9 parts by weight of the linear polyester resin; from about 5 to about 0.1 part by weight of ionically crosslinked copolymer and from about 5 to about 0.1 part by weight of polytetrafluoroethylene resin. The preferred compositions will include from about 99 to about 98.8 parts by weight of the linear polyester; from about 0.2 to 1 part by weight of the ionically crosslinked copolymers and from about 0.2 to 1 part by weight of polytetrafluoroethylene resin. The foamed linear polyester compositions of this process include flame retardant quantities of flame retardant agents and/or reinforcing amounts of reinforcing fillers. Useful flame retardants are disclosed in U.S. Patents 3,833,685, 3,334,154 and 3,671,487.

It is preferred that the flame retardant additive comprise a halogenated organic compound (brominated or chlorinated); a halogen-containing organic compound mixed with antimony oxide; elemental phosphorus or a phosphorus compound; a halogen-containing compound mixed with a phosphorus compound or compounds containing phosphorus-nitrogen bonds or a mixture of two or more of these. The linear polyesters have been employed for various thermoplastic molding applications and in the preparation of foamed structures.

The preferred halogen compounds for this process are aromatic halogen compounds such as halogenated polycarbonates, chlorinated benzene, brominated benzene, chlorinated biphenyl, chlorinated terphenyl, brominated biphenyl, brominated terphenyl, or a compound comprising two phenyl radicals separated by a divalent alkylene or oxygen group and having at least two chlorine or bromine atoms per phenyl nucleus. Especially preferred are the above-described halogenated polycarbonates, alone, or mixed with antimony oxide.

In general, the preferred phosphate compounds are selected from elemental phosphorus or organic phosphonic acids, phosphonates, phosphinates, phosphonites, phosphinites, phosphene oxides, phosphenes, phosphites or phosphates. Illustrative are triphenyl phosphene oxide. This can be used alone or mixed with hexabromobenzene or a chlorinated biphenyl and, optionally, antimony oxide.

1,4-Bis(β,γ,γ-Trichloroallyl)-2,3,5,6-Tetrachlorobenzene

W. Schmidt, E.N. Petersen and H. Richtzenhain; U.S. Patent 4,056,508; Nov. 1, 1977; assigned to Dynamit Nobel AG, Germany describe a flame-stabilized molding composition comprising a polyester and a flameproofing agent. The polyester is the condensation product of an aromatic dicarboxylic acid, acid halide or ester and an aliphatic diol having more than two carbon atoms in the chain. As a flameproofing agent, 1,4-bis(β,γ,γ-trichloroallyl)-2,3,5,6-tetrachlorobenzene of the formula

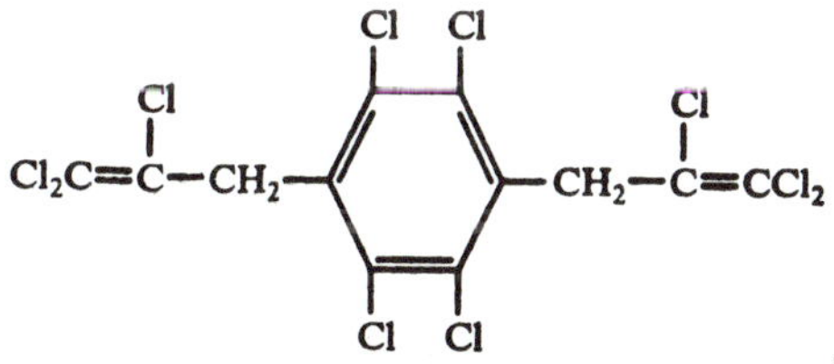

is used. This agent is present in the composition in an amount of 5 to 30 wt % based upon the weight of the polyester.

The polymer component of the molding composition consists wholly or mostly of a polyester, e.g., of polypropylene or polybutylene terephthalate, or of a mixture of the individual components. If desired, the polyesters can be blended with another polyester, e.g., polyethylene terephthalate, or with another polymer, such as ABS, provided the polymers are compatible with one another and with the flameproofing agent.

Especially good characteristics for fabrication and use are shown by those molding compositions whose polymer component consists wholly or mostly of polytetramethylene terephthalate having a reduced viscosity of 1.3 to 1.6.

The 1,4-bis(β,γ,γ-trichloroallyl)-2,3,5,6-tetrachlorobenzene used as a flameproofing agent is prepared in high yields from tetrachloro-p-xylene dichloride and trichloroethylene in the presence of anhydrous aluminum chloride.

The inclusion of 1,4-bis(β,γ,γ-trichloroallyl)-2,3,5,6-tetrachlorobenzene in a polyester molding composition in an amount of 20 parts per 100 parts of polyester together with 2.5 to 5 parts of antimony oxide per 100 parts of polyester renders the polyester self-extinguishing when subjected to a flame for a period of 10 seconds.

The molding compositions are particularly well suited for the preparation of injection molded articles. Articles having good characteristics are achieved by operating with a screw-type injection molding machine with a backflow blocking means (e.g., Monomat 150 S) and injecting into heated dies. In the case of molding compositions containing polybutylene terephthalate, the die temperature may be, for example, 50° to about 80°C.

Examples 1 through 6: Polybutylene terephthalate prepared from the condensation of dimethyl terephthalate and 1,4-butanediol having a reduced viscosity of 1.6 determined at 25°C in a 40:60 weight percent mixture of 1,1,2,2-tetrachloroethane and phenol utilizing 1 g of polyester dissolved in 100 ml of solvent, was charged into an extruder together with various amounts of 1,4-bis(β,γ,γ-trichloroallyl)-2,3,5,6-tetrachlorobenzene, as flameproofing agent. The various amounts used in Examples 1 through 6 are reported in the table below.

This particular double screw extruder provides a kneading and mixing action and forms strands of 2 to 4 mm thick as the polymeric mass emerges from the dye. The double screw extruder has various zones of heating. The polymer is heated at 210°C as it enters the extruder and undergoes various zones of heating as follows: 230°C; 240°C; 240°C; 230°C and it emerges from the nozzle or dye at 230°C. The strands formed were granulated and these granules were made into test specimens in a screw-type injection molding machine with an antibackflow. The mol temperature was between 50° to 60°C.

The various specimens made were subjected to a number of different tests. For testing of the flameproofing action, an independent inhouse test was developed and used, which permits a judgment of the performance of suspended specimens of plastic when a flame is applied to them under defined conditions. On the basis of the time of application of the flame, the afterburning time, the dripping, the flame propagation and flame temperature, and of the formation of gas, scores ranging from 1 (very good) to 5 (poor) are given.

Plastic specimens measuring 1.6 x 20 x 240 mm are suspended vertically from a suitable hanger in a square guard box open at the top and one side, and a flame is applied to their bottom ends from a nonincandescent, horizontal Bunsen burner flame 10 cm long directed vertically against the bottom edge of the specimen at a distance of 10 cm. The behavior of the specimen is observed and the scores from 1 to 5 are arrived at on the basis of the time of application of the flame, the afterburning time, the dripping, and other criteria. The scores are defined on the following page.

(1) After a flame application time of at least 10 seconds the specimen is extinguished immediately after removal of the flame; any drips that might fall do not burn.
(2) The specimen is extinguished a few seconds after removal of the flame; drips are extinguished as they fall.
(3) The specimen continues to burn after removal of the test flame, but shortly is extinguished; drips continue to burn briefly.
(4) The specimen continues to burn after removal of the test flame and is extinguished by the dripping off of the burning end; drips continue to burn for a longer time.
(5) The specimen continues to burn fully after removal of the test flame; drips likewise.

The scores given in the table are averages of 10 tests. For confirmation of the results thus obtained, the flame-inhibiting action was tested by means of the Oxygen Index test of ASTM D-2863-70. In the case of Examples 1 (standard), 4, 6 and 7, the UL 94 test was performed in order to confirm the results. The table below shows the results of the test for the various sample numbers. The antimony oxide was added during the passage of the molding mass through the double screw extruder.

Example No.	FPA[1]	Sb_2O_3	Dynamit Nobel Test Scores	Oxygen Index % O_2	UL 94
	Parts per 100 parts per Polyester				
Standard for comparison	0	0	5	23	failed
2	15	5	2	29	—
3	15	2.5	2	29	—
4	20	5	1	31	VE 0
5	20	2.5	1	31	—
6	25	5	1	32	VE 0
7	25	2.5	1	32	VE 0

[1]Flame proofing agent - 1,4-bis-(β,γ,γ-trichlorollyl)-2,3,5,6-tetrachlorobenzene

Phosphorus-Containing Heterocyclic Ring

E.L. Lawton and E.L. Ringwald; U.S. Patent 3,970,635; July 20, 1976; assigned to Monsanto Company describe a process whereby fire retardant qualities are incorporated into polyester fibers by adding to the polymer, prior to spinning into filaments, not more than 20%, based on the weight of the polymer, of a compound containing two 6-membered phosphorus-containing heterocyclic rings having carbon, oxygen and phosphorus in the heterocyclic ring. The compound has the following general formula:

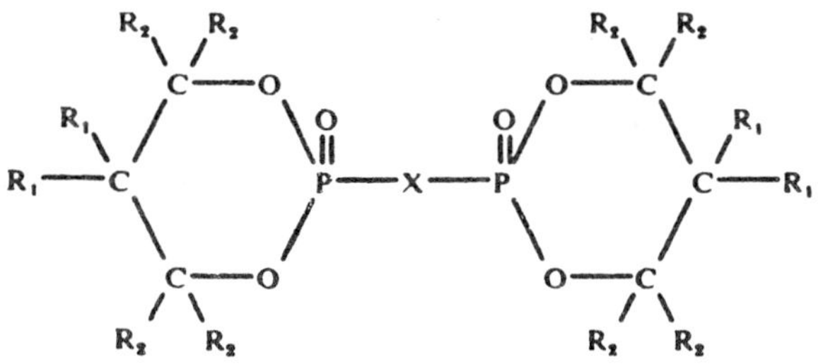

In the formula on the previous page R_1 is an alkyl group containing 1 to 5 carbon atoms; R_2 is a radical selected from the group consisting of H and an alkyl group containing 1 to 5 carbon atoms; and X is a divalent radical selected from the group which consists of

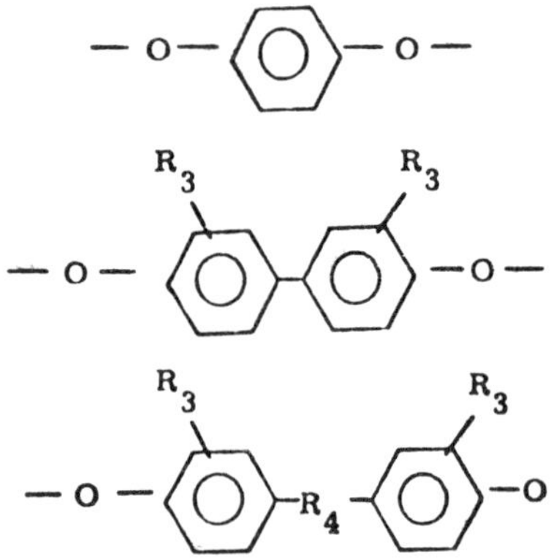

In the structures shown above, R_3 is a radical which is selected from the group consisting of hydrogen or a 1 to 5 carbon alkyl group, and R_4 is a radical selected from the group consisting of an alkyl radical having 1 to 10 carbon atoms, and SO_2.

These polyesters are relatively insoluble, chemically inactive, hydrophobic materials capable of being formed into filaments which can be cold-drawn to produce textile fibers of superior strength and pliability.

Example 1: Additive – A sample of the fire retardant additive was prepared as follows:

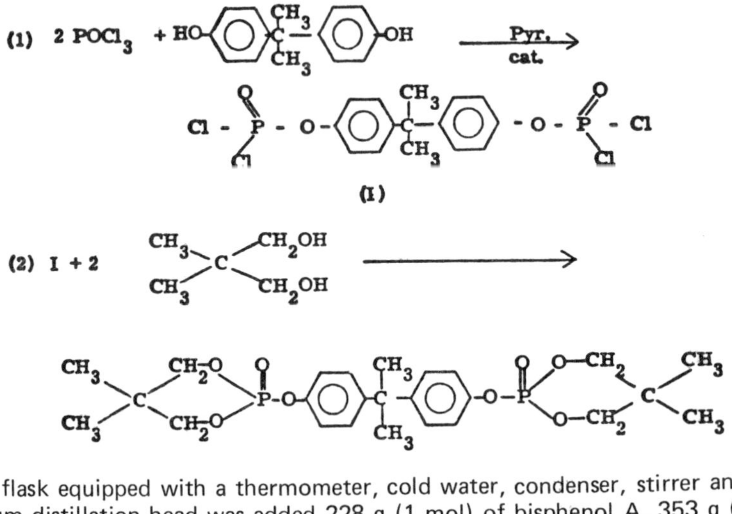

To a flask equipped with a thermometer, cold water, condenser, stirrer and vacuum distillation head was added 228 g (1 mol) of bisphenol A, 353 g (2.3 mol) $POCl_3$, 3 g of pyridine and 200 g toluene. This was heated to reflux over 1¾ hours and held there for 1¼ hours longer. HCl was evolved as an off-gas.

Excess $POCl_3$ was distilled off at 80 mm and 100°C. An additional 100 g of toluene was added to the reaction pot and the solution then cooled to room temperature. To this was added 208 g of neopentyl glycol. The reaction was exothermic and HCl gas was evolved. After this addition was completed, the reaction mix was heated to 98°C over 1¼ hours. It was then cooled to room temperature and the resulting crystalline material collected by filtration. The solid was washed with a solution of 15 g Na_2CO_3 in 300 ml hot water, followed by 600 ml of benzene and 600 ml acetone. The crystals were dried to a yield of 75%.

Example 2: A sample of a compound of the following structure was prepared as described in U.S. Patent 3,090,799.

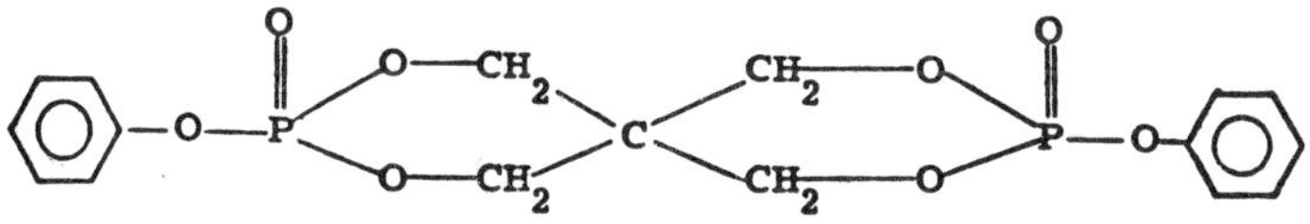

Example 3: A polyethylene terephthalate prepolymer having a specific viscosity (η_{sp}) of 0.043 in phenol/s-tetrachloroethane (60/40) at 0.5 g dl^{-1} and 25°C was polymerized at 265°C and approximately 1 mm Hg pressure until the polymer reached a solution specific viscosity greater than 0.28. The polymerization was performed in a 0.5 liter stainless steel autoclave. The phosphorus-containing additive was introduced into the molten polymer and mixed with a stirrer for 5 minutes. The molten polymer was then spun into a 10 filament bundle of fiber. The fiber was drawn over a heated pin at 80°C. Properties of the fibers were as shown in the following tables.

TABLE 1

Fiber Property	. . . . . .Weight Percent* of Example 2 Additive.				
	0	5.3	7.0	9.1	11.1
Draw ratio	4.9	3.7	4.9	4.4	3.9
Melting point, °C	256	250	249	247	249
Tenacity, g/denier	3.4	4.0	4.3	4.0	2.4
Ultimate elongation, %	31	42	10	24	38
Initial modulus, g/denier	114	94	138	138	82
η_{sp}	0.29	0.35	0.32	0.32	0.32
Chemstrand whiteness, W_C	82	82	75	85	74

$$\text{*Weight percent} = \frac{\text{weight additive}}{\text{weight additive + weight polymer}} \times 100$$

TABLE 2

Fiber Property	. . . . . .Weight percent of Example 1 Additive.				
	0	4.8	7.0	9.1	11.1
Draw ratio	4.9	5.5	5.5	5.5	4.8
Melting point, °C	256	251	252	248	247
Tenacity, g/denier	3.4	5.3	5.4	4.8	3.5
Ultimate elongation, %	31	14	12	12	28

(continued)

TABLE 2: (continued)

Fiber Property	0	4.8	7.0	9.1	11.1
	..... Weight percent of Example 1 Additive.....				
Initial modulus, g/denier	114	150	161	137	112
η_{sp}	0.29	0.32	0.31	0.32	0.29
Chemstrand whiteness, W_C	82	85	86	77	72

Tristimulus color was measured in a General Electric/Hardy Reflectance Spectrophotometer. Relative brightness is indicated directly by the value of Y. Chromaticity coordinates x and y are calculated from the reflectance measurements, and more particularly, as described at pages 9-10 of the Massachusetts Institute of Technology *Handbook of Colorimetry* (1936 - The Technology Press, Cambridge, Massachusetts). From the values x and y, purity (percent) and dominant wavelength (DWL in nm) is determined from chart 12a in the above-cited publication. Chemstrand whiteness (W_C) is calculated using the following equation: $W_c = 10\ (Y - 2p^2)^{1/2}$ where $Y = Y_{CIE}$ and p = purity.

Fibers containing either additive were reasonably stable to spinning conditions as shown by the following table in which each additive was added to the polymer at 5 wt % level and 265°C and then the mixture of additive and polymer is sampled at various times. The rate of specific viscosity change with time was recorded.

Time (min.)	Specific Viscosity No Additive	Example 1	Example 2
0	0.343	0.285	0.335
5	0.339	0.240	0.326
15	0.338	0.238	0.324
30	0.330	0.228	0.321
45	0.342	0.224	0.311
60	0.324	0.223	0.290
75	0.335	0.219	0.305

Figure 2.1 shows how the limiting oxygen index in polyester fibers is affected by the presence of phosphorus introduced by the additives of Examples 1 and 2.

FIGURE 2.1: EFFECT OF PHOSPHORUS CONTENT ON OXYGEN INDEX

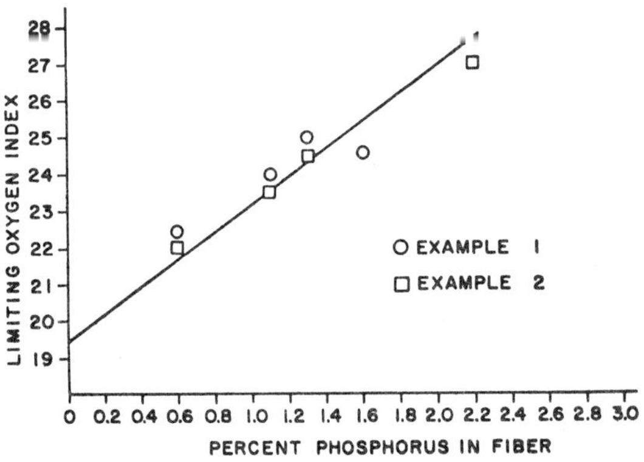

Source: U.S. Patent 3,970,635

Bis(Beta-chloroethyl) Vinyl Phosphonate/Maleic Anhydride Mixture

In a process described by *A.J. Spak and A.J. Yu; U.S. Patent 4,088,710; May 9, 1978; assigned to Stauffer Chemical Company* the gel time of unsaturated polyester thermosetting resins containing polymerizable vinyl monomers, such as styrene, and a vinyl phosphonate flame retardant, such as the bis(hydrocarbyl) vinyl phosphonates, which normally retards the gel time of such resins, is reduced by the addition of an effective amount for gel time reduction of maleic anhydride.

Example: This example illustrates the effect of bis(beta-chloroethyl) vinyl phosphonate (abbreviated as "Bis Beta") and maleic anhydride (MA) on commercially available styrenated polyester unsaturated resin. In addition to these two additives (when present) the polyester also contained 2 wt % of the polyester, of methyl ethyl ketone peroxide catalyst. All percentages are by weight based on the weight of polyester.

In the table, H_2O is measured as the percentage of water absorbed by the sample in 24 hours at room temperature, based on the weight of the dry sample. Water absorption or the moisture content of the material is intimately related to such properties as electrical insulation resistance, dielectric loss, mechanical strength, appearance and the dimensions of the sample. A lower value is generally desired for most applications. The weight change is measured as the percentage weight loss of a sample after immersion in acetone at room temperature for 24 hours.

The Limiting Oxygen Index (LOI) test is conducted by means of the procedure described by Fenimore and Martin in the November 1966 issue of *Modern Plastics*. It directly relates flame retardancy to a measurement of the minimum percentage concentration of oxygen in an oxygen-nitrogen mixture which permits the sample to burn. A higher LOI is indicative of a higher degree of flame retardancy and is more desirable for some applications.

Sample No.	Bis-Beta (%)	MA (%)	Gel Time (min)	H_2O Absorption (%)	Weight Change (%)	LOI	Flexural Strength* x 10^9 (dynes/cm^2)	Flexural Stiffness** x 10^{10} (dynes/cm^2)
1	0	0	55.3	+0.67	-2.2	20.4	2.1	9.4
2	10	0	>1,200	–	–	–	–	–
3	15	10	30.3	+0.3	-4.8	27.5	2.9	12.6
4	20	10	47.7	+0.41	-6.5	30.9	2.7	11.1

*The flexural strength of the sample is the maximum stress in the outer fibers of a specimen at the moment of crack or break. It is obtained by ASTM D 790-1966.

**This is a measure of the relative stiffness of plastic materials and is obtained by means of ASTM D 790.

The above data indicate that use of a bis(beta-chloroethyl) vinyl phosphonate-maleic anhydride mixture in a styrenated unsaturated polyester resin increases flammability resistance without seriously altering the other physical properties of the resin.

REACTIVE INTERMEDIATES

Perfluoroalkyl Esters

According to *R.H. Dettre and E.J. Greenwood; U.S. Patent 4,029,585; June 14, 1977; assigned to E.I. Du Pont de Nemours and Company* dry soil resistance and non-flame-propagating characteristics are insured in textile fibers by applying thereto an aqueous dispersion containing a perfluoroalkyl ester of a carboxylic acid of from 3 to 30 carbon atoms. After the dispersion is applied, the fibers are dried at between 120° and 170°C.

The fluorinated esters can be applied to synthetic thermoplastic fibers such as polyester fibers so as to leave from about 0.01 to 1.0% of the ester on the fibers, based on dry fiber weight. In one method of application an aqueous treating dispersion can be prepared as follows. The ester is liquefied by mixing with a small amount of volatile solvent such as methyl isobutyl ketone or the like, and the product dispersed in water containing a little emulsifying surfactant such as a tetraalkylammonium halide to make a composition containing, typically, about 10% ester.

This aqueous dispersion can be extended in water for application to a textile substrate such as a synthetic fiber carpet. Spray application, dipping and wringing, curtain coating or the like can be employed to coat the fibers uniformly with the dispersion, followed by drying at about 120° to 170°C. The fibers thus invested with a coating do not propagate a flame.

Incorporating Additives Into Melt Spun Fibers

R.T. Guthrie, J.L. Hirshman, S. Littman, E.L. Sukman and P.H. Ravenscroft; U.S. Patent 4,001,367; January 4, 1977; assigned to M&T Chemicals Inc. describe a process whereby additives are permanently incorporated into melt spun fibers by cold drawing the fibers under conditions that generate a network of interconnecting microvoids within the fiber. The microvoids are formed concurrently with the drawing of the fiber in a nonsolvent liquid or vapor medium in which the medium fills the microvoid network. The temperature of the medium is below the effective glass transition temperature of the fibers containing the medium. The additive is either present in the medium or is applied to drawn fibers containing the medium. Additives that impart flame retardancy to the fiber are applied by this process.

Compositions Containing Bis-Phenoxy Compounds

A.L. Anderson; U.S. Patent 3,971,763; July 27, 1976; assigned to Michigan Chemical Corporation describes plastic compositions containing polyesters and bis-phenoxy compounds having the formula

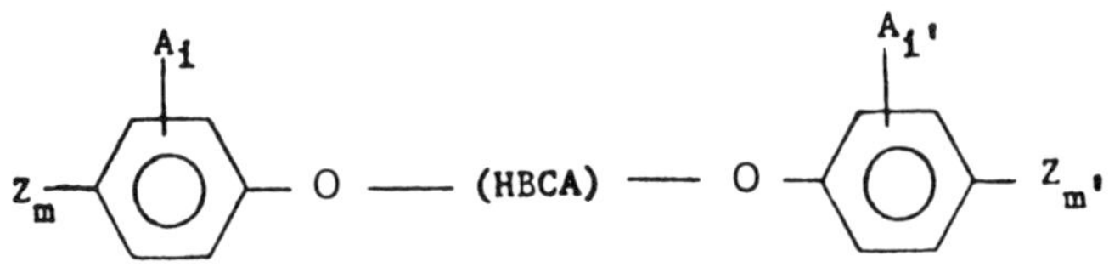

where Z is bromine or chlorine, m and m' are integers having a value of 1 to 5, i and i' are integers having a value of 0 to 2, HBCA is a halo-branched chain alkylene group having from 1 to 6 carbon atoms and A is cyano, nitro, lower alkoxy, lower alkyl, fluorine, dialkylamino, phenyl, halo-phenyl, benzyl or halo-benzyl. In the formula for the bis-phenoxy compounds shown on the previous page A, Z, i, i', m, m' and HBCA may be as listed in the following table:

Compound No.	Z	m	m'	A	i	i'	HBCA
1	Br	3	3	—	0	0	$CH_2C(CH_2Cl)H$
2	Br	3	3	—	0	0	$CH_2C(CH_2Cl)_2$
3	Br	5	5	—	0	0	$CH_2C(CH_2Cl)H$
4	Br	2	2	—CN	1	1	$CH_2CH_2C(CH_2Br)H$
5	Br	2	2	$—NO_2$	1	1	$CH_2C(CHCl_2)H$
6	Cl	2	2	$—OCH_3$	1	1	$CH_2C(CH_2Cl)H$
7	Br	3	3	$—OCH_3$	1	1	$(CH_2)_3C(CH_2Cl)_2$
8	Br	2	2	$—CH_3$	1	1	$CH_2C(CH_2Cl)H$
9	Br	2	2	F	1	1	$CH_2C(CH_2Cl)_2$
10	Br	2	2	$—N(CH_3)_2$	1	1	$CH_2C(CHBr_2)H$
11	Br	2	2	$—C_6H_5$	1	1	$(CH_2)_2C(CBr_3)H$
12	Cl	2	2	$—C_6H_3Br_2$	1	1	$CH_2C(CH_2Cl)H$
13	Br	2	2	$—CH_2C_6H_5$	1	1	$CH_2C(CCl_3)_2$
14	Br	2	2	$—CH_2C_6H_3Br_2$	1	1	$CH_2C(CH_2Cl)H$
15	Cl	3	3	$—C_6H_3Cl_2$	1	1	$CH_2C(CH_2Cl)H$
16	Br	3	3	F	1	1	$CH_2C(CCl_3)_2$
17	Cl	5	5	—	0	0	$CH_2C(CH_2Cl)H$
18	Br	4	4	—	0	0	$CH_2C(CH_2Cl)H$
19	Br	3	3	$—C_6H_2Br_3$	1	1	$(CH_2)_3C(CH_2Cl)_2$
20	Br	3	3	—	0	0	$CH_2C(CH_2Cl)HCH_3$
21	Br	2	2	—	0	0	$CH_2C(CCl_3)H$
22	Br	4	4	—	0	0	$CH_2C(CHBr_2)H$
23	Br	3	3	F	2	2	$CH_2C(CH_2Cl)H$
24	Br	1	1	$—C_4H_9$	1	1	$CH_2C(CH_2Cl)H$
25	Br	1	1	$—OC_4H_9$	1	1	$(CH_2)_3C(CH_2Cl)H$

The bis-phenoxy compounds are prepared by first reacting a halogenated phenol with a halogenated alkanol at elevated temperatures in the presence of a basic material such as alkali metal hydroxides, carbonates, bicarbonates, oxides, and hydrides to form an intermediate product. The preferred alkali metals are potassium and sodium. This product is then contacted with a halogenating agent such as $SOCl_2$, PCl_3, PCl_5, PBr_3, PBr_5, $(C_6H_5)_3PBr_2$ and the like as shown below, where X is halogen, preferably chlorine, and R is a C_3-C_6 alkylene.

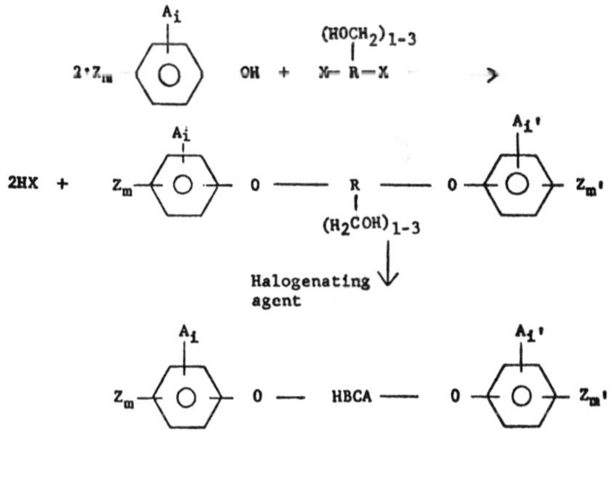

Preferably, the halogenation is carried out in a reaction medium such as an aromatic solvent. The following example illustrates the process.

Example: A linear saturated polyester material (Varlox 310) is utilized as the base resin in order to prepare 26 formulations (plastic compositions) designated Nos. 1-26 in the following table.

Flammability Data for Polyester Plastic Compositions Containing Bis-Phenoxy Compounds

Formulation No.	Bis-Phenoxy No.	Compound %	Enhancing Agent Sb_2O_3, %	Oxygen Agent %	UL 94
1.	—	0	0	22.0	SB
2.	1	20	0	23.5	SB
3.	1	20	10	29.0	SE-2
4.	3	20	0	23.0	SB
5.	3	20	10	33.0	SE-1
6.	4	20	0	24.0	SB
7.	4	20	10	28.0	SE-2
8.	5	20	0	23.5	SB
9.	5	20	10	29.5	SE-2
10.	6	20	0	23.5	SB
11.	6	20	10	27.5	SE-2
12.	8	20	0	24.0	SB
13.	8	20	10	28.0	SE-2
14.	12	20	0	24.5	SB
15.	12	20	10	29.0	SE-2
16.	14	20	0	24.5	SB
17.	14	20	10	29.5	SE-2
18.	16	20	0	23.5	SB
19.	16	20	10	31.0	SE-2
20.	17	20	0	23.5	SB
21.	17	20	10	29.5	SE-2
22.	20	20	0	24.5	SB
23.	20	20	10	30.5	SE-2
24.	23	20	0	23.0	SB
25.	23	20	10	28.5	SE-2
26.	25	20	10	27.0	SE-2
27.	—	0	0	18.0	SB
28.	1	20	0	2.5	SB
29.	1	20	10	27.5	SE-1
30.	3	20	0	22.5	SB
31.	3	20	10	32.0	SE-0
32.	4	20	0	20.5	SB
33.	4	20	10	28.0	SE-1
34.	5	20	0	21.0	SB
35.	5	20	10	28.5	SE-1
36.	6	20	0	20.5	SB
37.	6	20	10	28.5	SE-1
38.	8	20	0	21.0	SB
39.	8	20	10	28.0	SE-1
40.	12	20	0	21.5	SB
41.	12	20	10	29.0	SE-1
42.	14	20	0	22.0	SB
43.	14	20	10	29.5	SE-0
44.	16	20	0	21.0	SB
45.	16	20	10	30.5	SE-0
46.	17	20	0	21.5	SB
47.	17	20	10	29.0	SE-0

(continued)

Formulation No.	Bis-Phenoxy No.	Compound %	Enhancing Agent Sb_2O_3, %	Oxygen Agent %	UL 94
48.	20	20	0	22.0	SB
49.	20	20	10	31.0	SE-0
50.	23	20	0	21.0	SB
51.	23	20	10	28.5	SE-1
52.	25	20	10	26.5	SE-2

With the exception of formulation No. 1 (the control), the particular bis-phenoxy (and the antimony trioxide enhancing agent where indicated) is incorporated into the plastic by adding both to a Brabender mixer. The mixer is equipped with a pair of roller-type blades positioned within a head provided with heat transfer means.

The resultant mixture is heated to about 277°C; at this temperature, it is in a molten state. The percentages by weight of each component utilized in the respective formulations are listed in the abovementioned table. Each formulation is discharged from the mixer and upon cooling solidifies and is ground into chips. The chips are subjected to compression molding in a Wabash press by placing the chips between two platens, the bottom of which contains four equal size depressions 3 inches by 5 inches by one-eighth-inch deep. The top platen is then placed over the bottom platen and heat transfer means supplied to it in order to melt the chips and thus provide solid samples (after cooling) for testing.

An unsaturated polyester plastic material (Koppers 6101-25), a liquid product, is used as a base resin in order to prepare 26 formulations, designated Nos. 27-52 in the table. The particular bis-phenoxy (and antimony trioxide where indicated) is incorporated into the unsaturated polyester by mixing with 0.1 wt %, based on the total weight of the formulation, cobalt octoate and 2.0% methyl ethyl ketone peroxide to form a liquid mixture. This mixture is poured into the depressions and cured at 100°C under slight compression by the Wabash press. Formulation No. 27, without additives, is the control.

Portions of the solid samples of each respective formulation (Nos. 1-52) prepared according to the above procedures are then subjected to two different standard flammability tests, i.e., UL 94 and ASTM D-2863-70. The results of these flammability tests are shown in the table.

The results shown in the table demonstrate the effectiveness of these bis-phenoxys as flame retardants for polyesters. With reference to the saturated polyester-based formulations (Nos. 1-26), formulation No. 1 (the control) had an OI of 22.0 and UL 94 value of SB. The even-numbered formulations, employing the use of the particular bis-phenoxy compound, all show an increase in fire retardancy as measured by OI. While some formulations have a SB rating (UL 94), the individual UL rating covers a wide range and consequently, in these cases, the OI numbers are more indicative of increased flame retardancy. The use of an enhancing agent such as Sb_2O_3 to promote a cooperative effect between such agent and the bis-phenoxy compound is fully demonstrated via the results obtained from testing the odd-numbered formulations. The highest UL 94 ratings and significantly higher OI values are obtained when using an enhancing agent.

The results obtained using an unsaturated polyester, are basically similar to those obtained using the saturated polyester. Formulations Nos. 28-52 all demonstrated a significant increase in OI values as compared to the control, No. 27.

Alkylene Oxide Adduct of Tetrabromobisphenol A

S. Mizuno, R. Ishikawa and M. Miyazaki; U.S. Patent 4,085,086; April 18, 1978; assigned to Dainippon Ink and Chemicals, Inc., Japan describe a process for preparing a flame retardant thermoplastic polyester which comprises reacting a low molecular weight polyester having terminal hydroxyl groups with a polyfunctional isocyanate. The low molecular weight polyester is synthesized using as its acid component a compound selected from the group consisting of terephthalic acid and its lower alkyl esters of 1 to 4 carbon atoms and as its alcohol component 1,4-butanediol and a compound selected from the group consisting of an alkylene oxide adduct of 2 to 4 carbon atoms of tetrabromobisphenol A and an alkylene oxide adduct of 2 to 4 carbon atoms of tetrachlorobisphenol A. The thus-obtained flame retardant thermoplastic polyester can be used for various purposes but is especially useful as a molding material.

Example 1: 544 parts of tetrabromobisphenol A, 92 parts of ethylene oxide, 272 parts of toluene and 3 parts of caustic soda were charged to an autoclave, after which the autoclave was closed and the reaction was carried out for 6 hours at 120°C. After completion of the reaction, the reaction system was neutralized with phosphoric acid and washed with water. Toluene was then removed, and the reaction product was purified to obtain 2,2-bis[4-(2-hydroxyethoxy)-2,6-dibromophenyl] propane (i.e., an ethylene oxide adduct of tetrabromobisphenol A) having a melting point of 102°C.

A mixture of 100 parts of dimethyl terephthalate, 56 parts of 1,4-butanediol and 25 parts of the ethylene oxide adduct of tetrabromobisphenol A synthesized above was dissolved by heating at 130°C followed by the addition of 0.02 part of tetraisopropyl titanate as catalyst. The ester interchange reaction was initiated, and methanol was formed as a by-product.

The temperature of the system was then gradually raised, and at a point where the distillation of the methanol practically ceased the condensation reaction was carried out at a temperature of 240°C by reducing the pressure to about 10 mm Hg. Three hours after initiating the reduction of pressure, the system was returned to normal atmospheric pressure, and the resin formed was withdrawn. The low molecular weight modified polybutylene terephthalate had a hydroxyl value of 14, an acid value of 2, an intrinsic viscosity [η] of 0.35 and an average molecular weight of 7,000.

One hundred parts of this low molecular weight modified polybutylene terephthalate, 5.4 parts of diphenylmethane-4,4'-diisocyanate and 8 parts of antimony trioxide were placed in a stirrer-equipped reaction vessel heated at 240°C and stirred for 10 minutes under a stream of nitrogen. The resulting flame retardant thermoplastic polyester had a number average molecular weight of 25,000, a hydroxyl value of 2.5 and an acid value of 2. After comminuting this resin into granular form, it was injection molded into test pieces. The several tests for tensile strength (TS), flexural strength (FS), notched Izod impact strength (IS) and flame retardance were carried out, with the results shown in Table 1. The several tests in the case of this and subsequent examples were conducted in accordance with the the methods shown on the following page.

Tensile strength	ASTM D-638
Flexural strength	ASTM D-790
Impact strength	ASTM D-256
Flame retardancy	Underwriters' Laboratories, Inc. Subject 94 (UL 94)

Again, 100 parts of the low molecular weight modified polybutylene terephthalate, 5.4 parts of flaky diphenylmethane-4,4'-diisocyanate and 8 parts of antimony trioxide were mixed for 10 minutes in a tumbler, to which was then added 50 parts of 6-mm-long chopped glass fibers that had been treated with an aminosilane-type coupling agent, followed by mixing for a further 50 seconds.

This mixture was then fed to a vent-equipped 65-mm extruder provided with a full flight-type screw, the temperature of whose barrel was set at 250°C to prepare pellets of a flame retardant thermoplastic polyester (I-B) at such an extrusion speed that the dwell time of the starting material in the barrel would be 2 minutes. The number average molecular weight of the polymer was the same as that indicated above. Test pieces were prepared by injection molding these pellets, following which the test pieces were submitted to the several tests described hereinbefore, with the results shown in Table 1.

Again, 100 parts of the low molecular weight modified polybutylene terephthalate, 5 parts of flaky diphenylmethane-4,4'-diisocyanate, 5 parts of antimony trioxide and 2 parts of Epikote 1001 (a powdery epoxy compound of the glycidyl ether of bisphenol A type, epoxy equivalency 450, a product of Shell Company) were mixed for 10 minutes in a tumbler, to which was then added 50 parts of 6-mm-long chopped glass fibers which had been treated with an aminosilane-type coupling agent, followed by mixing for a further 50 seconds.

This mixture was fed to a vent-equipped 65-mm extruder provided with a full flight-type screw, the temperature of whose barrel was set at 250°C to prepare pellets of a flame retardant thermoplastic polyester (I-C) at such an extrusion speed that the dwell time of the starting material in the barrel would be 2 minutes. The number average molecular weight of the polymer was the same as that of the foregoing flame retardant thermoplastic polyester (I-B). 1.6 mm thick test pieces were prepared by injection molding these pellets. These test pieces were used, and the hereinbefore-described several tests were conducted, with the results shown in Table 1.

Control 1: A low molecular weight polybutylene terephthalate was synthesized as in Example 1 using 100 parts of dimethyl terephthalate and 56 parts of 1,4-butanediol as the starting materials. This low molecular weight polybutylene terephthalate had a hydroxyl value of 14, an acid value of 2 and a number average molecular weight of 7,000. Except that this low molecular weight polybutylene terephthalate was used instead of the low molecular weight modified polybutylene terephthalate used in Example 1 and that antimony trioxide was not added, the synthesis of a thermoplastic polyester (I-A') was otherwise carried out under identical conditions as in the case with the synthesis of the flame retardant thermoplastic polyester in Example 1. Similar tests as in the case of Example 1 were conducted on the resulting polyester. The results obtained were shown in Table 1.

Further, except that the foregoing low molecular weight polybutylene terephthalate was used instead of the low molecular weight modified polybutylene terephthalate used in Example 1 and that antimony trioxide was not added, the synthesis of a thermoplastic polyester (I-B') was otherwise performed under identical conditions as in the case with the synthesis of the flame retardant thermoplastic polyester in Example 1. The resulting polyester was tested as in Example 1, with the results shown in Table 1.

Example 2: Tetrachlorobisphenol A was used instead of tetrabromobisphenol A; by operating otherwise as in Example 1 an ethylene oxide adduct of tetrachlorobisphenol A was synthesized, after which the synthesis of a low molecular weight modified polybutylene terephthalate was carried out as in Example 1. This low molecular weight modified polybutylene terephthalate had a hydroxyl value of 12, an acid value of 3 and a number average molecular weight of 7,460.

One hundred parts of the so-obtained low molecular weight modified polybutylene terephthalate, 5 parts of flaky diphenylmethane-4,4'-diisocyanate and 4 parts of antimony trioxide were reacted as in Example 1 to obtain a flame retardant thermoplastic polyester (II-A). The so-obtained polyester had a number average molecular weight of 30,000. This polyester was tested as in Example 1, with the results shown in Table 1.

Again, 100 parts of the foregoing low molecular weight modified polybutylene terephthalate, 5 parts of flaky diphenylmethane-4,4'-diisocyanate, 4 parts of antimony trioxide and 50 parts of 6-mm-long chopped glass fibers which had been treated with a vinylsilane-type coupling agent were mixed as in Example 1 and thereafter processed in an extruder to obtain pellets of a flame retardant thermoplastic polyester (II-B). The so-obtained polyester had a hydroxyl value of 0.9, an acid value of 3 and a number average molecular weight of 29,000. This polyester was submitted to the same tests as in Example 1, with the results shown in Table 1.

Further, 100 parts of the foregoing low molecular weight modified polybutylene terephthalate, 5 parts of flaky diphenylmethane-4,4'-diisocyanate, 4 parts of antimony trioxide and 3 parts of Epikote 828 (an epoxy compound of the glycidyl ether of bisphenol A type, epoxy equivalency 190, a product of Shell Company) were mixed for 10 minutes in a tumbler, to which was then added 50 parts of 6-mm-long chopped glass fibers which had been treated with a vinylsilane type coupling agent. The so-obtained mixture was used, and by operating as in the case of the preparation of the flame retardant thermoplastic polyester (I-C) in Example 1 pellets of a flame retardant thermoplastic polyester (II-C) were prepared. This polyester was then tested as in Example 1, with the results shown in Table 1.

Example 3: 1,2-propylene oxide was used instead of ethylene oxide, and by operating as in Example 1 a 1,2-propylene oxide adduct of tetrabromobisphenol A was synthesized. Next, this adduct was used and a low molecular weight modified polybutylene terephthalate was synthesized as in Example 1. This low molecular weight modified polybutylene terephthalate had a hydroxyl value of 19, an acid value of 1 and a number average molecular weight of 5,600.

One hundred parts of this low molecular weight modified polybutylene terephthalate, 6 parts of flaky diphenylmethane-4,4'-diisocyanate and 12 parts of antimony trioxide were reacted as in Example 1 to obtain a flame retardant thermoplastic polyester (III-A). The so-obtained polyester had a hydroxyl value of 3.5, an acid value of 1 and a number average molecular weight of 25,000. This polyester was tested with the results shown in Table 1.

Again, 100 parts of the foregoing low molecular weight modified polybutylene terephthalate, 6 parts of flaky diphenylmethane-4,4'-diisocyanate, 12 parts of antimony trioxide and 50 parts of 6-mm-long chopped glass fibers treated with an aminosilane-type coupling agent were mixed as in Example 1 and thereafter processed in an extruder to obtain pellets of a flame retardant thermoplastic polyester (III-B). The number average molecular weight of this polymer was 26,000. The physical properties of the polymer were shown in Table 1.

Further, a mixture consisting of 100 parts of the foregoing low molecular weight modified polybutylene terephthalate, 5 parts of flaky diphenylmethane-4,4'-diisocyanate, 4 parts of antimony trioxide, 1.0 part of Epikote 154 (an epoxy compound of the glycidyl ether of novolak type, epoxy equivalency 180, a product of Shell Company) and 50 parts of 6-mm-long chopped glass fibers treated with an aminosilane-type coupling agent was used to prepare pellets of a flame retardant thermoplastic polyester (III-C) by operating as in the case of the preparation of the pellets of flame retardant thermoplastic polyester (I-C) in Example 1. The so-obtained pellets were tested in similar manner with the results shown in Table 1.

Example 4: A low molecular weight modified polybutylene terephthalate (hydroxyl value 11, acid value 0, number average molecular weight 10,180) was synthesized from 15 parts of the ethylene oxide adduct of tetrabromobisphenol A synthesized in Example 1, 100 parts of dimethyl terephthalate and 56 parts of 1,4-butanediol.

One hundred parts of this low molecular weight modified polybutylene terephthalate, 6 parts of flaky diphenylmethane-4,4'-diisocyanate and 12 parts of antimony trioxide were used, and by operating as in the case of the flame retardant thermoplastic polyester of Example 1 a flame retardant thermoplastic polyester (IV-A) was obtained. The so-obtained polymer had a hydroxyl value of 2.8, an acid value of 0 and a number average molecular weight of 40,000. This polymer was tested with the results shown in Table 1.

Again, 100 parts of the foregoing low molecular weight modified polybutylene terephthalate, 4 parts of flaky diphenylmethane-4,4'-diisocyanate, 10 parts of boron hydroxide and 50 parts of 6-mm-long chopped glass fibers treated with an aminosilane-type coupling agent were used, and by operating as in Example 1 pellets of a flame retardant thermoplastic polyester (IV-B) were obtained. The number average molecular weight of this polymer was 41,000. This polymer was tested with the results shown in Table 1.

TABLE 1

Flame Retardant Thermoplastic Polyester		TS (kg/ (cm²)	FS (kg/ cm²)	IS (kg . cm/cm)	Flame Retardancy (UL 94)*
	I-A	550	850	2	V-0
Example 1	I-B	1200	1700	8	V-0

(continued)

TABLE 1: (continued)

Flame Retardant Thermoplastic Polyester		TS (kg/(cm²)	FS (kg/cm²)	IS (kg . cm/cm)	Flame Retardancy (UL 94)*
	I-C	1400	2000	12	V-0
Control 1	I-A'	600	900	3	Burns with a flame
	I-B'	1300	1800	9	"
	II-A	650	900	3	V-1
Example 2	II-B	1150	1650	7	V-1
	II-C	1350	1900	10	V-1
	III-A	500	800	3	V-0
Example 3	III-B	1100	1600	7	V-0
	III-C	1300	1850	9	V-0
Example 4	IV-A	750	1000	5	V-0
	IV-B	1300	1800	8	V-2

*A 1.6 mm-thick test piece was used.

Modified Unsaturated Polyester Systems with Hypochlorite Ion

In a process described by *J.E. Aguirre; U.S. Patent 4,049,609; September 20, 1977; assigned to Sintetica Americana, Inc.* unsaturated polyester resin systems are modified to provide a curable, stable water-in-oil emulsion by reacting the resin system with a hydrolyzed source of hypochlorite ion, while maintaining during the reaction a pH of at least 4, whereby the reaction is predominantly one of chlorohydration.

The hypochlorite ion may be provided by addition of an aqueous solution of an inorganic salt of hypochlorous acid, e.g., by an aqueous solution of sodium, potassium, lithium, or calcium hypochlorite; or by an aqueous solution of a suitable chloramine; or the additive may comprise the reaction product of a chloramine with an aqueous solution of an inorganic salt of hypochlorous acid. The modified resin system may include over 50 wt % of water; and high quantities of conventional fillers may be added without disrupting the emulsion. The cured products yielded from the modified resin systems have excellent mechanical characteristics, and are highly flame retardant.

Unsaturated polyester resins have found application in myriad environments, including, e.g., in the fabrication of furniture and of similar structures, wherein the thermoset resin provides characteristics which may be highly simulative of wood, or of other materials commonly utilized in these objects. Similarly, such resins are widely utilized in the manufacture, as, e.g., by casting, of a large variety of other products including bowling balls, simulated marble, pistol grips, pearlescent buttons, etc.

Example 1: The treating solution used to modify the polyester resin system comprised a water solution of lithium hypochlorite having an available chlorine content of 5.0%. One part by weight of this treating solution was added to two parts by weight of a commercially available polyester resin system. The resin in particular was a resilient, high reactivity polyester resin system available from Marco Chemical Div. of W.R. Grace & Co. under the designation "GR-11015." The initial temperature of the system was approximately 20°C (ambient), and the mix of resin system and treating solution was stirred, with an almost immediate exotherm of about 15°C above the initial temperature. Hand mixing was continued until a uniform product was obtained. At this point in the process a marked change in resin color was noted. On the basis of the reaction conditions, it is believed that this change evidenced replacement of the double bonds in the unsaturated structure, by virtue of attachment of –Cl and –OH groups. This would effectively destroy the resonating structure of the molecules, with the consequent loss of color. The resultant product was a highly stable water-in-oil emulsion that showed good stability, i.e., no tendency to separate over any observed period. The viscosity of the modified product was virtually unchanged in comparison to the original resin system.

Example 2: After cooling, 60 wt % of calcium carbonate was added (as a filler) to the modified composition of Example 1, which was then catalyzed with 1% methyl ethyl ketone peroxide (based on resin system content) and utilized in the example for purposes of fabricating a table leg, by casting in a mold. A wood core was utilized in order to increase the strength of the part. For purposes of comparison, two parts by weight of the original resin system (GR-11015) containing 50% total weight of calcium carbonate as filler, was catalyzed with 1% methyl ethyl ketone peroxide and cast as just indicated in a table leg mold.

A wood core was similarly used to increase the strength of the resultant part. It was observed that the part prepared utilizing the composition of the process exhibited complete absence of cracking around the wood, as did as well the part based upon the prior art composition. Thus the composition of the process produced substantially equivalent results herein, i.e., in neither instance was cracking or warping observed. However, by means of the process, a third or more of the costly, unsaturated polyester resin was eliminated, i.e., a water solution of an inexpensive hypochlorite was effectively substituted for portions of the resin.

Where calcium carbonate is used as a filler, approximately 40 to 60 wt % of the total system is a preferable range of addition. Other fillers utilizable with the modified resin compositions of the process include glass bubbles, talc, pecan shell flour, glass fiber, cotton floc, etc. It may be noted in this connection that several of the aforementioned fillers, most notably the pecan flour, may be effectively utilized with the compositions of this process without substantially impairing the stability of the emulsion, whereas many prior art water-in-oil emulsion systems produced from polyester resins may not be effectively utilized with organic fillers. Thus, in this instance, typical additional ranges of 20 to 40% of the pecan shell flour may be effectively utilized with the compositions of the process, such as the composition of Example 1.

Example 3: In this example a conventional liquid polyester resin system, obtained from Marco Chemical Division of W.R. Grace & Co., under the product designation "GR-17060," was modified by the method of the process, and then utilized for preparing cured castings. The monomer present in this system is primarily divinylbenzene.

In particular, 66 parts by weight of the resin system was initially combined with 34 parts by weight of a 7% available chlorine, aqueous solution of sodium hypochlorite. This mix produced an exotherm of about 60°F. Slow stirring was used until full emulsification was obtained, i.e., to yield what appears as a single phase. A promoter was then prepared by adding 20 parts by weight dimethylaniline (DMA) to 100 parts by weight cobalt octoate, used as a 12% solution.

The promoter was incorporated into the modified resin system at three different addition levels, i.e., at 0.5%, at 1.0% and 1.5%, based on the resin content. In each instance, following the addition of the promoter, the milky white emulsion changed in color to a white medium tan, but retained the same consistency. All of the samples displayed the same viscosity as that of the unreacted, original resin system. Methyl ethyl ketone (MEK) peroxide was then used as a catalyst for curing of the three samples, in a concentration of 0.5% based on the resin content. No heating was necessary for the crosslinking reaction. The resultant samples were subsequently examined in order to determine such characteristics as impact strength, and flexibility. The data thereby yielded is set forth in the table below.

Samples	1	2	3
Polyester resin system parts by weight	66	66	66
NaOCl + H_2O	34	34	34
DMA/Cobalt (% based on resin system)	1.0	1.5	.5
Gel Time	5 min.	4.0 min.	9.5 min.
Impact Strength	Good	Good	Good
Curing Time (Green or Demold Cure)	14 min.	12 min.	25 min.
Flexibility 1/8" rod	Satis.	Satis.	Satis.

Alkylphosphine Oxide

U. Bollert, H.-J. Kleiner and F. Jakob; U.S. Patent 4,035,343; July 12, 1977; assigned to Farbwerke Hoechst AG vormals Meister Lucius & Bruning, Germany describe a process for the manufacture of flame resistant, synthetic, linear and phosphorus-containing polyesters from dicarboxylic acids or the esters with low molecular weight aliphatic alcohols, diols and phosphorus-containing compounds, which comprises adding, during the manufacture of the polyester, as phosphorus-containing compound a phosphine oxide of the general formula

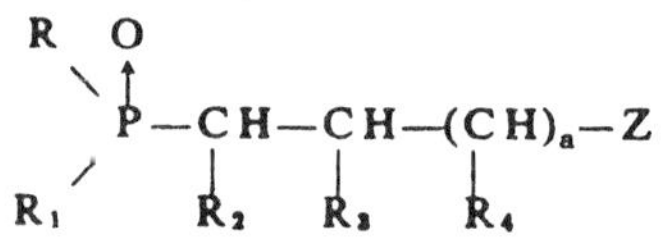

where the radicals R to R_4 represent organic radicals of which R_2, R_3 and Z may be substituted by 2 hydroxyl groups altogether, R_2, R_3 and R_4 may also represent hydrogen and a is zero or a whole number of from 1 to 4.

The bifunctional phosphine oxides added prior to, during or shortly before the end of polycondensation are prepared by the process described in Belgian Patent No. 771,302. In the final polyester the organic phosphorus compound is statistically incorporated in the macromolecule. Owing to the statistic distribution, the phosphine oxide units may sometimes also be bound at the end of the macromolecule. In order to ensure the desired flame resistance the polyester should contain at least about 0.5 wt % of phosphorus.

The preferred flame resistant polyesters are prepared with 2-dimethyl phosphine oxidomethyl-1,3-propanediol of the formula

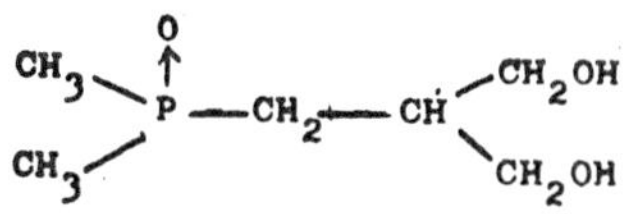

(a = zero)

They can be processed in usual manner to yield shaped structures such as filaments, fibers, sheets, injection and compression molded articles. The shaped articles made with the modified polyesters are characterized by very good permanent flame resistant and self-extinguishing properties.

The polyesters and the shaped articles made are used in all fields where easy inflammability is not desired. The fibers and filaments, for example, are preferably used for making canvas, carpets, curtains, and the like. The filaments may also be used as the one component of bicomponent filaments together with other polymers as the second component.

Example 1: One thousand grams of dimethyl terephthalate (DMT), 230 mg of $MnAc_2 \cdot 4H_2O$ and 740 ml of ethylene glycol were subjected under nitrogen to an ester interchange reaction. After termination of the methanol separation, 40 g of 2-dimethyl phosphine oxidomethyl-1,3-propanediol and 500 mg of Sb_2O_3 were added at 220°C. The reaction vessel was further heated, while being evacuated simultaneously, until an internal pressure of 1 mm Hg was reached at 250°C. The polycondensation was terminated at 0.2 mm Hg at 270°C. The product obtained had a relative viscosity of 1.64, measured with a 1% solution of the polymer in dichloroacetic acid at 25°C. The melting point was found to be 253° to 255°C, the phosphorus content 0.7% and the diglycol content 0.8%.

The polyester was polycondensed in the solid state until the relative viscosity had reached 1.98. The burning properties of the polyester were determined according to ASTM D635/68 using test bars of the dimensions 12.7·1.27 cm. The product could be classified in the group of self-extinguishing substances, while the same test bars of nonmodified polyethylene terephthalate belonged to the class of inflammable substances.

Example 2: Example 1 was repeated with the exception that 100 g of 2-dimethyl phosphine oxidomethyl-1,3-propanediol were used instead of 40 g of the compound. The polyester, which was not subjected to condensation in the solid state, had a relative viscosity of 1.65, a melting point of 240° to 241°C, a phosphorus content of 1.6% and a diglycol content of 0.9%.

The polyester was spun from the melt under the usual conditions and the filaments obtained were drawn in a ratio of 1:3.65. They had a strength of 3.1 g per dtex at an elongation at break of 22%. With the filaments a knitted hose was produced which was used for carrying out dyeing tests and a burning test. Deep shades were obtained with the following acid dyestuffs: Supranol fast red Br (Color Index No. 24,790); Alphanol fast blue (Color Index No. 62,155); and Lana pearl yellow 3 (Color Index No. 19,025).

For the burning test a piece of the knitted hose was burned on a semicircle tester according to DIN 54 331 over a carrier fabric of cotton. The flame extinguished at an angle of 100° to 110°C. A comparative knitted hose made from filaments of nonmodified polyethylene terephthalate burned completely.

Substituted Carbonyl Phosphonates

A process described by *K. Murayama and T. Kashihara; U.S. Patent 4,086,208; April 25, 1978; assigned to Toyo Boseki KK, Japan* provides a flame resistant polyester comprising a combination of an aromatic dicarboxylic acid or an ester-forming derivative thereof, a diol or an ester-forming derivative thereof, and a phosphorus compound, characterized in that during the preparation of the polyester at least one phosphorus compound having at least two, preferably at least three ester-forming functional groups is added so as to give a content of phosphorus in the range of about 500 to about 50,000 ppm in the resultant polyester. The phosphorus compound is selected from the group consisting of compounds having the following formulas:

(1)

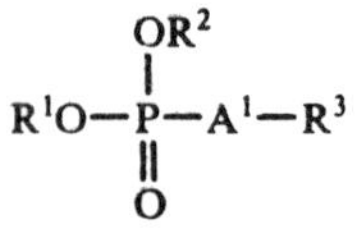

or

(2)

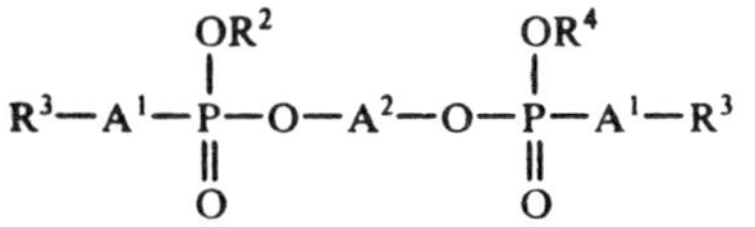

or

(3)

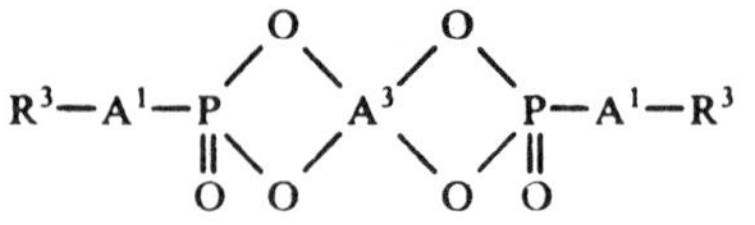

where R^1, R^2, and R^4 may be the same or different and represent hydrogen atoms or hydrocarbon groups having 1 to 18 carbon atoms which may contain at least one halogen atom, R^1 and R^2 being capable of forming a ring when taken together.

R^3 represents an ester-forming functional group selected from the group consisting of $-COOR^5$ and $-OR^5$, in which R^5 represents a hydrogen atom or a hydrocarbon group having 1 to 12 carbon atoms which may contain at least one hydroxyl group; A^1 and A^2 may the same or different and represent bivalent hydrocarbon groups having 1 to 12 carbon atoms which may contain at least one halogen atom; and A^3 represents a quadrivalent hydrocarbon group having 5 or 6 carbon atoms.

An aromatic polyester such as polyethylene terephthalate is one of the polymers which is now largely used as molded articles such as fiber, film, and board. Accordingly it has remained desirable to develop an aromatic polyester having flame resistance.

Preparation Example 1: Into a 2-liter, 4-necked flask equipped with a stirrer, a reflux condenser, a thermometer and a dropping funnel was placed a mixture of 5,500 g of dimethyl hydrogen phosphite and 4,300 g of methyl acrylate. With rapid stirring, 80 g of methanol containing 2 N sodium methoxide was introduced through the dropping funnel over 60 minutes. As the reaction proceeded it was necessary to absorb part of the reaction heat by surrounding the reaction vessel with a water bath, maintaining the temperature in the vessel below 80°C.

After the solution methoxide had been added, the contents were allowed to stand for 10 minutes. Afterwards, unreacted dimethyl hydrogen phosphite, methyl acrylate and others were distilled out under reduced pressures and there remained the reaction product. This crude product was further purified by distillation, yielding 8,100 g of methoxycarbonylethyl dimethyl phosphonate, chemical properties of which were found as follows: boiling point, 143° to 145°C at 10 mm Hg; n_D^{30}, 1.4327; and acid value, below 0.01 equiv. wt/mol.

Preparation Example 2: Preparation Example 1 was repeated except that a mixture of 5,520 g of diethyl hydrogen phosphite and 4,000 g of ethyl acrylate was used. The resultant reaction product was purified by distillation, yielding 8,000 grams of ethoxycarbonylethyl diethyl phosphonate, chemical properties of which were found as follows: boiling point, 152° to 154°C/10 mm Hg; n_D^{30}, 1.4287; and acid value, 0.01 equiv. wt /mol.

Example 1: Into a reaction vessel equipped with a stirrer, a condenser, a thermometer and an evacuating tube connected to a vacuum pump was placed a mixture of 194 parts of dimethyl terephthalate, 124 parts of ethylene glycol, 6.6 parts of ethoxycarbonylethyl diethyl phosphonate (obtained in Preparation Example 2), 0.12 part of potassium titanyl oxalate and 0.03 part of antimony trioxide. The reaction vessel was heated to a temperature of 150°C to initiate the ester exchange reaction. Further heat was applied to gradually increase the temperature, reaching 230°C after 150 minutes.

This application of heat enabled the theoretical amount of methanol to be distilled out. Then the reaction vessel was gradually heated up to 275°C over a period of 40 minutes, while it was slowly evacuated to 0.2 mm Hg. Under these conditions the condensation polymerization was carried out for 100 minutes, yielding the polymer. It was found that the polymer had an intrinsic viscosity of 0.60 and the phosphorus in the added phosphorus compound remained wholly (100%) in the polymer.

This polymer was melt spun at a temperature of 290°C and then drawn into flame resistant filaments. The flame resistance of drawn filament expressed as the number of ignitions was 5.5 times.

Comparative Example 1: Example 1 was repeated except that 3.6 parts of trimethyl phosphate was used in place of ethoxycarbonylethyl diethyl phosphonate. The catalyst for the ester exchange reaction, potassium titanyl oxalate, was immediately deactivated, and thereby the ester exchange reaction did not proceed at all. Consequently, no polyester was obtained.

Example 2: Example 1 was repeated except that 2.6 parts of methoxycarbonylethyl dimethyl phosphonate (obtained in Preparation Example 1) was used in place of ethoxycarbonylethyl diethyl phosphonate. There was obtained the

Example	Phosphorus compound: Name	Phosphorus compound: Amount (part)	Amount of Catalyst (part): Zinc acetate	Amount of Catalyst (part): Potassium titanyl oxalate	Amount of Catalyst (part): Antimony trioxide	Properties of Polyester: Intrinsic viscosity	Properties of Polyester: Ratio of remaining phosphorus (%)	Properties of Polyester: Content of diethylene glycol (mole %)	Properties of Polyester: Flame resis-tance (Times)
2	ethoxycarbonylethyl diethyl phosphonate	3.3	0.17	0.02	0.03	0.57	100	2.7	4.4
3	phenoxycarbonylethyl diethyl phosphonate	3.6	0.17	0.04	0.02	0.54	93	2.8	4.5
4	hydroxyethoxycarbonylethyl diethyl phosphonate	3.2	0.17	0.04	0.02	0.56	95	2.5	4.5
5	hydroxyethyl diisopropyl phosphonate	2.6	0.17	0.04	0.02	0.58	85	3.0	4.2
6	(a)	5.6	0.17	0.03	0.02	0.57	98	2.4	4.3
7	(b)	7.5	0.11	0.04	0.02	0.59	100	2.5	5.0

(a) 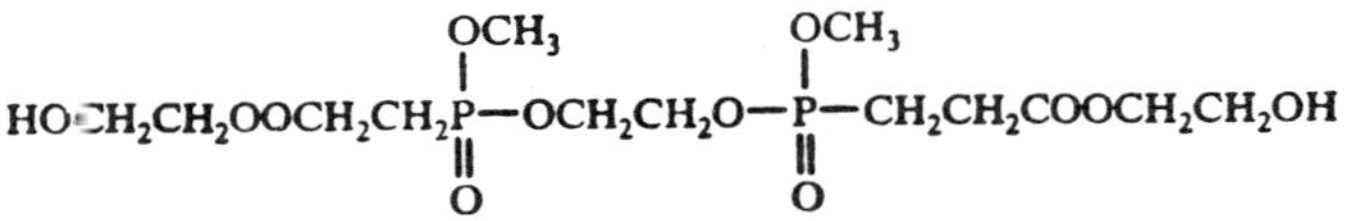

(b) 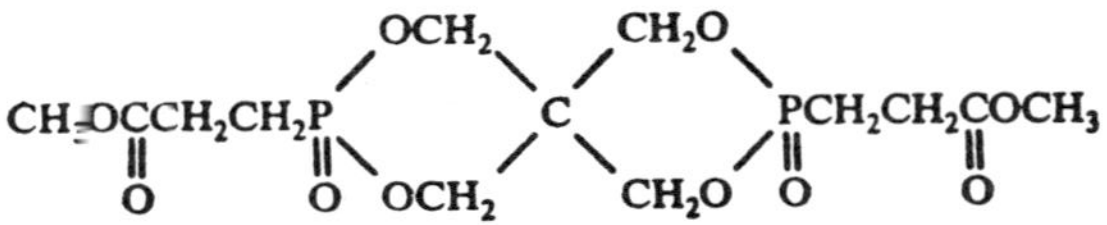

polymer, which was then melt spun and drawn into filaments in a similar manner as described in Example 1. The polymer has the following properties: intrinsic viscosity, 0.58; ratio of the phosphorus remaining in the polymer to that added, 99%. The properties of the drawn filament are: tenacity at break, 3.7 g/D; elongation at break, 27%; and flame resistance, 4.5 times.

Examples 2 to 7: Example 1 was repeated except that the phsophorus compound and catalysts shown in the table on the previous page were used in different amounts. Properties of the resultant polymers are also shown in the table.

Comparative Example 2: A mixture of 194 parts of dimethyl terephthalate, 124 parts of ethylene glycol, 0.06 part of zinc acetate and 0.06 part of antimony trioxide was subjected to ester exchange reaction in a similar manner as described in Example 1 and then 6.6 parts of ethoxycarbonylethyl diethyl phosphonate was added thereto. After the contents had been agitated in the nitrogen atmosphere at a temperature of 230°C for 20 minutes, the reaction vessel was heated up to 275°C, while the same was slowly evacuated to 0.2 mm Hg.

Under these conditions the condensation polymerization was carried out for 95 minutes. While the resultant polymer showed an intrinsic viscosity of 0.65, the ratio of the phosphorus remaining in the polymer to the added amount was as low as 30% and the flame resistance of the drawn filament made of this polymer was 3.5 times.

Aromatic Copolyester Composition

K. Hazama, I. Hirose, K. Yasue and D. Shinoura; U.S. Patent 4,064,196; Dec. 20, 1977; assigned to Unitika Ltd., Japan describe a resin composition comprising (1) an aromatic copolyester comprising terephthalic acid, isophthalic acid and a halogen-free bisphenol; and (2) a halogen-containing aromatic copolyester containing terephthalic acid, isophthalic acid and, at least in part, a bisphenol bearing a halogen substituent(s) as a comonomer.

The aromatic copolyesters can be prepared by several methods. For example, an interfacial polycondensation process can be used. In this case an aqueous alkaline solution of one or more bisphenols and a terephthaloyl dihalide-isophthaloyl dihalide mixture dissolved in an organic solvent which is immiscible with water are mixed and reacted. Suitable interfacial polycondensation processes which can be used are disclosed, for example, in W.M. Eareckson, *J. Polymer Sci.,* XL 399 (1959).

The following is a typical polycondensation. An aqueous alkali solution of a bisphenol is added to a terephthaloyl dihalide-isophthaloyl dihalide mixture, more preferably a terephthaloyl dichloride-isophthaloyl dichloride mixture dissolved in an organic solvent, or an organic solvent solution of a terephthaloyl dihalide-isophthaloyl dihalide mixture is added to an aqueous alkaline solution of a bisphenol. Alternatively, an aqueous alkaline solution of a bisphenol and an organic solvent solution of a terephthaloyl dihalide-isophthaloyl dihalide mixture can be simultaneously added to a reaction vessel.

Interfacial polycondensation takes place near the interface of the aqueous phase and the organic phase. However, since the aqueous phase and the organic phase are essentially immiscible, it is necessary to mutually disperse the phases. For this purpose an agitator or a mixer such as a Homo-mixer can be used.

The concentration of the terephthaloyl dihalide-isophthaloyl dihalide mixture dissolved in the organic solvent is usually from about 2 to about 25 wt %, based on the weight of the solution. The concentration of the bisphenol in the aqueous alkaline solution is also usually from about 2 to about 25 wt %, based on the weight of the solution.

The amount of the bisphenol and of the terephthaloyl dihalide-isophthaloyl dihalide mixture used (molar ratio) is preferably maintained equivalent. An excess of the terephthaloyl dihalide-isophthaloyl dihalide mixture is not desirable in the preparation of a high molecular weight aromatic copolyester having a logarithmic viscosity number greater than 0.50 as determined at a concentration of 1 g/100 ml in a solvent system of phenol-tetrachloroethane (6:4 by weight) at 25°C.

Typical examples of final products produced are films, monofilaments, injection molded materials such as machine parts, automobile parts, electrical parts, vessels and springs. The aromatic copolyester compositions find special use as engineering plastics for various uses which require their excellent properties.

Example: Preparation of Halogen-Free Aromatic Copolyester – 15.2 g of terephthaloyl dichloride and 15.2 g of isophthaloyl dichloride were dissolved in 462 g of methylene dichloride, and the solution temperature was adjusted to 10°C. Separately, an alkali aqueous solution consisting of 34.2 g of bisphenol A, 0.3 g of sodium dithionite, 0.17 g of trimethyl benzyl ammonium chloride, 12.6 g of sodium hydroxide, 0.56 g of p-methoxyphenol and 750 g of water was adjusted in temperature to 15°C. The two solutions were mixed with each other while stirring, and stirring was continued for 3 hours at room temperature.

Upon discontinuing the stirring and allowing the system to stand, a methylene dichloride phase dissolving the resulting polymer separated. This methylene dichloride phase was washed several times with pure water and then poured into acetone in a volume proportion of 1:1 to precipitate the polymer. The polymer precipitated was then dried. The logarithmic viscosity number thereof measured in a phenol-tetrachloroethane mixed solvent (6:4 by weight) was 0.65 at 25°C.

This powder was formed into chips and injection molded. As a result of combustion testing according to UL 94, a ⅛-inch-thick sample piece, a $\frac{1}{16}$-inch thick sample piece and $\frac{1}{32}$-inch-thick sample piece showed fire retardant standards of V-0, V-1, and V-1, respectively. The tensile strength, elongation at break, tensile impact strength and arc resistance of the polymers were 750 kg/cm², 70%, 600 kg-cm/cm² and 130 seconds, respectively.

Preparation of Halogen-Containing Aromatic Copolyester – 8.2 g of sodium hydroxide was dissolved in 460 ml of water, and, subsequently, 54.4 g of 2,2-bis-(4'-hydroxy-3',5'-dibromophenyl) propane and 0.27 g of trimethyl benzyl ammonium chloride were dissolved. To this solution (at 15°C) was added a solution (at 15°C) prepared by dissolving 10.1 g of isophthalic acid chloride and 10.1 g of terephthalic acid chloride in 230 cc of methylene dichloride, and the system vigorously stirred.

Immediately after initiating stirring, the mixture became emulsified. After 2 hours, the stirring was discontinued at 20°C, and the aqueous phase removed by adding 10 ml of a 0.1 N aqueous hydrochloric acid solution to the mixture while stirring at room temperature followed by allowing the resulting mixture to stand.

The methylene dichloride layer was then washed with water and the washed methylene dichloride layer was poured into hot water having a temperature of about 60° to 100°C to isolate the polymer. The resulting polymer had a logarithmic viscosity number of 0.45 (solvent: phenol/tetrachloroethane = 6/4 by weight, at 25°C), and the pyrolysis-initiating point (TGA; temperature elevation rate: 20°C/min) was 400°C.

The results of combustion testing according to the earlier mentioned UL standard on samples prepared by extrusion mixing the resulting bromine-containing aromatic copolyester with the halogen-free aromatic copolyester in various proportions, tensile strength, elongation at break and tensile impact strength are tabulated in the table.

No.	Compounding Proportion (wt%): Aromatic Copolyester	Bromine-containing Copolyester	Bromine Content of the Composition (wt %)	Combustion Test: 1/16 inch Rating	1/32 inch Rating	Tests on Physical Properties: Tensile Yield Strength (kg/cm^2)	Elongation at Break (%)	Tensile Impact Strength ($kg.cm/cm^2$)
1	100	0	0	V-1	V-1	750	70	600
2	99	1	0.47	V-1	V-1	770	70	605
3	97.5	2.5	1.2	V-0	V-1	760	70	590
4	95	5	2.4	V-0	V-0	750	65	570
5	90	10	4.8	V-0	V-0	740	64	560
6	85	15	7.2	V-0	V-0	735	63	540
7	80	20	9.6	V-0	V-0	740	60	520
8	75	25	11.8	V-0	V-0	700	50	500

The results of combustion testing show that when the bromine solvent was not less than 1 wt %, the samples met the standards of V-0 and V-1 (with a 1/16-inch-thick sample and a 1/32-inch-thick sample, respectively), and that the addition of the bromine-containing copolyester in a proportion of 25 wt % did not cause deterioration of other physical properties.

Superhalogenated Unsaturated Polyesters

P. Malfroid; U.S. Patent 3,979,368; September 7, 1976; assigned to Solvay & Cie, Belgium describes a process whereby superhalogenated unsaturated polyesters are produced by copolymerization of at least one α-epihalohydrin and maleic anhydride and optionally another cyclic organic anhydride, in a ratio of epihalohydrin to the sum of the anhydrides present of 1:1 to 2:1, using a titanium tetrahalide.

The concentration of titanium tetrahalide is not in itself critical and can vary within very wide proportions. It is nevertheless practical to keep it within certain limits so as to ensure that polyesters which have molecular weights and viscosities which enable them to be employed easily are produced. For this reason, 0.001 to 0.1 mol of titanium tetrahalide is generally used per mol of the mixture of monomers. The unsaturated polyesters have molecular weights of between 300 and 3,000, and are superhalogenated, that is, they possess a halogen content higher than that of the monomers incorporated into the polyester chain.

The halogen content arising from superhalogenation is generally between 1 and 35 wt %. Crosslinked polyester resins manufactured by using unsaturated polyesters according to the process have fire resistant properties which are markedly improved relative to those of the polymers produced by crosslinking polyetherester resins of the prior art, manufactured by employing epichlorohydrin and maleic anhydride; they are self-extinguishing. The halogenated unsaturated polyesters are particularly valuable in that they make it possible to manufacture crosslinked resins which combine several desired properties and especially self-extinguishability and resistance to chemical agents.

Examples 1 to 3: These examples, the particular characteristics of which are given in Table 1, relate to the manufacture of substantially alternating superchlorinated unsaturated polyesters by copolymerization of epichlorohydrin and maleic anhydride in the presence of $TiCl_4$. The manufacture is carried out in a 10-liter, 3-necked flask placed in a thermostatically controlled bath at 60°C and equipped with a glass stirrer with blades, a thermometer, a dropping funnel and a device for flushing with nitrogen. The apparatus is evacuated and then flushed with nitrogen three times so as to drive off all the air present in the flask and the dropping funnel.

Two liters of dichloroethane and amounts of maleic anhydride and $TiCl_4$ corresponding to those given in Table 1 are introduced successively into the flask. After the maleic anhydride has dissolved completely, epichlorohydrin is introduced gradually at a rate such that the temperature of the mixture does not exceed 80°C. Samples of the reaction mixture, removed at regular intervals, are investigated by infrared spectrometry. The reaction is considered to be complete when two consecutive spectra are identical. The reaction mixture is then cooled to ambient temperature and the catalyst is decomposed by treatment with dilute hydrochloric acid. The reaction product is filtered off, and finally washed with water until no trace of acidity remains, dried and evaporated.

The composition of the polyesters is determined from nuclear-magnetic resonance spectrograms which provide the molar ratio of epihalohydrin to anhydride and from the proportion by weight of carbon determined by elementary analysis. The halogen content arising from superhalogenation is calculated by the difference between the overall halogen content and the halogen content originating from the epihalohydrin. Comparison of Examples 1, 2, and 3 shows that the molecular weight and the degree of superhalogenation of the polyesters according to this process can be controlled and are essentially a function of the initial $TiCl_4$ content.

Examples 4 to 6: These examples relate to crosslinked resins manufactured by employing chlorinated unsaturated polyesters, after partial isomerization of the maleic double bonds. Three samples of chlorinated unsaturated polyester manufactured according to Example 1 were subjected to a noncatalytic thermal isomerization at 120°C for the periods of time given in Table 2.

The three samples of chlorinated unsaturated polyesters isomerized respectively to degrees of 20%, 45% and 70% were mixed with styrene in such a way as to produce mixtures containing 35 wt % of styrene. 1.5 wt % of a solution containing 50% of benzoyl peroxide, 0.25 wt % of a solution of cobalt octoate containing 1% of cobalt and 3% by weight of epoxidized soy oil were then added and this mixture was cast in the form of a sheet.

TABLE 1

Experiment No.	Starting constituents: Number of mols employed *: EPI	MA	$TiCl_4$	Degree of conversion % by weight	Composition of the polyester: Molar ratio EPI/MA	Chlorine content, % by weight: Total chlorine	Chlorine provided by $TiCl_4$	Characteristics of the polyester: Average molecular weight	Specific gravity, 20°C g/ml	Acid number, g of group/kg	Hydroxyl number g of group/kg
1	39	26	3.9	94.5	1.46	28.80	8.3	500	1.425	0.74	1.29
2	39	26	1.95	84.4	1.34	25.75	5.5	915	1.430	0.24	1.10
3	35.5	29.5	3.9	86.1	1.22	26.97	8.1	549	1.428	1.04	1.48

* Abbreviations used:
EPI = epichlorohydrin
MA = maleic anhydride

TABLE 2

Experiment No.	Isomerisation of the polyester: Duration	Degree of isomerisation, molar %	Mechanical properties of the crosslinked resins: Rockwell hardness	Flexural modulus ASTM D 790, kg/mm^2	Flexural strength ASTM D 790, kg/mm^2	Temperature of deflection under load °C ASTM D 648 (264 psi)
4	1 hr. 25 mins.	20	75	232	Does not break	+ 45
5	4 hrs. 25 mins.	45	88	289	7.2	+ 65
6	7 hrs. 45 mins.	70	92	304	6.6	+ 95

The sheet was then kept for 4 hours at ordinary temperature and cured thereafter by heating at 100°C for 6 hours in order to produce transparent products, with a very slight yellow color. The mechanical properties of the crosslinked resins are given in Table 2 and their fire resistance properties are given in Table 3.

TABLE 3:

	Fire-resistance of the crosslinked resins		
	ASTM Test D 635		
Experiment No.	Length burned, cm	Duration of non-supported burning, seconds	ISO Test R 181 according to Schramm
4	0	98	1.9
5	0	115	2.0
6	0.12	182	1.9

1,3-Di(Hydroxyalkyl)Benzimidazolones

A process described by *H. Batzer and J. Habermeier; U.S. Patents 4,021,506; May 3, 1977 and 3,997,511; December 14, 1976; both assigned to Ciba-Geigy Corporation* relates to unsaturated, linear polyesters, the acid component of which consists, to the extent of at least 20 mol percent, preferably at least 50 mol percent, relative to the total amount of dicarboxylic acid radicals, of unsaturated dicarboxylic acid radicals, and which are characterized in that they contain, as the flame retarding component, diol radicals of the formula

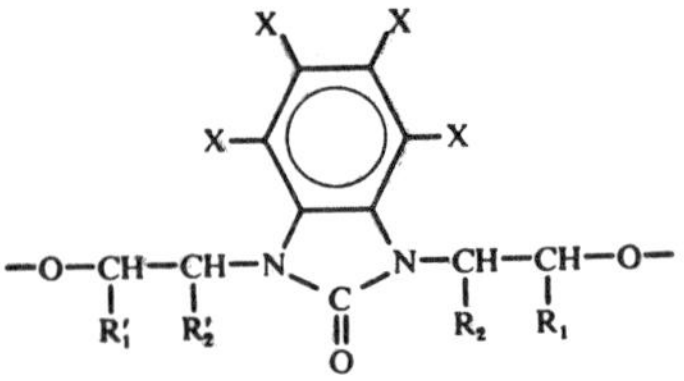

where X denotes bromine and/or chlorine atoms, R_1 and R_1' independently of one another each denote a hydrogen atom or the methyl, ethyl or phenyl group, and R_2 and R_2' independently of one another each denote a hydrogen atom or each, conjointly with R_1 or R_1', denote the trimethylene or tetramethylene radical.

In order that the polyesters may exhibit an adequate flame retarding property, the proportion of the chlorine-containing or bromine-containing diol component of the formula must, in the absence of synergists, be so regulated that the polyester resin has a chlorine content of at least 20 wt % and a bromine content of at least 15 wt %. The polyesters manufactured in accordance with this process, or mixtures with other polymerizable monomers, can be used in the protection of surfaces, in molding compositions and as casting resins, laminating resins, lacquer resins, powder resins and the like.

Preparation of the Chlorinated or Brominated Di(Hydroxyalkyl)Benzimidazolones: 1,3-Di(2-Hydroxyethyl)-4,5,6,7-Tetrabromobenzimidazolone – 333.3 g of 1,3-di-(2-hydroxyethyl)benzimidazolone (1.5 mols) are stirred in a reaction flask with

3.5 liters of water at 90°C until a clear solution is formed. 1,198 g of bromine (7.5 mols) are then added dropwise with gentle stirring over the course of 30 minutes. A yellowish precipitate settles out at once. After the dropwise addition, the reaction mixture is stirred more vigorously for a further 4 to 5 hours, in order to achieve more thorough mixing, and the internal temperature is raised to 92° to 96°C.

The hydrogen bromide formed in the reaction and the excess bromine vapors are fixed by means of 5% strength sodium hydroxide solution in a washing tower filled with active charcoal. The reaction mixture is subsequently cooled to 5° to 10°C and the product is isolated by filtration with suction. It is purified by being stirred with 5 liters of water and is again filtered off under strong suction. After being sucked dry, the product is dried in a vacuum cabinet at 100°C. 778.5 g (96.6% of the theoretic yield) of a practically colorless product melting at 266.2°C (Mettler "FP 51," heating rate 1°C/min) are obtained. 35 g of the crude product are purified by recrystallization from 150 ml of ethylene glycol. After drying at 150°C, pure white needles melting at 267.5°C are obtained.

Preparation of the Polyesters: Broming-Containing, Unsaturated Polyester Containing Approximately 40% of Bromine (Undiluted) – A mixture of 171.5 g of maleic anhydride (1.75 mols), 55.5 g of phthalic anhydride (0.375 mol) and 54.5 g of adipic acid (0.375 mol) is heated to 80°C under a stream of nitrogen in a laboratory glass apparatus equipped with a stirrer, a thermometer, a condenser and receiver for downward distillation and nitrogen inlet, and a mixture of 672.3 g of 1,3-di(2-hydroxyethyl)-4,5,6,7-tetrabromobenzimidazolone (1.25 mols), 146.0 g of diethylene glycol (1.375 mols) and 15.4 g of ethylene glycol (0.25 mol) is added.

This mixture is heated to 150°C under nitrogen over the course of 1 hour. The temperature is then raised uniformly over the course of 6 hours, with continual stirring, to 210°C. After heating for 4 hours at 210°C, a sample indicates an acid number of 122. After heating for a further 3 hours, the acid number has fallen to 12. The reaction mass is cooled to 180°C, 200 mg of hydroquinone are added and the whole is poured out onto a metal sheet to cool. A product which is solid at room temperature is obtained in a quantitative yield, having an acid number of 11.0 and a softening point (Kofler) of 62° to 64°C.

The unsaturated polyester thus obtained contains 37.4 wt % of stable bromine (calculated 38.61% of Br). On heating gently, the unsaturated polyester readily dissolves in styrene in most diverse mixing ratios, for example: A solution (1a) of 100 parts by weight of the polyester obtained above in 50 parts of styrene–

Density of the solution, g/ml at 20°C	1.399
Viscosity, cp at 30°C	644
Bromine content, wt %	25.7

Solution 1a is stable for a period of several weeks at temperatures between 5° and 25°C.

Example: 100 g of the styrene-containing polyester prepared in accordance with Solution 1a are homogeneously mixed with 0.4 g of 8% strength cobalt naphthenate and 1.5 g of 50% strength methyl ethyl ketone hydroperoxide (in dimethyl phthalate) are subsequently stirred in. After a brief treatment under vacuum, this homogeneous, clear mixture is cast into aluminum molds and is cured at room temperature (23°C). The resin is virtually completely cured after only 2 hours. The casting is removed from the mold and allowed to stand at room temperature for a further day and the following characteristics are measured:

Flexural strength (according to VSM* 77,103)	11.07 kg/mm^2
Deflection (according to VSM 77,103)	7.0 mm
Impact strength (according to VSM 77,105)	6.17 $cm\text{-}kg/cm^2$
Water absorption (4 days at 20°C)	0.22 percent
Dimensional stability (according to DIN 55,461)	61°C
Inflammability (according to CTM/20a)**	Grade 2 (Burning time: 1 second)

*VSM = Verein Schweizerischer Maschinenindustrieller
**CIBA-Testing Method/20a

Description of the Test According to CTM/20a – DIN standard bar (120 by 15 by 10 mm) of the plastic to be tested, clamped horizontally, is exposed for 1 minute to the flame of a Bunsen burner running on natural gas and inclined at an angle of 45° (burner opening 9 mm height of flame when burner is held vertically: 10 cm), so that the 15 mm wide surface of the sample is 3 cm above the upper edge of the burner and the front surface has a horizontal distance of 1 cm from the lower edge of the burner. Grade 0: sample burns for 15 to 60 seconds (or longer) after removing the flame. Grade 1: sample burns for 1 to 14 seconds after removing the flame. Grade 2: sample burns for 0 second, that is to say does not ignite. Grade 1 denotes, for example, that the bar continues to burn for not longer than 15 seconds after removing the flame. It is comparable with category 2 according to ISO/R (International Standards Organization/Recommendation) 1,210 (burning time 1 to 14 seconds).

Linear Thermoplastic Polyesters

A process described by *J. Habermeier; U.S. Patent 3,997,510; December 14, 1976; assigned to Ciba-Geigy Corporation* relates to flame resistant, linear thermoplastic polyesters having a relative viscosity of 1.10 to 4.0, measured at 30°C on a 1% strength solution consisting of equal parts of phenol and tetrachloroethane, and characterized by the general formula

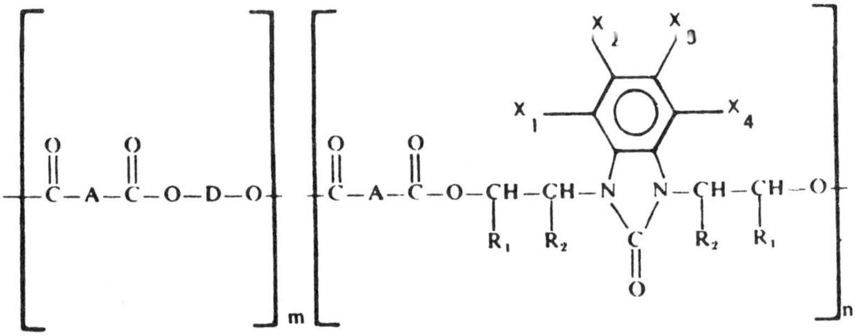

where A denotes a radical of the formulas

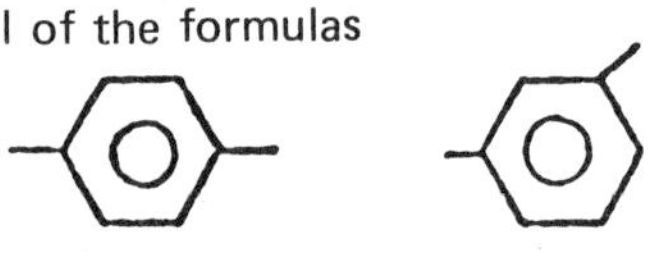

$-(CH_2)_2-$ and/or $-(CH_2)_4-$, D represents an aliphatic alkylene radical with 2 to 12, especially 2 to 10, carbon atoms and/or the 1,4-dimethylenecyclohexane radical, R_1 each denotes a hydrogen atom or the methyl, ethyl or phenyl group and R_2 denotes a hydrogen atom or together with R_1 denotes the tetramethylene radical, X_1 and X_4 independently of one another represent chlorine, a hydrogen atom or bromine atom and X_2 and X_3 represent a chlorine and/or bromine atom, m represents 0 or integers and n represents integers, and the molar fraction n/(n + m) determined by n and m can assume values from 0.005 to 1.

The polyesters of the formula are obtained by polycondensing terephthalic acid, isophthalic acid, succinic acid and/or adipic acid or their polyester-forming derivatives with di(hydroxyalkyl)benzimidazolone. The polyesters are particularly suitable for use as noninflammable "engineering plastic" materials, from which shaped articles, such as gear wheels, containers for chemicals or foodstuffs, machine components and apparatus components, sheets and films, and also semifinished goods which can be shaped by machining, are manufactured. The polyesters can also be used for coating articles, for example, in accordance with the known powder coating processes. The preparation of substituted di(hydroxyalkyl)benzimidazolones is described in the previous U.S. Patents 4,021,506 and 3,977,511.

Example 1: A mixture of 48.5 of pure dimethyl terephthalate (DMT) (0.25 mol), 26.9 g of 1,3-bis(2'-hydroxyethyl)-4,5,6,7-tetrabromobenzimidazolone (0.05 mol), 49.5 g of 1,4-butanediol (0.55 mol) and 0.04 g of tetraisopropyl orthotitanate is heated to 145°C under a N_2 atmosphere, while stirring.

The colorless homogeneous melt thereby produced is treated as follows: 2 hours/145° to 200°C/N_2 atmosphere/normal pressure, in the cource of which primarily methanol distills off, thereafter: 2 hr/200°C to 220°C/N_2 atmosphere/normal pressure, in the course of which the excess 1,4-butanediol begins to distill off, thereafter: 1 hr/220° to 235°C/N_2 atmosphere/760 mm Hg to 15 mm Hg; thereafter 1 hr/240°C/N_2 atmosphere/15 mm Hg to 0.2 mm Hg.

Following this, the hot melt is poured onto a metal sheet to cool, under a stream of nitrogen. 65 g of the polyester are obtained in the form of transparent, slightly cloudy lumps. The polyester, lumps of which do not burn when exposed to a flame, though it is still slightly inflammable in the form of fibers, displays the following properties:

Relative viscosity	1.86
Glass transition range (DSC)	67-79°C
Softening point (Kofler)	182°C
Polymer melting point T_m (DSC)	185°C
Decomposition point T_d (DSC)	309°C

In contrast, customary polybutylene terephthalate (PBT) is readily inflammable; furthermore, its glass transition temperature is only 24°C.

Example 2: The following mixture is polycondensed in accordance with the reaction conditions in Example 1; 48.5 g of pure DMT (0.25 mol), 40.35 g of diol A (0.075 mol), 49.50 g of pure 1,4-butanediol (0.55 mol) and 0.04 g of tetraisopropyl orthotitanate. A colorless copolyester is obtained, which is no longer inflammable even in the form of fine fibers. In addition to this noninflammability, it displays the properties shown on the following page.

Relative viscosity	1.74
Glass transition range (DSC)	83-97°C
Softening point	195°C

In related work, *J. Habermeier and H. Batzer; U.S. Patent 3,994,865; Nov. 30, 1976; assigned to Ciba-Geigy Corporation* detail linear, thermoplastic polyesters containing at least 50 mol percent of condensed terephthalic acid and at least 50 mol percent of condensed ethylene glycol or 1,4-butanediol, which have a relative viscosity of 1.1 to 2.0 measured at 30°C on a 1% strength solution in equal parts of phenol and tetrachloroethane, and which also contain 2 to 30 mol percent, incorporated by condensation, of a substituted di(hydroxyalkyl)benzimidazolone.

EPOXIES AND SILICONES

EPOXIES

Adducts of Cyclic Urea Derivatives and Dialkyl Phosphites

D. Porret, J. Habermeier and D. Baumann; U.S. Patent 3,971,813; July 27, 1976; assigned to Ciba-Geigy Corporation describe adducts of cyclic urea derivatives and dialkyl phosphites, containing glycidyl groups, for example adducts of a diglycidyl compound of hydantoins, substituted in the 5-position by alkyl groups or an alkylene group, and of diethyl phosphite or dimethyl phosphite. These adducts are used for the manufacture of flame retarding plastics. The following examples illustrate the process.

Example 1: 400 grams of 1,3-diglycidyl-5,5-dimethylhydantoin of epoxide content 7.7 epoxide equivalents/kg, 86.8 grams of diethyl phosphite (0.629 mol) and 0.2 ml of a 40% strength solution of tetramethylammonium chloride in methanol are stirred at 148° to 157°C reaction temperature (bath temperature 180°C). The addition reaction is followed by determining the epoxide content of samples withdrawn from the reaction mixture. After 60 minutes of reaction time the epoxide content is 5.80 epoxide equivalents/kg. The reaction is stopped after 260 minutes and the reaction product is freed of readily volatile constituents on a rotary evaporator at 110°C under a water pump vacuum.

Afterwards it is dried to constant weight at 110°C and 0.1 mm Hg. 436.8 grams of a yellow, clear, viscous resin having an epoxide content of 4.85 epoxide equivalents/kg and a phosphorus content of 3.77% P are obtained.

The IR spectrum shows the absence of the P-H band (there is a distinct P-H band of the reaction mixture, prior to the addition reaction, at 2,410 cm^{-1}) and the presence of the OH band at 2,880 cm^{-1}. The volatile constituents removed from the reaction product contain about 40% of ethanol, which is determined by fractional distillation through a rotating strip column. This ethanol is produced by transesterification of the hydroxyl groups, formed in the addition reaction, with the ethoxyphosphorono groups. This is also confirmed by analysis of the resin by gel chromatography which shows the presence of approximately 35% of a 1:1 adduct of 1,3-diglycidyl-5,5-dimethylhydantoin and diethyl phosphite and of approximately 65% of oligomers of molecular weight about 2,500.

Example 2: 100 grams of epoxide resin (manufactured according to Example 1) and 71.0 grams of hexahydrophthalic anhydride are converted to a clear, homogeneous melt at 120°C. This mixture is poured into aluminum molds of size 120 x 120 x 4 mm and 120 x 15 x 10 mm prewarmed to 120°C. The resin is cured in 2 hours at 120°C and 16 hours at 150°C. The moldings thus obtained display the following mechanical properties:

Flexural strength (VSM 77,103), kp/mm^2	13.6-14.7
Deflection (VSM 77,103), mm	7.1-9.7
Impact strength (VSM 77,105), $cmkp/cm^2$	12.8-15.0
Heat distortion point according to Martens (DIN 53,458), °C	139
Inflammability according to CTM 20, 6"	Level
Water absorption (4 days/20°C), %	0.88
Water absorption (1 hour/100°C), %	0.41

Description of the CTM 20 test used above is as follows. A horizontally clamped DIN standard bar (120 x 15 x 10 mm) of the plastic to be tested is exposed for 1 minute to the flame of a bunsen burner inclined at 45° and fed with town gas (burner orifice: 9 mm, height of flame with burner vertical: 10 cm), so that the 15 mm wide surface of the test specimen is 3 cm above the upper edge of the burner and the end face is at a distance of 1 cm, in the horizontal direction, from the lower edge of the burner. Level 1 means that after removing the flame the bar continues to burn for not more than 15 seconds. It is comparable with category 2 of ISO/R 1,210 (duration of burning 0-15 seconds).

Glass Fiber-Reinforced Compositions

L.M. Godfried; U.S. Patent 3,981,832; September 21, 1976; assigned to Fokker-VFW BV, Netherlands describe a process whereby glass fiber-reinforced epoxy resins are rendered fire-protecting by incorporating in their molding compositions a combination of three materials: a source of carbon, such as, for example, pentaerythritol, a source of phosphoric acid, such as, e.g., a water-insoluble nitrogen-containing polymer phosphate, and a source of noninflammable gases such as, e.g., melamine.

When the resins made from such compositions are exposed to high temperatures or even to an open fire, they will superficially give rise to the production of bubbles and foam, thus resulting in a multicellular foam layer at their surface and such a foam layer will isolate the underlying portions of the resins from combustion.

In the molding composition, the resin employed in most cases will be an epoxy resin from epichlorohydrin and diphenylol propane, but other epoxy resins from polyfunctional halohydrins and polyhydric phenols are also suitable.

The carbon source will generally be a material having many radicals capable of entering into an esterifying reaction with phosphoric acid, the material having a high carbon content and decomposing at higher temperatures than the phosphoric acid source. Typical examples include starch, casein and polyvalent alcohols such as pentaerythritol.

The phosphoric acid source may be any material that decomposes at elevated temperatures (lower than those of the carbon source), thereby generating phosphoric acid. Typical examples are ammonium orthophosphate, melamine phosphate and polyphosphoryl phosphate.

The source of noninflammable gases will generally be an amine or amide, such as, e.g., dicyanodiamide, urea, melamine and/or guanidine. These substances should be selected in such a way that they do not take part in the curing of the epoxy resin but remain in free state in the resin in order to be available for providing noninflammable gases at any time. The effect of these amines or amides may be increased by halogen atoms, such as bromine atoms, that have been built in chemically in the molecule of the epoxy resin or the curing agent because such halogen atoms may also serve as a source for noninflammable gases. The following examples illustrate the process.

Example 1: A mixture is made from 400 grams of Araldite F (an epoxy resin made from epichlorohydrin and diphenylol propane and having an epoxy value of 180 to 195; 130 grams of dipentaerythritol, 205 grams of Phos-Chek P/30 (a nitrogen-containing polymer phosphate), 190 grams of melamine and 75 grams of methyl ethyl ketone (diluent). The resulting noncured epoxy resin composition (1,000 grams) is a viscous opalescent product which is stable for 6 to 12 months at room temperature (20°C).

The composition is completed for molding by adding 10 parts by weight of Epikure RTU (a curing agent of the amine-adduct type, comprising about 90% of triethylenetetramine dissolved in about 10% of epoxy resin made from epichlorohydrin and diphenylol propane) to 100 parts by weight of the above mixture and thoroughly mixing these components with each other. The resulting molding composition has a pot life of at least 20 minutes at room temperature.

The molding composition is laminated with one or more layers of glass fiber reinforcement in such a way that the resulting laminate has an epoxy resin layer of at least 400 g/m^2 at one side, and the laminate is allowed to cure. After 6 hours at room temperature, the layer of epoxy resin is free from tackiness although the curing operation will last for at least 2 x 24 hours.

If the epoxy resin side of the resulting laminate is exposed to an open fire or to heat radiation sources having temperatures up to 1400°C it will spontaneously form an incombustible, insulating foam mass which insulates the remainder of the laminate against decomposition and combustion. The foaming surface is uniform and closed. The resulting foam mass is homogeneous.

Example 2: In the same way as described in Example 1, a mixture of uncured epoxy resin (epoxy value 180 to 190) with fire-protecting additives and a diluent is made. This mixture is stable for 6 to 12 months at room temperature.

A molding composition is made by adding 11 parts by weight of DDM curing agent (a curing agent based on diamino diphenyl methane and operative at elevated temperatures) to 100 parts by weight of the above mixture and thoroughly mixing these components with each other. The resulting molding composition has a pot life of about 1 week at room temperature.

The molding composition is applied to a glass fiber mat (30% or more by weight, based on the molding composition) and rolled into this mat, thus causing a good impregnation of the fibers with the molding composition. The resulting impregnated fiber mat is stored as a semifinished product.

Later on, a sandwich panel is made from two of these impregnated fiber mats by applying them as skin sheets onto a core of foam material, introducing the combination into a mold and curing the combination by means of heat and pressure. At 180°C, curing is completed in 10 to 20 minutes. The end product obtained from the mold has similar fire-protecting characteristics as the laminate of Example 1.

Epoxy Resins Cured with Phosphorus Trihalides

A process described by *K.M. Foley, R.H. Bell and F.P. McCombs; U.S. Patent 3,971,750; July 27, 1976; assigned to Owens-Corning Fiberglas Corporation* relates to a curable epoxy system consisting essentially of an uncured epoxy resin and a hardening agent of phosphorus trihalide where the uncured epoxy resin is represented by the structure:

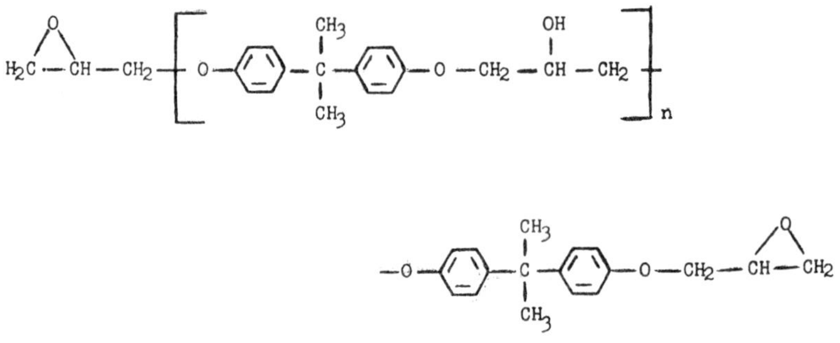

where n ranges from essentially zero to about 30 and where the uncured epoxy resin and hardening agent have a ratio of epoxy equivalent weight to halogen equivalent weight ranging from 3:1 to 1:3. These higher molecular weight homologs are derived from the simplest epoxy resin, 2,2-bis[4-(2',3'-epoxypropoxy)-phenyl]propane. These epoxies have found use in various types of adhesives. They are known for their excellent strength and chemical resistance. The following examples illustrate the process.

Example 1: This example demonstrates the curing of an epoxy resin with phosphorus trichloride. 200 grams of epoxy resin DER 332 and 2,2-bis[4-(2',3'-epoxypropoxy)phenyl] propane (1.15 epoxy equivalents, epoxy equivalent weight of 174) were placed in a 500 ml resin pot equipped with a mechanical stirrer, addition funnel and reflux condenser. 100.3 ml of phosphorus trichloride (1.15 mol, 3.45 phosphorus-chlorine equivalents) was rapidly added to the pot with stirring. The reaction warmed and began to reflux. Refluxing stopped after the reaction consumed all the phosphorus trichloride. The reaction was allowed to cool and a viscous liquid with 0.287 phosphorus-chlorine equivalents and a phosphorus-chlorine equivalent weight of 155.7 was obtained. The product was a mixture with the major component being:

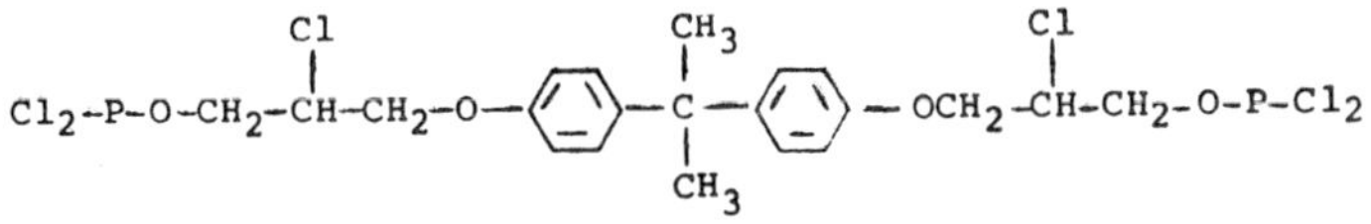

50 grams of DER 332 (0.287 epoxy equivalents, epoxy equivalent weight of 174) was then added to 44.7 grams of the above product in a 250 ml beaker. The result was a hard polymeric product similar to that obtained when the DER 332 was cured with an aliphatic amine such as DEH 24.

Example 2: This example demonstrates the fire resistance properties of the cured resin of Example 1. A sample of that product 1 x 1 in, ¼ in thick was placed in the flame of a Fisher burner until it appeared that combustion had occured. The flame of the sample immediately extinguished when it was removed from the flame of the Fisher burner. The sample was self-extinguishing.

Example 3: This example demonstrates the curing of an epoxy resin with phosphorus oxychloride. 50 grams of epoxy resin Araldite 6005 (0.2703 epoxy equivalents, epoxy equivalent weight of 185) with 8.25 ml of phosphorus oxychloride (0.0901 mol, 0.2703 phosphorus-chlorine equivalents) and 0.75 ml (0.00846 mol) of pyridine catalyst was mixed in a beaker. After approximately 2 minutes, the reaction mixture got hot, darkened in color, and set up to give a hard polymeric product.

Example 4: The procedure of Example 3 was repeated except that only 0.05 ml (0.000564 mol) of pyridine was used. The beaker was sealed from the atmosphere. The beaker did not become warm. After 4 days, the contents of the beaker were hard, indicating a hard polymeric product.

Unsaturated Amic Acid Additive

A process described by *A. Takahashi, M. Wajima and R. Tada; U.S. Patent 3,984,373; October 5, 1976; assigned to Hitachi, Ltd., Japan* relates to an epoxy resin composition containing an N,N'-unsaturated amic acid compound represented by the general formula, $HOCO{-}R_2{-}CONH{-}R_1{-}Y$. R_1 is a divalent group having 2 or more carbon atoms, R_2 is a divalent group having a carbon-to-carbon double bond, and Y is

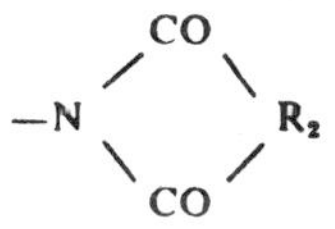

or $-NHCO{-}R_2{-}COOH$.

Examples of the N,N'-unsaturated amic acid amides include 1-(maleamic acid)-2-maleimido-ethylene, 1-(maleamic acid)-6-maleimido-hexamethylene, 1-(maleamic acid)-3-maleimido-phenylene, and 1-(maleamic acid)-4-maleimido-phenylene.

The N,N'-unsaturated bisamic acids represented by the formula,

$$HOCO{-}R_2{-}CONH{-}R_1NHCO{-}R_2{-}COOH$$

are N,N'-ethylene-bismaleamic acid, N,N'-hexamethylene-bismaleamic acid, N,N'-m-phenylene-bismaleamic acid, N,N'-p-phenylene-bismaleamic acid, N,N'-4,4'-diphenylmethane-bismaleamic acid, N,N'-4,4'-diphenylether-bismaleamic acid, N,N'-4,4'-diphenylsulfone-bismaleamic acid, etc.

The amide group and carboxyl group of the amic acid may be in either trans- or cis-configuration.

The epoxy resins to be used are those having in the molecule more than one vicinal epoxy group on average, such as bisphenol A type epoxy resins which are condensation products of bisphenol A and epichlorohydrin, resorcinol-type epoxy resins, tetrahydroxyphenylethane-type epoxy compounds having the formula,

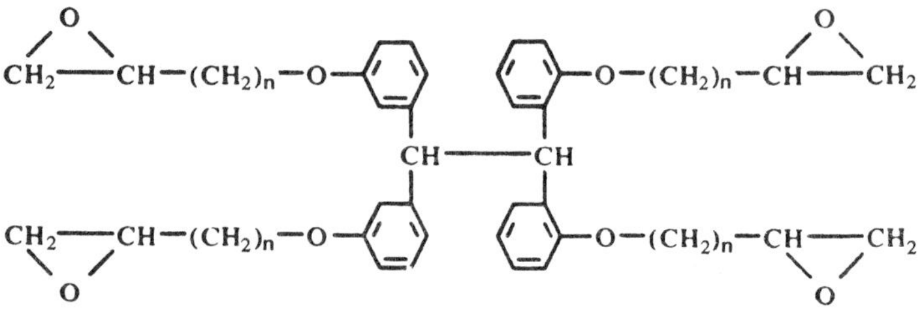

where n is greater than 1, novolac-type epoxy resins, cycloaliphatic-type epoxy resins, and halogenation products of these resins.

The curing agents include polycarboxylic acid anhydrides such as maleic anhydride, dichloromaleic anhydride, dodecenylsuccinic anhydride, pyromellitic anhydride, methyltetrahydrophthalic anhydride, benzophenone tetracarboxylic anhydride, phthalic anhydride, etc.

Owing to the unsaturated amic acid compound the cured product obtained from the components is superior in flame retardance to that obtained from ordinary epoxy resins alone (excluding those which have been endowed with flame retarding property, such as halogenated epoxy resins). The following examples illustrate the process. Parts are by weight unless otherwise specified.

Example 1: 50 parts of N,N'-4,4'-diphenylmethanebismaleamic acid and 50 parts of cresol-novolac-type epoxy resin (ECN-1280, CIBA; epoxy equivalent, 233) were dry-blended to obtain a resin composition. The composition was cured by heating at 180°C for 90 minutes. The weight loss on heating and the temperature at the beginning of weight decrease of the cured product were as shown in the following table.

Example 2: 55 parts of N,N'-4,4'-(diphenyl ether)bismaleamic acid and 50 parts of phenol-novolac-type epoxy resin (DEN-438, Dow Chemical Co.; epoxy equivalent, 180, liquid resin) was obtained in a manner similar to that in Example 1.

		Example No. 1	Example No. 2
Gel time at 180°C. (sec.)		150	170
Curing conditions	Temperature (°C.)	180	180
	Time (min.)	90	100
Temp. at beginning of wt. decrease (°C.)		400	395
Loss on heating at 500°C. (%)		23	25
Flexural strength (%)	100°C.	100	100
	150°C.	90	85
	180°C.	82	78
	250°C.	60	51

Metal Clad Laminate

A process described by *H. Fujiwara and H. Suzuki; U.S. Patents 4,021,403; May 3, 1977; and 4,020,225; April 26, 1977; both assigned to Maruzen Oil Co. Ltd., Japan* relates to a thermosetting resin composition having excellent flame resistance and thermal stability comprising an epoxy resin and a halogenated polyhydroxystyrene (PHS). A metal clad laminate comprising at least one planar base material having impregnated in it the above described flame resistant thermosetting resin composition and at least one metal foil laminated on it is also described.

The halogenated polyhydroxystyrene used can be a halogenated polyhydroxystyrene having a broad range of mean degrees of polymerization. However, from the standpoints of the curing operation and the properties of the hardened products it is preferable to use a halogenated polyhydroxystyrene having a mean degree of polymerization of from about 10 to 150.

The halogenated polyhydroxystyrene is prepared by reacting a polyhydroxystyrene with a conventional halogenating agent such as molecular chlorine, bromine, or iodine or further N-bromosuccinimide or chloramine T. Preferred halogen atoms in the halogenated polyhydroxystyrene are bromine, chlorine or a mixture.

The epoxy resin blended with the halogenated polyhydroxystyrene can be any one of the epoxy resins which are generally used and mainly comprise compounds having at least two epoxy groups per molecule. These epoxy resins can further include any of the saturated or unsaturated aliphatic, cycloaliphatic, aromatic or heterocyclic epoxy resins. These include polyglycidyl ethers and esters, cycloaliphatic polyepoxides (epoxy type resins), basic polyepoxy compounds (nitrogen-containing epoxy resins) and diene-type epoxy resins.

The composition of this process can be used for various purposes and in various forms, such as a coating composition, a molding, a laminate, etc., as well as a base material for a metal clad laminate. For example, a flat coating film of the composition can be formed on a support by dissolving the composition in an appropriate solvent and coating the support with the solution of the composition followed by drying and heating or alternatively by placing a powder of the composition on a support and compressing and heating the powdery composition. The following example illustrates the process.

Example: Copper-clad laminates clad on both surfaces and copper-clad laminates clad on only one surface were prepared using the following three kinds of resin compositions. Brominated poly-p-hydroxystyrene having a mean molecular weight of 13,000 and a bromine content of 51.7% or brominated poly-p-hydroxystyrene having a mean molecular weight of 10,900 and a bromine content of 41.5% was mixed with an epoxy resin such as a bisphenolic epoxy resin, Epikote 1001, a novolac type epoxy resin, Epikote 154, and an alicyclic epoxy resin, CX-221 so that the phenolic hydroxyl group/epoxy equivalent was 1.

After adding 1 phr of p-dimethylaminobenzaldehyde to the mixture, the resultant mixture was dissolved in a definite amount of acetone to provide a varnish. A glass cloth was immersed in each of the varnishes and then the impregnated cloth was allowed to stand for 4 hours at room temperature. Thus three kinds of the resin-impregnated base materials, that is, prepreg cloths, were prepared. Nine sheets of each kind of the prepreg cloth were laminated and a copper foil of a thickness of 35 microns was placed on one side or on both sides of the laminate followed by press-molding of the laminates by heating under pressure under the conditions as shown in Table 1 to provide copper clad laminates of a thickness of 1.6 mm. The laminates were then postcured for 2 hours at 170°C.

When the properties of the copper clad laminates thus obtained were measured, the results shown in Table 2 were obtained. A solder float was measured for the double-surface copper clad laminates and the peeling strength in a direction perpendicular to the plane of the laminate and other properties were measured for the one-surface copper-clad laminates. In this case, however, the volume resistivity and the flame retardant property were measured for the laminate after stripping off the copper foil or foils.

TABLE 1

Sample No.	..... Resin Composition (parts by weight) ...... Br-PHS	Epoxy Resin	Accelerator (phr)	Acetone (parts by weight)
1	34.0 (bromine content, 51.7%)	Epikote 1001, 66.0	1	95
2	53.7 (bromine content, 41.5%)	Epikote 154, 46.3	1	100
3	60.0 (bromine content, 41.5%)	CX-221, 50.0	1	75

Sample No.	Heating and Pressing Conditions	Resin Content of Laminate (wt %)
1	160°-170°C/50 kg/cm²/15 min	37
2	150°-160°C/50 kg/cm²/30 min	40
3	160°-170°C/50 kg/cm²/20 min	35

TABLE 2

	.........Sample Number......... 1	2	3
Solder float (280°C), min	10+	30+	5+
Peel strength in the direction perpendicular to the plane (kg/cm)	1.74	1.50	1.63
Peel strength after solder float, (kg/cm)	1.70	1.45	1.58
Volume resistivity after boiling 2 hr (Ωcm)	4.2×10^{13}	5.3×10^{12}	3.5×10^{13}
Flame retarding properties*			
JIS C-6481			
Burning time (sec)	3	2.5	2.0
Burning distance (mm)	11	7	7
UL 492			
Burning time (sec)	3 (max. 6)	2 (max. 4)	2 (max. 3)

*Passed UL 94 (V-0).

Intumescent Coatings Containing 4,4'-Dinitrosulfanilide

P.M. Sawko and S.R. Riccitiello; U.S. Patent 4,061,579; December 6, 1977; assigned to U.S. Administrator of the National Aeronautics and Space Administration describe an intumescent composition, which comprises a mixture of 4,4'-dinitrosulfanilide as the intumescent agent in a polymer binder mixture of a chlorinated polyolefin, a bisphenol A epoxy resin and a rubber-like amine hardener.

In the preparation of the composition, from 35 to 60 weight percent of the dinitrosulfanilide component based on the total weight of the composition is combined with a binder system. Accordingly, the components of the binder system amount to from 80 to 30 weight percent of the intumescent agent. The binder system must be compatible with the intumescent agent and should also have the characteristics of softening or decomposing below the activation temperature of the intumescent agent, of being processable with the intumescent agent and environmentally stable as a coating.

Moreover, in order to provide a more fire resistant and thermally effective system, the binder system should be nonflammable or self-extinguishing when exposed to a flame and, upon anaerobic thermal decomposition, should show no or very little exothermic behavior. A binder which satisfies these conditions is a chlorinated polyolefin of the formula:

$$CH_3 \!-\!\!\!+\! CH_2\!-\!CCl_2]_y\!-\![CHCl\!-\!CHCl\!-\!\!\!+_x COOH$$

where the ratio y:x is such that the chlorine content of the compound ranges from 58 to 62%. The amount of polyolefin component in the overall composition generally ranges from 15 to 60 weight percent. Two deficiencies of the binder namely inflexibility and the low residue stability and adhesion of the binder during thermal decomposition of the applied intumescent coating, are overcome by blending a flexible epoxy resin copolymer with the chlorinated polyolefin. Suitable epoxy resins for the epoxy component include the bisphenol A types.

A hardener component is necessary for the binder system in order to crosslink the epoxy resin. A rubber type amine hardener is generally used because it maintains the desired degree of flexibility in the resulting cured composition.

The present composition can be applied to any suitable substrate by any convenient means. Suitable substrates are those which do not decompose below the intumescent temperature of this composition and include such metals as steel, aluminum and the like; plastics, wood, cardboard and the like. The composition is therefore especially applicable in protecting metal parts which potentially could be exposed to heat or fire such as in aircraft, spacecraft, ordnance and the like. The following examples illustrate the process.

Example 1: A 26.9 part by weight amount of 4,4'-dinitrosulfanilide, 32.9 parts by weight of chlorinated polyolefin CP-310-6, having a 60% chlorine content, 1.0 part by weight of a bisphenol A epoxy resin, Epon 828 having average molecular weight of 389 and an epoxide equivalent of 185-192 in 28.8 parts by weight of toluene were homogeneously dispersed in a pebble jar mill for 4 hours or in a high speed kinetic dispersion unit for 20 minutes. To the dispersion were added 10.4 parts by weight of methylene chloride to complete Mixture A of the coating composition. To Mixture A was added Mixture B which was a mixture of 4.0 parts by weight of Hycar ATBN (amine terminated butadiene-acrylonitrile) and 4.0 parts by weight of toluene.

The coating composition thus prepared is a reactive coating with a pot life of 4 to 8 hours. The solution was sprayed with a Devilbiss Type MBC, No. 96 aircap spray gun with a pressure pot of 2 to 3.5 x 10^4 pascals (3-5 psi) and air pressure of 2.8 to 3.2 x 10^5 pascals (40-45 psi), onto cold rolled steel substrates to a dry film thickness of 60 mils and then air dried to a tough, adherent coating. The coatings were exposed to an atmosphere of 38°C, and 90% relative humidity for 150 days and then heated in an oven for 20 minutes at 1250°C to intumesce the coatings. The linear expansion of the intumesced coatings was 13 to 19 times the original thickness of the coated film. The coatings exhibited toughness when abraded with a Taber Abrader. An average weight loss of 111 milligrams occurred when abraded with a CS-10 wheel for 1,000 cycles and a 500 gram load. The coated film had a density of 1.38 g/cc.

The thermal protection afforded the substrate by the intumescent coating was determined by the combustion of a JP-4 fuel source impinging on the coating surface. (All coatings tested were air dried for one week prior to testing.) Heat at a rate of 11 to 12 x 10^4 watts/square meter was applied to a 7.62 cm x 7.62 cm x 0.159 cm cold rolled steel panel, at the center of which was a 2.54 cm diameter hole into which was cemented (epoxy novolac/polyamide adhesive, 60/40) a 2.36 cm diameter by 0.159 cm thick cold rolled disc. Time-temperature histories of the calorimeter disc were obtained at the prescribed heating rate. An average dry coating thickness of 62 mils was used for the thermal testing.

Example 2: Another composition in which the intumescent agent was present in a concentration of 40% was prepared by mixing 17.6 parts by weight of 4,4'-dinitrosulfanilide, 39.7 parts by weight of chlorinated polyolefin, CP-310-6, and 1.1 parts by weight of epoxy resin, Epon 828, in 30.8 parts by weight of toluene. This dispersion was prepared as described in Example 1. To the dispersion was added 10.7 parts by weight of methylene chloride to complete formulation of Mixture A of the composition. To Mixture A was added Mixture B, which comprises a mixture of 4.5 parts by weight of Hycar ATBN and 4.5 parts by weight of toluene.

The thermal protection offered by the composition was determined according to the procedure described in Example 1. An average dry coating thickness of 60 mils was used for the thermal testing.

Hardenable Crosslinked Resins

A process described by *N. Vollkommer and W. Trautvetter; U.S. Patent 4,031,163; June 21, 1977; assigned to Dynamit Nobel AG, Germany* relates to a hardenable cross-linkable resinous composition comprising a crosslinked polymer having:

(1) Between 22 and 76 units of a residue of a nuclearly chlorinated or brominated xylyleneglycolbisallylether;

(2) Between 11 and 54 units of a residue of maleic acid anhydride; and

(3) Between 6 and 57 units of a residue of an epoxy compound.

In accordance with this process it has been discovered that when a hardenable resinous composition containing nuclearly brominated or chlorinated xylyleneglycolbisallylethers and maleic acid anhydride is initially preset and thereafter cured, the resultant composition can be improved with respect to hydrolysis resistance and resistance to caustic soda if the composition also includes an epoxy compound. It can also be found that by including an epoxy compound in the resinous composition the preliminary setting and subsequent curing process can be conducted conveniently and no special steps are required to control the crosslinking polymerization.

Cast or formed articles of the crosslinked polymer can be prepared by initially melting the three component composition, for example, at a temperature of 60° to 80°C. Thereafter, the initiator and additionally, should the occasion arise, the accelerator can be added and the composition can be formed into the desired shape. Hardening is thereafter effected in the usual manner. The resin forming components can also be melted in the presence of conventional additives, such as stabilizers, inert fillers, flame-proofing agents and the like, if desired. If desired, the additives can also be added to the melted components.

The mixtures in which xylyleneglycolbisallylethers of a high degree of chlorination or bromination in the nucleus are used, can be fabricated into flameproof products without the addition of flameproofing agents in most cases. For example, the resin prepared on the basis of 356 weight-parts of tetrachloroxylyleneglycolbisallylether, 196 weight-parts of maleic acid anhydride and 250 weight-parts of bisphenol A diglycidyl ether, and hardened, is self-extinguishing in spite of a content of only about 18 weight-percent of organically bound chlorine. In the UL test (test for flame resistance by a method standardized by the Underwriters Laboratories) it achieves a rating of SE 1. The following examples illustrate the process.

Example 1: In a 2 liter three-necked flask equipped with stirrer and reflux condenser 750 ml of allyl alcohol were placed. While a current of nitrogen was passed through, 88 grams (2.2 mols) of sodium hydroxide were added portionwise, with stirring, and dissolved at approximately 50°C. At the same temperature, 312 grams (1 mol) of 2,3,4,6-tetrachloro-m-xylylenedichloride was added.

The mixture was heated to 130°C (bath temperature) and refluxed for 4 hours. Then 580 ml of allyl alcohol was distilled off, the cooled residue was diluted with 500 ml of acetone, and the reaction product was precipitated by pouring the acetone solution into water. It was suction filtered, washed with water to free it of NaCl, and reprecipitated from acetone in water to refine it. After drying, 321 grams of 2,3,4,6-tetrachloro-m-xylyleneglycolbisallylether was obtained. Yield: 90% of the theory. MP: 55° to 56°C.

By the same procedure, 309 grams of 2,3,5,6-tetrachloro-p-xylyleneglycolbisallyl ether was prepared from 312 grams (1 mol) of 2,3,5,6-tetrachloro-p-xylylene dichloride. Yield: 87% of the theory. MP: 86° to 88°C.

Example 2: 71 grams of 2,3,4,6-tetrachloro-m-xylyleneglycolbisallylether, 40 grams of maleic acid anhydride and 50 grams of bisphenol A diglycidyl ether was melted together at 70° to 80°C, and 3.28 grams (2% of the weight of the resin mixture) of dibenzoyl peroxide paste (a 50% paste) was stirred into the melt. The clear, fluid melt was poured into a mold and set for 1.5 hours at 85°C and 1 hour at 100°C. Then the temperature was increased over a period of 1 hour to 200°C and the casting was cured at this temperature for 2 hours. After removal of the mold, a transparent, honey-yellow plate 4 mm thick was obtained, which had the following characteristics:

Ball impression hardness, 10 sec.	1950 kp/cm^2
Vicat temperature	>220° C
Thermal stability of shape,	
Martens*	183° C
ISO/R 75 A	201° C
Bending strength	1173 kp/cm^2
Impact toughness*	12 cmkp/cm^2

* Tested on the small standard bar.

In the UL test, the casting proved to be self-extinguishing (Rate SE 1) even though the material contained approximately 18% organically bound chlorine by weight.

The infrared spectrum showed, in addition to an anhydride function that is still present (C=O valence vibration at 1,830 cm^{-1}), a strong ester-carbonyl absorption at 1,715 cm^{-1} indicating the reaction of the epoxide groups with the anhydride groups.

The weight loss of the set resin as measured on the thermal scale (air atmosphere; heating rate 8°C/min) amounted to 1% at 293°C, 5% at 334°C and 10% at 354°C.

Small standard bars were soaked in water at 100°C and in normal aqueous soda solutions at 100°C. After 14 days of soaking in hot water, the surface of the small standard bars showed no alteration. The weight increase was 3.34%, the thermal stability of shape on the basis of ISO/R 75 A was 186°C and the impact toughness was 13.1 $cmkp/cm^2$.

After 48 hours of standing in the hot caustic soda lye the surface of the small standard bars still showed no alteration. The weight increase was 7.4%, the thermal stability of shape according to ISO/R 75 A was 167°C and the impact toughness was 13.7 $kp\text{-}cm/cm^2$. Roughening and scaling of the surface did not occur until it remained in a hot caustic soda solution for 72 hours. This was combined with a loss of weight from the standard bars.

Example 3: This is included for purposes of comparison (without epoxy modification). 71 grams of 2,3,4,6-tetrachloro-m-xylyleneglycolbisallylether and 39 grams of maleic acid anhydride were mixed together and melted at 90°C. 0.1 weight percent of dibenzoyl peroxide paste (50% dibenzoyl peroxide) was added, and the clear melt was poured into a mold preheated to 70°C. The mold was immersed into an oil bath heated to 80°C to set the composition. As soon as incipient jelling was signaled by an increase in the viscosity of the casting resin, approximately one hour later, the mold was immersed in an oil bath of a temperature of 50°C for external cooling to prevent an excessive temperature rise due to the exothermic setting process, which would result in a decomposition of the resin causing it to foam up.

If the resin has jelled after another hour at 50°C, the temperature is increased to 80°C and maintained at that level for 2 hours, and then the resin is cured for 2 hours at 140°C to complete the crosslinking copolymerization. The mold was opened and a transparent plate 4 mm thick was obtained having the following characteristics:

Ball indentation hardness, 10 sec	1,850 kp/cm^2
Vicat temperature	>200°C
Martens thermal stability of shape*	136°C
ISO/R 75 A thermal stability	158°C
Bending strength	570 kp/cm^2

*Tested on standard small bar.

In the UL test, the material proved to be self-extinguishing (Rating SE 1). The weight loss determined on the thermal scale (air atmosphere; heating rate 8°C per minute) was 1% at 285°C, 5% at 326°C and 10% at 345°C.

Small standard bars were soaked in water at 100°C and in normal aqueous caustic soda solution at 100°C. While the surface of the castings crazed after only 48 hours in hot water, the bars had completely disintegrated after 17 hours of standing in the soda lye.

The superiority of the resin formula containing bisallylether, maleic acid anhydride and epoxy over the one consisting of bisallylether and maleic acid anhydride is thus clearly shown by Example 2. A substantial improvement in thermal stability of shape, bending strength and impact toughness is attained, plus a decided improvement in resistance to hydrolysis.

SILICONES

Arylorganobromosilicones

A process described by *T.J. Swihart; U.S. Patents 4,042,563; August 16, 1977; and 4,031,122; June 21, 1977; both assigned to Dow Corning Corporation* relates to a method for enhancing the flame-retardant characteristics of a substrate which comprises adding about 0.1 to 5.0% by weight of an arylorganobromosilicone compound having the general formula

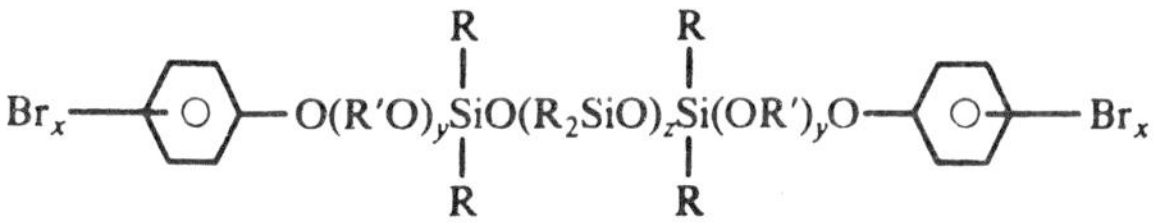

where x has an average value of from 2 to 3, R' is selected from the group consisting of ethylene, propylene and butylene radicals, y has a value of from 1 to 10, R is selected from the group consisting of hydrocarbon radicals containing from 1 to 18 carbon atoms, fluorinated hydrocarbon radicals containing from 1 to 18 carbon atoms, chlorinated hydrocarbon radicals containing from 1 to 18 carbon atoms and cyanohydrocarbon radicals containing from 1 to 18 carbon atoms, and z has a value of from 0 to 25. However, the best flame-retardant properties can generally be obtained when z has a value of from 0 to 8.

These compounds are prepared by the reaction of a bromoarylcarbinol with an acetoxy endblocked siloxane in the presence of a suitable catalyst. The arylorganobromosilicone compounds can be applied to various substrates, particularly cellulosic substrates, to improve their flame retardant properties. Examples of such substrates include wood, jute, cotton, or hemp in the form of plywood panels, rope or textiles, for example, as well as substrates made of nylon, polyamides, rayon or the like.

The following examples illustrate the process. All parts and percents referred to here are by weight and all viscosities are measured at 25°C unless otherwise specified.

Example 1: To a three-necked flask equipped with stirrer, condenser, dropping funnel, ammonia sparger and heating mantle there are added 188 grams of

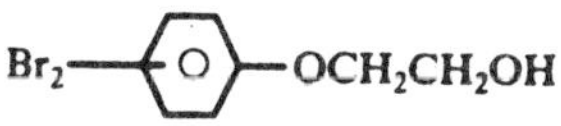

(composed of mono-, di-, and tribromo-substituted aromatic rings averaging two bromine atoms per molecule) and 200 ml of a methoxy endblocked polyethylene glycol. The mixture was saturated with ammonia and then while the mixture was heating at 90°C, 106 grams of $CH_3COO[(CH_3)_2SiO]_3OCCH_3$ was slowly added. The amount of acetoxy endblocked siloxane was determined by following the reaction by gas-liquid chromotography (GLC) analysis. The heating completes the reaction more quickly. The salt (ammonium acetate) was washed from the mixture using two water washings. This caused little, if any, hydrolysis according to GLC analysis. The mixture was then stripped on a Rinco separator to obtain a solid product having a melting point of 26° to 30°C and the following structure:

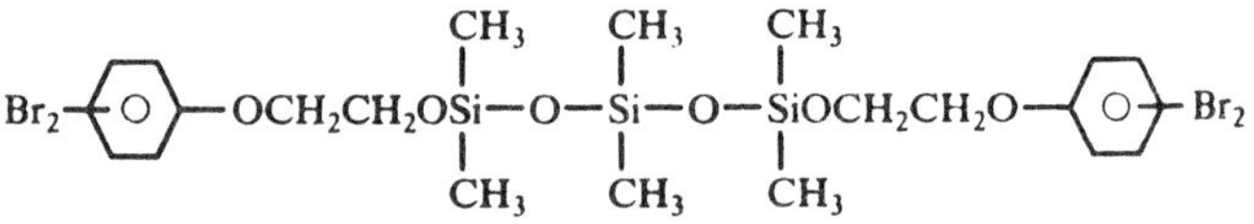

Example 2: The product of Example 1 was tested for flame resistant characteristics using the following tests.

Manifold Test — This test is described in Federal Test Methods No. 791A, Method 6053 "Manifold Ignition Test." The test fluid is preheated to 450±5°F and then dropped at the rate of 100 drops per minute from a dropping funnel whose outlet is five inches above the center of the manifold. The surface temperature of the manifold is varied from 600° to 1200°F in 100° increments. The minimum temperature at which the fluid burns is recorded.

Kim-Wipe Test — In this test a Kim-Wipe tissue is soaked with the test fluid. A bunsen burner is then held to the tissue for five seconds followed by cooling for five seconds, a total of four consecutive heating and cooling cycles being used.

Aluminum Dish Test — About seven grams of the test fluid is placed in a small aluminum dish or cup in this test. The fluid is heated continuously with a bunsen burner to see if it will ignite or burn.

In the Manifold Test the test fluid ignited at about 1080°F. In the Kim-Wipe test the tissue did not burn. In the aluminum Dish Test the test material did not burn after the flame was removed.

Organobromosilicone Fluids

P.L. Brown; U.S. Patent 4,031,121; June 21, 1977; assigned to Dow Corning Corporation describes organobromosilicone fluids having the general formula

$$(BrCH_2)_3CCH_2O(R_2SiO)_xCH_2C(CH_2Br)_3$$

where each R is selected from the group consisting of hydrocarbon radicals, fluorinated hydrocarbon radicals, chlorinated hydrocarbon radicals and cyanohydrocarbon radicals, all radicals containing 1 to 18 carbons and x has a value of from 1 to 20. These fluids are useful for improving the fire resistance of cellulosic substrates.

The organobromosilicone fluids can be applied to various substrates, particularly cellulosic substrates to improve their flame retardant properties. Examples of such substrates include wood, jute, cotton, or hemp in the form of plywood panels, rope or textiles, for example, as well as to substrates made of nylon, polyamides, rayon or the like. The following examples illustrate the process.

Example 1: To a three-necked flask equipped with stirrer, condenser, dropping funnel, ammonia sparger and heating mantle there was added 162.3 grams of $(BrCH_2)_3CCH_2OH$ and 200 ml of toluene. The mixture was saturated with ammonia and then 81 grams of $CH_3COO[(CH_3)_2SiO]_3OCCH_3$ was slowly added while heating the mixture at 90°C. The heating completes the reaction more quickly. The salt (ammonium acetate) was washed from the mixture using two water washings. Then the toluene and most of the low boiling components were removed by stripping over a short, open column at a head temperature of about 130°C and a pot temperature of about 240°C at 0.25 mm of mercury pressure using a nitrogen sweep. The resulting product,

$$(BrCH_2)_3CCH_2O[(CH_3)_2SiO]_3CH_2C(CH_2Br)_3$$

was dark in color with a viscosity of 73.6 cs and a refractive index at 25°C of 1.5071. The product had a purity of greater than 95% by gas-liquid chromatography analysis.

The infrared analysis spectrum shows the presence of $Si(CH_3)_2$ at 2,806 and 1,260 cm^{-1}, OCH at about 1,414 and 1,427 cm^{-1}, SiOSi and SiOC at 1,020 to 1,130 cm^{-1}. The SiOSi and SiOC region suggests that the degree of polymerization is greater than 3, but the hydrogen nuclear magnetic resonance analysis shows that the proton ratios are correct for the trimer structure.

Example 2: The procedure of Example 1 was repeated except that the reactants were 124 g of $(BrCH_2)_3CCH_2OH$, 250 g of $CH_3COO[(CH_3)_2SiO]_{\sim 15.4}OCCH_3$ and 350 ml of toluene. The product, $(BrCH_2)_3CCH_2O[(CH_3)_2SiO]_{\sim 15.4}CH_2C(CH_2Br)_3$, had a viscosity of 43.5 cs and a refractive index at 25°C of 1.4392.

Example 3: The products of Examples 1 and 2 were tested for flame-resistant characteristics using the Manifold Test, the Kim-Wipe Test and the Aluminum Dish Test as described in U.S. Patent 4,042,563. The results of these tests are shown in the following table.

	Test		
Test Fluid	Manifold	Kim-Wipe	Aluminum Dish
Example 1	~1040° F.	No ignition	No ignition after flame from burner removed.
Example 2	~1020° F.	Burns slowly	Burns with low flame after burner removed.

Polyphosphazenes and Silicone Polymer Blends

A process described by *R.L. Dieck and E.J. Quinn; U.S. Patent 4,026,839; May 31, 1977; assigned to Armstrong Cork Company* relates to foamed blends of polyphosphazenes and silicone polymers. The blends are useful as fire retardant coatings, sheets, films or flexible or semi-rigid foams.

The polyphosphazenes employed in the blends of this process are characterized by repeating

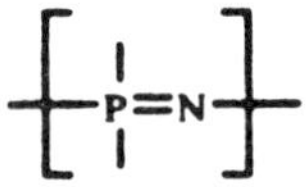

units which contain substituted aryloxy substituents (preferably substituted in the para position) on the phosphorus atoms in nonregular fashion and which can be represented by the following formulas:

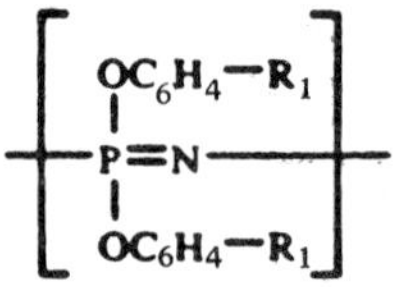 , 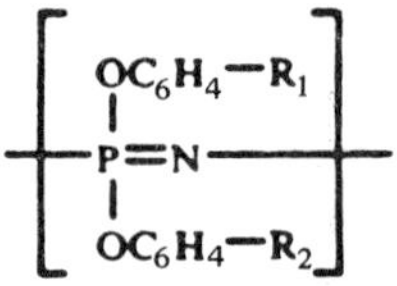 and 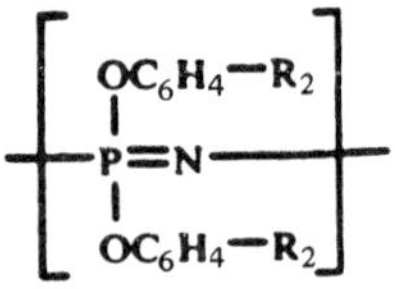

where R_1 and R_2 are the same or different and are hydrogen, a C_{1-10} linear or branched alkyl radical, or a C_{1-4} linear or branched alkoxy radical substituted on any sterically permissible position on the phenoxy group, with the proviso that when R_2 is alkoxy and when copolymers are to be prepared, R_1 and R_2 are different. Examples of R_1 and R_2 include ethoxy, methoxy, isopropoxy, n-butoxy, methyl, ethyl, n-propyl, isopropyl, sec-butyl, tert-butyl, tert-pentyl, 2-ethylhexyl and n-nonyl.

The above described polymers may be crosslinked and/or cured at moderate temperatures (for example, 200° to 350°F) by the use of free radical initiators, for example, peroxides, using conventional amounts, techniques and processing equipment.

The silicone elastomers which can be utilized in the blends are poly(organosiloxanes), preferably completely soluble in benzene and consisting essentially of repeat units of structure

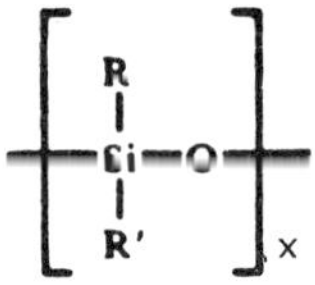

where R and R', which may be the same or different, are hydrocarbyl radicals or substituted hydrocarbyl radicals and x is an integer. The average molecular weight of the poly(organosiloxane) may vary widely, but is preferably within the range 40,000 to 200,000. Poly(organosiloxanes) having molecular weights outside this range may be used if desired.

Preferred poly(organosiloxanes) are phenylvinylpolydimethylsiloxanes or methylvinylpolydimethylsiloxanes which contain minor amounts of vinyl groups, i.e., having up to about 25 mol percent of organosiloxane units containing vinyl radicals.

The blends may be used to prepare foamed products which exhibit excellent fire retardance and which produce low smoke levels, or essentially no smoke when heated in an open flame.

The following formulation and method were employed in Examples 1 and 2.

	Parts by Weight
Cup A	
Polyphosphazene	5
Silicone elastomer	5
Cup B	
1,1'-azobisformamide	3
Oil treated urea (activator)	2.3
Magnesium oxide	0.5
Zinc stearate	0.5
Coumarone-indene resin (Cumar P-10)	0.2
2,5-dimethyl-2,5-di-tert-butyl-peroxyhexane	0.5
Benzoyl peroxide (78% active)	0.5
Alumina trihydrate	varied

Polymers were blended on a two-roll mill with one roll at 120° to 140°F and the other at ambient conditions. Samples in Cup A were blended for 15 minutes to ensure homogeneous mixing. The ingredients in Cup B were added to the polymer blend on the research mill. This was allowed to mix for an additional 15 minutes. The unexpected blend was then precured in a press for 12 minutes at a temperature of 220°F and pressure of 2,000 psi. Finally, the precured pad was expanded in a circulating air oven for 30 minutes at 300°F.

Example 1: Using the formulation above $[NP(OC_6H_5)(OC_6H_4\text{-4-sec-}C_4H_9)]_n$ and a phenyl vinyl polydimethylsiloxane (Silastic 675U) were blended with the ingredients in Cup B and 2.5 grams of alumina trihydrate. Silastic 675U is a low temperature serviceability silicone rubber which when press molded (100 parts to 13 parts of 50% 2,4-dichlorobenzoyl peroxide, five minutes at 116°C) has the following properties in a 0.075 inch thick slab: cure, 24 hours at 250°C; durometer hardness, Shore A, 75+5, tensile strength, psi min, 650; elongation, % maximum, 20; shrinkage, percent, 2.85±0.35. The mixture was foamed and resulted in a flexible salmon pink foam. Rohm and Haas XP2 Smoke Chamber tests on the sample gave the maximum smoke density value of 23; SV/g = 13.

Example 2: Using the formulation above, $[NP(OC_6H_5)(OC_6H_4\text{-4-sec-}C_4H_9)]_n$ and Silastic 675U (both described in Example 1) were blended with the ingredients in Cup B and 5.0 grams of Hydral-710. The mixture was foamed and resulted in a flexible, salmon pink foam. Rohm and Haas XP2 Smoke Chamber tests on the sample gave the values: maximum smoke density = 14; SV/g = 7.

The following general formulations were employed in Examples 3 and 4:

	Parts by Weight
Cup A	
$[NP(OC_6H_4\text{-4-}OCH_3)(OC_6H_4\text{-4-sec-}C_4H_9)]_n$	x
Silicone elastomer (Silastic 675U)	y
Cup B	
Alumina trihydrate	10
1,1'-azobisformamide	2
Oil treated urea	0.5

(continued)

	Parts by Weight
Magnesium oxide	0.5
Zinc stearate	1
Coumarone indene resin (Cumar P-10)	0.2
2,5-dimethyl-2,5-di-tert-butyl peroxyhexane	0.6
Benzoyl peroxide (78% active)	0.2
Dicumyl peroxide	0.1

The method employed in the following examples is the method described in conjunction with Examples 1 and 2.

Example 3: Using the method and formulation above, where x = 8.0 grams and y = 2.0 grams, the polymers and ingredients were blended and formed. The resulting foams were flexible and salmon pink in color. NBS Smoke Chamber testing in the flaming mode resulted in the value D_m(corr) = 121; SV/g = 9; Oxygen Index (O.I.) = 39.9.

Example 4: Using the method and formulation above, where x = 7.0 grams and y = 3.0 grams, the polymers and ingredients were blended and foamed. The resulting foams were flexible and salmon pink in color.

Silicate Containing Adhesive Composition

J.A.L.F. Chollet; U.S. Patent 4,066,463; January 3, 1978 describes a silicate-containing flame resistant adhesive composition comprising (1) an inorganic component consisting, with respect to the weight of the total composition, of (a) 20 to 90 weight percent of a concentrated aqueous alkali metal silicate solution, (b) 5 to 25 weight percent of a clay and 2 to 7 weight percent of deflocculated asbestos fibers; (2) an organic component present as a 30 to 70% aqueous solution of a carboxymethylcellulose, a starch ether, a dextrin or mixtures thereof. This organic component is present in an amount (dry weight) of 0.2 to 2 wt % of the total composition.

The process for the preparation of the composition is carried out in the following manner. The silicate solution is charged in a suitable container and is then rotated in a given direction by means of a suitable stirring device, at a speed of preferably between 500 and 2,000 rpm.

The aqueous organic component solution is prepared separately and is then intimately dispersed in the stirred silicate solution. To effect the dispersion, the organic solution is added portionwise, preferably near the bottom of the container, while rotating it in a direction opposite to that used to rotate the silicate solution; this may be achieved, for example, by means of a perforated dispensing head rotating at a speed of 250 to 1,000 rpm (preferably at a speed of 500 rpm).

On completion of the dispersion, the clay is added, with continued stirring, and the deflocculated asbestos fibers are then added and thoroughly dispersed in the liquid mixture.

When a mineral filler is to be added, this is added to the dispersion prior to addition of the clay; the liquid silicone, when used, is added after the asbestos and the optional swelling agents are added last. All the process steps are conducted at ordinary temperature (preferably >5°C).

The resulting composition may be stored in conventional pots, drums or closed containers without setting to a mass during a period of time of at least six months. The following examples are given to illustrate the process.

Example 1: The two following adhesive compositions are prepared according to the above described method:

	Composition 1 (wt %)	Composition 2 (wt %)
Sodium silicate (38°/40°Bé)	85	76
Kaolin	7	16
Deflocculated asbestos fibers	4	4
Solvitose CL* (50% aqueous solution)	2	2
Methyl siliconate (liquid silicone)	2	2
	100	100

*Starch ether sold by Doitteau, France

Both compositions exhibit outstanding tack and, after setting, are classified "flame resistant" and "fireproof" according to the usual standard specifications.

The tensile strength is found to be 11 to 15 kg/cm^2 at normal temperature when black sheet iron is bonded to concrete; after 7 days at 40°C and 95% relative humidity, the tensile strength is 11 to 18 kg/cm^2.

Agglomerated wood fiber boards were bonded to concrete with the above compositions; the agglomerate breaks at 8 to 9 kg/cm^2 at normal temperature but the bonding plane remains intact.

Plywood was bonded with the composition: the plywood was found to break between 12 and 14 kg/cm^2, but the bonding plane remained intact.

Compositions exhibiting analogous properties were obtained by replacing the starch ether with carboxymethylcellulose or dextrin.

Example 2: Using the above described process, the following coating composition was prepared:

	Weight Percent
Potassium silicate (42°Bé)	41
Kaolin	13
Deflocculated asbestos fibers	4
Crude perlite	40
Solvitose CL (50% aqueous solution)	1
Methyl siliconate	1
	100

A flame resistant and fireproofing coating is obtained after setting.

One of the applications of the composition is the bonding of various articles, for example the bonding of (a) pieces of sheets made of aluminum, of black sheet iron, of galvanized sheet metal or of zinc-plated sheet metal, of glass, of cement, etc., on (b) wood, plywood, particle- or fiberboards, polystyrene foam, phenol resins, polyurethanes, and the like. The following bondings may also be achieved: wood-wood, wood-polystyrene, carpet-floor, cement-wood, and the like.

Another highly important application is the production of flame resistant and flame-proofing coatings on a variety of substrates such as wood, plywood, particle- or fiberboards, expanded polystyrene, phenol foams, polyurethanes, cardboard-reinforced plaster, and the like.

Intumescent Silicone Base Composition

In a process described by *R.D. Pratt, P.F. Proffitt and G. Webb; U.S. Patents 4,065,394; December 27, 1977; 4,052,526; October 4, 1977; and 3,983,082; September 28, 1976 all assigned to General Electric Company* a silicone resin base fire retardant system having a temperature capability at least to about 450°F (230°C) combines intumescent characteristics with a crusting and charring capability through use of a combination of an oxide of iron and a phosphate of potassium for crusting, along with a charring material and a blowing agent, in a silicone resin base to allow for adequate expansion.

Such an intumescent mixture includes, by weight 2 to 4 parts ferric oxide; 35 to 40 parts potassium orthophosphate; 15 to 25 parts tripentaerythritol; and 6 to 9 parts melamine.

This mixture in the range of about 70 parts by weight is mixed with about 55 parts by weight of the silicone resin. The process is particularly useful in connection with metallic materials or alloys based on elements selected from Al, Mg and Ti.

The silicone resin used in one evaluation was prepared by catalyzing about 55 parts by weight of resin with about ½ part by weight of a catalyst commercially available as Thermolite 12 material. The catalyzed resin was extended with a solvent to a consistency suitable for spraying. To this thinned, catalyzed resin was added about 70 parts by weight of an intumescent mixture. This mixture of powder ingredients having a maximum particle size of about 60 mesh size U.S. Standard Sieve consisted of, by weight, about 2.5-3.5 parts ferric oxide, about 36-38 parts potassium orthophosphate, about 7-8 parts melamine and about 20-25 parts tripentaerythritol.

The total material which included the intumescent mixture and the catalyzed silicone resin, appropriately thinned, was sprayed in thin layers, for example of at least about 0.01 inch (0.25 mm), in repeated layers on Al alloy specimen panels, typical of such metals or alloys based on Al, Mg or Ti. In this series of examples, the alloy was one consisting nominally, by weight, of 4.4% Cu, 1.5% Mg, 0.6% Mn with the balance Al. In this series of examples, an average total thickness of about 0.04 inch (1 mm) was attained.

After application of the material to the surface of the metallic specimen panels, the panels were tested by positioning the panels horizontally over a flame with the coating facing downward and exposing the coated side of the specimen to about a 2000°F (1090°C) direct flame from a Meeker burner with an equivalent heat influx intensity of about 6,000 Btu/hr for a duration of 15 minutes. The following table includes typical test data for such an evaluation.

In the following tests, the conditions remained stable after about 5 minutes and no burn through of the specimens occurred at the conclusion of the 15 minute tests.

2000°F Fire Tests*

	Time (Min:Sec)	Alloy Panels Backside Temp. (°F)
Specimen A; 0.039" Coating	1:45	1000
	2:30	1100
	3:00	950
	5:00	740
Specimen B; 0.044" Coating	0:26	600
	1:45	800
	3:30	850
	4:30	700

*Aluminum alloy panels, 0.025" thick.

PHENOLICS AND OTHER THERMOSETS

PHENOLIC AND AMINE-FORMALDEHYDE RESINS

Sulfanilamide-Formaldehyde Intumescent Compositions

S.H. Roth; U.S. Patents 3,988,266; October 26, 1976 and 3,991,021; Nov. 9, 1976; both assigned to Cities Service Company describe a water-insoluble amine-aldehyde copolymer intumescent agent having a number average molecular weight of 350 to 2,000 and consisting essentially of repeating units corresponding to the formula:

(1)

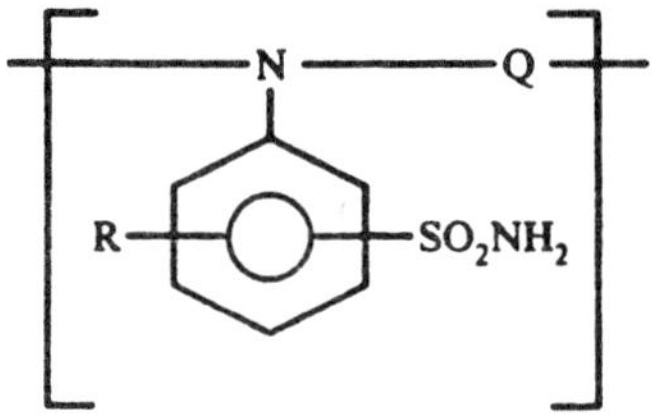

Q is the aldehyde residue and R is hydrogen, alkyl, haloalkyl, alkoxy, haloalkoxy, halo, nitro, acetamido, or sulfonamido. This is mixed with 10 to 50% by weight of a synergist corresponding to the formula:

(2)

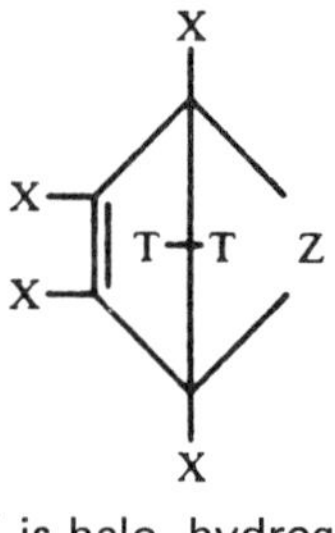

where X is chloro or bromo, T is halo, hydrogen, or a hydrocarbyl or hydrocarbyloxy group, and Z is a monocyclic divalent saturated polybromocycloalkylene group containing 5 to 16 carbon atoms and 2 to 6 bromine atoms, the valences being on adjacent carbon atoms.

The synergist may be prepared by reacting a polyhalocyclopentadiene with a suitable multiunsaturated cycloaliphatic hydrocarbon to from a 1:1 Diels-Alder adduct, and then adding bromine to the residual unsaturation and, if desired, introducing additional bromine by substitution bromination.

For use in protecting a substrate from heat and fire, the intumescent agent/synergist compositions may be applied by electrodeposition, spraying of powdered intumescent agent/synergist composition onto an adhesive substrate, etc. It is usually preferred to compound the intumescent agent and synergist with a binder, e.g., nitrocellulose, and optionally also with one or more of the other ingredients used in intumescent compositions, e.g., stabilizers, dispersing agents, pigments, driers, biocides, antifoamers, thickeners, protective colloids, fillers, blowing agents, etc.; disperse the composition in a suitable liquid medium, e.g., water or a solvent or solvent mixture; and apply the coating composition thus formed to the substrate to be protected.

The following examples illustrate the process. Quantities mentioned are quantities by weight, and the sulfanilamide-formaldehyde copolymer employed as an intumescent agent is a copolymer which has been prepared in accordance with U.S. Patent 3,922,251 so that the sulfanilamide units correspond to the formula:

Example 1: Control – Place 1 g of sulfanilamide-formaldehyde copolymer in an aluminum pan having a diameter of 57 mm and a height of 16 mm. Flame the sample with propane torch. The copolymer intumesces to form a foam volume of 1⅓ pans.

Example 2: Control – Repeat Example 1 except for replacing the sulfanilamide-formaldehyde copolymer with a compound corresponding to the formula:

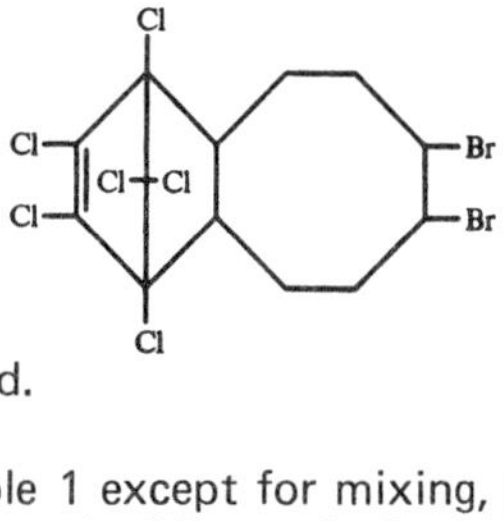

No intumescence is observed.

Example 3: Repeat Example 1 except for mixing, respectively, 0.05, 0.1, 0.2, 0.3, and 1.0 g of the compound of Example 2 with 1 g of the sulfanilamide-formaldehyde copolymer of Example 1 before flaming. The compositions intumesce to form foam volumes of 1½, 2, 4, 2¼, and ⅔ pans, respectively. In each case, the foam has excellent char structure.

Example 4: Control – Repeat Example 1 except for replacing the sulfanilamide-formaldehyde copolymer with a compound corresponding to the formula:

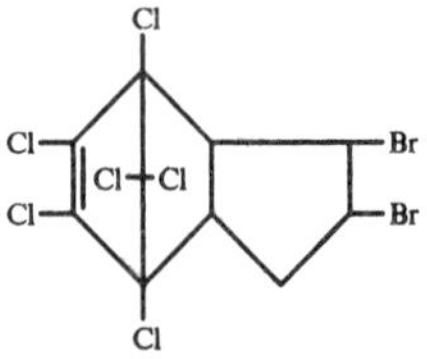

No intumescence is observed.

Example 5: Repeat Example 1 except for mixing, respectively, 0.1, 0.2, 0.4, 0.5 and 1.0 g of the compound of Example 4 with 1 g of the sulfanilamide-formaldehyde copolymer of Example 1 before flaming. The compositions intumesce to form foam volumes of 1⅔, 3, 2, 2, and about 1, respectively. In each case, the foam has excellent char structure.

The preceding examples show that the synergists increase the foam volume when used at the appropriate concentrations, even though they may actually decrease the foam volume when used at higher or lower concentrations. Similar results are observed when the sulfanilamide-formaldehyde copolymer and/or the above specified synergists are replaced by materials taught to be their equivalents in the specification. The following example demonstrates the inability of other halogenated compounds to be substituted for the synergists of this process.

Example 6: Control – Repeat Example 1 except for mixing, respectively, 0.1, 0.2, 0.3, and 0.5 g of decabromodiphenyl oxide with 1 g of the sulfanilamide-formaldehyde copolymer of Example 1 before flaming. The compositions intumesce to form foam volumes of not more than ¼ pan in each case.

In related work, *S.H. Roth and W.S. Stoy; U.S. Patents 4,003,863, January 18, 1977 and 4,003,864; January 18, 1977; both assigned to Cities Service Company* describe intumescent amine-aldehyde copolymer compositions having improved UV stability. These are obtained by incorporating 0.5 to 75% by weight of a polyol polyacrylate or polyol polymethacrylate into an intumescent composition comprising a low molecular weight, water-insoluble amine-aldehyde copolymer consisting essentially of repeating units which correspond to Formula (1) of the previous work by Roth.

Phosphate Ester Derivatives of ortho-Novolaks

A process described by *B.F. Dannels and A.F. Shepard; U.S. Patent 4,010,144; March 1, 1977; assigned to Hooker Chemicals & Plastics Corporation* relates to polymer compositions and to a process of producing phenolic novolak resins having a high phosphorus content.

The phosphorus used is suitably in the form of P_2O_5, P_2S_5, or P_2O_3. In terms of P_2O_5, amounts between 20 and 45% by weight may be reacted with novolak resins. The starting novolak resin that is required contains largely o,o'–rather than o,p'–linkages between the phenol nuclei. A method of making such o,o'-linked novolak resins is described in U.S. Patent 3,425,989.

Generally, when mixtures of P_2O_5, P_2S_5 and P_2O_3 are used, the moieties present on the reacted novolaks can be represented as follows:

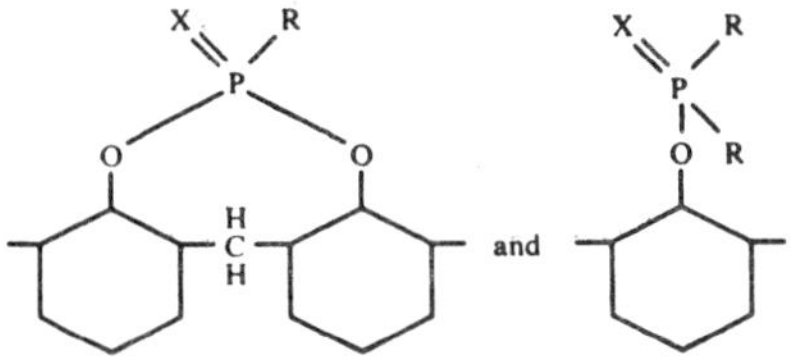

where X is O or S and each R is independently selected from the group of XH and H. In the reacted novolaks, at least 90% of the alkylidene linkage is ortho to ortho, and at least 25% of the phenolic hydroxyl groups are esterified.

The novolak resin phosphate esters of this process are useful in the fabrication of flame retardant paper, paper products, laminates, fibrous insulation, foams and molded articles.

Example 1: 312 parts by weight of a novolak resin containing 90% o,o'-linkage and having a molecular weight of 350 was heated to 130°C. 140 parts by weight of P_2O_5 was then added in small increments. As the viscosity of the reaction mixture increased, the temperature was raised so that the mixture remained fluid. The last portion of P_2O_5 was added at a temperature of 220°C. The reaction mixture became so viscous that it was not practical to add additional P_2O_5. The mixture was allowed to cool and the resulting product was a dark colored brittle resin. The product was essentially completely soluble in warm 10% NaOH indicating the absence of a high degree of crosslinking.

Example 2: In a similar manner 312 parts by weight of a randomly linked novolak with a molecular weight of 500 was reacted with P_2O_5. After 75 parts by weight had been added, the mixture became so viscous that further addition was impossible. The product was not soluble in hot 10% NaOH nor in common organic solvents, indicating a crosslinked gel.

Example 3: The free phenol content of the ortho-novolak resin used in Example 1 was reduced to less than 0.5% by distilling phenol off under reduced pressure. 595 parts by weight of the resulting novolak, which now had a molecular weight of 550, was heated to 180°C and P_2O_5 slowly added. After the addition of 130 parts by weight, the reaction mixture became too viscous to continue. This product dissolved in warm 10% NaOH. At the point where 110 parts by weight of P_2O_5 had been reacted, the reaction mixture was sufficiently fluid to be easily stirred and aptly suited for molding applications.

Example 4: 590 parts by weight of dephenolated novolak used in Example 3 was dissolved in 830 parts of chlorobenzene. The mixture was heated to 90°C and 261 parts by weight of P_2O_5 were added over a 1 hour period. The reaction temperature was then increased to 132°C and held for an additional 6 hours. It was noted that a trace amount of sticky solid adhered to the walls of the reactor.

The solution was decanted into 6,000 parts by weight of hexane. A light pink solid precipitated in the form of a fine powder.

Mixture of Acids and Glycol as Curing Agent

D. Simic, H.-U. Zupancic and M. Cherubim; U.S. Patent 4,001,148; January 4, 1977; assigned to Deutsche Texaco AG, Germany describe a process for the production of flame resistant phenolic resin foam plastics where an aqueous alkaline-condensed phenol-formaldehyde resin solution is mixed with an expanding agent and an acidic curing agent, the mixture is subjected to foaming at temperatures of 15° to 30°C and is permitted to cure. The curing agent utilized in the process of this method has the following composition:

8 to 70 vol % of a strong inorganic or organic acid,
10 to 40 vol % of a glycol, and
20 to 80 vol % of an about 75 to 85 wt % phosphoric acid.

An especially useful curing agent is a mixture comprising phenolsulfonic acid, ethylene glycol, and phosphoric acid in the above volume percentages. It has been found that the glycol, especially ethylene glycol, component of the curing agent prevents the formation of cracks in large foam blocks. As phenolic resins, resoles may be used, i.e., alkaline-condensed phenol-formaldehyde resins, having a mol ratio of phenol to formaldehyde ranging from 1:1.1 to 1:2.0. The following examples illustrate the process.

In Examples 1 and 2, an aqueous phenol-formaldehyde resin solution was employed which was formed by condensing phenol and formaldehyde in a mol ratio of 1:1.64, with sodium hydroxide solution to a viscosity of 3,500 cp at a solids content of 73 wt %. In Example 3, an aqueous phenol-formaldehyde resin solution of similar viscosity and solids content was employed, with the mol ratio of phenol to formaldehyde being about 1:1.52.

The curing composition (i.e., Curing Composition A) used in Examples 1 through 3 had the following composition:

10 parts by volume phenolsulfonic acid, 60 wt % (density 1.27 g/cm^3),
20 parts by volume ethylene glycol (density 1.13 g/cm^3),
70 parts by volume phosphoric acid, 85 wt % (density 1.70 g/cm^3).

Example 1: To 120 kg of phenol formaldehyde resin (mol ratio 1:1.64) 2.8 vol % of pentane and 25 vol % of Curing Composition A were added under intensive stirring. This mixture was then poured into an open 1 m^3 mold at 20°C (cold foaming). After 1 minute the reaction mixture started to foam and after 20 minutes foaming was ended. After being cured for approximately 60 minutes, the finished foam block was removed from the mold. The resulting foam block had a density of 0.10 g/cm^3. After tempering (a heat treatment for 48 hours at 80°C) the foam block was cut into plates and these plates were tested for flame resistance. The results are shown in Tables 1 and 2.

Example 2: In accordance with Example 1, 10 vol % pentane and 25 vol % Curing Composition A were added under stirring to 55 kg phenol-formaldehyde resin (mol ratio 1:1.64), and subjected to foaming in an open 1 m^3 mold at 20°C. After the foam had been cured for about 75 minutes, it was removed from the mold. The resulting foam block had a density of 0.04 g/cm^3. Like the foam of Example 1, it was tested for flame resistance and the results obtained are shown in Tables 1 and 2.

Example 3: To 55 kg of phenol-formaldehyde resin (mol ratio 1:1.52) 10 vol % pentane and 25 vol % of Curing Composition A were added under intensive stirring. The mixture was poured into an open 1 m^3 mold maintained at 60°C in a hot chamber (hot foaming) and foamed. In about 60 minutes foaming was completed and after being cured for 180 minutes the foam block was removed from the mold. The resulting foam block had a density of 0.04 g/cm^3. This foam block was then tested in the same manner as in Examples 1 and 2 and the results are reported in Tables 1 and 2.

Flame-Resistance Test (1) (Laboratory Method) — Flame resistance was tested by fastening a 4 cm-thick sample with a surface area of 25 x 25 cm, 10 cm above a Bunsen burner and exposing it to the flame until it was burnt through. All tested foam plastics of this process exhibited the following characteristics: they burnt while exposed to the flame, but when removed, burning ceased by itself.

Flame Resistance Test (2) (Technical Method) — Foam plates with measurements of 100 x 19 x 2.5 cm were hung in a fire pit and exposed to blazing from below. After 10 minutes blazing was ended and the samples were removed. The following criteria were evaluated: height of flame, temperature of flue gas (must not exceed 250°C), and length of the nondestructed sample portion (must not be less than 15 cm).

The foam plastics obtained in Examples 1-3 were subjected to the above tests and, for comparison, foam plastics cured with hydrochloric acid-ethylene glycol and foam plastic cured by phenolsulfonic acid were also tested.

TABLE 1: RESULTS OF FLAME RESISTANCE TEST (1)

	Density (g/cm^3)	Time Required for Burning Through (min)	Remarks
Comparative sample cured with hydrochloric acid-ethylene glycol	0.04	3	burning does not cease by itself
Comparative sample cured with phenolsulfonic acid	0.04	8	burning does not cease by itself
Example 1	0.10	150	burning ceases by itself
Example 2	0.04	38	burning ceases by itself
Example 3	0.04	42	burning ceases by itself

TABLE 2: RESULTS OF FLAME RESISTANCE TEST (2)

	Height of Flame (cm)	Temperature of Flue Gas (°C)	Length of Nondestructed Sample Portion (cm)
Comparative sample cured with hydrochloric acid-ethylene glycol	100	270	0

(continued)

TABLE 2: (continued)

	Height of Flame (cm)	Temperature of Flue Gas (°C)	Length of Nondestructed Sample Portion (cm)
Comparative sample cured with phenolsulfonic acid	85	220	0
Example 1	30	113	28
Example 2	50	116	18
Example 3	50	116	20

Phenolic-Resol Resin Solution

A. Franz and S. Kopnick; U.S. Patent 4,074,015; February 14, 1978; assigned to Dynamit Nobel AG, Germany describe a process for producing nonflammable laminated materials which includes the steps of impregnating a cellulose-containing material with an impregnating solution containing a phenolic-resol resin solution, at least one plasticizer for the resin and optionally flameproofing agents.

The impregnated material is subsequently dried to effect preliminary condensation of the resin and thereafter superimposed layers of the impregnated and preliminary condensed material by the application of heat and pressure are cured. The phenol-resol resin solution used contains 50 to 80% by weight of an ammonia alkaline-condensed phenol-resol resin with a solids content of 65 to 80% by weight and with a molecular weight distribution determined by gel chromatography of: dimers 20 to 26%; tri- and tetramers 6 to 12%; penta- and oligomers 24 to 36% expressed in percent per unit area; and hexamethylenetetramine in amounts of 1 to 15% by weight based on the weight of the impregnating solution.

The cellulose-containing material is preferably paper in the form of sheets or fibrous webs. The cellulose-containing material can, however, also be in the form of fleeces; it is also possible to replace this material partially or completely by fabrics, mats, or papers made from synthetic substances which can likewise form fibers, or by spun glass fibers or glass wool. Examples of the synthetic substances are polyesters or polyamides.

The phenol-resol resins utilized are selected from conventional products. The characteristic factor for the resins to be employed is their molecular weight distribution, determined with the aid of gel chromatography (i.e., GPC which means gel permeation chromatography).

The laminated materials used in the examples listed in the table were manufactured as follows: A commercial phenol-resol resin (T55 produced and sold by Dynamit Nobel) having the above indicated molecular weight distribution and a 75% solid resin content is mixed under agitation with a 40% by weight slurry of hexamethylenetetramine in water at room temperature. Subsequently, the plasticizer, methanol as the solvent, as well as optionally the flameproofing agent are mixed. A clear solution is obtained which is used as the impregnating solution in this form.

Soda kraft paper (180 g/m^2) is impregnated with this impregnating solution by dipping. The dipping time is approximately 30 seconds. The impregnation can also be effected by coating, application by a doctor knife, or other conventional methods. Thereafter the impregnated paper is dried in a furnace at temperatures of about 170°C and preliminarily condensed.

At the outlet of the impregnating machine, the material is cut into sheets. The sheets carry a resin application of 120 parts of solid resin substance on 100 parts of paper substance. In order to obtain desired laminate properties this resin application may vary from 30 to 300 parts of solid resin substance on 100 parts of paper substance preferably from 40 to 200. Respectively six of these sheets are combined and heated under a heating press for 60 minutes under a pressure of 100 kp/cm^2 to 165°C. The laminates have a nominal thickness of 1.5 mm.

The properties of these laminates are listed in the following table. Examples 1 and 3 are laminates according to this process, while Example 2 is a comparison product without the addition of hexamethylenetetramine. This product when compared, primarily with Example 1, demonstrates that the afterburning time could be considerably reduced by adding hexamethylenetetramine, with the electrical and mechanical characteristics remaining almost the same.

Comparative Example 2 shows furthermore that the mode of operation of this process results in a considerably higher impregnating efficiency, and that the proportion in solids of deleterious substances is considerably lower. This amount of deleterious substances is the quantity of solid particles emitted during impregnation or drying into the exhaust air. The process as conducted in accordance with this description thus is also considerably less polluting or requires smaller devices for collecting and/or subsequently combusting the damaging substances.

Example 4 is likewise a comparative example, where the product was prepared with the use of a mixture of phenol-resol resin and cresol-resol resin according to the conventional double impregnating method. In spite of its high content of flameproofing agents, this product does not fulfill the conditions of flameproofing class V1.

If, for reasons of facilitating the punching of the product, a plasticizer content is employed that is 15 to 25% by weight, and yet the burning classification V0 is desired in the final product, this can be accomplished by a minor addition of known flameproofing agents. Example 3 serves to demonstrate this situation.

		. . . . Example Number	. . . .		
	Standard	1	2	3	4
Burn test classification	UL Subject 94	94 V0	94 V1	94 V0	94 HB
Afterburning time, sec	UL Subject 94	1-8	5-22	2-9	10-50
Insulation resistance, $10^{10}\ \Omega$	DIN 7735	27	25	24	28
Water absorption, mg	DIN 7735	23	24	25	24
Flexural strength transverse, kp/cm^2	DIN 7735	1,350	1,350	1,420	1,300
Punchability 60°C, symbol	DIN 53 488	2.0	2.0	2.5	2.0
Pentabromodiphenyl ether, pbw		0	0	5	10
Hexamethylenetetramine, pbw		10	0	10	0
Phenol-resol resin, pbw		60	70	65	24
Cresol-resol resin, pbw		0	0	0	46

(continued)

		Example Number			
	Standard	1	2	3	4
Tricresyl phosphate, pbw		30	30	20	30
Methanol, pbw		5	5	5	100
Water, pbw		25	25	25	0
Soda kraft paper (180 g/cm^2), pbw		80	80	80	80
Impregnating efficiency, kg/hr		180	100	180	100
Deleterious substance, quantity in solid particles, kg/hr		6	10	6	10
Deleterious substance, quantity in the solvent, kg/hr		5	5	5	100
Curing time, min		60	75	60	80

Chloral Hydrate and Sodium Bromide for Urea-Formaldehyde Foam

M. Widmann; U.S. Patent 3,979,341; September 7, 1976; assigned to Borden Products Limited, Canada describes a flame retardant foaming system which includes a foaming agent solution and a resin solution. The foaming agent solution includes from 1.6 to 4.8 mmol of dibutyl naphthalene sulfonic acid as surfactant, from 5 to 20 mmol of phosphoric acid as catalyst and foam intumescent agent, from 0 to 45.5 mmol of resorcinol as foam anticollapsing agent, from 9.1 to 54.6 mmol of chloral hydrate as fire retardant agent and sufficient water to make a solution of about 5% solids concentration.

The resin solution includes urea-formaldehyde resin produced by reacting from 1.0 to 2.4 mols of formaldehyde per mol of urea, from 6.5 to 500 mmol polyethylene glycol as foam charring component, from 0.3 to 1.2 mmol of calcium chloride as foam uniform cell size component, from 3 to 310 mmol of dicyandiamide as foam hydrolysis resistance component, from 2 to 30 mmol of sodium bromide as foam aging component, from 0 to 150 mmol of furfuryl alcohol as formaldehyde odor masking component and sufficient water to prepare a solution of 40 to 45% solids concentration. One part by weight of the foaming agent solution is frothed and then from 0.9 to 1.5 parts by weight of the resin solution is sprayed on the froth to obtain foam.

Being a two-component system, a specially designed gun is used to produce the foam by foaming the foaming agent in a foaming chamber with the aid of pressurized air and then uniformly spraying the foamed agent with the aqueous solution of urea-formaldehyde resin. Since this is an air froth system, there is no noticeable exothermic reaction and no postexpansion of the foam after it is injected into a cavity. Because it is initially a liquid foam, it will entirely fill any shaped cavity.

This foam has been used with excellent results as thermal insulation between precast concrete cladding or marble curtain wall and the supporting structures of multistory complexes or any similar porous cavity constructions. The ease of application permits its use as a sound absorber in existing walls to effectively reduce sound transmission. It has also been used to provide both sound insulation and fire protection for laundry and garbage incinerator chutes.

Chloral hydrate is a fire retardant. To achieve instantaneous fire retardancy in the foam, sodium bromide, or any other water-soluble alkali metal halide, is used. In the absence of sodium bromide, the foam must be aged for a period of one week to several months before it achieves the required fire retardancy.

Example 1: Preparation of Chloral Hydrate – Chloral hydrate is prepared by loading 50 parts by weight of trichloroacetaldehyde (chloral) into a reactor provided with cooling coils. The cooling water is turned on and water is slowly added to the reactor at a rate to reach but not exceed 70°C. A great volume of offensive fumes is evolved during the reaction. A total of 50 parts by weight of water is added to the reactor. The exotherm diminishes after about 20% of the water is added, at which time the remainder of the water is added faster. After addition of water is completed, contents of the reactor are cooled to room temperature and stored in a stainless steel tank.

Example 2: Preparation of the Foaming Agent Solution –

	Parts by Weight
Resorcinol	1.5
Water	89.3
Chloral hydrate (Example 1)	3.7
BASF foaming agent 514	5.5

The BASF foaming agent 514 includes 1.1% of dibutyl naphthalene sulfonic acid, 1.6% phosphoric acid and 2.5% water, all amounts being based on the weight of the foaming agent solution.

Resorcinol is dissolved in water and chloral hydrate is added to the solution. A check must be made at this point to make sure that all components have been dissolved. BASF foaming agent is added and the solution is agitated very carefully to avoid foaming. The foaming agent solution is loaded into a drum with fused polyethylene liner.

Gardner color of the foaming agent solution is 4 to 5, its pH is 1 to 1.5 and its shelf life is 1 year.

Example 3: Preparation of the Urea-Formaldehyde Resin –

	Parts by Weight
(1) Formaldehyde, 50.0% solution	34.569
(2) Water	16.374
(3) Triethanolamine	0.076
(4) Aqua ammonia, 28% (26°Be) solution	0.525
(5) Urea prills	16.706
(6) Formic acid, 30% solution	0.027
(7) Polyethylene glycol 200 EX (Dow Chemical)	4.640
(8) Formic acid, 30% solution	0.066
(9) Caustic soda, 50% solution	0.046
(10) Urea prills	6.669
(11) Water	11.668
(12) Furfuryl alcohol	1.061
(13) Water	6.497
(14) Dicyandiamide	0.50
(15) Sodium bromide	0.070
(16) Calcium chloride, anhydrous	0.033
(17) Water	0.464
(18) Caustic soda, 50% solution	0.009

Formaldehyde (1) is loaded into a thoroughly cleaned reactor followed by water (2), triethanolamine (3). After the heat has been turned on and solution temperature reaches 60°C, aqua ammonia (4) and urea prills (5) are added. This will cause the temperature to drop off temporarily. The solution is heated to reflux and held at a mild reflux for 10 minutes following which the heat is turned off. Formic acid (6) and polyethylene glycol (7) are added and the pH of the solution is adjusted to about 4.6 or 4.7 with formic acid (8).

Adjustment of pH at this point must be made slowly because too much acid added at once will cause a violent exotherm, thus condensation will proceed at an uncontrollable rate. The solution is again heated to maintain a mild reflux at about 95° to 98°C.

Dilutability of the resin is checked every 15 minutes until 100% is reached and then it is checked more frequently until dilutability of 280 to 350% is reached. Dilutability is determined by placing 2 cc of the resin at 25°C into a graduated cylinder and water at 25°C is slowly added until clouding occurs. One hundred times the amount of water divided by 2 gives percent dilutability.

After the desired dilutability has been achieved, the resin solution is immediately neutralized with caustic soda (9) and pH is adjusted to 7.5 to 7.9 with additional caustic soda, if needed. Urea (10) is added and the resin solution is held for 5 minutes following which, water (11) and furfuryl alcohol (12) are also added and mixed for 5 minutes. After additional water (13) and dicyandiamide (14) are introduced into the solution, cooling of the solution is commenced at such a rate as to reach 25°C in 50 minutes.

If the solution is held for more than 5 minutes following the urea addition, or if it is cooled to 25°C slower than in 50 minutes, a resin is obtained which produces foam with a high shrinkage factor and the resin itself will have a short shelf life. If the solution is held for less than 5 minutes or if it is cooled to 25°C faster than in 50 minutes, the resin will not initially produce light foam but must be aged for a period of about several weeks to produce desirable foam. It appears that during this period, the resin is artificially aged to develop the desired properties.

Sodium bromide (15) is dissolved in the calcium chloride solution (16 and 17) and this solution is added to the batch. Stirring is continued for 30 minutes, pH of the finished resin solution is adjusted to the range of 7.4 to 7.7 with caustic soda (18) and the resin solution is discharged into a drum lined with a fused polyethylene liner.

The calcium content of the resin solution has to be adjusted to the desired value after determining hardness of water used for its preparation. Hardness determination of water is made by preparing a fresh solution of 60.0 g of 0.01 molar disodium ethylenediaminetetraacetate solution, 10.0 g of a buffer for water hardness solution (ammonium chloride-hydroxide buffer with magnesium ethylenediaminetetraacetate complex) and about 0.2 g of an indicator which is a 1:1 mixture of Eriochrome black T and sodium chloride.

By means of a graduated 10 cc pipette, 7 cc of this solution is added to 20.0 g of the resin solution and thoroughly mixed. If the mixture is of a blue color, more calcium chloride must be added. If the mixture is red, the maximum calcium level has been exceeded. Solids content of the resin solution is 40 to 45%, specific gravity is 1.148 to 1.150, viscosity is 10 to 20 cp (1/60) and calcium ion content is 115 to 125 ppm.

After polyethylene glycol (7) has been added, condensation should be completed in about 60 minutes. If more time is required to complete condensation, the resulting resin will produce foam which will exhibit high shrinkage. Storage temperature of the resin solution should be above 12°C.

Example 4: Preparation of Foam – Foaming agent solution is introduced into a foaming chamber through a nozzle which is 1.2 mm in diameter. The foaming chamber is 6.7 cm long by 13 mm ID and contains a fine mesh screen of 100 mesh, disposed at the bottom of the foaming chamber, which serves the purpose of mixing the foaming agent with pressurized air. Air with the other components is also introduced into the foaming chamber and forms a froth with the other components. The froth is conveyed out of the foaming chamber into a coupling and then into a mixing chamber in a hose. The coupling is 2 cm long by 7 mm ID and the hose is about 1.7 cm ID and about 6 feet long.

The resin solution is sprayed onto the froth in the mixing chamber through a 1.2 mm spray nozzle to produce foam which is stable, flame retardant, has low shrinkage, long life expectancy and exceptional thermal and acoustical insulation properties. A special adapter is used to produce a resin solution spray in the form of a solid cone. As the foam is conveyed through the hose, additional mixing takes place to ensure a homogeneous foam composition with uniform air cells.

Aminoformaldehyde-Furfuryl Alcohol Foams

P.L. Wolff and J.A.G. Gent; U.S. Patent 4,016,111; April 5, 1977; assigned to Koninklijke Emballage Industrie Van Leer BV, Netherlands describe a method of preparing foamed solid aminoformaldehyde-furfuryl alcohol products by polymerizing a liquid furfuryl alcohol containing aminoformaldehyde resin composition in the presence of an acid.

As a starting material, a resin composition containing 15 to 90% by weight of total free and bound furfuryl alcohol and having an urea-formaldehyde molar ratio of 1:1 to 1:5 and at least 10 parts by weight of a boron compound per 100 parts by weight of the resin is used. The boron compound is present in an amount sufficient so that the foam product will pass the muffle furnace test at 500°C, and has no second exotherm at 200°C. The boron compound can be selected from the group consisting of boric acid, boron oxide, borax, boric acid esters, boron tribromide, boron trichloride, boron nitride, boron phosphate, boron trifluoride etherate, boron trifluoride methanol, boron trifluoride monoethylamine, complexes of boric acid with polyhydroxyl compounds, and mixtures of these. The foamed products find utility as thermal insulators.

Example 1: A series of foams was prepared using the formulations (pbw) stated in Table 1. The foams were tested according to NEN Test No. 1076, part C. The test results are stated as the time elapsed before a flash-over occurs at an energy input of 0.4 cal/cm² sec. The requirement for a class 1 rating in the test is that this time should exceed 15 minutes. In addition, results of muffle furnace tests and second exotherm tests are stated.

TABLE 1

Foam No.	1.1	1.2	1.3	1.4	1.5
UFF resin 1)	60	60	60	100	100
Furan resin 2)	40	40	40		
Silicone surfactant (DC 193)	1	1	0.8	1	1
Boric acid	30	30	25	35	35
Borax	—	—	—	2	3
Blowing agent (Freon 113)	6	6	6	6	1
Fyrepruffe 3)			25		
Catalyst:					
Orthophosphoric acid, 89%	9.1	7.2	—	9.7	9.4
Polyphosphoric acid	—	—	10	—	—
Isopropanol	—	—	6	—	—
1,4-Butanediol	2.6	—	—	—	—
Methanol	—	2	—	—	—
Water	2.6	2.0	—	4	5.2
Glucose	9.7	10.2	--	10.7	10.4
Muffle furnace test 500° C, flash +/–	+	+	+	—	—
2nd exotherm at 200° C	450	227	442	206	206
NEN 1076 part C test, time to flash-over (min., sec.) at 0.40 cal/cm² sec.	8.40	13.50 and 15.30	12.35	>18	>25
at 0.5 cal/cm² sec.				>15	
at 0.6 cal/cm² sec.				>15	

1) Borden TN1
2) Synphorm P 490
3) A commercial fire retardant containing ammonium phosphate and ammonium sulphate.

Foam 1.1 does not pass the muffle furnace test nor the NEN 1076 test although it contains 30 parts boric acid. This is consistent with the high amount of furan resin used in the preparation of the foam, which makes it difficult to obtain a complete cure, as is evident from the high temperature reached in the second exotherm test.

Foam 1.2 is made with another catalyst, and the second exotherm is now much lower. This foam shows a very brief flash in the muffle furnace test and is almost passing the NEN test.

Foam 1.3 shows that conventional fire retardant cannot produce the desired effect.

Finally foams 1.4 and 1.5 show the superior performance in the NEN 1076 test of urea-formaldehyde-furfuryl alcohol (UFF) foams to which sufficient boric acid has been added to make them pass the muffle furnace test.

Example 2: Foamed materials were prepared using the formulations (pbw) given in Table 2. The resin blend was pumped to a mixing chamber in which it was intimately mixed with the catalyst. The mixture was poured into a mold and allowed to expand. The foams were postcured at 70°C for 15 minutes (foam 2.1), alternatively at 60°C for 2 hours (foams 2.2-2.5). Samples were cut and tested in the second exotherm test and the muffle furnace test as described above. Finally, the foams were tested in the flash-over test NEN 1076, part C. Test results are given in Table 3.

TABLE 2

Foam No.	2.1	2.2	2.3	2.4	2.5
Borden TM1 a)	57.7	70	70		
Borden FRD 511 a)		30	30	100	
Borden FRD 530 a)					100
Furan resin	34.0				
Furfurylalcohol	8.3				
Glucose	40				
Boric acid		20	10		20
DC 193 b)	0.2				1
Pluronic f 88 c)		1	1		
Dowfax a 11 d)				4	
Freon 113 e)	2.5	6	6		4
Freon 11 e)		4	4		5
Dimethoxymethane	1.25				
Dichloromethane	1.25				
Pentane				10	
Catalyst:					
Orthophosphoric acid, 45 pct		6	6		5
Tetraphosphoric acid	7.8				
Isopropanol	5.8				
Ammonium chloride, 25 pct				4	

a) commercial UFF resins available from Borden Chemical Co., UK.
b) silicone surfactant from Dow.
c) surfactant from BASF, Wyandotte.
d) surfactant from Dow.
e) halogenated hydrocarbons from du Pont.

TABLE 3

Foam No.	2.1	2.2	2.3	2.4	2.5
Muffle furnace test at					
550° C flash/burn, sec.		16–32	8–118	6–87	10–121
T max. in ° C		570	595	690	600
515°C flash/burn, sec.		—	22–120	5–183	14–123
T max. in ° C		530	585	690	575
508° C flash/burn, sec.	flash + burn	—	89–132	19–157	—
T max. in ° C		535	565	630	560
495° C flash/burn, sec.			99–131		
T max. in ° C		570			
2nd exotherm at 200° C max. temp.	435	200	210		

(continued)

TABLE 3: (continued)

Foam No.	2.1	2.2	2.3	2.4	2.5
NEN 1076 performance	class 2	class 1	class 2	class 2	class 1
Time to flash over at					
0.2 cal/cm² sec.				more than 16 min.	
0.4 cal/cm² sec.	6 min. 45 sec.	more than 16 min.	7 min. 25 sec.	1 min.	more than 16 min.

Foam 2.1 contains no boron compound and fails both in muffle furnace test and in the NEN 1076 test.

Foam 2.2 contains 20 parts of boric acid and passes the muffle furnace test both at 525°C and at 508°C and hence also the NEN 1076 test with a very high performance.

Foam 2.3 contains 10 parts of boric acid which with this resin composition gives failure both in the muffle furnace and in the NEN 1076 test.

From the combined results of these two foams, it is concluded that only a slight increase in boric acid is required to bring foam 2.3 to pass the tests.

Foam 2.4 is made from a resin composition free of furan resin, but even then the fire performance in the absence of a boron compound is very poor.

Foam 2.5 is made from another urea-formaldehyde furfuryl alcohol resin and contains 20 parts boric acid. It passes the muffle furnace test at 508°C and as expected also the NEN 1076 test.

It appears from the results both in Example 1 and Example 2, that there exists a good correlation between the muffle furnace test and the NEN 1076 test.

Cured Resol Foams Employing Premixed Catalyst Combinations

R.L. Greer; U.S. Patent 4,075,139; February 21, 1978; assigned to The Upjohn Company describes a process for the preparation of a phenolic foam from the reaction of a phenol-aldehyde resole resin, a blowing agent and an acidic catalyst. This comprises employing a premixed catalyst combination as follows: (a) an aromatic sulfonic acid; and (b) a phosphonate compound having the formula

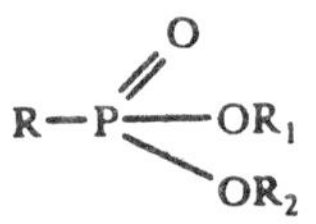

where R, R_1 and R_2 are independently selected from the class consisting of alkyl having from 1 to 8 carbon atoms, aryl having from 6 to 12 carbon atoms, and aralkyl having from 7 to 13 carbon atoms.

The phenolic foams are utilized in the production of preformed thermal insulation panels, such as roof and wall insulation, and particularly as the core insulant in prefabricated portable booths, kiosks, and the like. Further utility for the foams is found as sound insulation to deaden noise transmission through ceilings, floors or multistory buildings, and various acoustical panels. The foams are also used in the floral trade in plant potting, as fire barriers in shipping and storage of pyrotechnic devices, as electric battery separators, and as decorative coatings in the building trade.

Example: The phenolic foams A, B, and C of this example were prepared by blending the reactants and proportions described in the following table at ambient room temperature (about 20°C). The blending procedure consisted of first mixing the resole resin (BRL-2760) with the surfactant (L-5320) using a high speed drill press motor equipped with an efficient agitator blade. The Freon blowing agents were blended into the mixture followed by the boric acid antipunking agent, ethylene glycol, and benzyltrimethylammonium chloride, where they were called for.

The premixed catalyst combinations were prepared by mixing in a separate container first the dimethyl methylphosphonate (DMMP) with the surfactant (TMN-10) followed by the mixed sulfonic acid combination of p-toluenesulfonic acid and xylenesulfonic acid in equal proportions by weight which caused an immediate and substantial exotherm accompanied by a gradual increase in viscosity. Stirring of catalyst ingredients was continued by means of a suitable stirrer for about 15 minutes. The amine polyurethane catalyst Niax ES was added last where it was called for.

The premixed catalyst combination was then thoroughly stirred into the resin blend for about 30 to 60 seconds. A sufficient amount of the ingredients was employed in each foam preparation so that, upon completion of stirring in the catalyst, approximately 2,000 g of each reaction mixture was poured into a 12 inch x 12 inch x 12 inch cardboard box which had been preheated to about 140°F. Each sample was allowed to exotherm and rise freely to form a solid block of foam about a cubic foot in volume. Postrise curing conditions for the foams are described in the following table. The foam blocks did not split or crack.

Foam B contained the antipunking agent boric acid. Both Foam A and Foam C, the latter which also lacked the additional agents including the antipunking agent, were characterized by the same absence of punking as Foam B and excellent flame resistance including the lack of any smoke formation. Foam C was also characterized by a FSR = 25 in a test which is comparable to the ASTM E-84 Tunnel Test.

Flammability characteristics are determined using the following procedures.
(a) Flame resistance (FR) is determined qualitatively by positioning the center of a 4" x 4" x 1" piece of the foam sample in the blue oxidizing flame of a Meeker burner about 2" above the top of the burner and noting the burning characteristics of the sample while at the same time observing the smoke which is being generated.

(b) The test for punking is performed by holding a 6" x 6" x 2" sample of the foam above the same blue flame described above and about 2" above the burner top so that the flame impinges on the middle of one 6" face of the sample for about 3 minutes. The sample is then removed from the flame and its burned side superimposed immediately on the charred face of another similarly sized sample of the same foam which had been allowed to cool. The test is considered positive for punking when the heat radiation from the hot piece of foam causes both pieces to ignite or glow.

(c) The 4 ft Tunnel Test is a minaturized adaptation of the ASTM E-84 Tunnel Test and the FSR values obtained from a calibrated 4 ft tunnel are considered comparable to those obtained in the ASTM E-84 test.

	Foam A	Foam B	Foam C
Resin ingredients (parts by weight)			
BRL-2760[1]	100	100	100
L-5320[2]	4	4	4
Freon-11[3]	6	6	6
Freon-113[4]	6	6	6
Boric acid	–	15	–
Ethylene glycol	5	5	–
Benzyltrimethylammonium chloride solution[5]	3	3	–
Catalyst (parts by weight)			
Catalyst I[6]	20	–	–
Catalyst II[7]	–	20	–
Catalyst III[8]	–	–	20
Foam rise characteristics (min:sec)			
Cream	1:30	<18:00	2:00
Gel	13:30	–	–
Rise	16:00	<30:00	17:00
Foam curing conditions	15 min/130°-140°F then stand overnight at room temperature	1 hr/150°F	1 hr/110°F
Foam properties			
Core density (pcf)	1.61	2.38	2.29
Closed cells (%)	–	4.0	3.6
Flammability characteristics			
Flame resistance (FR)	Excellent	Excellent	Excellent
Smoke formation	None	None	None
Punking	None	None	None
4′ Tunnel, flame spread rating (FSR)	38	–	25

[1] Proprietary liquid phenol-formaldehyde resole resin (Union Carbide Corp.); viscosity is 2,000 to 6,500 cp at 25°C, pH is 6.0 to 6.3.

[2] Proprietary silicone surfactant (Union Carbide Corp.).

[3] Trichloromonofluoromethane (Union Carbide Corp.).

[4] Trichlorotrifluoroethane (Union Carbide Corp.).

[5] 40% by weight in methanol (Miles Laboratories, Sumner Division).

[6] A premixed catalyst combination comprising the following ingredients in parts by weight: 2 parts dimethyl methylphosphonate (DMMP) (Stauffer Chemical); 2 parts TMN-10, a nonionic surfactant prepared from the reaction of about 10 mols of ethylene oxide with trimethylnonanol (Union Carbide Corp.); 3 parts of a combination of para-toluenesulfonic acid and xylenesulfonic acid in equal parts by weight (Ultra TX Acid, Witco Chemical Co.); and 1 part of Niax Catalyst ES, a proprietary silicone amine polyurethane catalyst (Union Carbide Corp.).

[7] A premixed catalyst combination comprising the following ingredients in parts by weight: 2 parts DMMP, 2 parts TMN-10, 6 parts of Ultra TX Acid (described above), and 1 part Niax Catalyst ES.

[8] A premixed catalyst combination comprising the following ingredients in parts by weight: 2 parts DMMP, 2 parts TMN-10, and 6 parts Ultra TX Acid (described above).

OTHER THERMOSETS

Isocyanurate Resins

S. Kamiyama, N. Nagai and H. Miyamoto; U.S. Patent 4,087,414; May 2, 1978; assigned to Mitsubishi Gas Chemical Company, Inc., Japan describe a process for producing a flame retardant, thermosetting isocyanurate resin by polymerizing a polyfunctional isocyanate compound with a polyfunctional epoxy compound at a ratio of 0.05 to 0.2 equivalent of the epoxy to 1 of the isocyanurate compound at a pressure of 1 to 800 kg/cm^2 (gauge) at a temperature of 100° to 250°C in the presence of an organic isocyanate polymerization catalyst.

Polyfunctional isocyanate compounds utilized are xylylene diisocyanate, tolylene diisocyanate, naphthylene diisocyanate, diphenylmethane diisocyanate, polymethylene polyphenyl isocyanate, and a combination of an aliphatic polyfunctional isocyanate compound such as hexamethylene diisocyanate and the like and a polyfunctional isocyanate compound containing an aromatic nucleus.

Preferable polyfunctional epoxy compounds are diglycidyl ether of butanediol, diglycidyl ether of polyethylene glycol, diglycidyl ether of isopropylidenediphenol, polyglycidyl ether of phenol-formaldehyde novolac, polyglycidyl ether of cresol-formaldehyde novolac, and diglycidyl ester of linoleic dimer acid.

The organic isocyanate polymerization catalysts used in this process may be a conventional organic isocyanate trimerization catalyst, for example, metal salts of carboxylic acids such as sodium acetate, zinc octylate, etc., alkali metal or alkaline earth metal alkoxides such as sodium methoxide and the like, alkali metal or alkaline earth metal phenoxides such as potassium phenoxide, tertiary amines such as triethylamine, triethylenediamine, dimethylaminomethylphenol, methylmorpholine, pyridine, etc.

The isocyanurate resins produced are infusible by heat and insoluble in solvents, and have excellent flame retardancy, heat resistance, heat stability, chemical resistance, adhesivity, mechanical properties and electrical properties. In particular, two of the outstanding features of the isocyanurate resins are their excellent flame retardancy as well as excellent dielectric properties. The isocyanurate resins may be used for producing compression molding articles, casting articles, and laminating articles and adhesives.

The flame-retarding effects of the resin were evaluated according to the UL Standard 94 of the United States, using test pieces of 3.2 mm thickness. Material classed 94V-0 and 94V-1 shall not have any specimens which burn with flaming combustion for more than 10 seconds and 30 seconds, respectively, after each application of the test flame, and shall not have a total flaming combustion time exceeding 50 seconds and 250 seconds, respectively, for the 10 flame applications for each set of 5 specimens.

Example 1: Diphenylmethane diisocyanate (Millionate MR, the isocyanato equivlent, 136; viscosity, 200 cp/25°C) and diglycidyl ether of bisphenol A (Epikote 828, epoxy equivalent, 190; viscosity, 15,000 cp/25°C) were mixed with each other in various ratios given in the following table. To 100 parts of the resulting mixture, 1.0 part of N-methyl morpholine was added, and the mixture was heated at 70°C and preliminary polymerization was carried out to obtain a viscous prepolymer having an isocyanato conversion of 30 to 40%. 200 parts of finely powdered silica was added to the prepolymer, which was kneaded with a roll and finely pulverized to prepare a molding material.

The molding material thus obtained was placed in a mold, and the polymerization was started under conditions such that the temperature was 150°C and the gauge pressure was 500 kg/cm² (gauge), and then the temperature was raised to 200°C over a period of 30 minutes and remained at the same temperature for a further 2 hours to effect the polymerization reaction. The isocyanurate resin thus obtained was brown and opaque. The following table shows the properties of the resins obtained.

. . . . Starting Materials		. . . . .Properties of Resin							
Diphenyl-methane Diisocyanate (parts)	Diglycidyl Ether of Bisphenol A (parts)	Epoxy Equivalent/ Isocyanato Equivalent	Density (g/cm³)	Vicat* SP (°C)	Weight Loss on Heating at 200°C for 100 hr (%)	Burning Test by UL . . .Standard 94 . . . Flaming Combustion Time (sec)	Class of Flammability	Electric Properties (10 kHz) Dielectric Constant (ε)	Dielectric Tangent (tan δ)
93.5	6.5	0.05	1.84	204	0.8	6	94V-0	2.8	0.0027
87.7	12.3	0.10	1.83	240	0.7	14	94V-0	2.8	0.0028
82.7	17.3	0.15	1.82	254	0.6	32	94V-0	2.9	0.0030
78.2	21.8	0.20	1.82	261	0.5	70	94V-1	3.0	0.0032

*The Vicat softening point was measured in accordance with DIN 57302.

Comparison Example 1: The same procedure as in Example 1 was repeated except that diphenylmethane diisocyanate alone was used without using a diglycidyl ether of bisphenol A to obtain isocyanurate resin. The resin had a Vicat softening point of 186°C and a weight loss on heating of 1.1%, and had further a poor heat stability and had gas foams over the whole surface.

Comparison Example 2: The same procedure as in Example 1 was repeated except that the diphenylmethane diisocyanate and diglycidyl ether of bisphenol A were used in amounts of 70.5 parts and 29.5 parts (epoxy equivalent/isocyanato equivalent, 0.30), respectively. The resulting isocyanurate resin had a total flaming combustion time of 160 sec and a maximum flaming combustion time of 42 sec and further it falls in class 94V-2. In addition, the product had a dielectric constant (ε) of 3.2 and a dielectric tangent (tan δ) of 0.0036.

Example 2: 83.4 parts of diphenylmethane diisocyanate (Millionate MR) and 16.6 parts of the polyglycidyl ether of phenol-formaldehyde novolac (Epikote 154, epoxy equivalent, 180; viscosity, 50,000 cp/52°C) were mixed with each other (epoxy equivalent/isocyanato equivalent, 0.15). 0.5 part of N,N-dimethylbenzylamine was added to the mixture, which was heated at 80°C and preliminarily polymerized to obtain a viscous prepolymer having an isocyanato conversion of 38%. The prepolymer was diluted with methyl ethyl ketone and impregnated into glass fiber sheets, which were heated at 120°C to obtain prepregs.

The prepregs thus obtained were piled one upon another and laminated for 2 hours under the conditions such that the temperature was 180°C and the pressure 50 kg/cm² (gauge), and a laminated article having a resin content of 36% was obtained. The properties of the laminated article were tested similarly to Example 1. As the result of the test, the laminated article was found to have a density of 1.80 g/cm^3, a weight loss on heating of 0.6%, a total flaming combustion time of 72 seconds and to be classified to 94V-1 according to UL Standard 94.

Prepreg Resin Laminates

C.L. Wright and H.H. Beacham; U.S. Patent 4,034,136; July 5, 1977; assigned to FMC Corporation describe a process whereby certain polychlorinated and polybrominated aromatic hydrocarbon compounds can be incorporated in thermosetting polymer compositions to make the compositions flame retardant without injuring the desirable physical and electrical properties of the thermosetting compositions when in the thermoset state. This result is achieved by adding at least 1 mol of a polyunsaturated monomer with carbon to carbon unsaturation to the thermosetting composition for each 3 mols of polychlorinated or polybrominated aromatic hydrocarbon added to the thermosetting composition.

These thermosetting compositions generally comprise: (a) 100 parts by weight of a blend of about 10 to 70 parts of a polyunsaturated monomer with carbon to carbon unsaturation, with 90 to 30 parts of a polymer of the class consisting of a polyunsaturated polymer with carbon to carbon unsaturation and polyphenylene ether polymers, and (b) about 5 to 50 parts by weight per 100 parts of (a) of a polychlorinated or polybrominated aromatic hydrocarbon containing at least 50% by weight of combined chlorine or bromine and whose molecular weight is greater than 200, in an amount not in excess of 3 mols of polyhalogenated aromatic hydrocarbon for each mol of the polyunsaturated monomer in (a).

The resin compositions of this process can be employed in the usual process for manufacture of fibrous reinforced thermoset resin laminates using the prepreg technique. A polymerizable liquid, i.e., a monomer such as diallyl ortho-phthalate and the like, a polyhalogenated aromatic hydrocarbon, a polyphenylene ether polymer, and, where desired, a polymer containing carbon to carbon unsaturation such as an allylic prepolymer or a reactive type polyester resin, a catalyst and modifying ingredients such as dyes, pigments, fillers, glass coupling agents, inhibitors and so forth are mixed together and used to impregnate a fibrous nonwoven mat or a woven fabric; where glass mats or fabrics are used, it may be desirable to have the glass treated with a glass coupling agent.

The use of some solvent is usually required in order to reduce the viscosity level of the resin composition to make it suitable for application to the mat or fabric with conventional, commercial saturating or impregnating equipment.

It is not necessary to dissolve the polyphenylene ether polymer in the impregnating compositions. Simple uniform dispersion of the polyphenylene ether polymer powder in the solvent-monomer-polymerizable resin mixtures suffices. Prepregs are generally most economically processed with 30 to 60 parts of the resin composition dispersed in 70 to 40 parts of a suitable solvent such as acetone, methyl ethyl ketone, methyl isobutyl ketone, toluene, xylene, chloroform, methylene chloride, trichloroethylene, perchloroethylene and mixtures of these and other solvents known to be useful in preparing allylic prepregs.

The mat or fabric is impregnated with the solvent solution and then dried to remove the solvent. After impregnation and drying of the impregnated fabric, the laminate is laid up and cured with heat and mild pressure using cure cycles and conditions similar to those used in curing the wet lay-up type laminates. Roving, including glass roving, is similarly preimpregnated for processing by filament winding techniques into pipe, other cylindrical shapes and hollow tapered and conical shapes. Products made by filament winding are generally cured at about 150°C in 60 minutes. The fiber content of the prepreg laminates varies from about 15 to 40% by weight for low density fibers and up to 55 to 75% of the total weight of the cured laminate for glass mat or glass fabric laminates. The fiber content of filament wound constructions such as pipe, when made from impregnated glass roving, is generally 70 to 80% of the total weight of the cured product.

Reinforced laminates of fibrous materials such as glass cloth, glass mats, synthetic fiber, cloth mats, paper and the like can be copper-clad to produce copper-clad laminates with excellent electrical properties to be used in preparing printed circuits. Copper-clad laminates can be prepared by coating copper foil with a composition of this process containing a polyphenylene ether polymer and then baking the coated copper foil at 160°C for about 15 minutes.

The baked resin coated foil is then placed on resin impregnated fibrous materials such as glass cloth which has been impregnated with the resin compositions, unsaturated polyester resins, diallylic phthalate prepolymers and the like, and then the laminate is pressed at 50 to 2,000 psi at 100° to 170°C for at least 5 minutes to convert the resinous materials to the thermoset state. If desired, the coated copper foil can be laminated to almost any previously cured base. Dicumyl peroxide is the preferred catalyst for producing the copper-clad laminates. The resulting copper-clad laminate has excellent adhesion of the copper to the base material and has excellent electrical properties.

Those compositions which contain polyphenylene ether resins as a portion of the system are useful as thermosetting coatings. They adhere strongly to a wide variety of substrates involving metals, ceramic bases and other resinous substances while providing excellent electrical insulating characteristics. They may be used as conformal coatings for electronic devices or as primer coats to bond additional electrical insulation to the devices. Many molding compounds used in the encapsulation of electronic components do not form good bonds to the devices and to the wire leads, thereby permitting access of moisture. When such devices or leads are first coated with the compositions of this process, and then subjected to mold encapsulation, hermetically sealed systems are obtained which resist moisture even on prolonged exposure to boiling in water or high pressure steam.

Example 1: Diallyl phthalate prepolymer, prereacted polychlorinated aromatic compound with monomer, molding compounds were prepared. The prereacted polychlorinated aromatic hydrocarbon and monomer reaction products were prepared in 3½ lb batches. Chlorinated biphenyls, containing 68% combined chlorine (Aroclor 1268), and a diallylic phthalate monomer were heated together in a ratio of 62 parts of chlorinated biphenyls to 38 parts of diallylic phthalate monomer with 2 parts by weight of dicumyl peroxide per 100 parts of Aroclor plus monomer.

The reaction products were made in a 1 gallon sigma blade mixer equipped with a steam jacket. The steam to the jacket was at 160°C and the internal temperature of the blade and side walls was about 120°C. Although accurate reaction time measurements could not be made, reasonably short reaction times of about 10 minutes were observed in preparing these reaction products. The reactants were highly fluid up to the point of gelation, from which point the reaction proceeded rapidly to a brittle stage, breaking up thereafter into small lumps.

The prereacted materials were ball milled for a 2 hour period in a 1½ gallon jar mill to produce 1,000 g of ground product. Reaction products were prepared using diallyl phthalate and diallyl isophthalate monomer and are designated Reaction Product No. 1 and No. 2 respectively in Table 1. The reaction products were mixed with diallyl phthalate prepolymer, diallyl isophthalate prepolymer, some catalyst, monomer and fillers to make molding compounds. The composition details and physical properties of samples molded from the molding compounds are found in Table 1. For comparison, the table includes qualifying specifications for diallyl phthalate molding compounds for military use under MIL M-14F and MIL-19833.

TABLE 1: MOLDING COMPOUNDS

	Compound No. 1	Compound No. 2	Mil M-14F Specification Requirements
Formulation (parts by weight)			
Diallyl phthalate prepolymer (Dapon 35)	74	—	—
Diallyl isophthalate prepolymer (Dapon M)	—	70	—
Diallyl phthalate monomer	3	—	—
Diallyl isophthalate monomer	—	3	—
Reaction product No. 1*	26	—	—
Reaction product No. 2**	—	30	—
tert-Butyl perbenzoate	3	3	—
Calcium stearate	2	2	—
Antimony oxide	15	18	—
Wollastonite	30	20	—
Glass (¼" strands)	90	100	—
Properties			
Ignition time (sec)	93	98	90***
Burning time (sec)	78	76	90†
Heat deflection temperature (°C)	175	270	160***
Izod (ft lb) impact	0.66	0.60	0.30***
Compressive strength (psi)	24,620	26,500	16,000***
Flexural strength (psi)	11,462	11,200	9,000***
Modulus of elasticity (psi)	1.70×10^6	1.72×10^6	—
Water absorption (%)	0.12	0.11	0.40†
Specific gravity	1.902	1.897	—
DC $10^3/10^6$††	4.50/4.39	4.38/4.24	4.6/4.4†
DC $10^3/10^6$ (wet)	4.52/4.38	4.41/4.29	4.7/4.5†
DF $10^3/10^6$†††	0.005/0.005	0.005/0.005	0.009/0.015†
DF $10^3/10^6$ (wet)	0.006/0.005	0.006/0.005	0.013/0.017†
Volume resist. (ohm), 720 hr at 70°C, 100% RH	1.0×10^{12}	0.8×10^{12}	5×10^9***

*62/38 reaction product of chlorinated biphenyl with diallyl phthalate monomer.
**62/38 reaction product of chlorinated biphenyl with diallyl isophthalate monomer.
***Minimum.
†Maximum.
††Dielectric constant.
†††Dissipation factor.

Example 2: A portion of the diallyl phthalate monomer-chlorinated biphenyl reaction product, Reaction Product No. 1 of Example 1, was incorporated into a diallyl phthalate polyester molding compound. Test specimens were molded from the compound for 3 minutes at 150°C at 2,000 psi. The composition of the molding compound, the physical and electrical properties of the molded test specimens and the MIL-M-14F specification requirements for MAI 30 diallyl phthalate polyester molding compounds are shown in Table 2.

TABLE 2: MAI-30 (DAP-POLYESTER) COMPOUNDS

		MIL M-14F Specification Requirements
Formulation (parts by weight)		
Diallyl phthalate monomer	35	–
Uncut isophthalate polyester resin, reactive type	35*	–
Reaction product No. 1	30**	–
Glass (¼" strands)	57	–
Hydrated alumina	48	–
Wollastonite	48	–
Antimony oxide	18	–
Calcium stearate	2	–
tert-Butyl perbenzoate	2	–
Physical properties		
Compressive strength (psi)	21,200	20,000***
DC $10^3/10^6$	4.95/4.54	6.3/6.2†
DC $10^3/10^6$ (wet)	5.15/4.63	6.4/6.4†
DF $10^3/10^6$	0.010/0.009	0.015/0.012†
DF $10^3/10^6$ (wet)	0.012/0.011	0.017/0.015†
Flexural strength (psi)	16,700	14,000***
Flexural modulus (psi)	2.0×10^6	–
Heat deflection temperature (°C)	272	200***
Impact (ft lb)	4.37	3.0***
Water absorption (%)	0.17	0.5†
Ignition time (sec)	105	100***
Burning time (sec)	17	75†
Rockwell hardness (M)	108	–
Specific gravity	1.93	–
Volume resistivity (ohm cm)	1.2×10^{15}	–
Surface resistivity (ohm)	2.8×10^{15}	

*Equimolar amounts of maleic anhydride and isophthalic were heated together with a 10% excess of diethylene glycol. Water was removed azeotropically and the heating, at about 200°C, was continued until the acid number of the polyester resin fell below 6. Excess diethylene glycol was removed under vacuum and the polyester resin recovered.

**62/38 Reaction product of chlorinated biphenyl with diallyl phthalate monomer from Example 1.

***Minimum.

†Maximum.

Example 3: Wet lay-up or vacuum bag laminates were prepared using a diallylic phthalate prepolymer, a diallylic phthalate monomer and chlorinated terphenyl (Aroclor 5460). The wet lay-up was prepared by making a liquid blend of the monomer, chlorinated terphenyl and diallylic phthalate resin and catalyst. The liquid polymerizable resin composition was poured onto several layers of woven glass fabric.

After impregnation, the laminates were placed in a vacuum bag and evacuated to remove air from the polymerizable liquid resin mixture. The lay-up was laminated while maintaining the vacuum at pressures of 15 psi or less, as in the very early stages of cure it was very difficult to maintain pressure on the polymerizable liquid. Full cure was obtained after about 15 minutes at 150°C as measured by hot hardness, hot strength, and solvent resistance in boiling acetone. This laminate construction was also cured in a vacuum bag by exposure to a 3 MeV (million electron volt) electron beam in multiple passes of 5 Mrad each until a room temperature Barcol hardness of 60 was developed. The polymerizable liquid details and the physical properties of the cured laminates are shown in Table 3.

TABLE 3: VACUUM BAG LAMINATES

Formulation (parts by weight)		
Diallyl phthalate prepolymer (Dapon 35)	50	–
Diallyl isophthalate prepolymer (Dapon M)	–	50
Diallyl phthalate monomer	20	–
Diallyl isophthalate monomer	–	20
Chlorinated terphenyl (Aroclor 5460)	30	30
Dicumyl peroxide	2	2
Physical properties		
Burning time (ASTM D229, Part 1) (sec)	16	27
Specific gravity	2.104	2.040
Percent resins	28.0	32.7
Rockwell hardness (M)	117	121
Flexural strength at 25°C (psi)	94,000	101,700
Flexural strength at 150°C (psi)	25,410	6,050*
Modulus at 25°C (psi)	4.08×10^6	3.86×10^6
Modulus at 150°C (psi)	2.49×10^6	1.47×10^6 *
Compressive strength (psi)	49,110	48,740
Tensile strength (psi)	63,460	62,110
Shear (psi)	3,710	2,040
Flexural, 2 hr H_2O/100°C (psi)	93,390	88,810
Modulus, 2 hr H_2O/100°C (psi)	4.18×10^6	3.26×10^6
Flexural at 155°C, aged 1 day (psi)	101,900	–
Aged 4 days	97,660	–
Aged 10 days	93,900	–
Flexural at 180°C, aged 1 day (psi)	–	95,400
Aged 4 days	–	98,870
Aged 10 days	–	88,580
DC $10^3/10^6$	4.94/4.85	4.72/4.65
DC $10^3/10^6$ (wet)	5.00/4.84	4.81/4.65
DF $10^3/10^6$	0.0054/0.0049	0.0068/0.0040
DF $10^3/10^6$ (wet)	0.0088/0.0057	0.0091/0.0048
Volume resistivity (ohm cm)	4.17×10^{15}	3.56×10^{15}
Surface resistivity (ohm)	5.08×10^{14}	2.36×10^{14}

*Test run at 260°C.

Example 4: A low pressure laminate containing diallyl isophthalate monomer and prepolymer, a chlorinated terphenyl and polyphenylene ether polymer was prepared in which the following formulation was used to saturate glass cloth from which the laminate was prepared.

Ingredients	Parts by Weight
Polyphenylene ether resin (Grade 691-111)	25
Chlorinated terphenyl containing 60% combined chlorine (Aroclor 5460)	15

(continued)

Ingredients	Parts by Weight
Diallyl isophthalate prepolymer (Dapon M)	30
Diallyl isophthalate monomer	30
Dicumyl peroxide (recrystallized)	3
Acetone	130

The diallyl isophthalate prepolymer and polyphenylene ether polymer were ball-milled together until 70% of the product was -200 mesh (U.S. Standard Sieve Series 1940). The chlorinated terphenyl-diallyl isophthalate monomer and dicumyl peroxide were dissolved in the acetone solvent and the ball-milled powder was added to this solution to make up a saturating mixture.

Woven glass cloth, type 181, was impregnated with this mixture and allowed to dry at least 48 hours at ambient temperature. The dry impregnated cloth was cut into 12 by 12 inch squares, stacked 18 plies deep with the warp yarns parallel. The preimpregnated lay-up was made in a flat bed press for 30 minutes at 80°C at contact pressure, 30 minutes at 120°C at 300 psi and 1 hour at 160°C at 300 psi. The laminated sample had a resin content of 44% and was found to be nonburning in ASTM Flammability Test D-229, Part 1. The following physical properties were obtained on the laminated sample:

Laminate Physical and Electrical Properties

Resin content	44%
Specific gravity	1.752
Rockwell hardness (M)	120
Compressive strength, psi	50,400
Tensile strength, psi	43,700
Flexural strength, psi at 23°C	67,300
Flexural strength, psi at 200°C	13,700
Flexural modulus at 23°C, psi	3.0×10^6
Flexural modulus at 200°C, psi	1.3×10^6
Tensile shear strength, psi	3,530

POLYPROPYLENE AND POLYETHYLENE

PHOSPHORUS-CONTAINING ADDITIVES

Cured Intumescent Flame Retardant

H.W. Bost; U.S. Patent 4,072,643; February 7, 1978; assigned to Phillips Petroleum Company describes a process whereby flame retardant polyolefin compositions are prepared by blending polypropylene, polyethylene and a rubbery ethylene-propylene copolymer, and a cured intumescent flame retardant. These compositions are characterized as having a low brittleness temperature and an acceptable balance of properties.

The cured intumescent flame retardant is a highly crosslinked polymer comprising the following ingredients: phosphorus oxide, pentaerythritol, and melamine. Optionally, a color improver may also be added. Examples of suitable color improvers include 2-ethylhexanol, 4-tetradecanol, 1-octadecanol, triethylene glycol, ethylene glycol, cyclohexanol and the like.

To prepare the cured intumescent flame retardant, the above ingredients are combined in any suitable manner, at a temperature and for a time sufficient to provide a composition which, when incorporated into a thermoplastic resin, will impart flame retardant properties to that resin. Generally, preparation of the flame retardant can be conveniently carried out as follows:

(1) The phosphoric acid and phosphorus pentoxide are mixed, allowing an autogenous increase in temperature resulting from the exothermic reaction. The reaction period can vary broadly from about 5 to 20 minutes.

(2) The pentaerythritol, the color improver, and about 5 to 15% of the melamine are added, with stirring to the mixture produced in (1), at a temperature in the range of 75° to 150°C for about 0.5 to 8 hours, preferably about 110° to 130°C for about 0.5 to 4 hours.

(3) The balance of the melamine is added, with stirring, to the mixture produced in (2). The melamine can be added continuously or intermittently over a period of about 0.1 to 3 hours or more, at a temperature of about 180° to 300°C, preferably about 0.1 to 1.5 hours at about 200° to 250°C.

(4) The reaction mixture obtained from (3) is cured by continued heating, with stirring, at a temperature of about 200° to 300°C for about 1 to 5 hours, preferably 240° to 270°C for about 2 to 3 hours.

(5) The reaction mixture from (4) is cooled, then pulverized to a suitable particle size.

Exclusion of oxygen from the reaction mixture during the curing (4) and the cooling and pulverizing (5) generally results in a further improvement in color quality of the finished flame retardant. Oxygen can be excluded by maintaining an inert atmosphere, such as nitrogen or helium, or by maintaining a reduced pressure over the reaction mixture.

Example 1: Preparation of the Cured Intumescent Flame Retardant (IFR) – Phosphoric acid (0.4 mol) was added in a slow stream to phosphorus pentoxide (1.0 mol) with stirring in a 2-liter flask. A moderate flow of nitrogen was kept over the mixture at all times. A mixture of pentaerythritol (0.6 mol), melamine (0.12 mol) and color improver, was added in small increments to the mixed acid. This mixture was stirred while the temperature was raised to and held at about 120°C in a fluidized sand bath for about 3 hours. Additional melamine (1.08 mols) was added in small portions with stirring as the temperature was raised over a period of about 1 hour to 250°C. The temperature was held at 250° to 265°C for 2 to 3 hours. During this time as crosslinking continued in the product the mixture became very viscous and difficult to stir.

When the sample was adequately cooled, generally after standing overnight, the flame retardant was removed from the flask and pulverized in a Waring blender. The material was screened through an 80-mesh sieve and was then ready for compounding into the polymer blend.

Example 2: Preparation of Polymer Blends – Polypropylene having a melt flow of about 12 g/10 min and a density of about 0.908 g/cc, polyethylene having a melt index of about 18 g/10 min and a density of about 0.955 g/cc, and an amorphous ethylene/propylene copolymer (EPR) having a Mooney viscosity range of about 55 to 70 were blended together on a hot roll mill in the amounts shown in the table below. The hot roll mill temperature ranged from 160° to 185°C, generally 168° to 178°C. The cured intumescent flame retardant was then added to the blend. After the flame retardant had been completely added to the blend, milling was continued for about 4.5 minutes to disperse the flame retardant throughout the polymer blend.

The following tests were conducted: Melt Flow (melt index), ASTM, D 1238-73, condition E for polyethylene, condition F for polypropylene; Flexural Modulus, ASTM, D 790-71; Izod Impact Strength, ASTM, D 256-72a; and Brittleness Temperature, ASTM, D 746-73. The UL-94, Vertical Burning Test and the Oxygen Index Test were performed for flammability.

Polymer Blend Plus Cured IFR

Run No.	Blend Component, wt %. PP	PE	EPR	IFR (php)	UL-94 Rating	LOI (%)	Melt Flow	Brittleness Temperature (°F)	Flexural Modulus (psi x 10^{-3})	Izod Impact Strength, Notched (ft lb)	Other Ingredients (php)
1	42.5	30	27.5	None	Fail	–	3.5	-59	84	5.7	–
2	40	35	25	A 40	V-1	24.3	4.8	-13	197	4.5	–
3	42.5	30	27.5	A 40	V-0	26.2	2.9	7	139	4.3	–
4	42.5	30	27.5	A 40	V-0	27.8	3.6	-13	162	4.7	–
5	42.5	30	27.5	A 42.5	V-0	29.3	3.7	-21	177	5.2	–
6	42.5	30	27.5	A 45	V-0	27.3	3.6	-21	212	5.0	–
7	45	30	25	A 30	V-1	25.7	5.5	18	183	4.1	–
8	45	30	25	A 50	V-0	26.5	4.7	13	204	3.3	–
9	47.5	32.5	20	A 30	V-1	23.5	6.7	31	203	3.2	–
10	42.5	30	27.5	A 38*	V-0	26.5	2.4	11	144	4.2	–
11	42.5	30	27.5	A 40	V-0	26.0	5.3	-13	157	5.0	BCE-6Br 3 php**
12	45	30	25	A 35	V-0	25.4	5.8	25	184	4.2	BCE-6Br 3 php**
13	45	30	25	A 40	V-0	25.4	5.9	21	201	3.8	BCE-6Br 3 php**
14	45	30	25	A 35	V-0	25.4	6.3	33	214	3.4	TPPO 5 php***
15	45	30	25	B 35	V-0	23.5	5.6	21	259	5.5	Sb_2O_3 2.5 php
16	45	30	25	B 40	V-0	26.0	6.1	26	205	5.5	Sb_2O_3 2.5 php
17	45	30	25	B 45	V-0	25.4	5.8	33	233	3.3	Sb_2O_3 2.5 php
18	42.5	30	27.5	B 42.5	V-0	25.0	3.3	33	154	4.5	–
19	42.5	30	27.5	B 42.5	V-0	27.5	3.1	25	142	3.9	–
20	42.5	30	27.5	B 45	V-0	26.5	3.6	34	164	3.9	–
21	42.5	30	27.5	C 45	V-0	27.0	3.7	38	159	3.9	–
22	76.5†	8.5††	15	D 30	Fail	22.4	1.9	>RT	204	1.0	–
23	76.5†	8.5††	15	D 30	Fail	23.3	0.5	>RT	194	2.7	–
24	45†	30††	25	E 30	Fail	27.0	0.7	-24	148	8.5	–

*IFR contains only 0.037 mol of octadecanol.
**BCE-6Br bis(cyclohexenyl)ethylenehexabromide.
***TPPO triphenylphosphine oxide.
†Melt flow is 5.0.
††Melt index is 0.15.

Results of testing of the flame retarded polymer blends are given in the table above. IFR-A contains 0.055 mol octadecanol; IFR-B contains 0.055 mol octadecanol and 0.1 mol triethylene glycol; IFR-C contains 0.1 mol triethylene glycol; IFR-D had no other components; and IFR-E contained 0.053 mol stearic acid.

Examination of the above data shows that the flame retarded compositions of runs 2 through 14 have acceptable balances of properties. In contrast, the polymer blend of run 1, having no flame retardant, has a low flexural modulus, thus upsetting the desired balance of properties and, of course, is quite flammable.

For comparison, the polymer blends of runs 22 through 24 were prepared using polypropylene and polyethylene having melt flows outside the scope of this process. Additionally, the component ratios of the blends of runs 22 and 23 are outside the scope of this process. The flame retarded blends of runs 22 through 24 fail the flammability test and have unacceptable melt flows. The blends of runs 22 and 23 additionally have brittleness temperatures above room temperature and unacceptable impact test values.

In related work, *H.W. Bost; U.S. Patent 4,026,810; May 31, 1977; assigned to Phillips Petroleum Company* describes color improvements of a cured intumescent flame retardant by incorporating a suitable alcohol, carboxylic acid or amine into the composition prior to curing the composition.

Red Phosphorus

C.F. Raley, Jr., U.S. Patents 4,086,192; April 25, 1978; 4,012,343; March 15, 1977; both assigned to The Dow Chemical Company describes an olefin polymer composition containing a halogenated organic moiety, and an amount of elemental phosphorus having a specific gravity greater than 2, sufficient to reduce the propensity of the olefin polymer composition to propagate burning or combustion after the removal of a small scale ignition source.

Elemental phosphorus refers to the colored allotropic forms of elemental phosphorus having specific gravities greater than 2.0. Specifically the term includes red phosphorus having specific gravity of about 2.34, black phosphorus having specific gravity of about 2.70, and violet phosphorus having specific gravity of about 2.36. Also included are mixtures of the red, black and/or violet forms. Red phosphorus which is usually a brownish red powder which is insoluble in water and carbon disulfide and is both nonpoisonous and nonluminous is preferred. Preferred are powders having particle size less than about 200 microns, especially those less than about 150 microns.

Best results are obtained with about 0.5 to about 20 parts by weight of elemental phosphorus per hundred weight parts of olefin polymer, preferably from about 1 to about 10 parts, and an amount of halogenated organic moiety such that the atomic ratio of phosphorus to halogen is from about 4:1 to 1:6.

Compounds having a halogenated organic moiety are chemically inert to the elemental phosphorus at temperatures from ambient up to the temperatures employed in incorporating the phosphorus and the halogenated organic compound into the olefin polymer and in fabricating the polymer to a desired shape. It is further required that the halogenated organic compounds are substantially inert to the olefin polymer at such temperatures. Preferably the halogenated organic compounds have chlorine and/or bromine in concentrations ranging from about 25 to 97 weight percent based on halogenated organic compound, especially from 50 to 95 weight percent.

Examples of halogenated compounds include chlorinated paraffins containing from about 50 to 80 weight percent chlorine, particularly those having chlorine content of about 60 to 75 weight percent and from 18 to 40 carbon atoms in the molecular chain; chlorinated polyethylenes particularly those having chlorine content above 30 weight percent preferably from about 50 to 80 weight percent, and chlorinated castor oil or chlorinated tall oil. The following example is illustrative of an olefin polymer composition.

Example: One hundred parts of granular polyethylene having a melt index of 2 decigrams per minute as determined by ASTM D-1238-65T(E) together with 6.2 parts of elemental red phosphorus having a specific gravity of 2.20 and an average particle size of $<$75 microns, and 10 parts of chlorinated paraffin wax containing about 70% chlorine (P:Cl~1:1) are compounded thoroughly on a hot roll mill at temperatures about 140° to 150°C, and then reintroduced, such that the entire mass is plastified and apparently homogeneous. The resultant mixture is cooled, ground into small particles and fed into a plastic extruder at a rate corresponding to 7 pounds of the mixture per hour. The plastic extruder has a barrel enclosing a 1¼ inch diameter screw equipped with a mixing head similar in design to that described in U.S. Patent 2,453,088.

A sealing plate is positioned midway of the screw and an inlet to the barrel is provided adjacent to the sealing plate for feeding volatile organic blowing agent into the barrel and into contact with the polyethylene. The sealing plate forms a constricted passageway between the rim of the plate and the bore of the barrel so that flow of the heat-plastified polyethylene through the constricted passageway by pressure of the flights of the screw forms an effective plastic seal against countercurrent flow or leakage of the volatile organic compound from the extruder.

The heat-plastified polyethylene formation is forwarded under pressure of the screw around the sealing plate and into the second section of the barrel of the extruder where it is mixed with 1,2-dichlorotetrafluoroethane as a foaming agent fed at a rate of 1.6 lb/hr. The resulting mixture is blended under pressure, e.g., 700 psig, principally by action of the mixing head on the screw of the extruder, into a uniform gel composition and is brought to a temperature of 190°C, passed through a cooling section to bring about a gel temperature of 100° to 110°C, and is then forced through a discharge orifice of ¼ x ⅛ inch cross-section into the atmosphere. The extruded material is allowed to expand freely in the air.

The polyethylene foam product is a cellular oval rod having a cross-section of 21 x 9 mm and is composed of substantially uniform fine individually-closed thin-walled cells. The foam has a density of about 2.0 to 3.0 lb/cu ft and physical properties, e.g., impact strength and tensile strength, comparable to those of polyethylene foam containing no additive.

The foamed product is tested for its burning characteristics by the following procedure: a sample (21 mm x 9 mm x 23 cm) of the foamed product is inclined at a 30° angle below horizontal while in a draft-free enclosure; a gas flame approximately 1 inch high is applied to the lower end of the sample for 5 seconds which causes ignition; the flame is then removed and the time for the ignited sample to extinguish itself is measured. There are up to 30 ignitions, up to 10 for each of 3 samples depending on burning time. The times required for the foam samples to extinguish after removal of the flame (average extinction times) are in the range of 1 to 4 seconds.

For purposes of comparison, polyethylene foam is prepared and tested under similar conditions except for using 8 parts per 100 parts of polymer of antimony trioxide and 8 parts of chlorinated paraffin wax containing about 70% by weight of chlorine. The foam obtained by this procedure extinguishes after 11.2 seconds.

Bis[Bis(Carbamoylalkyl or Carboxyalkyl) Phosphonylmethyl] Alkylbenzenes

J.A. Hoffmann; U.S. Patents 4,009,207; February 22, 1977; 3,976,685; Aug. 24, 1976; both assigned to American Cyanamid Company describes compounds having the formula:

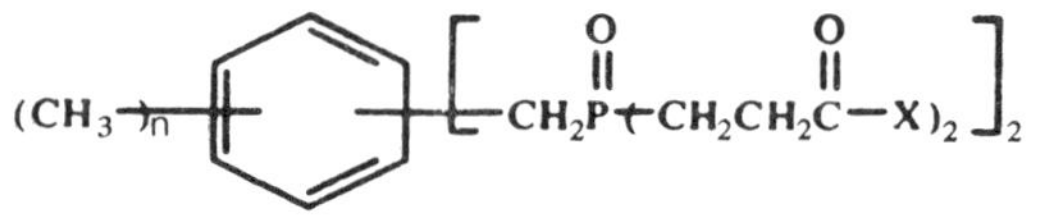

wherein X is OH or NH_2 and n is a whole, positive integer of from 1 to 4, inclusive. These compounds have been found to be effective in flameproofing polyolefins, particularly polypropylene.

Among the preferred carboxyethyl and carbamoylethyl compounds are the following: 1,4-bis[bis(2-carboxyethyl)phosphonylmethyl]-2-methylbenzene; 1,4-bis[bis(2-carboxyethyl)phosphonylmethyl]-2,5-dimethylbenzene; 2,4-bis[bis(2-carboxyethyl)phosphonylmethyl] 1,3,5-trimethylbenzene; and 1,4-bis[bis(2-carbamoylethyl)phosphonylmethyl]-2,3,5,6-tetramethylbenzene. These compounds may be readily prepared by reacting bis(2-carboxyethyl)phosphine oxide or bis(2-carbamoylethyl)phosphine oxide with the appropriately substituted halomethyl benzene.

The flame retardant compounds may be utilized as such or in conjunction with various flame retardant additives such as ammonium polyphosphates in the ratio of carboxy or carbamoyl derivative to ammonium polyphosphate of 1:1 to 1:3. Additionally, to the ammonium polyphosphate-carbamoyl or carboxy derivative mixture may be added a metal oxide such as titanium dioxide in amounts ranging from about 0.5 to 5.0% by weight, based on the weight of the polymer. These metal oxides perform synergistically with the ammonium polyphosphate and carboxy or carbamoyl derivative to minimize dripping of the polymer to which they are added when it is burning and before it extinguishes itself.

Example 1: Preparation of 1,4-Bis[Bis(2-Carboxyethyl)Phosphonylmethyl]-2,3,5,6-Tetramethylbenzene – To 1 liter of concentrated hydrochloric acid are added 103 parts of 1,4-bis[bis(2-cyanoethyl)phosphonylmethyl]-2,3,5,6-tetramethylbenzene. The reaction mixture is refluxed for 24 hours and 800 ml of hot water are then added. The resultant reaction mixture is cooled to 25°C and the product is filtered, washed with water and dried. The yield is 137 parts of 1,4-bis[bis(2-carboxyethyl)phosphonylmethyl]-2,3,5,6-tetramethylbenzene having a melting point of 265° to 267°C.

Example 2: Preparation of 1,4-Bis[Bis(2-Carbamoylethyl)Phosphonylmethyl]-2,3,5,6-Tetramethylbenzene – 109 parts of the 1,4-bis[bis(2-carboxyethyl)phosphonylmethyl] 2,3,5,6-tetramethylbenzene of Example 1, 600 ml of methanol, 100 ml of benzene and 1 ml of concentrated sulfuric acid are heated at reflux until complete solution is achieved (4 hours) and heating is then continued for an additional 5 hours. The resultant solution is then cooled and 15 parts of sodium bicarbonate are added. The solution is stirred for 30 minutes and filtered. The filtrate is cooled in an ice/acetone bath stirred for an additional 1½ hours, filtered, washed with methanol and dried. 111 parts of 1,4-bis[bis(2-carbomethoxyethyl)phosphonylmethyl]-2,3,5,6-tetramethylbenzene are recovered of 149° to 150°C MP.

40 parts of the carbomethoxyethyl compound are then mixed with 300 ml of concentrated, aqueous ammonia. The mixture is stirred for 2 hours at 10°C and then allowed to come to room temperature. Stirring at ambient temperature for 48 hours, filtering, washing with water and drying, results in the recovery of 30 parts of 1,4-bis[bis(2-carbamoylethyl)phosphonylmethyl]-2,3,5,6-tetramethylbenzene having a melting point of 322°C dec.

Example 3 through 15: To 80 parts of unstabilized polypropylene are added, by milling at 175°C, 0.1 part of pentaerythrityl tetrakis(3,5-di-t-butyl-4-hydroxyphenyl)propionate, 0.25 part of distearylthiodipropionate (as primary and secondary stabilizers), 10.0 parts of the product in Example 1, 10.0 parts of ammonium polyphosphate and 1.0 part of titanium dioxide. The resultant compounded polypropylene composition is molded at 200°C into test bars 6" x 0.5" x 0.125" and the bars are tested for flammability according to Underwriter's Test (UL)-94 (vertical). The flammability is rated according to the following definitions:

- V-0: flame extinguishes in 0 to 5 seconds; nondripping or, if dripping, the drippings do not ignite surgical cotton.
- V-1: flame extinguishes in 6 to 25 seconds; nondripping or, if dripping, the drippings do not ignite surgical cotton.
- V-2: flame extinguishes in 0 to 25 seconds; drips and ignites surgical cotton.

The results are described in the table below, along with other test results utilizing other compounds and combinations of materials and flame retardant additives. Comparative results are indicated by the letter (c).

The flame retardant compounds are described by reference to the following structural formula:

$$(X-\overset{\overset{O}{\|}}{C}-CH_2CH_2)_2-\overset{\overset{O}{\|}}{P}-R-\overset{\overset{O}{\|}}{P}-(CH_2CH_2\overset{\overset{O}{\|}}{C}-X)_2$$

EX.	X	R	FLAME TEST	COMPAT-IBILITY	EXTRACT-ABILITY
3	OH	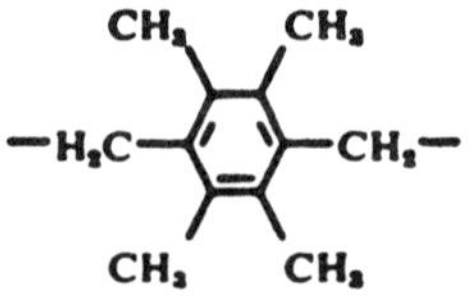	V-1	+	+
4	OH	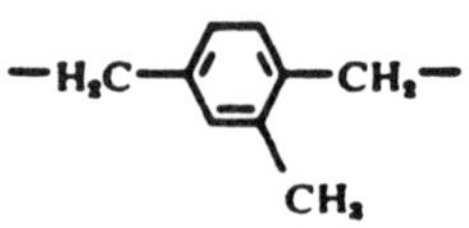	V-2	+	+
5	OH	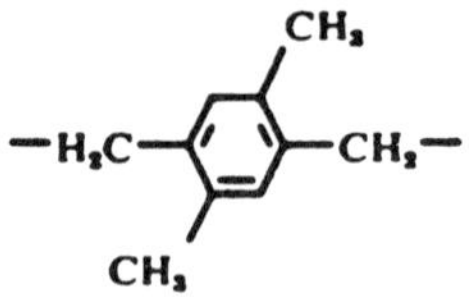	V-2	+	+
6	OH	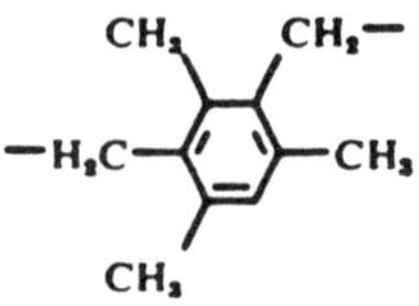	V-2	+	+
7	NH_2	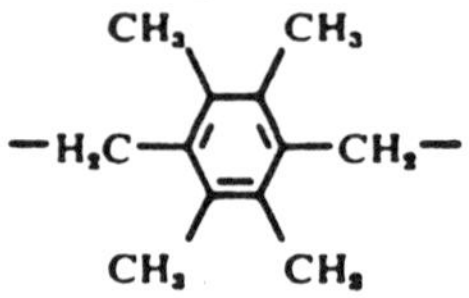	V-1	+	+
8	NH_2	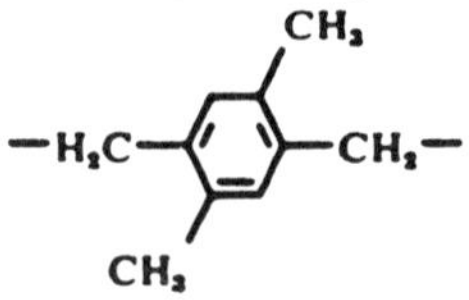	V-1	+	+
9	NH_2	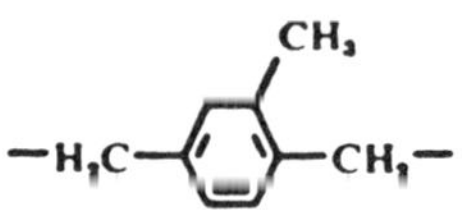	V-2	+	+
10	NH_2	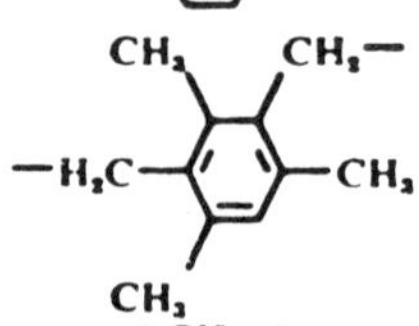	V-2	+	+
11c	OH	$-(CH_2)_2-$	V-1	–	–
12c	NH_2	$-(CH_2)_2-$	V-0	–	–
13c	OCH_3	$-(CH_2)_2-$	V-2	–	–
14c	OCH_3	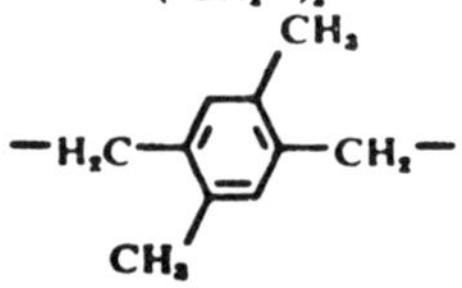	V-1	–	–

(continued)

EX.	X	R	FLAME TEST	COMPAT-IBILITY	EXTRACT-ABILITY
15c	OCH_3	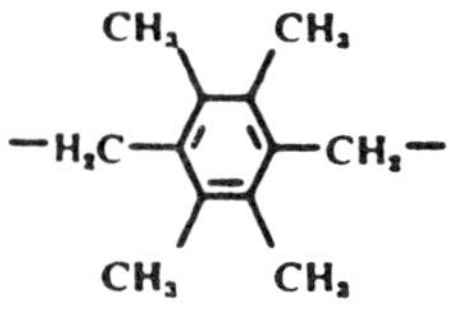	–	–	–

(+) = compatible and nonextractable
(–) = incompatible and extractable

Phosphorylated Novolak Resin

B.T. MacKenzie, Jr., and E.V. Wilkus; U.S. Patent 4,085,162; April 18, 1978; assigned to General Electric Company describe a process whereby a high level of resistance to flame and combustion is achieved in polyolefin compounds comprising ethylene-containing polymers by including in them the combination of phosphorylated novolak-type phenol-formaldehyde and triallyl cyanurate. The polyolefin compositions comprise ethylene-containing polymers such as polyethylene, copolymers of ethylene and other polymerizable materials. The process is especially useful in providing flame-resistant polymeric electrical insulations for metallic conductors.

The phosphorylated novolak phenol-formaldehyde component of this process consists of the reaction product of phenol-formaldehyde condensation novolak resin and triphenylphosphate. For example, a phenol-formaldehyde condensation novolak resin (1,696 g) of the formula:

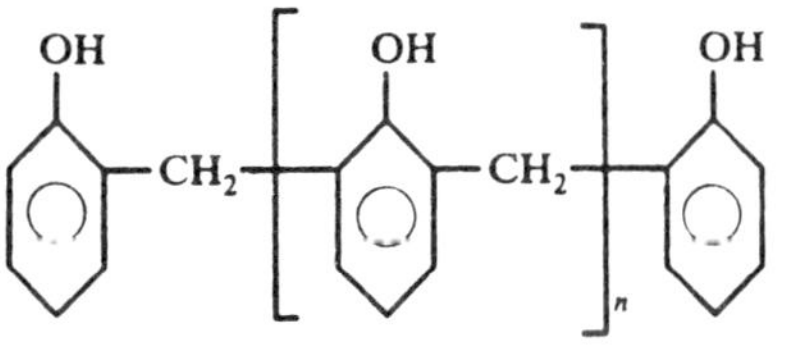

is reacted with triphenylphosphate (5,740 g) by heating in the presence of about 0.1 g of magnesium catalyst at a temperature of about 280°C for a period of approximately 3 hours. The phosphorylated novolak phenol product comprises:

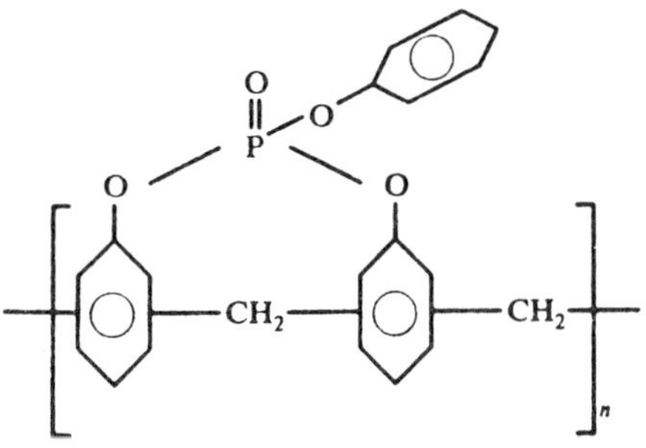

Suitable amounts for effective flame resistance in polyolefins comprise about 5 to 20 parts by weight of the phosphorylated novolak phenol-formaldehyde, and about 0.5 to 5 parts by weight of the triallyl cyanurate, per 100 parts by weight of the polyolefin compound.

The formulations of the examples of this process and for the comparative control standards are all given in parts by weight of the enumerated ingredients. The compositions of all examples and all control standards were prepared and evaluated identically. That is, all ingredients, except the peroxide curing agent, were initially combined and mixed on a hot two-roll mill at a temperature of about 195°F for approximately five minutes, then in each case the peroxide was subsequently added and dispersed by mixing for about three minutes.

The examples and the control standards are as follows with Control A and Examples 1 through 4 comprising like laboratory slab specimens, and Control B and Example 5 comprising specimens of wire coated with the respective compositions at a thickness of about 50 mils. Curing was effected by subjecting the slab specimens to 300°F in a press for 60 minutes and the control wire specimen to steam at 400°F for two minutes.

	Slab Specimens					Wire Specimens	
	Control	Examples				Control	Ex.
	A	1	2	3	4	B	5
Ingredients							
Polyethylene	100	100	100	100	100	100	100
Calcined clay (treated with 1% silicone tetramer)	50.5	50.5	50.5	50.5	50.5	50.5	50.5
Carbon black	5	5	5	5	5	5	5
Antioxidant, Flectol H (polymerized trimethyldihydroquinoline)	1.75	1.75	1.75	1.75	1.75	1.75	1.75
Di-α-cumyl peroxide	3.3	3.3	3.3	3.3	2.2	3.3	2.2
Phosphorylated novolak phenol-formaldehyde condensate (PN)	–	–	10	10	10	–	10
Triallyl cyanurate (TAC)	–	0.9	–	0.9	2.2	–	2.2
Properties							
Percent toluene extractables	7.8	6.7	15.9	9.1	–	–	–
Oxygen index slab	0.191	0.204	0.221	0.236	0.240	–	–
	weak char, drips	weak char, drips	strong char, no drip	strong char, no drip	strong char, no drip	–	–
Wire insulation	–	–	–	–	–	0.190	0.224
	–	–	–	–	–	weak char, drips	strong char, no drip
Copper mirror test, percent removed after 16 hr at 175°C	0	0	0	0	0	0	0
Surface bloom	none	none	none	none	none	none	none
Original, quick rise dielectric strength in water at room temperature, volts/mm	–	–	–	–	–	725	894

The relative resistance to flame or combustion of each of the examples and of the standards is demonstrated by the Oxygen Index Test (ASTM D-2863-70). The test was run upon specimens of the polymeric materials of the Examples and Control Standards in the form of a test slab or as an insulation covering on wire.

The copper mirror test which measures the potential corrosive properties of materials on metal, is as follows: A given sample of the material for evaluation is placed in a test tube together with a copper glass mirror comprising vacuum deposited copper of a known weight. The copper mirror is suspended in the test tube above the sample and the tube is sealed. The lower end of the sealed tube is then heated at 175°C for 16 hours in an oil bath or other suitable means while its upper end containing the copper mirror is maintained below 60°C. After heating for 16 hours, the copper mirror is removed and examined for evidence of corrosion, and if 50% or more of the copper is removed from the mirror, the material is classified as corrosive with respect to copper. Results are reported as the percentage of copper removed.

Phosphorus Oxide/Melamine/Polyol Flame Retardant Systems

According to a process described by *D.G. Brady; U.S. Patent 4,010,137; Mar. 1, 1977; assigned to Phillips Petroleum Company* a phosphorus-containing flame retardant is produced when suitable amounts of at least one phosphorus oxide compound of the formula $(P_2O_5)_{0.5-1.0}(H_2O)_{0.0-1.5}$, melamine, and at least one saturated open-chain polyol containing from 5 to 15 carbon atoms and from 4 to 8 hydroxyl groups are combined and heated under such conditions that the resulting product when incorporated into a synthetic resin will impart flame resistance to it.

Phosphorus oxide compounds include phosphorus pentoxide, orthophosphoric acid, and condensed phosphoric acids. Examples of suitable polyols include sorbitol, mannitol, pentaerythritol, dipentaerythritol, and tripentaerythritol.

In Table 1 several methods are shown in which phosphorus-containing systems are prepared and added to polypropylene, namely Embodiments 1, 2, 3 and 4. There are two subembodiments under the heading of Embodiment 1.

TABLE 1: PHOSPHORUS-CONTAINING FLAME RETARDANT ADDITIVES AND SUITABLE MOLDING TEMPERATURES (SMT)

Embodiment (1)	Embodiment (2)*	Embodiment (3)**	Embodiment (4)***
(a) Ammonium polyphosphate (AP) and dipentaerythritol (DPE) (b) Melamine pyrophosphate (MP) and at least one of pentaerythritol (PE), dipentaerythritol (DPE) and tripentaerythritol (TPE)	Same additives as (1); thermal degassing of additives prior to formulation in polypropylene distinguishes (2) over (1)	Reaction product of H_3PO_4, urea, dipentaerythritol and melamine	Reaction product of H_3PO_4, P_2O_5, dipentaerythritol and melamine
SMT 350° to 400°F	SMT 350° to 450°F	SMT 350° to 475°F	SMT 350° to 475°F

*The additives are heated at 175° to 275°C for 0.5 to 4 hr to cause a foaming reaction. After this thermal degassing, the residue is used to formulate a nonburning, nondripping polypropylene.

**The reaction product is heated at 190° to 260°C for 0.5 to 4 hr before formulating in polypropylene (integral part of additive synthesis procedure).

***The reaction product is heated to 190° to 280°C for 0.5 to 6 hr before formulating in polypropylene (integral part of additive synthesis procedure). Pentaerythritol or tripentaerythritol can be used in place of dipentaerythritol in (4).

In Embodiment I, it is advantageous to coat the ammonium polyphosphate with about 1 to 3 weight percent stearylamine prior to compounding in polypropylene. This procedure alleviates an incompatibility problem in the polypropylene. The use of uncoated ammonium polyphosphate results in blooming in the molded polypropylene which exhibits a tacky surface with a dull appearance. No incompatibility problem was encountered in Embodiments 1(b), 2, 3 and 4.

The thermal degassing of the phosphorus-containing additive can be carried out on a blend of compounds as described in Embodiment 2 or the thermal degassing can be a step in a synthetic procedure as described in Embodiments 3 and 4. The phosphorus-containing flame retardant additives which are thermally degassed according to the process generally correspond to one or more of the compositions described in Embodiments 2, 3 and 4 of Table 1.

Embodiment 1 primarily shows the flame-retardant effectiveness of a phosphorus-containing additive system which has not been degassed. Foaming, i.e., void formation, is such a severe problem of Embodiment 1 in which the additive system has not been a priori degassed, that molding must be carried out at an impractically low temperature of about 350° to 400°F if the problem is to be minimized. This can cause other problems.

The increase in suitable molding temperatures in Embodiment 2 over Embodiment 1 illustrates that Embodiment 2 yields nonburning, nondripping polypropylene formulations which possess significantly less tendency to foam. Therefore, these formulations can be molded at correspondingly higher temperatures. The higher molding temperatures of Embodiments 3 and 4 illustrate that these systems yield polypropylene samples with less tendency to foam during molding. In general, molding temperatures of approximately 450° to 500°F are considered desirable or necessary.

Embodiment 1: The additives shown in Table 1 under Embodiment 1(a) and 1(b) were blended in polypropylene on a three-inch roll mill for 5 minutes at 320° to 350°F. Samples for testing were cut from sheets of the formulation that had been molded at 350°F in picture frame molds. Results are summarized in Table 2.

TABLE 2: EVALUATION DATA ON FLAME RETARDANT POLYPROPYLENE SAMPLES (EMBODIMENT 1 OF TABLE 1)*

Run No.	Additive (php)**	ASTM D635***	ASTM D2863†	UL-94	Ratio (AP/DPE)
Contol	none	B	~18	–	–
1	AP (30)	SE	ND	not SE-0	5:1
	DPE (6)	–	–	–	–
2	AP (40)	SE	ND	not SE-0	5:1
	DPE (8)	–	–	–	–
3	AP (40)	SE-NE	28.7	SE-0	3.33:1
	DPE (12)	–	–	–	–
4	AP (40)	NB	29.5	SE-0	2.5:1
	DPE (16)	–	–	–	–

(continued)

TABLE 2: (continued)

Run No.	Additive (php)**	ASTM D635***	ASTM D2863†	UL 94	Ratio (AP/DPE)
5	AP (30)	NB	ND	SE-0	1.87:1
	DPE(16)	–	–	–	–
6	AP (20)	NB	ND	SE-0	1:1
	DPE (20)	–	–	–	–
7	AP (40)	SE	ND	SE-0	2.5:1
	DPE (16)	–	–	–	–
	asbestos (10)	–	–	–	–
8	AP (40)	NB-SE	ND	SE-0	2.5:1
	DPE (16)	–	–	–	–
	glass fiber (10)	–	–	–	–
9	MP (30)	NB-SE	31	SE-0	1.87:1
	DPE (16)	–	–	–	–

*Formulations containing ammonium polyphosphate coated with 1 to 3 wt % stearylamine exhibited no incompatibility in molded polypropylene. The stearylamine in chloroform was contacted with ammonium polyphosphate. Removal of chloroform solvent on a rotary evaporator gave the coated ammonium polyphosphate suitable for use in Embodiment 1(a).

**AP, MP and DPE represent respectively, ammonium polyphosphate, melamine pyrophosphate and dipentaerythritol.

***B, SE and NB represent respectively, burning, self-extinguishing and nonburning.

†ND represents not determined. The numbers represent oxygen index as defined in ASTM D2863.

In Table 2, UL-94 refers to an Underwriters Laboratory test method described in Mod. Plastics, October 1970, p. 96. In accordance with this test, the SE-0 rating indicates that a vertically positioned molded specimen contacted for 10 seconds with a flame extinguishes within 5 seconds after removal of the flame on two successive flame applications and does not drip flaming particles capable of igniting a cotton ball positioned 12 inches below the molded specimen. Molded specimen sample slabs were 5' x ½" x ⅛".

The results of runs 1 through 9 demonstrate the flame retardant effectiveness of Embodiment 1. Note SE-0 ratings on UL-94 test for runs 3 through 9.

The amount and nature of the additives of Embodiments 1 and 2 generally suitable for use in polypropylene are in the following approximate ranges.

Additive	Broad (php)	Preferred (php)
Ammonium polyphosphate (AP) or melamine pyrophosphate (MP)	10-70	20-40
Dipentaerythritol (DPE)	5-30	10-20
Fillers such as glass fiber, TiO_2 or asbestos	0-15	1-10
Weight ratio AP/DPE or weight ratio MP/DPE	14:1-0.25:1	4:1-1:1
Total loading of components	30-90	40-60
Weight percent stearylamine based on ammonium polyphosphate for coating	0.5-5	1-3

Embodiment 2: The operability of Embodiment 2 is substantiated by the thermal degassing of a melamine pyrophosphate/dipentaerythritol blend. The thermal degassing can be carried out at 175° to 275°C until a foaming reaction has substantially subsided. The resulting residue can be formulated with polypropylene and molded at 350° to 450°F without foaming. The molded samples are nonburning and nondripping.

The following procedure is a representative thermal degassing treatment: An 8 g sample of melamine pyrophosphate was mixed with 4 g dipentaerythritol and the mixture was pulverized. The resulting powder was placed in a 200 ml round bottomed flask equipped with stirrer and nitrogen inlet tube. The mixture of reactants was degassed by heating the flask in an oil bath up to a temperature of 260°C over a 1.75 hours time interval. The resulting light tan solid was ground to a powder and used as a flame retardant in polypropylene. (See Runs 10 and 11 in Table 3. In Run 12 a different ratio of MP/DPE was used.)

TABLE 3: EVALUATION DATA ON POLYPROPYLENE SAMPLES BASED ON DEGASSED MP/DPE ADDITIVE (EMBODIMENT 2)

Run No.	Level of Additive (php)	(MP/DPE) Weight Ratio*	ASTM D635-63	Observations on Molded Polypropylene Samples
10	45	2:1	NB-SE	Good char; nondripping; foams a little at 475°F in 5 min
11	45	2:1	SE	No foam at 425°F; nondripping
12	45	1.67:1	NB	Stable at 475°F for 5 min (no foam); nondripping

*Prior to thermal degassing treatment of MP/DPE blend.

The results in Table 3 indicate that in Embodiment 2 suitable molding temperatures (SMT) in the range of 450°F (350° to 450°F) which are desirable can be used. The thermal degassing procedure of Embodiment 2 thus significantly reduces the tendency of formulated polypropylene to foam during molding. The samples of Embodiment 1 were molded at approximately 350°F to avoid the foaming problem.

Embodiment 3: The operability of Embodiment 3 is demonstrated by the thermal degassing of the reaction product derived from H_3PO_4, urea, dipentaerythritol, and melamine. The product from the reaction of H_3PO_4, urea, dipentaerythritol is isolated as a white solid which is subsequently heated at an elevated temperature with melamine to complete the preparation and degassing of the flame retardant additive.

HALOGEN-CONTAINING ADDITIVES

Brominated Propyl Amides

H. Braus and J.R. Woltermann; U.S. Patent 4,092,292; May 30, 1978; assigned to National Distillers and Chemical Corp., describe a flame retarded composition comprising a solid polymer and a flame retardant amount of a halogen-containing amide having the formula:

$$(C_nH_{2n+1-y}X_y)_{2-m}\overset{H_m}{\overset{|}{N}}-\overset{O}{\overset{\|}{C}}-R$$

where R is lower alkoxy; lower alkoxy carbonyl;

$$-\overset{H_m}{\overset{|}{N}}(C_nH_{2n+1-y}X_y)_{2-m}$$

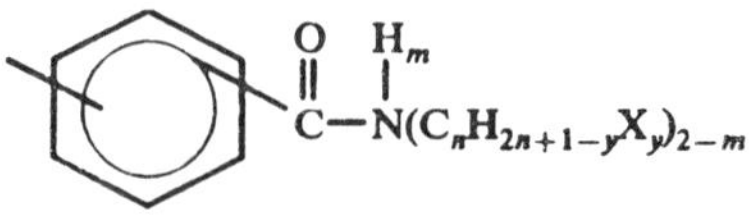

or

$$-\overset{O}{\overset{\|}{C}}-\overset{H_m}{\overset{|}{N}}(C_nH_{2n+1-y}X_y)_{2-m}$$

where X is chlorine or bromine; m is 0 or 1; n is 1 to 6; and y is 1 to 13.

The flame retardant compositions may comprise a homogeneous or inhomogeneous intermixture of polymer and agent, the latter being incorporated intimately in a compounding operation, applied to the surface as a coating, or absorbed into a porous, e.g., foamed article. Preferably, molded articles comprise an intimate mixture of flame retardant agent and polymer, to a level of 10 to 20% by weight of flame retardant. Typically, a flame retardant amount, for example, 2 to 5 weight percent up to 10 to 15% or more of the flame retardant will be introduced to the polymer. Solid polymers, such as polypropylene and polyethylene may be treated with the flame retardant compounds of this process.

Example 1: N,N-bis(β,γ-dibromopropyl)ethyloxamate was prepared by the bromination of N,N diallylethyloxamate as follows: Ethyloxalate, 146.1 g (1.0 mol), was charged into a 3-necked flask equipped with a mechanical stirrer, dropping funnel, condenser and drying tube. Diallylamine, 97.2 g (1.0 mol) was slowly added with constant stirring. There was a moderate exotherm during the addition until approximately one-half the amine was added, at which time the temperature declined. When the addition was complete, the bath was heated to 80° to 90°C for 4 hours and then cooled by standing overnight.

The product was then fractionated and vacuum distilled and the portion boiling at 105° to 106°C at 1 mm Hg was collected. The product was submitted for mass spectra determination which showed a mass of 197 and then expected fragments of the desired product, N,N-diallylethyloxamate.

The N,N-diallylethyloxamate, 77.3 g (0.39 mol), dissolved in 100 ml of chloroform, was charged to a one liter, 3-necked flask equipped with a stirrer, thermometer, dropping funnel condenser and drying tube. The solution was cooled to 4°C and bromine, 124.7 g (0.78 mol), was slowly added. When all the bromine was added, the chloroform was distilled at atmospheric pressure. The remaining chloroform was removed under reduced pressure (water pump). The residue was taken up in benzene, filtered and evaporated under a stream of N_2 (MP 160°C, with decomposition). Mass spectra analysis confirmed the desired compound. N,N-bis(β,γ-dibromopropyl)ethyloxamate (30 g) is dissolved in perchloroethylene (180 ml) using Triton X-100 (5 g) which is an octylphenoxypolyethoxy ethanol.

Example 2: A series of flame retardant agents in accordance with the process was formulated in a Banbury mixer at 300° to 320°F with polypropylene copolymer, comprising a minor proportion of copolymerized ethylene (Pro-fax 1823 Hercules, MI 0.4, density 0.901-0.902) at a 10% level by weight, molded into test plaques, and performance evaluated with results as follows:

	Control	Flame Retardant Agents 1	2	3	4
Oxygen Index		25.3	25.0	25.0	24.9
Avg. 1st burn, sec.	Burns to Consumption	0	0	0	0
Avg. 2d burn, sec.		0	0	0	0
Avg. afterglow, sec.		38.0	40.4	55.8	45.6

Flame retardant agents
1. N,N'-bis (β,γ-dibromopropyl) ethyl oxamate
2. bis(tetrabromodipropyl) isophthalamide
3. tetra(β,γ-dibromopropyl) urea
4. N,N'-bis(β,γ-dibromopropyl) oxamide

All materials were formulated identically, with 5% antimony oxide (based on polymer) and a stabilizer system comprising 0.5% Mark 645 (a proprietary, tin stabilizer), 0.3% Irganox 1010 ⟨tetrakis[3-(3,5-di-t-butyl-4-hydroxyphenyl)propionoxymethyl] methane⟩ and 0.3% dilaurylthiodipropionate.

The control burned to consumption upon application of flame, whereas all of the flame retarded materials were self-extinguishing on first and second application of flame, and exhibited a relatively short afterglow. There was no drip of the flame retarded materials, as distinct from the control.

Bisphenoxy Compounds

A. L. Anderson; U.S. Patent 4,065,429; December 27, 1977; assigned to Velsicol Chemical Corporation describes plastic compositions containing polyolefins and bisphenoxy compounds having the formula:

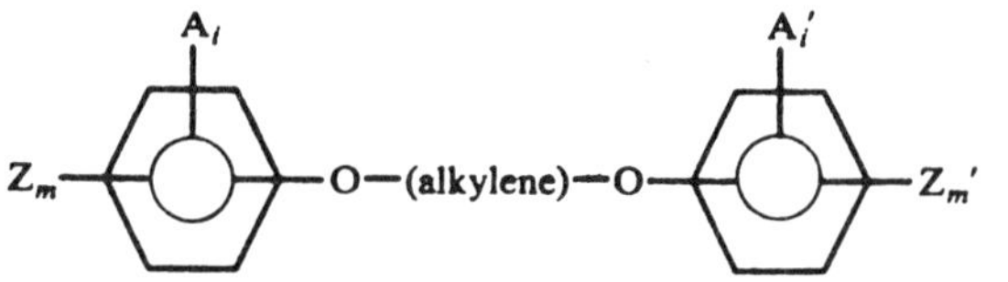

where Z is bromine, m is an integer having a value of 1 to 5 and m' is an integer having a value of 0 to 4, i is an integer having a value of 0 to 2 and i' is an integer having a value of 1 to 5, alkylene is a straight or branched chain alkylene group having from 1 to 6 carbon atoms and A is chlorine.

Examples illustrating the process are as stated in the section under polyesters.

Crosslinked Vinyl Halide Polymers

J.I. Jin and P. Kraft; U.S. Patents 4,049,747; September 20, 1977; and 4,022,849; May 10, 1977; both assigned to Stauffer Chemical Company disclose that crosslinked copolymers of the vinyl halides, particularly of vinyl chloride, can be blended with burning thermoplastics so as to yield fire retardant polyblends which are devoid of any detrimental effects upon their physical properties.

The fire retardant additives of this process are thus seen to comprise crosslinked copolymers of:

(1) One or more vinyl halides selected from the group consisting of vinyl chloride, vinyl bromide and vinyl fluoride with the use of vinyl chloride being preferred because of its lower cost and more widespread commercial availability; and

(2) One or more polyfunctional ethylenically unsaturated monomers, i.e., one or more monomers containing two or more polymerizable, ethylenically unsaturated bonds including, for example, allyl methacrylate, divinyl benzene, diethylene glycol dimethacrylate, ethylene glycol dimethacrylate, trimethylolpropane trimethacrylate, diallyl adipate, and 2-vinyl oxyethyl ether.

Optimum results are, however, obtained by the use of allyl methacrylate since its use leads to a higher degree of conversion of the vinyl chloride monomer to the form of polymeric moieties as well as a more extensive degree of crosslinking as indicated by the presence of little, if any, methanol extractable residues in the resulting crosslinked copolymer.

The polyblends comprising blends of any of the above described thermoplastic polymers with one or more of the crosslinked vinyl halide copolymer additives may be utilized in any of the coating, impregnating and especially in the molding applications known to those skilled in the art wherein it is desired to provide fire retardancy to the resulting end product.

For example, these compositions may be used for preparing such diverse items as calendered films, blow molded bottles, extruded and blown films, extruded and shaped articles such as panels, tubes, sheets, rods, fibers and particularly with polypropylene and other polyolefin fibers, and in carrying out such processes as injection molding, fluidized bed coating, electrostatic powder spraying and rotational coating, etc.

Example: This example illustrates the preparation of a crosslinked vinyl halide copolymer additive and its use in the preparation of a fire retardant polyblend with polypropylene. Ninety parts of vinyl chloride and ten parts of allyl methacrylate are suspension polymerized in a solution of 0.5 parts of methyl cellulose in 300 parts of deionized water which also contains 0.5 parts of t-butyl peroxypivalate as an initiator. The latter mixture is stirred at 60°C for 10 hours while in a nitrogen atmosphere. The resulting crosslinked 90:10 vinyl chloride-allyl methacrylate copolymer is isolated by filtration followed by washing with water and drying at 70°C for 24 hours. The copolymer is obtained in a yield of 98% and contains 50%, by weight, of chlorine.

A polyblend containing polypropylene along with 30% by weight of the above described crosslinked copolymer is prepared by milling a mixture of those two resins for 10 minutes on a two-roll mill in which the front roll is at a temperature of 185°C and the back roll is at 130°C. As shown in the following table, the resulting blend, as compared with unmodified polypropylene, displays reduced flammability upon having its limiting oxygen index (LOI) determined by means of ASTM D-2863, the latter procedure also being described by Fenimore and Martin in the November, 1966 issue of *Modern Plastics.*

Flammability of Polypropylene Polyblends

Polyblend Compositions*	LOI Value
0/100	17.6
30/100	21.5
30/100 + 6% antimony trioxide**	23.8

*Parts crosslinked copolymer/parts polypropylene.
**As based on the weight of the polypropylene

The above data reveal the improved fire retardancy of the polyblends as well as the fact that the use of a relatively low concentration of antimony trioxide in combination with the fire retardant crosslinked copolymer additive of this process effects an additional increase in the fire retardancy displayed by one of the polyblends of this process.

Comparable fire retardant polyblends are obtained upon blending this crosslinked copolymer additive with various burning thermoplastics including polymethyl methacrylate, polystyrene, polyethylene, cellulose acetate, nylon 66, polyphenylene oxide, polycarbonate and polyurethane resins. Moreover, excellent fire retardant polyblends are also provided by the use of the following crosslinked vinyl halide copolymers:

(1) A 95:5 vinyl chloride-allyl methacrylate copolymer;

(2) A 45:45:10 vinyl chloride:vinyl bromide:allyl methacrylate copolymer;

(3) A 90:10 vinyl chloride:ethylene glycol dimethacrylate copolymer;

(4) A 95:5 vinyl chloride:2 vinyloxyethyl ether copolymer;

(5) A 90:5:5 vinyl bromide-allyl methacrylate copolymer; vinylidene chloride copolymer; and

(6) An 80:20 vinyl chloride-ethyl diallylphosphonate copolymer.

PHOSPHORUS AND HALOGEN-CONTAINING ADDITIVES

Intumescent Adhesives of Shellac, Chloroparaffin and Monoammonium Phosphate

J.D. Marshall and M.C. Kuklies; U.S. Patent 4,058,643; November 15, 1977; assigned to Mobil Oil Corporation describe a fire retardant laminate construction comprising a plastic backing film adhesively secured to an insulation batting material. The adhesive material comprises an intumescent coating, containing one or more fire retardant additives, which will foam and char at high temperatures to form a solid multicellular, fire retardant insulating layer which prevents wicking of the plastic material into the insulation batting. In general, intumescent coatings foam at high temperatures to form an insulating layer. Such coatings contain at least four basic compounds including: (1) an acid source, preferably phosphoric; (2) a carbonaceous residue source; (3) a blowing agent; and (4) a solvent.

The acid source includes such material as monoammonium phosphate, diammonium phosphate, ammonium polyphosphate, melamine phosphate, guanylurea phosphate, urea phosphate, ammonium sulfate and ammonium borate. The ammonium phosphates have been found particularly effective acid source materials for purposes of the present coating compositions.

As the carbonaceous residue source, shellac provides an excellent foamed char residue when it is employed in the fire retardant adhesive coating compositions. Shellac exhibits an ability to maintain its adhesive properties with very high loadings of other additive compounds which may be present in the adhesive coating composition. Shellac does not drip or migrate from the flame source, and thereby, eliminates undesirable wicking of the plastic material, such as polyethylene, into the insulation batting leaving the flame retardant additives in the coating behind.

Further, it has been found that shellac-based adhesives do not degrade or lose their holding power over extended periods of time which is an essential feature, for example in holding and maintaining insulation materials in a fixed location in building construction. Shellac-based adhesives are nontoxic, nonfuming and do not produce objectionable smoke when they are subjected to combustion temperatures.

The compositions of the process may be applied to backing materials employing plastic films such as polyolefins including, for example, polyethylene. It has been found that polyethylene films, while flammable by nature, may be modified to make them suitable for employment as backing for insulation.

Example: The formulations were prepared from commercially available 5 pound cut shellac (i.e., five pounds of shellac per gallon of alcohol). The shellac was measured into a beaker and manually agitated while the other ingredients were added. Although the order of addition is not critical, the chlorowax and other components will settle if the solution is not continuously agitated.

The coating formulation employed was about 10% chlorowax, 57.7% of the five pound cut shellac, and 26.3% monoammonium phosphate. This mixture is applied with a brush to uniformly coat one side of a test material, allowed to dry, and subsequently subjected to ignition testing.

The test material used for screening this formulation was ⅛ inch sheet of foamed polystyrene such as is commercially sold for institutional food trays. One side of the 4 x 10 inch test sheet was coated with the fire retardant formulation in a continuous pattern and dried for 3 hours in a 140°F circulating air oven.

Ignition was tested by attempting to light the material with a match along the bottom (4 inch) side of the coated test material. When match ignition failed a propane torch was applied to the bottom of the sample (which is hanging in a vertical plane) for 10 seconds and withdrawn. Following this test, if ignition still was unsuccessful the torch was applied and the torch was raised with the char front to determine if ignition could be started with more flame exposure. In the case of the above formulation, all of these ignition tests failed to start ignition that would support itself when the flame or torch was removed.

A coating formulation that passed the initial ignition test was then tested as a simulated product where polyethylene film was brush coated with the formulation then manually adhered to a 1½" thick fiber glass batting the same size as the test films. This laminate construction was then dried overnight in an oven and subjected to the series of the flammability tests described above progressing from match ignition through following the char front up the surface of a vertical sample with a propane torch. All tests failed to provide a self-supporting flame within 2 seconds of removal of the source of ignition.

The third series of test samples used a discontinuous pattern or grid of lines of the retardant/adhesive composition. The grid pattern allows a heavy bead of adhesive to be formed every ½" to 1½" which effectively stops the flame front while the heat from the flame also shrinks the polyolefin films away from the flame faster than the flame propagation velocity. The film tends to shrink to the nearest retardant bead and not to wick into the film in this method of adhering the samples.

In preparing the third series of test samples, a glass rod was used to transfer the shellac retardant/adhesive formulation from the beaker into irregular wave-like grids of nonintersecting lines of the formulation on the surface of polyethylene film. The film in this case contained a fire retardant additive, Flameguard, at a 5% by weight loading. A fiber glass batting was then pressed onto the film and it adhered at the retardant adhesive composition grid lines. The sample was dried overnight in a 140°F oven prior to the series of tests. Self-supporting ignition again did not occur.

A number of formulations were examined to determine the optimum formulations. Two successful compositions are as follows: (a) 50% shellac and 50% monoammonium phosphate, and (b) 57% shellac, 2% chloroparaffin, 20% monoammonium phosphate and 21% starch. Results were that samples prepared as above could not burn or failed to ignite.

Aromatic Phosphine Oxides

J.J. Mrowca; U.S. Patent 4,064,104; December 20, 1977; assigned to E.I. Du Pont de Nemours and Company describes aromatic phosphine oxide compounds which are useful as flame retardants in synthetic polymers and textiles. These compounds have the formulas as shown below.

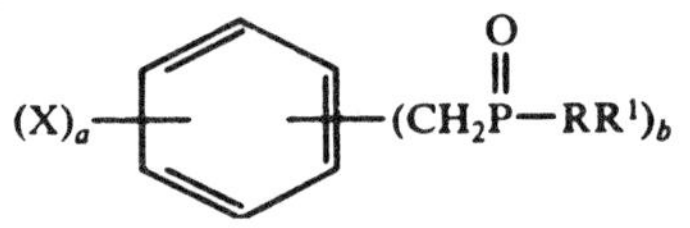

In the formula R and R[1], individually, are alkyl, aryl or aralkyl, each of up to 10 carbon atoms having up to five chlorine, bromine or fluorine atoms; X is chlorine, bromine, OR, SR, SOR, SO_2R, SO_2NH_2, SO_2NHR or SO_2NR_2, R being as defined above; a is 1 to 4; b is 2 to 5; and a + b is 3 to 6.

The following examples illustrate the synthesis and use of aromatic phosphine oxides.

Example 1: 2,4,5,6-Tetrabromo-1,3-Xylylenebis(Diphenylphosphine Oxide) –

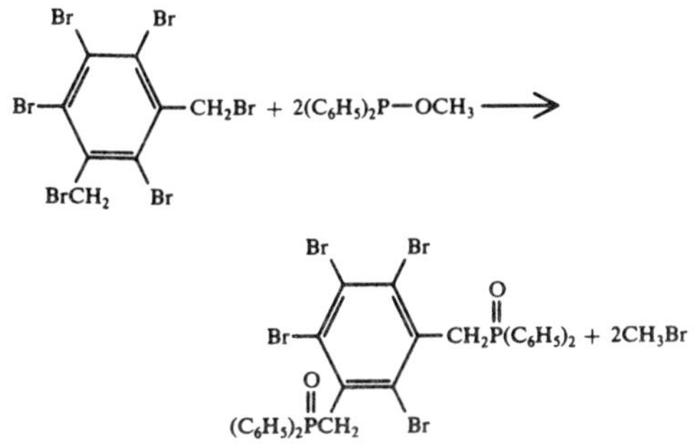

To a stirred, refluxing mixture of 10.00 g of α,α',-2,4,5,6-hexabromo-m-xylene in 100 ml of o-dichlorobenzene under nitrogen was added dropwise a solution of 15.00 g of methyl diphenylphosphinite in 50 ml of o-dichlorobenzene. After the addition was complete, the mixture was refluxed for 5 hours and cooled. The white solid was filtered, washed with benzene and hexane, and recrystallized from dimethylsulfoxide to give 7.36 g of 2,4,5,6-tetrabromo-1,3-xylylenebis(diphenylphosphine oxide) as white crystals, MP, 251° to 254°C.

Example 2: 2,5-Dibromo-1,4-Xylylenebis(Diphenylphosphine Oxide) –

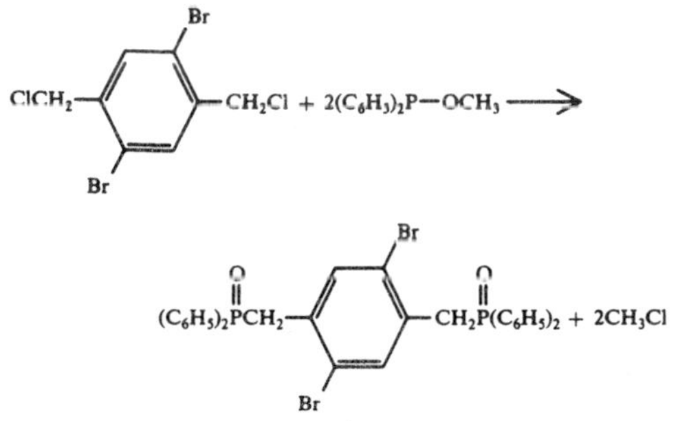

To a stirred, refluxing mixture of 10.00 g of 1,4-bis(chloromethyl)-2,5-dibromobenzene in 150 ml of o-dichlorobenzene under N_2 was added dropwise a solution of 15.00 g of methyl diphenylphosphinite in 50 ml of o-dichlorobenzene. After the addition was complete, the mixture was refluxed for 5 hours and cooled. The white solid was filtered, washed with benzene and hexane, and dried at 100°C and 0.1 mm for 18 hours to give 19.10 g of 2,5-dibromo-1,4-xylylenebis(diphenylphosphine oxide), MP, 301° to 303°C.

Example 3: 2,3,5,6-Tetrachloro-1,4-Xylylenebis(Diphenylphosphine Oxide)

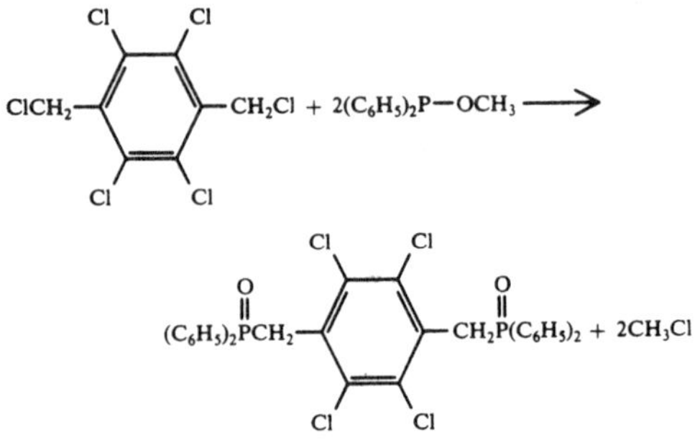

To a stirred, refluxing mixture of 10.00 g of α,α'-2,3,5,6-hexachloro-p-xylene in 150 ml of o-dichlorobenzene under nitrogen was added a solution of 20.00 g of methyl diphenylphosphinite in 50 ml of o-dichlorobenzene. After the addition was complete, the mixture was refluxed for 5 hours and cooled. The white solid was filtered, washed with benzene and hexane, and recrystallized from 750 ml of o-dichlorobenzene to give 18.50 g of 2,3,5,6-tetrachloro-1,4-xylylene-bis(diphenylphosphine oxide) as white crystals, MP 339° to 341°C dec.

The use of these compounds as flame retardants for polyolefins has been demonstrated by following the Underwriters Laboratory procedure (UL-94) in the following manner: Commercial polypropylene (28 g) was fluxed on a 2-roll mill (Junior Lab 3 x 8 inch rolls), 12 g of the flame-retardant additive was added to the resin band, and the mixture milled about 6 minutes, during which time five-to-ten cuts were made to enhance blending. The resin sheet was ground up in a Wiley mill. The resulting powder was molded into 5 x ½ to ⅛ inch flex bars using a laboratory scale injection mold.

The bars (five from each batch) were subjected to flammability testing. The bars were supported vertically. Each sample was ignited twice, I_1 and I_2, for ten seconds with a Bunsen burner. After each ignition, there was no continued burning, i.e., the polymer was self-extinguishing. Even though all dripped molten polymer to cotton batting 12 inches below, in only one sample of a series of five (using the compound of Example 3) did the cotton ignite. Compounds used in the test included those of Examples 1, 2 and 3. In contrast to these results, when 1,2-xylylenebis(diphenylphosphine oxide) was used in the same manner as a control, three of the five samples ignited the cotton. The data are presented in the following table.

Ex. No.	Sample No.	Time of I_1	Burn Time	Time of I_2	Burn Time	Drip	Ignite Cotton
		..........seconds..........					
1	1	10	–	10	–	yes	no
	2	10	–	10	–	yes	no
	3	10	–	10	–	yes	no
	4	10	–	10	–	yes	no
	5	10	–	10	–	yes	no

(continued)

Ex. No.	Sample No.	Time of I_1	Burn Time	Time of I_2	Burn Time	Drip	Ignite Cotton
		. . . . seconds.			. . . .		
2	1	10	–	10	–	yes	no
	2	10	–	10	–	yes	no
	3	10	–	10	–	yes	no
	4	10	–	10	–	yes	no
	5	10	–	10	–	yes	no
3	1	10	–	10	–	yes	yes
	2	10	–	10	–	yes	no
	3	10	–	10	–	yes	no
	4	10	–	10	–	yes	no
	5	10	–	10	–	yes	no
control	1	10	–	10	–	yes	yes
	2	10	–	10	–	yes	yes
	3	10	–	10	–	yes	no
	4	10	–	10	–	yes	no
	5	10	–	10	–	yes	yes

Tris(2,4,6-Tribromophenyl) Phosphate

According to *I. Touval; U.S. Patent 4,021,406; May 3, 1977; assigned to M & T Chemicals Inc.,* the resistance to both flaming and glowing combustion exhibited by polymer compositions where the flame retarding agent is comprised at least in part of a halogen-containing organic compound in combination with an antimony oxide is significantly increased by including specified halogenated phosphates, thiophosphates or thiophosphoryl chlorides as part of the flame retarding agent. Optionally, the composition may also include up to 50% of sulfur, based on the weight of the phosphorus compound.

The halogen-containing organic compounds of this process contain between 2 and 20 carbon atoms and preferably contain two or three fused carbocyclic ring structures, one or more of which may be of the bicyclic type as represented by the formula:

and contain a plurality of chlorine or bromine atoms to the extent that halogen atoms may occupy all of the valences not satisfied by carbon-carbon bonds. The hydrocarbon radicals may be alkyl, cycloalkyl, aralkyl, aryl or alkaryl. Alternatively, the radical may contain one or more carbon to carbon double bonds. Various functional groups, e.g., hydroxyl, carboxy and carboxylate radicals may also be present on the molecule.

A preferred class of halogen-containing organic compounds are Diels-Alder reaction products of brominated or chlorinated cyclopentadiene with hydrocarbons, carboxylic acids or carboxylic acid anhydrides containing a carbocyclic ring structure and one or two carbon-carbon double bonds. The carbocyclic compounds are represented by cyclopentadiene, maleic anhydride and 1-cyclohexene-3,4-dicarboxylic acid.

The concentration of total flame retarding agent, i.e., combination of antimony oxide, halogen-containing organic compound, halogenated phosphorus compound and, optionally, sulfur can be between 3% and 50% by weight of the substrate, depending upon the organic halogen source. As a general rule, a smaller amount of a bromine-containing compound will be required than a chlorine-containing compound to obtain an equivalent level of flame retardant performance. The following examples demonstrate the efficacy of the flame retarding agents of this process.

Example 1: This example demonstrates the flame retardancy imparted to polypropylene using tris(2,4,6-tribromophenyl) phosphate in combination with antimony trioxide and various halogen-containing organic compounds as the flame retarding agent.

Test samples were prepared by combining the indicated amounts of flame retardant with 100 parts by weight of a commercial grade of powdered polypropylene on a two-roll mill. The resultant sheet was cut into pellets approximately 0.125" (0.317 cm) in diameter. The pellets were injection molded at a temperature between 218° and 234°C to form bars measuring 6 inches x 0.5 inch x 0.125 inch (15 cm x 1.3 cm x 0.317 cm).

The data obtained from the UL 94 test are disclosed in Tables 1 and 2. Samples 1 and 6 in Table 1 did not contain any phosphate and were employed as controls (CTRL). A class 0 rating in accordance with the UL 94 test was only achieved using the halogenated aromatic phosphate in combination with antimony oxide and the halogen-containing organic compound.

TABLE 1

	Sample Number						
	1 (CTRL)	2	3	4	5	6 (CTRL)	7
Polypropylene	100	100	100	100	100	100	100
Antimony trioxide	17	17	—	17	—	17	17
Dechlorane*	50	50	—	—	50	—	—
Chloran**	—	—	—	—	—	50	50
TTBPP***	0	5	5	5	—	0	5
UL 94 class	1	0	fails†	fails†	fails†	fails†	0
Average burning time††	7	2	—	—	—		<1
Average afterglow time†††	50	7	—	—	—	—	0

*Perchloropentacyclodecane.

**
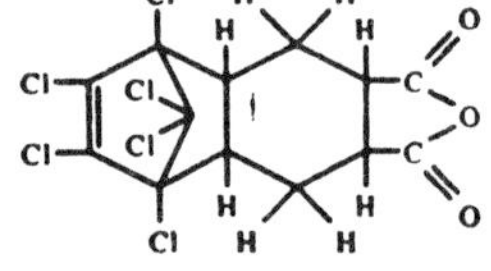

***Tris(2,4,6-tribromophenyl) phosphate.

†Samples were too flammable to qualify in any UL class.

††Elapsed time (sec) between ignition of sample and extinguishing of flame (6 trials).

†††Elapsed time (sec) between extinguishing of flame and disappearance of glowing, as evidenced by red or orange-red coloration (6 trials).

TABLE 2

	Sample Number 8	9	10	11
Polypropylene, parts	200	200	200	200
Antimony trioxide, parts	40	40	40	34
Chlorinated Diels-Alder adduct, parts*	40	80	80	100
TTBPP, parts	40	40	20	20
Flameout time, sec	–	2	4	2
Afterglow time, sec	–	20-60	20-60	6
UL 94 rating	2	1	1	0

*A 2:1 mol ratio Diels-Alder reaction product of hexachlorocyclopentadiene and 1,5-cyclooctadiene (Dechlorane Plus).

Example 2: This example demonstrates the increase in efficiency of the flame retardant achieved by the incorporation of sulfur in combination with a suitable phosphorus compound. The test samples were prepared as described in Example 1 and rated for flammability as described in UL 94 procedure. The composition and flame retardancy ratings are summarized in Table 3 where all concentrations are expressed in parts by weight.

TABLE 3

	Sample 1	Sample 2
Polypropylene	225	225
Perchloropentacyclodecane	100	100
Antimony trioxide	33.4	33.4
Tris(2,4,6-tribromophenyl) phosphate	16.7	15.6
Sulfur	–	1.1
Flameout time, sec	5.5	3.0
Afterglow time, sec	21.0	7.0
UL 94 rating	SE-1	SE-0

PHOSPHORUS-CONTAINING REACTANTS

Acrylic Acid Esters

F. Linke, W. Dürsch and H.-J. Kleiner; U.S. Patent 4,072,776; February 7, 1978; assigned to Hoechst AG, Germany describe a process for the flameproofing of fiber material, which comprises applying (meth-)acrylic acid-alkyl-esters of the general formula:

$$R_2(O)_n-\underset{\underset{O}{\|}}{\overset{\overset{R_1}{|}}{P}}-(O)_mZ-O-CO-CR_3{=}CH_2 \qquad (1)$$

in which n and m are 0 or 1 and (n+m) is 0 or 1, and R_1 stands for a saturated or unsaturated aliphatic hydrocarbon radical having from 1 to 8, preferably from 1 to 4 carbon atoms. The radical may also be substituted, preferably monosubstituted by chlorine, bromine, the nitrile group or the radicals $-OR_4$ or $-COOR_4$.

R_2 has the same meaning as R_1 and, for n = 0 and m = 1, may also be a radical of the formula:

(2)
$$CH_2{=}CR_3{-}CO{-}O{-}Z^1{-}O{-}\underset{\underset{\displaystyle O}{\|}}{\overset{\overset{\displaystyle R_1}{|}}{P}}{-}Z^2{-}$$

or

(3)
$$X{-}CHR_5{-}CH_2{-}O{-}\underset{\underset{\displaystyle O}{\|}}{\overset{\overset{\displaystyle R_1}{|}}{P}}{-}Z^2{-}$$

Z, Z^1, Z^2 represent straight-chain or branched (C_1-C_3)alkylene radicals which may also be substituted, preferably monosubstituted by chlorine, bromine or the hydroxy group, X stands for a hydrogen, a chlorine, or a bromine atom, a hydroxy or a methyl group. R_3 represents a hydrogen atom or a methyl group. R_4 is a hydrogen atom or an alkyl group having from 1 to 8 carbon atoms. R_5 is a hydrogen atom, the methyl or the chloromethyl group. The above (meth-)-acrylic acid-alkyl-esters are dissolved in an inert solvent and are subjected to a polymerization by using radicals. These monomers are generally well soluble in water and can be applied onto the fiber material in an aqueous finishing bath.

The good polymerization capability of the acrylic acid esters makes the polymerization possible without protection by an inert gas and in the presence of atmospheric oxygen, on the textile substrates under conditions which are appropriate for textile materials, even on a large substrate surface.

The polymers thus obtained impart a good flameproof finish to many different textile fabrics. The amounts required for this purpose are generally in the range of from 100 to 500 parts by weight of monomers of the formula 1, or of mixtures of such monomers, calculated on 1,000 parts by weight of a textile substrate.

In order to increase the permanence, particularly the fastness to washing, of the flameproof finish obtained, it is advantageous to add further comonomers to the monomers of the formula 1, especially those with a crosslinking effect. Examples of these comonomers are acrylamide, N-methylol-acrylamide, methylene-bisacrylamide, N-methylol-methylene-bisacrylamide, N,N'-dimethylol-methylene-bisacryl-amide, N-formamidomethyl-acrylamide, divinylbenzenes, triallylcyanuric acid esters, imides and anhydrides of acrylic acid and methacrylic acid, and 1,3,5-trisacryloyl-hexahydro-1,3,5-triazine.

A 200 g/l sample of a 40% aqueous plastic dispersion of a copolymer of ethylacrylate/acrylonitrile/N-methylol-acrylamide in a ratio of 6:3:1 to be used in the following examples is made up.

Example 1: A needle felt on the basis of 100% polypropylene fibers having a weight of 1,000 grams per square meter was impregnated on a two-roll padding mangle with an aqueous impregnation solution containing 350 g/l of dimethylphosphinyl-methyl-methacrylate; 90 g/l of a 50% aqueous solution of N,N'-dimethylol-methylene-bisacrylamide; 5 g/l of potassium persulfate; and 200 g/l of the 40% plastic dispersion. The squeezing-off effect was 105%. The drying was carried out for 25 minutes at 130°C.

The needle felt on the basis of polypropylene fibers thus finished had a good flameproof quality. The flameproof effect lasted for several washings at a temperature of from 40° to 50°C and several shampooings. The feel was very elastic and was not hardened. The needle felt showed a very good retention of shape and could be easily rolled up.

The flameproofing test was effected according to the US Motor Vehicle Safety Standard No. 302 (US-MVSS 302). A test sample which had been finished only with 200 g/l of the 40% plastic dispersion mentioned in the example continued to burn, after removal of the flame source, with a very strong flame front, and the flame covered a measuring distance of 10 cm with a burning rate of 2 minutes and 10 seconds.

The test sample finished with the flameproof agent did not go on burning after removal of the flame source, but only showed an afterburn effect of 10 seconds. After 6 shampooings a period of afterburning of 15 seconds was detected, and after this time the flame had gone out.

Example 2: A needle felt floor covering consisting of 100% polypropylene fibers having a weight of 1,050 grams per square meter was treated on a two-roll padding mangle with an aqueous impregnation bath containing 380 g/l of the compound of the formula:

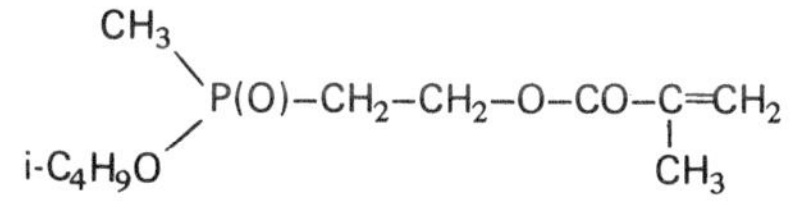

90 g/l of a 50% aqueous solution of N,N'-dimethylol-methylene-bisacrylamide; 5 g/l of potassium persulfate; and 150 g/l of plastic dispersion. The squeezing-off effect was 100%. The drying was effected for 20 minutes at 135°C.

The needle felt thus impregnated showed a very good retention of shape and an improved fastness to rubbing, as compared against a needle felt which had been impregnated only with the plastic dispersion. It could be reeled off very well. The flameproof effect on the polypropylene needle felt was surprisingly good. It lasted for several washings at a temperature of 50°C and several shampooings.

The flameproofing test was effected according to DIN 54332 (Determination of the rate of flame spreading for textile materials, which is a good approach to MVSS 302). After removal of the flame source, the needle felt did not burn any longer. After 5 shampooings, an afterburn effect of 10 seconds was detected. Then the flame went out. Similar results were obtained when the compound of the formula below was used.

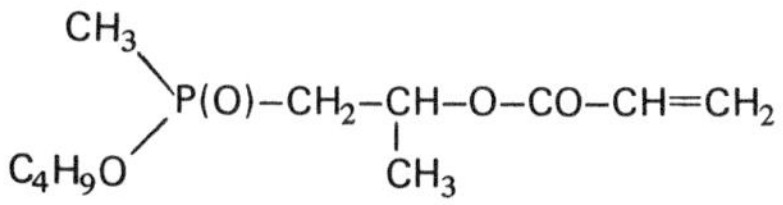

OTHER ADDITIVES

Iron Oxides and Bromobiphenyls

B.T. MacKenzie, Jr., and S. Rothenberg; U.S. Patent 4,022,945; May 10, 1977; assigned to General Electric Company describe a fire retarding system comprising combinations of iron oxides with halogen-containing materials which had heretofore been used in fire retarding systems with antimony compounds. For example, conventional halogenated hydrocarbons for flame resistance comprise chlorinated diphenyl alkanes, and their brominated or other halogenated equivalents such as hexabromobiphenyl, decabromobiphenyl or decabromobiphenyl oxide.

With the fire retarding systems of this process, typical amounts of the oxides of iron for dispersion through a polymeric composition comprise approximately 5 to 10% by weight, based upon the total weight of the combustible components of the overall polymeric composition. The proportions of the halogen-containing material range from 15 to 35% by weight of overall bromine content.

The following examples and data derived from a comparative evaluation of several fire retarding systems and ingredients, including the system of this process comprising the replacement of antimony trioxide with smaller proportions of iron oxides, when applied to common crosslink cured polyethylene compositions with a typical bromine-containing hydrocarbon, demonstrate the relative effectiveness of such systems.

The polymeric composition employed consisted of the cured product of the following curable compound formulation given in relative parts by weight: low density polyethylene, 500 parts; antioxidant, polymerized 1,2-dihydro-2,2,4-trimethylquinoline, 5 parts; curing coagent, triallyl cyanurate, 5 parts; and peroxide curing agent, di-α-cumyl peroxide (Di Cup R, Hercules Company), 17.5 parts.

To provide the halogen-containing material for the fire retarding systems evaluated within the given polyethylene composition, 100 parts by weight of a brominate flame retardant was combined with the polyethylene formulation.

The flame resistance of the polyethylene composition containing 100 parts by weight of a brominated flame retardant as a standard, and the base composition including an oxide were each measured by the Oxygen Index Test (ASTM D-2863-70). The bromine-containing materials added to the given base composition providing the standard, and the oxygen index values of the standard alone and including the added iron oxides in the amounts specified are all given in the following table of the fire retarding systems and their relative resistance to flame. All amounts of ingredients are in relative parts by weight, and in each instance the base composition is present in 627.5 parts by weight, in the relative proportions of ingredients set forth above.

	. . . Standards . . .		. . Examples. . .	
	A	C	1	2
Polyethylene	500	500	500	500
Antioxidant	5	5	5	5

(continued)

	Standards A	Standards C	Examples 1	Examples 2
Curing coagent	5	5	5	5
Peroxide curing agent	17.5	17.5	17.5	17.5
Hexabromobiphenyl	100	–	100	–
Decabromobiphenyl oxide	–	100	–	100
Fe_2O_3 (red oxide)	–	–	32.5	32.5
Oxygen Index	0.198	0.200	0.212	0.289

The polyethylene compositions of the standard and the additives of each example were all prepared and heat cured with the peroxide curing agent in an identical manner. That is all ingredients of the formulation, except the peroxide curing agent were first blended by mixing at a roll temperature of about 200°F for a period of about 10 minutes on a two roll rubber mill, then the peroxide curing agent was added and mixing continued at a temperature of 200°F for 3 minutes. Equal quantities of the blended composition of the standard and each sample were molded at about 1,300 psi and 350°F for about 45 minutes, thereby crosslink curing the polyethylene with the peroxide agent, and forming a test specimen measuring 4⅔" x 4⅔" for evaluation of its flame resistance by the oxygen index test procedure.

The method of rendering polymeric materials or compositions resistant to flame and combustion, comprises the addition and distribution of the components of the fire retarding system comprising a particulate oxide of iron with a halogen-containing hydrocarbon throughout the mass of the polymer by any suitable mixing or compounding technique or apparatus.

Magnesium Hydroxide

H. Yui, S. Moriwaki, Y. Ichikawa, and S. Ohi; U.S. Patent 4,067,847; assigned to Mitsubishi Petrochemical Co., Ltd., Japan describe a self-extinguishing resin composition which comprises 55 to 25 wt % of a thermoplastic resin having a melt index of 10 to 0.1 g/10 min and 45 to 75 wt % of magnesium hydroxide wherein the magnesium hydroxide has a specific surface area of less than 45 m^2/g. Use of this special magnesium hydroxide prevents the formation of a silver-colored streak after extrusion-molding or injection-molding of the composition.

The surface area of magnesium hydroxide to be used in a specific case is dependent upon the melt index of the thermoplastic resin used. The following are the ranges of the optimum values:

Optimum Specific Surface Area of Magnesium Hydroxide (m^2/g)	MI of Thermoplastic Resin (g/10 min)
Extrusion molding:	
38–45	more than 8
40–33	8–4
36–28	4–1
32–25	1–0.1
Injection molding:	
30–32	more than 10
25–18	10–6
22–10	6–3
less than 20	3–1

These values are especially suitable for the thermoplastic resin, polypropylene. As can be seen, the optimum specific surface area decreases with decreasing melt index of the resins. The following examples illustrate the use of magnesium hydroxide at various specific surface areas.

Magnesium Hydroxide: Preparation No. 1 – In a 2 liter flask, 1 liter of a bittern (magnesium content of 3.25 wt %) was charged, and 270 ml of an equivalent amount of an aqueous solution of sodium hydroxide was added dropwise to the bittern with stirring during 60 hours. The temperature in the flask was maintained at 70°C. After the addition, the mixture was allowed to stand for 4 hours. The precipitate was separated, dried and crushed. The specific surface area measured was 34 m^2/g.

Preparation No. 2 – In a 2 liter flask, 500 ml of the bittern of Preparation 1 was mixed with 1.5 wt % of a slurry of magnesium hydroxide having a solid content of 32 wt %. The specific surface area of the magnesium hydroxide in the solid component was 63 m^2/g. This was produced by adding dropwise 17.5 ml of calcium hydroxide containing an equivalent amount relative to 500 ml of sea water (magnesium content of 0.13 wt %) during 12 hours.

Afterwards the mixture was kept for 4 hours and the supernatant liquid was separated. 270 ml of an equivalent amount of calcium hydroxide was added dropwise to the mixture with stirring during 96 hours. The temperature in the flask was maintained at 80°C. After the addition, the mixture was allowed to stand for 4 hours. The precipitate was separated, dried and crushed. The specific surface area measured was 19 m^2/g.

Preparation No. 3 – In a 2 liter flask, 500 ml of sea water (magnesium content of 0.13 wt %) was charged, and 17.5 ml of an equivalent amount of an aqueous solution of calcium hydroxide was added dropwise to the sea water with stirring during 6 hours at 50°C. After the addition, the mixture was allowed to stand for 4 hours. The precipitate was separated, dried and crushed. The specific surface area measured was 85 m^2/g.

Preparation No. 4 – Preparation 3 was repeated except for using a bittern (magnesium content of 3.25 wt %) instead of sea water to produce magnesium hydroxide. The specific surface area measured was 17 m^2/g.

Example 1: 40 wt % of polypropylene (MI = 8 g/10 min), 60 wt % of magnesium hydroxide having a specific surface area of 19 m^2/g (Preparation 2) and a lubricant of calcium stearate in a ratio of 1.8 wt parts to 100 wt parts of the total components were blended and pelletized to obtain a composition having an MI of 1.5 g/10 min. Pellets of the composition were injection-molded at 250°C under a primary pressure of 820 kg/cm^2 by an injection-molding machine. The moldability of the composition, the appearance of the molded product and extrusion moldability were all good. On the other hand, magnesium hydroxide having a specific surface area of 34 m^2/g (Preparation 1) was used in the same formulation and the mixture was blended and pelletized under the same conditions to obtain a composition having an MI of 0.05 g/10 min.

When the pellets of the composition were used, the injection moldability was slightly inferior (a primary pressure of more than 950 kg/cm^2 was required) and a silver-colored streak appeared on the surface, deteriorating the appearance, although the extrusion moldability was good. The combustibility of the molded product was 94 V-0. On the other hand, magnesium hydroxide having a specific surface area of 85 m^2/g or 47 m^2/g (Preparations 3 and 4) was used in the same formulation and the mixture was blended and pelletized under the same conditions. A composition having an MI of less than 0.01 g/10 min was obtained and was difficult to use for either injection molding or extrusion molding.

Example 2: Magnesium hydroxide, having various specific surface areas and polypropylene, propylene-ethylene copolymer, high density polyethylene or polystyrene were blended and pelletized. Injection-molding and extrusion-molding were conducted using pellets of each composition. When using magnesium hydroxide having a specific surface area of more than 45 m^2/g, both the injection-molding and the extrusion-molding were difficult to perform. When using magnesium hydroxide having a specific surface area of less than 45 m^2/g (34 m^2/g and 19 m^2/g) injection-molding could be performed. When magnesium hydroxide, having a specific surface area of less than 20 m^2/g (19 m^2/g) was used, no silver-colored streaks were found on the surface of the molded product, giving a good appearance. The combustibility of the products was self-extinguishing.

Boric Acid Esters

H. Mayerhoefer, W. Müeller, and U. Sollberger; U.S. Patent 4,021,464; May 3, 1977; assigned to Sandoz Ltd., Switzerland describe procedures for the synthesis of boric acid esters and their use as flame retardants for organic combustible materials, in particular, polypropylene and polyethylene. These compounds are represented by the following formula:

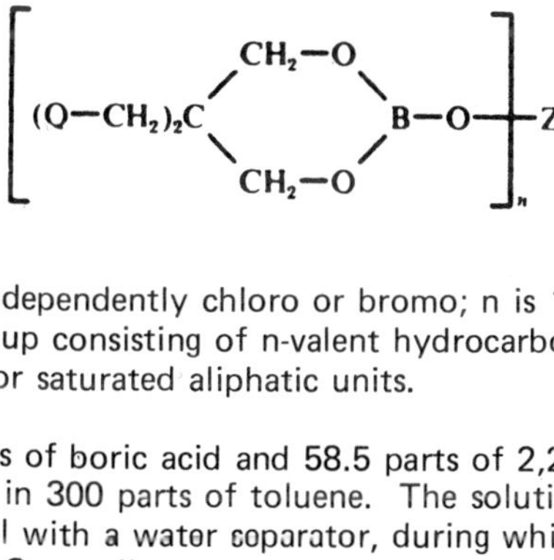

in which each Q is independently chloro or bromo; n is 1, 2, 3 or 4, and Z is selected from the group consisting of n-valent hydrocarbon groups which consist of aromatic and/or saturated aliphatic units.

Example 1: 3.7 parts of boric acid and 58.5 parts of 2,2,2-(tris-bromomethyl)-ethanol are dissolved in 300 parts of toluene. The solution is boiled for 6 hours with reflux in a vessel with a water separator, during which time 3.2 parts of water are separated. On cooling to room temperature, a white crystalline product settles out. Recrystallization from toluene yields 52 parts of the product with MP 136° to 138°C, which is equivalent to a yield of 83%.

Example 2: A solution of 52.4 parts of 2,2-(bis-bromomethyl)-1,3-propanediol and 12.35 parts of boric acid in 350 parts of benzene is boiled with reflux in a vessel with a water separator. In about 3 hours 8 parts of water are separated, at which time 13.4 parts of monoethanolamine are slowly added. In the next two hours a further 1.5 parts of water are separated. On cooling to room temperature 54.2 parts of white crystalline product, MP 161° to 163°C, settled out. This is equal to a yield of approximately 82%.

Example 3: 9 parts of ethylene glycol are slowly added to 35 parts of boron trichloride, with subsequent stirring for 1 hour at room temperature under nitrogen. At this point 50 parts of toluene are added and then, with cooling, 73 parts of 2,2-(bis-bromomethyl)-1,3-propanediol in portions. The reaction mixture is stirred for 2 hours at room temperature and freed from solvent by the application of vacuum. 84 parts of a colorless oil are obtained, which amounts to a yield of 96%.

Example 4: A solution of 8.16 parts of boric acid and 51.9 parts of 2,2-(bis-bromomethyl)-1,3-propanediol in 240 parts of benzene is boiled for 5 hours with reflux in a vessel with a water separator. During this time, 7.1 parts of water are separated. The solvent is removed by vacuum distillation and the oily residue freed from residual solvent in a high vacuum at 80°C by the introduction of nitrogen. 52 parts of an almost colorless, highly viscous oil are obtained, which is equal to a yield of 98%.

Example 5: The compound of Example 1 or 4 is dissolved in trichloroethylene and applied from exhaust baths to a polypropylene fabric weighing 186 grams per square meter. The fabric is freed from trichloroethylene in a vacuum drying oven at 80°C for about 2 hours.

Example 6: Polypropylene powder is intimately mixed with 8% of the boric acid ester of Example 4 by adding a 5% solution of the ester in carbon tetrachloride to the polypropylene powder in a rotating vacuum drum and running the drum to evaporate the solvent. The weight ratio is calculated so that after evaporation of the carbon tetrachloride, the polypropylene contains 8% of the boric acid ester. After regranulation, the melt is spun at 270°C as continuous filament which is cold drawn and twisted to yarn. The yarn is converted into knitted fabric on a circular knitting machine.

Hydrotalcites

S. Miyata, T. Hirose and N. Iizima; U.S. Patent 4,085,088; April 18, 1978; assigned to Kyowa Chemical Industry Co., Ltd., Japan describe a fire retarding thermoplastic resin composition comprising (a) a thermoplastic resin, and (b) a fire retarding amount of a hydrotalcite having a specific surface area, determined by the BET method, of not more than 30 m²/g.

Preferred hydrotalcites used are compounds of the following formula:

$$Mg_{1-x}Al_x(OH)_2 \cdot A_{x/n}^{n-} \cdot mH_2O$$

where x is a number of more than 0 but up to 0.5, A^{n-} represents an anion having a valence of n, preferably a divalent anion such as CO_3^{2-}, or SO_4^{2-}, and m is a positive number.

The hydrotalcites used may be treated with an anionic surface active agent to form solid particles of hydrotalcite coated with the surfactant. The coating can be performed by contacting the hydrotalcite with anionic surfactants. For example, an aqueous solution of an anionic surfactant in a concentration of not more than 20 millimoles per liter is mixed with solid particles of hydrotalcite under conditions which ensure their intimate contact, for example, by agitating them sufficiently, or by hydrothermal treatment at 120° to 250°C to form a solid powder of hydrotalcite coated with the anionic surfactant. This contacting operation causes the surfactant to be chemically adsorbed on the surface of the solid particles of the hydrotalcite, and can lead to more improved properties when the hydrotalcite is incorporated in thermoplastic synthetic resins. The amount of the anionic surfactant adsorbed to the solid particles of the hydrotalcite is such that a coating having a thickness larger than a monolayer but up to a triple layer can be formed.

Specific examples of the surface active agent are sodium stearate, potassium behenate, potassium stearate, sodium oleate, potassium oleate, sodium palmitate, potassium palmitate, sodium laurate, potassium laurate, sodium dilaurylbenzenesulfonate, potassium octadecylsulfate, sodium laurylsulfonate, or disodium 2-sulfoethyl α-sulfostearate.

Example: One kilogram of hydrotalcite having the composition formula

$$Mg_{2/3}Al_{1/3}(OH)_2(CO_3)_{1/6}\cdot 1/2H_2O$$

and having a BET specific surface area of 12 m^2/g was poured at one time into 20 liters of an aqueous solution of sodium oleate in a concentration of 5 millimoles per liter at about 40°C. The mixture was stirred thoroughly for about 30 minutes to cause the sodium oleate molecules to be chemically adsorbed to the surface of the hydrotalcite crystals, followed by dehydration, washing with water, dehydration, and drying. The adsorption cross-sectional area of one molecule of sodium oleate was 46 $(A)^2$. Hence, the amount of sodium oleate required for monolayer adsorption was 43 millimoles per kilogram. Therefore, the sodium oleate in an amount corresponding to 2.3 times (5 x 20/43) the monolayer thickness was contacted with the hydrotalcite in the above procedure. 120 parts by weight of the hydrotalcite so treated and 100 parts by weight of polypropylene were melt-kneaded, and molded.

The properties of the molded articles were evaluated and the results are as follows: injection pressure (the relative value of the injecting pressure to that used for molding the resin alone), 80; izod impact strength (ASTM D256, notched), 4.2 kg-cm/cm; tensile elongation (ASTM D638), 25.0%, fire retardancy (UL standards 94VE), V-1 or V-2; and appearance of the surface of molded article—no flash, more lustrous than the base resin.

POLYSTYRENE AND ABS

HALOGEN-CONTAINING ADDITIVES

Polyhalogenated Cycloalkanes in Expandable Styrene Polymers

E. Nintz, R. Schick, H. Horacek and E. Stahnecker; U.S. Patent 3,975,327; August 17, 1976; assigned to BASF AG, Germany describe a process whereby particulate, expandable styrene polymers containing a flameproofing organic halo compound as expanding agent and also a mixture of from 2.5 to 30% by weight of a hydrocarbon of from 3 to 6 carbon atoms and from 97.5 to 70% by weight of monofluorotrichloromethane, based on the total weight of the mixture serving as expanding agent, show good processability and exhibit self-extinguishing properties even during processing.

Suitable flameproofing agents are bromo compounds such as tetrabromobutane, dibromoethylbenzene, pentabromomonochlorocyclohexane, brominated organic monomers and organic phosphates. Preferred compounds are the brominated oligomers of butadiene and of isoprene, having a degree of polymerization of from 2 to 2,000. These compounds may be completely or partially brominated. Typical representatives are for example 1,2,5,6-tetrabromocyclooctane, 1,2,5,6,9,10-hexabromocyclododecane and brominated polybutadiene. The flameproofing agents are usually employed in amounts of from 0.4 to 1% by weight of the total weight of the polymer. It may also be advantageous to supplement the flameproofing agents by synergistic substances such as di-t-butyl peroxide, dicumyl peroxide and poly-p-diisopropylbenzene.

Suitable expanding agents are, for example, normally gaseous or liquid halo hydrocarbons and hydrocarbons of from 3 to 6 carbon atoms which are nonsolvents for the styrene polymer and have boiling points below the softening point of the polymer. Due to its good solubility in styrene polymers, a suitable halo hydrocarbon is monofluorotrichloromethane. Suitable C_{3-6} hydrocarbons are, for example, propane and the isomers or isomer mixtures of butane, pentane and hexane.

The prefoamed polymer particles obtained from particulate expandable self-extinguishing styrene polymers may be used for packaging purposes in place of wood-wool or as an upholstery material. The prefoamed particles may also be sintered by heating in molds which are not gas-tight when closed to form foamed articles having dimensions corresponding to those of the cavity of the mold used.

Example 1: A mixture of:

Water of hardness 15°dH	480 kg
Sodium pyrophosphate	360 g
Styrene	405 kg
t-Butyl perbenzoate	1.19 kg
Dibenzoyl peroxide	330 g
Dicumyl peroxide	820 g
1,2,5,6,9,10-Hexabromocyclododecane	2.65 kg

is polymerized in a 1,000 liter pressure vessel equipped with stirrer and purged with nitrogen, for 3 hours at 90°C, 1 hour at 100°C, 1 hour at 110°C and then 5 hours at 115°C, with stirring. During the heating period from 90° to 115°C, 15 kg of a 10% w/w aqueous polyvinyl pyrrolidone solution having a K value of 90 are added to the reaction mixture. After a polymerization time of 3.5 hours, an expanding agent mixture consisting of 33 kg of monofluorotrichloromethane and 8.2 kg of pentane is added to the reaction over 15 minutes. The expandable polystyrene produced in this way is prefoamed in conventional prefoamers to form material having a density of 12 g/l.

The foamed particles are placed in a coarse mesh brass basket having a capacity of approximately 0.5 liter and are subjected to a burning test. This test consists in applying a 5 to 6 cm long Bunsen burner flame to the prefoamed particles for 3 seconds. After removal of the flame, the foamed beads burn for a further 0.6 to 0.9 second.

Example 2: A mixture of:

Water of hardness 15°dH	480 kg
Sodium pyrophosphate	360 g
Styrene	385 kg
Acrylonitrile	21 kg
t-Butyl perbenzoate	1.11 kg
Dibenzoyl peroxide	440 g
Dicumyl peroxide	820 g
Pentabromomonochlorocyclohexane	4.1 kg

is polymerized in a 1,000 liter pressure vessel equipped with stirrer and purged with nitrogen, for 4 hours at 85°C, 2 hours at 100°C, 1 hour at 110°C and 4 hours at 115°C. After a polymerization time of 1 hour at 85°C, 13 kg of a 10% w/w aqueous polyvinyl pyrrolidone solution having a K value of 90 are added to the reaction mixture, and after a polymerization time of 4 hours, an expanding agent mixture consisting of 31 kg of monofluorotrichloromethane and 8.2 kg of pentane is added to the reaction mixture over 15 minutes. The resulting expandable styrene copolymer is prefoamed to form a material having a density of 12 g/l, which is subjected to the burning test described in Example 1. The prefoamed beads exhibit a burning time of from 0.8 to 1.2 second.

Example 3: An expandable styrene polymer is manufactured in a manner similar to that described in Example 1 except that 3.5 hours after the reaction mixture has reached a reaction temperature of 90°C an expanding agent mixture consisting of 29 kg of monofluorotrichloromethane and 10.3 kg of pentane is added over 15 minutes. The prefoamed beads have a density of 12 g/l and a burning time of from 1.2 to 1.5 seconds.

Example 4: An expandable styrene polymer is manufactured by a method similar to that described in Example 1 except that 3.5 hours after the reaction mixture has reached a reaction temperature of 90°C 30 kg of a mixture consisting of 75% by weight of n-pentane and 25% by weight of isopentane are added. The prefoamed product has a density of 12 g/l and burns away completely when the source of ignition is removed.

The following table shows that shaped articles made from expandable styrene polymers as described in Examples 1 through 3, are self-extinguishing immediately after manufacture and have a burning time of less than 3 seconds. By contrast, shaped articles made from expandable polymers as described in Example 4 are still flammable after a storage time of 1 week.

Ex. No.	Foaming Rate in Prefoaming* (sec)	Density After Steaming** (g/l)	Steam Pressure in Mold (atm gage)	Steaming Time (sec)***	Degree of Interwelding (%)†	Sag of Total Sides (%)††
1	34	10.0	0.5	50	50	2.0
			0.4	90	45	1.5
2	37	11.0	0.5	50	60	2.3
			0.4	90	45	2.0
3	33	10	0.5	50	60	2.5
			0.4	90	40	1.8
4	33	10	0.5	50	60	1.4
			0.4	90	30	1.0

*At steam pressure 0.3 atm gage. Initial weight of sample 9 kg, desired final density 15 g/l.

**For 6 min measured at atm pressure.

***The steaming time is the time from the moment at which the specified steam pressure is achieved in the block mold to the moment of closing the steam inlet valve.

†The degree of fusion is the ratio of the number of torn beads to the total number of beads in the surface of a fracture produced by force.

††Sag of the sides is measured 24 hr after the block has been removed from the mold. It is determined by measuring the thickness of the block in the center of one of the large sides at right angles thereto. The difference between the inside dimension of the mold at the point and the thickness of the molding at the same point, converted to percent of the inside dimension of the mold, gives the degree of sag.

In related work, *K. Hinselmann, K. Penzien, R. Schick, M. Walter, H. Weber and H. Wirth; U.S. Patent 4,086,191; April 25, 1978; assigned to BASF AG, Germany* provide a process for manufacturing expandable styrene polymers containing from 0.0001 to 0.2% by weight, based on the styrene polymer, of a nonhomogeneously distributed nucleating agent for expandable polystyrene. This comprises: treating expandable polystyrene particles at a temperature above their softening point with organic bromine nucleating compounds which contain at least 3 bromine atoms per molecule bonded to an aliphatic or cycloaliphatic structure and limiting the time available for the diffusion of the bromine compound into the expandable polystyrene particles such that the bromine compound is only present in a thin peripheral zone of the particles whereas the interior of the particles is free from the bromine compound.

After foaming the expandable polystyrene particles, the foamed particles show a coarse cellular structure of from 4 to 7 cells per millimeter in the interior while the thin peripheral zone has a fine cellular structure which is from 0.1 to 3 millimeters thick, the peripheral zone having a negligible thickness compared to the total cross-sectional area of the foamed particle.

2,5-Tribromomethyl-1,3,4-Thiadiazoles

H. Fleig, H. Hagen, H. Naarmann and K. Penzien; U.S. Patent 4,049,622; Sept. 20, 1977; assigned to BASF AG, Germany describe a process whereby thermoplastic molding compositions based on styrene polymers contain, as a flameproofing agent, an organic bromine compound of the general formula:

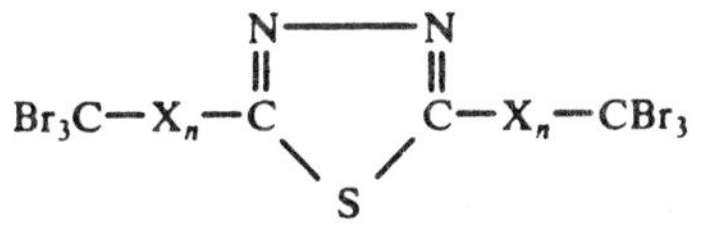

where X is a hydrocarbon radical and n is 0 or 1. The compositions may be used for the manufacture of self-extinguishing moldings, e.g., foams.

Preferred organic bromine compounds to use are those of the above thiadiazoles which contain more than 40% by weight of bromine. These compounds are particularly suitable since they are of low volatility, have only a slight plasticizing action, if any, and do not have an objectionable odor. The bromine compounds are suitably used in such amounts that the molding compositions contain not less than 0.1%, and not more than 5% of bromine. Preferably, they contain from 0.5 to 3% by weight of bromine.

The molding compositions may also contain further components, e.g., fillers, colored pigments, lubricants, plasticizers, antistatic agents, antioxidants, stabilizers or compounds which assist foaming.

The self-extinguishing moldings manufactured from the molding compositions are tested as follows: The test is carried out with moldings of size 0.1 x 10 x 30 cm in the case of unfoamed compositions and with moldings of size 0.5 x 15 x 40 cm in the case of foamed compositions; the moldings are held for 5 seconds in a gas flame of 40 mm height, and the flame is then removed with a gentle movement. The time the molding takes to self-extinguish after it has been removed from the flame is a measure of its nonflammability. Molding compositions which have been rendered insufficiently nonflammable, if at all, burn away completely after removal from the flame.

Examples 1 through 17: In each case, 30 parts of a styrene polymer and 0.3 part of an organic bromine compound are dissolved in 100 parts of methylene chloride. 3 parts of pentane are added to the solution. The solution is then cast on a glass plate and the methylene chloride is allowed to evaporate at room temperature. At the same time the pentane remains homogeneously distributed in the mixture. The film thus obtained is foamed in steam at 100°C and is dried under reduced pressure at 35°C for 12 hours. The film sheets obtained are tested for their nonflammability by the method described above. The results are shown in the table. Azodiisobutyronitrile (y) or dicumyl peroxide (a) were employed as synergistic agents. HBCD is hexabromocyclododecane, a conventional flameproofing agent. Compounds (I) and (II) are shown below.

(I)

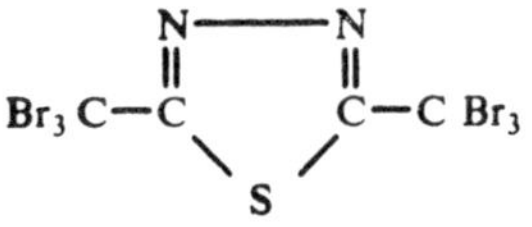

(II)

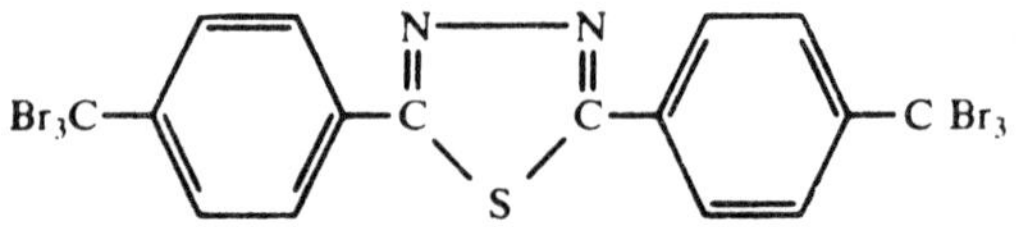

	Bromine compound		Synergistic agent		Extinction time
Example	Type	Amount	Type	%	seconds
1	HBCD		—	—	12.0
2	"	1%	a	0.05	3.5
3	"	1%	a	0.1	3.0
4	"	1%	a	0.5	2.0
5	"	1%	y	0.05	4.9
6	"	1%	y	0.1	4.5
7	"	1%	y	0.5	2.0
8	I	1%			0
9	I	0.8%			0
10	I	0.6%			0
11	I	0.4%			0
12	II	1%			0
13	II	0.8%			0
14	II	0.6%			0
15	II	0.4%			1.0
16	I	1%	a	0.05	0
17	I	1%	HBCD	0.2	0

When using (I) or (II), only slight dripping of flaming particles was observed, while in Comparative Experiments 1 through 7 and 16, very severe dripping was observed. Examples 8 through 17 are in accordance with this process.

Example 18: Example 8 was repeated, using a copolymer of 75% of styrene and 25% of acrylonitrile. The extinction time was 0 second while when using 1% of HBCD and 0.3% of synergistic agent (a) it was 3.5 seconds.

Example 19: A mixture of 100 parts of a styrene polymer which has been manufactured by polymerizing 95 parts of styrene in the presence of 5 parts of polybutadiene, and 1.5 parts of bis-tribromomethylthiazole (I) is extruded in an extruder, using a sheet die, to give a 1 mm thick sheet. The latter is tested for its nonflammability. It self-extinguishes 1.0 second after removing the flame. If, by way of comparison, a sheet which contains 1.5 parts of HBCD and 0.5 part of synergistic agent (a) is tested, it self-extinguishes after 8 seconds.

Diadducts of Halogenated Cyclopentadienes

I. Gordon, J.J. Duffy and N.W. Dachs; U.S. Patent 4,000,114; December 28, 1976;

assigned to Hooker Chemicals & Plastics Corporation describe a fire retardant polymeric composition comprising a high impact thermoplastic polymer and an effective fire retardant proportion of decabromodiphenyl oxide and a diadduct of a halogenated cyclopentadiene and a polyunsaturated cyclic compound. The high impact thermoplastic polymers used are the styrenic based polymers, ABS and high impact polystyrene. The diadducts of halogenated cyclopentadienes useful in the fire retardant polymer composition have the formula:

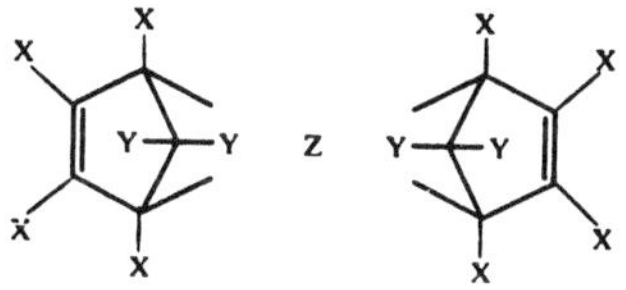

where X is selected from chlorine, bromine and fluorine, Y is selected from chlorine, bromine, fluorine, alkyl of 1 to 10 carbon atoms, alkyloxy where the alkyl group contains 1 to 10 carbon atoms, haloalkyl and haloalkyloxy where the alkyl groups contain 1 to 10 carbon atoms and halo is chloro, bromo or fluoro; Z is a tetravalent cyclic radical having at least 4 carbon atoms which may be substituted by alkyl groups of 1 to 6 carbon atoms, chlorine, bromine or fluorine.

The alkyl and alkoxy radicals preferably have 1 to 6 carbon atoms. Z is preferably a tetravalent cyclic hydrocarbon of 5 to 18 carbon atoms or a tetravalent heterocyclic radical of 4 to 18 carbon atoms and preferably has 1 to 5 cyclic structures. When Z is a plurality of cyclic structures, they are fused, that is, share carbon atoms.

The foregoing diadducts are generally prepared by the Diels-Alder reaction of 1 mol of a polyunsaturated cyclic compound, such as a polyunsaturated cycloaliphatic compound or polyunsaturated heterocyclic compound, and 2 or more mols of a polyhalogenated cyclopentadiene of the formula:

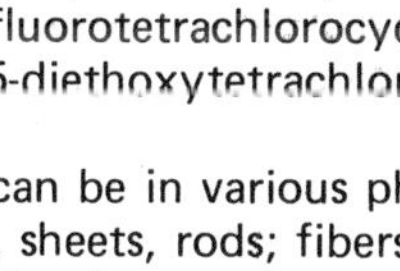

where X and Y have the same definition as previously stated. Illustrative of the polyhalogenated cyclopentadienes suitable for use in preparing the diadducts are hexachlorocyclopentadiene, 5,5-dimethoxytetrachlorocyclopentadiene, hexabromocyclopentadiene, 5,5-difluorotetrachlorocyclopentadiene, 5,5-dibromotetrachlorocyclopentadiene and 5,5-diethoxytetrachlorocyclopentadiene.

The polymers of this process can be in various physical forms, such as shaped articles, for example, molding, sheets, rods; fibers, coatings, films and fabrics. Such materials find use in applications such as television cabinets and other appliances.

Examples: Formulations were prepared by melt blending fire retardant additive compounds and either an ABS polymer or a high impact polystyrene on a two roll mill, followed by injection molding the mixed formulations, in the proportions shown in Tables 1 to 4. The properties of each molded specimen were determined and the values are also shown in Tables 1 to 4.

TABLE 1

	A	B	C	D	E	F	G	H	I	J	K
Formulation, parts by wt											
ABS polymer (Blendex 101)	100	80	80	80	80	80	80	80	80	80	80
Antimony trioxide	–	5	5	5	5	5	5	5	3	3	3
Diadduct of hexachloro-cyclopentadiene and furan	–	15	–	12	10	7.5	5	3	11	6	4
Decabromodiphenyl oxide	–	–	15	3	5	7.5	10	12	6	11	13
Oxygen index	19	24.5	25.5	26±1	27±1	27.5	27±1	27±1	26±1	27±1	26±1
ASTM D-635	B	SE	SE	SE	SE	SE	SE	SE	SE	SE	SE
UL-94 ⅛" specimen	–	V-1	V-2	V-0	V-0	V-0	V-0	V-0	V-0	V-0	V-0
UL-94 1/16" specimen	–	NA	NA	NA	NA	NA	NA	NA	NA	NA	NA
Izod impact strength ft lb/in notch	8.6	3.3	2.8	4.4	4.6	7.4	4.2	4.0	4.5	4.3	4.0
Heat distortion											
at 264 psi, °C	69	70	69	67	68	67	67	68	69	68	68
at 66 psi, °C	83	84	83	83	83	84	83	85	85	84	84
Tensile strength											
at yield	5,431	5,034	5,397	4,988	5,046	5,136	5,158	5,199	5,056	5,128	5,189
at break	4,056	3,950	4,150	3,819	3,819	3,874	3,949	3,819	3,814	3,796	3,904
Flexural strength, psi	8,656	8,532	8,859	8,034	7,952	8,052	8,257	8,357	8,191	8,129	8,441
Flexural modulus x 10^5	2.71	2.94	2.93	2.54	2.50	2.53	2.64	2.59	2.57	2.55	2.70

Note: SE is self-extinguishing; NA is not acceptable; and B is burning.

TABLE 2

	A	B	C	D	E	F	G
Formulation, parts by weight							
ABS polymer (Blendex 101)	100	80	80	80	80	80	80
Antimony trioxide	–	5	5	5	5	5	5
Diadduct of hexachlorocy-clopentadiene and 1,5-cyclooctadiene	–	15	–	10	7.5	5	3
Decabromodiphenyl oxide	–	–	15	5	7.5	10	12
Oxygen index	19	24.5	25.5	26±1	26±1	26±1	26±1
ASTM D-635	B	SE	SE	SE	SE	SE	SE
UL-94 ⅛" specimen	NA	NA	V-2	NA	V-0	V-0	V-0
UL-94 1/16" specimen	NA	NA	NA	NA	NA	NA	NA
Notched Izod strength, ft lb/in notch	8.6	1.1	2.8	2.1	2.5	2.7	2.9
Heat distortion							
at 264 psi, °C	69	71	69	69	68	70	70
at 66 psi, °C	83	87	83	86	84	85	84
Tensile strength							
at yield	5,431	5,425	5,397	5,367	5,400	5,338	5,311
at break	4,056	4,223	4,150	3,959	4,015	3,982	3,991
Flexural strength, psi	8,656	9,097	8,857	8,610	8,627	8,512	8,492
Flexural modulus x 10^5	2.71	2.14	2.93	2.82	2.75	2.72	2.70

TABLE 3

	A	B	C	D	E	F	G	H	I	J
Formulation, parts by wt										
High impact poly-styrene (Styron 492)	100	80	80	80	80	80	80	80	80	80
Antimony trioxide	–	5	5	5	5	5	5	5	3	3
Diadduct of hexachloro-cyclopentadiene and furan	–	15	–	12	10	7.5	5	3	11	6

(continued)

TABLE 3: (continued

	A	B	C	D	E	F	G	H	I	J
Formulation, parts by wt										
Decabromodiphenyl oxide	–	–	15	3	5	7.5	10	12	6	11
Oxygen index	17.6	26	28	26	28	30	27	28	26	24
ASTM D-635	B	SE	SE	SE	SE	SE	SE	SE	SE	SE
UL-94 ⅛" specimen	NA	V-1	V-0	V-1	V-1	V-1	V-1	V-0	V-0	V-0
UL-94 1/16" specimen	NA	V-2	V-0	V-2	V-1	V-2	V-2	V-2	V-2	V-2
Izod impact strength, ft lb/in notch	1.7	0.8	1.1	1.3	1.3	1.3	1.3	1.3	1.3	1.2
Heat deflection										
at 264 psi, °C	75	71	76	74	74	74	73	74	74	73
at 66 psi, °C	86	84	85	83	83	84	83	85	83	84
Tensile strength										
at yield	4,425	–	–	4,756	4,827	4,728	4,869	4,949	4,777	4,832
at break	4,443	4,728	4,997	4,601	4,702	4,584	4,779	4,896	4,672	4,755
Flexural strength, psi	8,721	8,403	9,147	9,754	9,705	9,715	9,911	10,227	9,823	10,040
Flexural modulus x 10^5	3.30	3.98	4.01	3.82	3.81	3.74	3.86	3.86	3.71	3.79

TABLE 4

	A	B	C	D	E	F	G	H
Formulation, parts by weight								
High impact polystyrene (Styron 492)	100	80	80	80	80	80	80	80
Antimony trioxide	–	5	5	5	5	5	5	3
Diadduct of hexachlorocyclopentadiene and 1,5-cyclooctadiene	–	15	–	10	7.5	5	3	4
Decabromodiphenyl oxide	–	–	15	5	7.5	10	12	13
Oxygen index	17.6	24	28	28	26	26	26	28
ASTM D-635	B	SE	SE	SE	SE	SE	SE	SE
UL-94 ⅛" specimen	NA	V-1	V-0	V-0	V-1	V-0	V-0	V-0
UL-94 1/16" specimen	NA	V-2	V-0	V-2	V-0	V-2	V-0	V-2
Izod impact strength, ft lb/in notch	1.7	0.5	1.1	0.7	1.1	1.0	1.3	1.3
Heat distortion								
at 264 psi, °C	75	79	76	79	76	75	75	72
at 66 psi, °C	86	89	85	87	86	87	85	82
Tensile strength								
at yield	4,425	–	–	4,960	–	–	4,963	–
at break	4,443	4,968	4,997	5,029	4,892	4,975	4,908	4,814
Flexural strength, psi	8,721	9,235	9,147	8,770	10,297	8,721	10,391	8,766
Flexural modulus x 10^5	3.3	4.21	4.01	4.20	3.88	3.97	3.90	3.96

From the data in Table 1, it is seen that composition B containing the furan diadduct had an impact strength of 3.3 and the composition C containing the oxide had an impact strength of 2.8, but the compositions containing a mixture of the two additives have impact strengths of 4.0 up to 7.4, compared to 8.6 for the base polymer. The compositions containing a mixture of additives had a fire retardancy rating of V-0 as measured by the Underwriter's Laboratory Test No. 94 as compared to a rating of V-1 and V-2 for the compositions containing the individual additives.

From the data in Table 2, it is seen that composition B containing the diadduct of 1,5-cyclooctadiene had an impact strength of 1.1, but when this additive was blended with the oxide, higher impact strengths were achieved. The fire retardancy as measured by the Underwriter's Laboratory Test-94 was better for the blends compared to the individual additives.

The data in Tables 3 and 4 show that the improvement in properties that prevailed for the additive blends in ABS in Tables 1 and 2, also prevail for the additive blends of the process when used in high impact polystyrene.

Styrene-Butadiene-Styrene Block Copolymer

J.L. Snyder; U.S. Patent 4,038,237; July 26, 1977; assigned to Shell Oil Company describes a fire resistant polymer composition suitable for the coating of electrical conductors for use in high temperature service. This composition comprises:

(1) 100 parts by weight of a block copolymer having at least 2 polymer end blocks A and at least one polymer mid block B, each block A being a nonelastomeric polymer block of a monoalkenyl arene having an average molecular weight of between 2,000 and 125,000 and each block B being an elastomeric hydrogenated polymer block of a conjugated diene, having an average molecular weight between 10,000 and 250,000;

(2) 0 to 200 parts by weight of an extending oil;

(3) 10 to 300 parts by weight of a polymeric α-olefin;

(4) 5 to 100 parts by weight of antimony trioxide;

(5) 25 to 200 parts by weight of a chlorinated hydrocarbon; and

(6) 50 to 300 parts by weight of a calcium sulfate filler.

Compositions employing calcium sulfate in place of other fillers such as clay, silicates, and calcium carbonate have much improved fire retardant properties.

Typical block copolymers include:

(1) Polystyrene-hydrogenated polybutadiene-polystyrene,

(2) Polystyrene-hydrogenated polyisoprene-polystyrene,

(3) Poly(α-methyl styrene)-hydrogenated polybutadiene-poly(α-methyl styrene), and

(4) Poly(α-methyl styrene)-hydrogenated polyisoprene-poly(α-methyl styrene).

Petroleum fractions as well as olefin oligomers may be used as extending oils. The petroleum derived extending oils include relatively high boiling materials having number average molecular weights between 30 and 1,400. The oligomers may be polypropylenes, polybutenes, hydrogenated polyisoprene, hydrogenated polybutadiene, or the like having average molecular weights between 350 and 10,000.

The polymeric α-olefin component is preferably isotactic polypropylene, which is an essentially crystalline polypropylene. The solid isotactic polypropylene has weight average molecular weight of between 200,000 and 300,000. The chlorinated hydrocarbon employed is preferably a fire retardant compound derived from hexachlorocyclopentadiene (HEX). One such compound is the Diels-Alder product made by adding 2 mols of hexachlorocyclopentadiene to 1 mol of the stable cis-isomer of 1,5-cyclooctadiene. The structure of the adduct is:

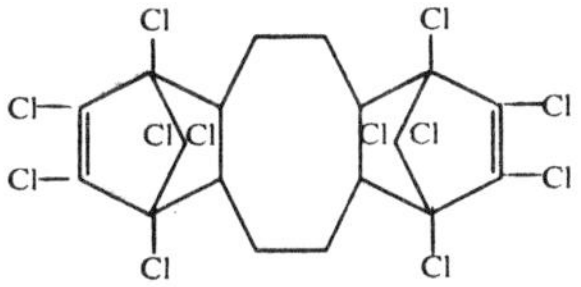

One method to prepare the compound is described in U.S. Patent 3,385,819. This is available under the name Dechlorane Plus 25.

As the inorganic filler, anhydrous $CaSO_4$ is preferred. Other additives such as pigments, extenders, antioxidants, stabilizers, and other fire retardants may also be added. The additives that can be incorporated should be selected from those whose electrical properties are such that they will not materially reduce or impair the electrical properties of the composition.

The various components may be combined in any suitable manner and may then suitably be employed as wire coatings. These compositions are especially superior for use as insulators for electrical conductors not only because of their superior electrical properties, but also for their high degree of adherence to metallic surfaces without special treatment of the surface. The electrical conductors usually employed with the instant compositions are metallic wires such as copper, silver, aluminum, and alloys thereof.

Example: Four wire coating compositions were formulated and extrusion coated on an 18 AWG bare copper wire (0.030 inch coating). In each case, the block polymer employed was a selectively hydrogenated styrene-butadiene-styrene block copolymer (SEBS). Other common ingredients included a rubber compounding oil, polypropylene resin, antioxidant package [Plastonox 2246 (2,2'-methylene-bis(4-methyl-6-tertiary-butylphenol) and Plastonox DLTDP (dilauryl thiodipropionate)], and two fire retardants; antimony trioxide and Dechlorane Plus 25 chlorinated hydrocarbon.

In composition number 1, a talc filler was employed, while in composition number 2, a clay filler was employed. The filler in composition number 3 was Camel-Carb, a calcium carbonate filler. In composition number 4, the filler was anhydrous $CaSO_4$.

These various wire coatings were subjected to a standard vertical flame test according to the method of UL-83. Only the $CaSO_4$-filled composition was self-extinguishing, the other three formulations were not self-extinguishing. The various compositions in parts by weight are presented in the table below.

Composition No.	1	2	3	4
Component				
Block copolymer	100	100	100	100
Oil	100	100	100	100
Polypropylene resin	25	25	25	25
Filler, type	Talc	Clay	$CaCO_3$	$CaSO_4$
amount	150	150	250	175
Antioxidant				
Plastonox 2246	2	2	2	2
Plastonox DLTDP	0.5	0.5	0.5	0.5

(continued)

Composition No.	1	2	3	4
Fire retardants				
Sb_2O_3	30	30	30	30
Dechlorane +25	90	90	90	90
Test Results				
	Burns	Burns	Burns	Self-Extinguishing

Halogen Derivatives of Stilbene in Expandable Polystyrene

A process described by *F. Sonnenberg; U.S. Patent 4,024,092; May 17, 1977; assigned to Foster Grant Co., Inc.* relates to alkenyl aromatic polymer compositions having a relatively high limiting oxygen index value as measured by the Oxygen Index Method (ASTM D 2863-70). They are prepared by incorporating into the polymer composition a minor but effective amount of an alpha, beta-bromine or chlorine derivative of stilbene.

These compositions may contain a mixture of these halogenated stilbenes with other conventional flame retardants and/or synergists. Polymer forms particularly susceptible to oxygen index improvement include expandable polystyrene compositions containing a small amount of blowing agent which when heated above the boiling point of the blowing agent and above the softening point of the polystyrene are caused to expand to a porous or cellular mass. Examples of suitable stilbenes include stilbene bromide, stilbene dibromide, stilbene dichloride, and stilbene tetrachloride. The following example illustrates a suitable method for the preparation of stilbene dibromide.

Example 1: A solution of 45 g (0.25 mol) of trans-stilbene in 750 ml of ether is prepared in a 1-liter, 3-necked round-bottom flask fitted with a reflux condenser, an efficient mechanical stirrer, and a dropping funnel. To the well stirred solution there is added 13.8 ml (43 g; 0.27 mol) of bromine, during the course of 10 minutes. A solid begins to separate in 5 minutes, but stirring is continued for 1 hour. The product is collected on a Buchner funnel and washed with ether until it is white. The yield of stilbene dibromide, melting at 235° to 237°C, is 65.8 to 69.1 g (77 to 81%). The product contains about 47% by weight bromine.

It has been found that relatively modest amounts of the stilbene derivatives provide dramatic increases in the oxygen index values of polymeric materials in which they are incorporated. For example, the use of stilbene dibromide in expandable polystyrene compositions at impregnation levels as low as 0.5% by weight increases the oxygen index of such compositions from the low twenties to values in the order of 30. Polymers particularly susceptible to oxygen index enhancement include homopolymers of styrene, α-methyl styrene, ar-vinyl toluene, ar-vinyl xylene and ar-ethylvinyl benzene, and copolymers or one or more of the above.

The advantages of oxygen index enhancement is best realized in expandable alkenyl aromatic resins which contain minor amounts in the order of 3 to 10% by weight of a foaming agent in homogeneous dispersion. Suitable foaming agents include volatile organic compounds which may be liquids or gases at ordinary temperatures and pressures, and include such materials as pentane, hexane heptane, cyclopentane and other compounds having a boiling range in the order of about 15° to 100°C.

The enhancement of oxygen index in expandable polymers is further realized by the inclusion of minor amounts of synergists in the composition. Such additives are effectively employed at levels of from 0.1 to 2% based on polymer weight. The amount of stilbene halide which may be used is from 0.1 to 10% based on polymer weight. In expandable polystyrene, less than 2% based on polymer weight, e.g., from 0.5 to 2% by weight, has proven highly effective. Typical additives which may be employed in addition to the stilbene halides include the oxides and sulfides of antimony, arsenic and bismuth, the preferred species being antimony oxide.

Example 2: This example illustrates the preparation of a control sample of expandable beads which does not contain the additives of this process. Polystyrene beads having an average particle size of less than 50 U.S. Mesh were impregnated with foaming agent according to the following procedure:

50 ml of distilled water and 50 g of the polystyrene beads were introduced into a pressure-tight 10 oz bottle. Also added were 0.0025 g of calcium hydroxide, 1.5 g of an aqueous slurry containing tricalcium phosphate as a suspending agent at a level of 37.5% solids and 5.85 ml of normal pentane. After the bottle was sealed air tight, it was tumbled in a heated oil bath at 110°C for a period of 4 hours. After this time, the contents were cooled and the pressure was released. The contents of the bottle were removed, washed, centrifuged and dried.

Example 3: Example 2 was repeated using the same formulation and impregnation techniques except that the bottle charge included 0.75 g of stilbene α,β-dibromide (1.5% by weight beads) in the form of a finely divided powder and 1.0 ml of a 1% aqueous solution of dodecylbenzene sodium sulfonate as a dispersing aid.

Example 4: Example 2 was repeated using the same formulation and impregnation techniques except that the bottle charge included 0.75 g of stilbene, α,β-dibromide powder, 1.0 ml of a 1% aqueous solution dodecylbenzene sodium sulfonate, and 0.25 g of dicumyl peroxide (0.5% by weight beads).

Example 5: Example 2 was repeated using the same formulation and impregnation techniques except that the formulation included 0.75 g of finely divided stilbene α,β-dichloride and 0.25 g of dicumyl peroxide.

Example 6: Example 2 was repeated using the same formulation and impregnation techniques except that the formulation included 0.25 g of finely divided stilbene α,β-dibromide (0.5% by weight beads) and 0.05 g of dicumyl peroxide (0.1% by weight beads).

Examples 7 through 9: Three different samples of expandable beads were prepared using the same formulation and impregnation techniques as Example 2 and including a 1.5% by weight beads charge of various prior art flame retardants in combination with a 0.5% by weight beads charge of dicumyl peroxide. The various flame retardants employed were as follows:

Example	Flame Retardants
7	1,2,3,4-Tetrabromobutane (about 85% Br)
8	Hexabromobenzene (about 89% Br)
9	Pentabromophenoxybenzene (about 71% Br)

Example 10: Polystyrene beads having an average particle size of less than 50 U.S. Mesh were impregnated according to the following procedure:

7,500 ml of distilled water and 7,500 g of the polystyrene beads were introduced into a pressure-tight 3 gallon pressure autoclave. Also added were 200 g of an aqueous slurry containing tricalcium phosphate as a suspending agent at a level of 35% solids, 112.5 g of stilbene α,β-dibromide powder (1.5% by weight beads), 1.0 g of dodecylbenzene sodium sulfonate, 37.5 g of dicumyl peroxide (0.5% by weight beads), and 937 ml of normal pentane. After the autoclave was sealed, stirring was commenced and the contents heated to about 225°F for a period of 6 hours. After this time, the contents were cooled and the pressure was released. The contents of the autoclave were removed, washed, centrifuged and dried.

Oxygen index values on the various samples prepared according to Examples 2 through 10 were obtained as follows: The impregnated beads were treated with steam and hot water on a screen to form a prefoamed bead having a density of approximately 1 pcf. This material was then aged for about 24 hours and placed in a mold and expanded into a block by the application of steam for approximately 5 minutes, followed by cooling. Sections having dimensions of about ½" x ½" x 6" and suitable for testing in accordance with ASTM 2863-70 were cut from the inner portions of the expanded block. These sections were also about 1 pcf density.

Comparative oxygen index values on foamed samples prepared from the beads of Examples 2 through 10 are reported in the following table.

Sample	Ex. 2	Ex. 3	Ex. 4	Ex. 5	Ex. 6	Ex. 7	Ex. 8	Ex. 9	Ex. 10
Oxygen Index	23	31.7	36.2	28.1	29.1	35.9	25.7	27.0	41.5

As can be seen from the table, all samples exhibited an improvement in limiting oxygen index (LOI) value as opposed to the control sample of Example 2. Note, however, that the beads of Examples 3 through 6 and 10 are superior in terms of LOI when compared with Examples 8 and 9. Examples 3 and 4 compare quite favorably with Example 7 when one considers that stilbene dibromide contains only about 47% by weight bromine, whereas the 1,2,3,4-tetrabromobutane contains about 85% by weight bromine. The very high oxygen index value achieved in Example 10 illustrates the improved impregnation efficiency of a larger scale impregnation process as opposed to the smaller scale process described for Example 4.

The derivatives of stilbene may also be successfully employed to enhance the oxygen index of other polymers or waxy materials, particularly, polyolefins such as polyethylene or polypropylene, polyamides, polyurethanes, polyesters and like materials which can be processed at temperatures below about 240°C. They may also be used in polymeric adhesives or coating compositions, particularly those that are not processed at high temperatures.

Brominated Polydiisopropyl Arylenes

H.H.J. Oosterwijk, U.E. Wiersum, E.P. Magré and J.H. van Dijk; U.S. Patent

4,049,726; September 20, 1977; assigned to Akzona Incorporated describe brominated polydiisopropyl arylene compounds which can be used for imparting fire retardance properties to polymers such as polystyrene. These compounds contain bromine as a substituent on the aliphatic radicals and have the following formula:

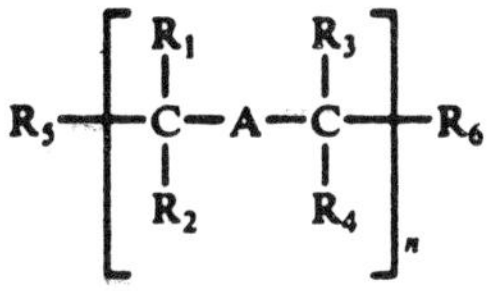

A is a thienylene group or an arylene group such as a substituted or unsubstituted phenylene, biphenylene, naphthylene, or anthracylene group, R_1 through R_4 a methyl group substituted or not with bromine, R_5 and R_6 a bromine atom or a hydrogen atom, and n a whole number in the range of 2 to 10, the number of bromine atoms per molecule in the aliphatic groups being at least 2 and the total bromine content being not more than 70% by weight.

It is preferred that A should be a phenylene group which may, together with the brominated or nonbrominated isopropyl groups be included in the structural unit by para or meta substitution. In addition, the phenylene group may contain one or more substituents such as an alkyl group with 1 to 4 carbon atoms such as methyl, ethyl, propyl, butyl, isobutyl, a chlorine atom or a bromine atom and the like.

The bromine content of the polymeric end product should generally be in the range of 0.5 to 2% by weight. The compounds provided by this process may be incorporated to advantage in many types of polymers. Particularly favorable results are obtained when they are contained in polypropylene and polystyrene polymers or copolymers.

By polymer compositions having a basis of polystyrene are to be understood here both polystyrene and copolymers thereof, the latter containing at least 50% by weight of the styrene monomer. Examples of other monomers that may be incorporated in the copolymers according to the process are: α-methylstyrene, acrylonitrile, methacrylonitrile, esters of acrylic acid or methacrylic acid with alcohols having 1 to 8 carbon atoms, esters of fumaric acid with alcohols having 1 to 8 carbon atoms, vinyl pyridine, N-vinyl compounds such as N-vinyl carbazole, butadiene or also small quantities, for instance 0.001 to 1 and preferably 0.001 to 0.1% by weight of divinyl benzene.

The flame retardance in the following examples was determined by means of the limiting oxygen index (LOI), measured in conformity with ASTM D 2863-70, unless indicated otherwise.

Example 1: Preparation of Brominated Polydiisopropyl Benzene — Into a three-necked flask provided with a stirrer, a thermometer, a dropping funnel and a reflux condenser, were charged 100 g poly-m-diisopropyl benzene. To this substance were added, in the presence of UV light, 108 g bromine over a period of 3 hours at 85°C. After the mixture had been stirred for 4 hours at 85°C, it was cooled and brought into 200 ml benzene and washed acid-free with a 10% by weight solution of $NaHCO_3$. The product was dried on magnesium sulfate, after which the benzene was evaporated off. The yield was 150.3 g, the bromine content being 33.3% by weight.

By nuclear magnetic resonance measurement (NMR) it could be established that the bromine was exclusively contained in the aliphatic groups, upon further analysis the product was found to be composed as follows: 60% by weight dimer, 21% by weight trimer and 19% by weight tetramer and higher oligomers.

Example 2: Preparation of Brominated Polydiisopropyl Benzene – Into the three-necked flask of Example 1 were charged 98 g poly-m-diisopropyl benzene together with 100 ml carbon tetrachloride and 1 ml pyridine. After the mixture had been heated to 70°C, 2 ml water and a few crystals of iodine were added. Subsequently, 250 g bromine were added dropwise over a period of 7 hours, after which the reaction mixture was stirred for 3 more hours at 78°C.

The reaction mixture was concentrated by evaporation under reduced pressure (2 mm Hg) at about 50°C, as a result of which a viscous liquid was obtained, from which after dissolving in petroleum ether and cooling down 171.1 g solid matter could be isolated. The bromine content of the resulting product was 46.7% by weight. Here too, it was found by nuclear magnetic resonance measurement that the bromine was exclusively contained in the aliphatic groups. The composition of the product was as follows: 60% by weight dimer, 21% by weight trimer and 19% by weight higher polymer.

Example 3: Fire Retardant Polystyrene – An amount of polystyrene beads together with successively 1, 3 and 5% by weight of the brominated polydiisopropyl benzene of Example 1, was dissolved in dichloromethane (1 part of solid substance in 5 parts of solvent). After the mixture had been poured onto a glass plate and air dried, a film was obtained which was dried for 2 more hours in vacuo at 100°C. After the film had been cooled, it was ground into a coarse powder in a cross beater mill.

The powder obtained was molded into rectangular test specimens measuring 150 x 6.5 x 3 mm at a temperature of 150°C and over a period of about 1½ minutes. Of the best specimens obtained, the values of the limiting oxygen index (LOI) were determined. The results are shown in Table 1.

TABLE 1

Run	% by weight of brominated poly-m-diisopropyl benzene	LOI-value
1	0	18.3
2	1	25.8
3	3	27.8
4	5	29.7

Example 4: Fire Retardant Polystyrene – The preparation of fire retardant polystyrene was carried out in the same way as in Example 3, except that this time use was made of the brominated poly-m-diisopropyl benzene of Example 2, which has a bromine content of 46.7% by weight. The results of these experiments are listed in Table 2 on the following page.

TABLE 2

Run	% by weight of brominated poly-m-diisopropyl benzene	LOI-value
1	1	26.3
2	3	28.9
3	5	31.6

Trifluoromethanesulfonates

C.E. Reineke; U.S. Patent 4,038,245; July 26, 1977; assigned to The Dow Chemical Company describes a char-forming composition of matter comprising (A) a polymeric component that comprises (1) polymerized ethylenically unsaturated carboxylic anhydride in an amount sufficient to provide at least 1 meq of anhydride moiety per gram of the char-forming composition, and (2) polymerized monovinylidene aromatic monomer in an amount sufficient to provide at least 1 meq of reactive aromatic moiety per gram of the char-forming composition, and (B) a char-forming amount of a stable ester of the formula:

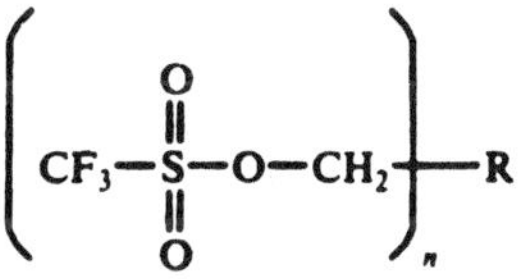

where n is a positive integer and R is an n-valent tertiary hydrocarbyl radical or an inertly substituted n-valent tertiary hydrocarbyl radical.

Polymerized ethylenically unsaturated carboxylic anhydride means an ethylenically unsaturated carboxylic anhydride (A) which has undergone addition polymerization with itself or with another ethylenically unsaturated monomer to form a normally solid thermoplastic polymer or copolymer, and (B) which contains at least one anhydride moiety of the formula:

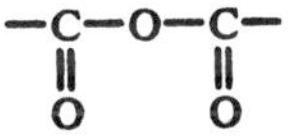

Most common of the carboxylic anhydrides are the α,β-ethylenically unsaturated carboxylic anhydrides which include maleic anhydride, phenylmaleic anhydride, citraconic anhydride, itaconic anhydride, chloromaleic anhydride, bromomaleic anhydride, etc.

The monovinylidene aromatic monomer is a compound of the formula:

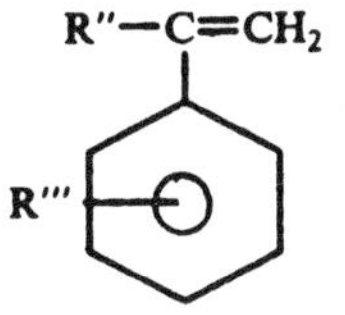

where R'' is H or lower alkyl and R''' is hydrogen, bromine, chlorine, an alkyl radical of from 1 to 6 carbon atoms or a haloalkyl radical having from 1 to 6 carbon atoms and from 1 to 3 atoms of bromine or chlorine. Examples of such preferred species are styrene, α-methylstyrene, ar-methylstyrene, ar-(t-butyl)styrene, ar-chlorostyrene, ar-bromostyrene, most preferably styrene.

The stable trifluoromethanesulfonates used are referred to as triflate esters. The term "stable" connotes both thermal stability and low volatility. A triflate ester is deemed stable (a) if it is as thermally stable as 2,2-dimethylpropyl trifluoromethanesulfonate, preferably as stable as 3-bromo-2,2-bis(bromomethyl)propyl trifluoromethanesulfonate, and (b) if it has a boiling point of at least about 60°C, preferably at least 100°C, at a pressure of 3 mm of mercury.

As a general rule, the triflate ester is present in an amount sufficient to provide from 0.1 to 0.4 meq of triflate ester moiety,

$$-CH_2-O-\overset{\overset{\displaystyle O}{\|}}{\underset{\underset{\displaystyle O}{\|}}{S}}-CF_3$$

per gram of char-forming composition.

Example: Preparation of Additive A – A solution of 25.5 g (0.09 mol) of trifluoromethanesulfonic anhydride in methylene chloride is added dropwise, with stirring, over a 1½ hour period to a solution containing 26.0 g (0.08 mol) of 3-bromo-2,2-bis(bromomethyl)-1-propanol and 6.3 g (0.08 mol) pyridine in 75 ml of methylene chloride, at 0°C. The reaction product is recovered by:

(1) washing the reaction mixture with 100 ml of cold distilled water in a separatory funnel and draining off the methylene chloride solution layer;

(2) drying the methylene chloride layer over anhydrous sodium sulfate;

(3) filtering out the sodium sulfate; and

(4) vacuum distillation of the methylene chloride.

The product recovered is a colorless liquid exhibiting a boiling point of 102°C at 0.03 mm of mercury and a nuclear magnetic resonance (NMR) spectrum in agreement with that expected for 3-bromo-2,2-bis(bromomethyl)propyl trifluoromethanesulfonate (A). A 33.5 g portion of the product is recovered, representing a yield of 91.5%.

Preparation of Additive B – For comparative purposes a quantity of 3-bromo-2,2-bis(bromomethyl)propyl acetate (B) is prepared by reacting the 3-bromo-2,2-bis(bromomethyl)-1-propanol with acetic acid.

Preparation of Char-Forming Compositions: Polystyrene compositions and styrene/maleic anhydride copolymer compositions, comprising 75 mol percent styrene and 25 mol percent maleic anhydride, each containing the aforementioned triflate ester are prepared by blending the ester into the polystyrene, or the styrene/maleic anhydride copolymer, with a Brabender mixer at 180°C or 220°C, respectively. In the same manner, compositions containing the aforementioned acetate and polystyrene, or the styrene/maleic anhydride copolymer, are prepared. The resulting compositions are molded into test bars and subjected to a modified version of the Underwriters Laboratory Vertical Burning Test for Classifying Materials 94V-0, 94V-1, or 94V-2. Such tests demonstrate the ignition properties or burning characteristics of the polymer compositions when exposed to small-scale ignition sources. The test procedure is as follows: Test specimens, 5.0 inches in length by 0.50 inch in width with a maximum thickness of 0.50 inch, are ignited in a draft-free environment while supported vertically lengthwise by a holding clamp attached to the upper one quarter of the specimen with the lower end 12 inches above a horizontal layer of dry absorbent surgical cotton having a maximum free-standing thickness of one-quarter inch. Ignition is performed using a Bunsen burner adjusted to produce a three-quarters inch blue flame. The flame is placed centrally under the lower end of the test specimen for 10 seconds and is then withdrawn. When the flaming of the specimen ceases, the flame is placed immediately under the specimen for 10 more seconds and is again withdrawn. The parameters observed for classification are:

(a) Duration of flaming following the first flame application,

(b) Duration of flaming following the second flame application,

(c) Duration of flaming and glowing following the second flame application,

(d) Whether or not the specimens burned up to the holding clamp,

(e) Whether or not the specimens dripped flaming particles which ignited the surgical cotton.

Results indicative of a 94V-0 rating are achieved by meeting the following criteria:

(a) None of the specimens burn with flaming combustion for more than 10 seconds after each flame application,

(b) The specimens exhibit no more than an average 10 seconds total flaming combustion for the 2 flame applications combined,

(c) None of the specimens burn with flaming or glowing combustion up to the holding clamp,

(d) None of the specimens drip flaming particles that ignite the cotton below the specimens,

(e) None of the specimens exhibit glowing combustion persisting longer than 30 seconds after the second flame removal.

Results indicative of a 94V-2 rating are achieved by meeting the following criteria:

(a) None of the specimens burn with flaming combustion for more than 30 seconds after each flame application,

(b) The specimens exhibit no more than an average 50 seconds total flaming combustion for the 2 flame applications combined,

(c) None of the specimens burn with flaming or glowing combustion to the holding clamp,

(d) Only briefly burning flaming drips, some of which ignited the cotton below the specimen are observed,

(e) None of the specimens exhibit glowing combustion more than 60 seconds after the second flame removal.

Test results are summarized in the table on the following page.

Sample	Polymer	Additive Type, wt %		Bromine wt %	Halogen wt %	UL-94 Rating	Average Time of Flaming Combustion (sec)	Visible Char Formation
1*	Styrene/maleic anhydride copolymer**	none	–	–	–	sustained combustion	–	–
2	Styrene/maleic anhydride copolymer**	A	5	2.6	3.3	94 V-2	5	yes
3*	Styrene/maleic anhydride copolymer**	B	5	3.3	3.3	94 V-2	11	no
4	Styrene/maleic anhydride copolymer**	A	10.3	5.4	6.7	94 V-0	<1	yes
5*	Styrene/maleic anhydride copolymer**	B	10.3	6.7	6.7	94 V-0	6	no
6	Polystyrene	A	10.3	5.4	6.7	94 V-2	3	no
7*	Polystyrene	B	10.3	6.7	6.7	94 V-2	10	no

*Samples 1, 3, 5 and 7 are not embodiments of the process.
**75:25.

As is apparent from the table above, the samples containing styrene/maleic anhydride copolymer and the triflate esters form char upon combustion while the samples containing styrene/maleic anhydride copolymer and the corresponding acetate ester do not. Neither the acetate nor the triflate ester induce char formation in styrene homopolymer.

Bisphenoxy Compounds

A.L. Anderson; U.S. Patent 3,971,759; July 27, 1976; assigned to Michigan Chemical Corporation describes plastic compositions containing polystyrenes and bisphenoxy compounds having the formula:

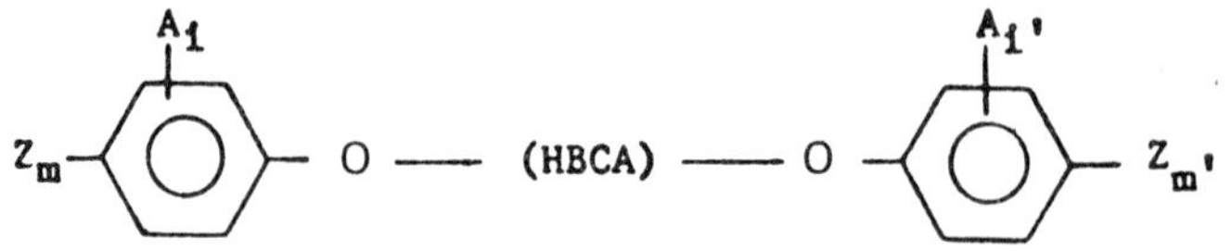

wherein Z is bromine or chlorine, m and m' are integers having a value of 1 to 5, i and i' are integers having a value of 0 to 2, HBCA is a halo-branched chain alkylene group having from 1 to 6 carbon atoms and A is cyano, nitro, lower alkoxy, lower alkyl, fluorine, dialkylamino, phenyl, halo-phenyl, benzyl or halobenzyl. These compounds and examples have been described previously in the chapter on Polyesters (see page 94).

Interpolymers of Styrene, Maleic Anhydride and Tribromophenyl Acrylate

Y.C. Lee; U.S. Patent 4,048,263; September 13, 1977; assigned to Monsanto Company describes rubber modified fire retardant polymer compositions comprising an interpolymer of styrene, maleic anhydride and a copolymerizable brominated comonomer such as tribromophenyl acrylate, which have been formulated with a halogenated fire retardant additive, a metal oxide and a smoke suppressant.

The styrene-type monomers used include styrene, α-methylstyrene, and halogenated styrenes. The amount of styrene-type monomer used is from 55 to 85% by weight based on the weight of the total polymer composition.

The amount of maleic anhydride used is preferably 7 to 20% by weight based on the total polymer composition. The preferred brominated monomers include tribromophenyl acrylate, tribromophenyl methacrylate, tribromoneopentyl acrylate and tribromoneopentyl methacrylate.

The amount of brominated monomer used is selected so as to provide an interpolymer with a bromine content of at least 6% by weight. In order to obtain this minimum bromine level, it will be necessary to use brominated monomer in amounts in the range of from 10 to 20% by weight when using tribromo monomers.

Suitable halogen fire retardant additives used are diaromatic or aromatic aliphatic ethers. The most preferred compound is decabromodiphenyl ether. The amount of halogenated fire retardant additives is in the range of 4 to 12% by weight in order to provide additional halogen to the compositions.

Examples of synergists used in combination with the halogen in order to obtain improved fire retardant properties include metal oxides such as Sb_2O_3, Bi_2O_3, MoO_3, SnO_2, WO_3, and the like. The preferred metal oxide is Sb_2O_3. The amount of metal oxide used will be from 3 to 12% by weight, based on the total weight of the formulated composition.

The polymers may be further modified with smoke suppressants in order to enhance fire retardant properties and/or smoke evolution properties. Examples of smoke suppressants would include hydrates, carbonates and borates such as alumina trihydrate, aluminum hydroxide-sodium carbonate hydrate, magnesium carbonate, hydrated magnesia, hydrated calcium silicate, hydrated calcium borate, calcium carbonate, and magnesium borate. One of the preferred smoke suppressants is dawsonite [$NaAlCO_3(OH)_2$].

The polymers also contain from 5 to 20% by weight of a synthetic or natural rubber component. Examples of the rubber components include polymers and copolymers of butadiene, isoprene, chloroprene, as well as other rubbers such as nitrile rubbers, epichlorohydrin rubbers, acrylate rubbers, natural rubbers, etc. The preferred rubber components are epichlorohydrin rubbers, polychloroprene rubbers and chlorinated polyethylene.

These compositions, which have a UL-94 rating of at least V-1, are especially useful for use in radio and television cabinets, furniture, appliance housings and other similar applications.

Metal Oxide-Polystyrene Dispersion Used in Molding Compositions

A process described by *R.A. Barkhuff, Jr. and W.E. Sprenkle, Jr.; U.S. Patent 4,069,288; January 17, 1978; assigned to Monsanto Company* relates to a method for preparing fire retardant styrenic molding compositions which contain halogen fire retardant additives and a metal oxide such as antimony oxide.

Preferred styrenic polymers for use in the process are those high impact polystyrene (HIPS) compositions which are obtained by blending polystyrene with from 3 to 9% by weight of a diene type rubber. Preferably, the rubber is grafted with styrene or a styrene copolymer composition. Examples of diene rubbers used include butadiene, isoprene, chloroprene, etc.

The fire retardant additives are those which contain one or more bromine or chlorine atoms such as aromatic bromine compounds, brominated furan-maleic anhydride adducts, 4-bromoalkyl or tribromoneopentyl esters, chlorinated cyclopentadiene and derivatives of these.

The amount of brominated fire retardant additives used is selected to provide a bromine content in the range of from 6 to 10% bromine by weight in order to provide sufficient bromine to the composition to obtain the desired degree of flame retardancy. When using the relatively less efficient chlorinated additives, the amounts are increased to provide 10 to 15% of chlorine to the molding composition.

The antimony oxide used functions as a synergist in improving the efficiency of the halogen additive in imparting flame retardancy to the abovementioned polymers. The amount of oxide used is in the range of from 4 to 12% by weight based on the total weight of the formulated composition.

In addition to the grafted rubber component used to modify the polystyrene composition, a second rubber component may optionally be used. The purpose of such rubber is to improve impact strength and to help disperse the metal oxide component during the intensive mixing step. In addition, the second rubber component may help to control dripping of the polyblend under flaming conditions.

The second rubber component is preferably a block copolymer or a graft copolymer used in the amounts of 2 to 6% based on the total weight of the polyblend composition. The following process steps are used in preparing the composition.

The metal oxide component is blended with a minor amount of styrenic polymer component and optionally one of the rubber components. The blend is then subjected to high intensity mixing and heated at or above the fusion temperature of the styrenic polymer component but below the decomposition or degradation temperature of the polymer component. The preferred temperatures at which the polymer is heated are in the range of 160° to 230°C.

High intensity mixing of the heated blend may be obtained using conventional high intensity mixers such as Banbury mixers, roll mills, Pfaudler mixers, continuous intensive mixers, twin screw extruders, etc. Mixing is continued until the metal oxide is uniformly dispersed throughout the molten polymer. When using a Banbury mixer, the mixing time required is about 1 to 6 minutes, preferably 2 to 4 minutes after fusion of the polymer component occurs.

The blend is then comminuted to the desired particle size using pelletizers, dicers, knives, grinders, choppers, and the like. The comminuted particles are then blended with the other ingredients used to prepare the polyblend compositions. These ingredients include the balance of the styrene matrix polymer, the grafted diene rubber which is used to modify the styrene matrix polymer, the second rubber component, the halogen fire retardant additive, lubricants, stabilizers and other such additives.

The polystyrene-metal oxide dispersion and the other components of the molding composition are then uniformly mixed using conventional mixers such as tumblers, drums, rotary mixers, conical blenders, Kenics blender, ribbon blenders and the like. The blended ingredients are then compounded at temperatures in the range of 260° to 290°C. The temperatures referred to above are the stock temperatures of the polyblend. The impact strength of molded test specimens increases with increasing compounding temperatures in the preferred range.

The ingredients that form the molding composition are compounded by heating and mixing. The compounding is carried out in extruders, Banbury mixers, mill rolls, Farrell continuous mixers and the like. The preferred compounding method is in an extruder. The compounded material is then comminuted to the desired size using the methods described above to form the molding composition.

Additive Composition of Magnesium Bentonite and Chlorowax for ABS Resins

R.K.S. Mehta and P. Weiss; U.S. Patent 4,070,315; January 24, 1978; assigned to General Motors Corporation describe a fire resistant acrylonitrile-butadiene-styrene (ABS) copolymer composition.

The fire retarding additive consists of the combination of magnesium bentonite and a halogenated organic compound which when dispersed in the resin and combusted will yield a hydrogen halide. The magnesium bentonite may be characterized as having at least about 2% by weight of an exchangeable magnesium cation, a volume swell in water of at least about 50% and a cation exchange capacity of at least 50 meq per 100 g. All suitable bentonites will have similar volume swells in water and ion exchange properties.

Example 1: The fire resistant ABS composition was prepared by first forming magnesium bentonite and then uniformly blending the bentonite and Chlorowax, a chlorinated paraffin, into an ABS resin.

The magnesium bentonite was formed by dissolving about 9.6 g of magnesium chloride (a magnesium salt) in about 100 ml of distilled water. As this solution was stirred in a Waring blender, about 100 g of sodium bentonite (Thixo-Jell #1) was added. A slurry formed within about a minute and started to thicken. After 2 minutes, the stirring was stopped and the contents were allowed to stand for about 2 hours. The product was then transferred to a shallow pan, freeze-dried and powdered to about 100 mesh size.

Thermogravimetric analysis of the magnesium bentonite showed that the material was hydrated. It had retained an overall moisture content of 15% by weight and this water was given off gradually as the temperature increased. It continued to be released until the temperature of the sample was about 700°C, a temperature well above the temperature of most polymer fueled fires.

The ABS resin was compounded with the magnesium bentonite and Chlorowax S70 on a Getty hot roll mill, and compression molded into slabs about a quarter inch thick. The slabs were then cut into bars having dimensions of 5" x 0.5" x 0.25", and flammability tests were performed on these bars.

Two standard methods of testing flame retardant polymers were used. The first is the oxygen index test which is described in ASTM D-2863-74 and which measures the minimum percentage of oxygen necessary to support a flame on the sample. The other test was an internal Fisher Body test, TM 32-12, which is a slight modification of the flammability test described in Federal Motor Vehicle Safety Specification 302. In this test, the sample is held horizontally within a metal cabinet and exposed to a 1½ inch flame of a Bunsen burner. The burning is timed and reported in inches per minute.

Table 1 shows the oxygen index and burn rate of several compositions (in parts by weight) prepared within the procedures described.

TABLE 1

Ingredients and Properties	Compositions 1	2	3
ABS resin	112	112	112
Magnesium bentonite	–	15	–
Sodium bentonite (Thixo-Jell #1)	–	–	37
Chlorowax S70	–	15	–
Oxygen index	18.1	22.1	19.4
Burn rate (in/min)	2.0	SE*	–

*Self-extinguishing.

The first ABS composition is the control which contains no fire retarding additive. The second composition contains 15 parts of the magnesium bentonite and 15 parts of Chlorowax. In this composition, the oxygen index has increased by 4 points and the resin has become self-extinguishing. The third example is included to show the effect of a high concentration of the sodium bentonite on the fire resistance of a polymer sample.

Example 2: In accordance with the procedures in Example 1, an additional series of experiments were run using bentonite clays which contain various cations including hydrogen, alkali metals and alkaline earth metals. The results which are shown in Table 2, indicate that all 6 bentonite compositions in combination with Chlorowax, significantly and unexpectedly increase the fire retardant properties of ABS resins. In addition, from this data, it is evident that certain bentonite compositions are more effective than others. In view of all 6, however, it does not appear that there is a difference in kind, but rather a difference in degree of effectiveness. (Amounts of ingredients are in parts by weight.)

TABLE 2

Ingredients and LOI	Compositions 1	2	3	4	5	6	7	8
ABS	112	112	112	112	112	112	112	112
Chlorowax	–	15	15	15	15	15	15	15
Mg bentonite	–	–	15	–	–	–	–	–
H bentonite (T-300)	–	–	–	15	–	–	–	–
H bentonite (T-30)	–	–	–	–	15	–	–	–
Ca bentonite (T-60)	–	–	–	–	–	15	–	–
Na bentonite(Thixo-Jell #1)	–	–	–	–	–	–	15	–
Sr bentonite	–	–	–	–	–	–	–	15
LOI	18.1	19.5	22.1	21.8	21.2	21.2	21.0	21.0

Example 3: In accordance with the procedures described above, a series of experiments were run to determine the effect of bentonite and Chlorowax individually on the oxygen index of ABS. The results of this investigation are set out in Table 3, and from that it is evident that the combined effect of the hydrogen bentonite and Chlorowax is significantly greater than what one would expect by simply adding the increase provided by each ingredient individually. (Amounts of ingredients are in parts by weight.) This set of experiments suggests synergism is at work and clearly shows that the additive combination provides unexpected results.

TABLE 3

Ingredients and Properties	Compositions 1	2	3	4
ABS	112	112	112	112
Hydrogen bentonite (T 300)	–	15	–	15
Chlorowax S70	–	–	15	15
Oxygen index	18.1	18.6	19.5	21.8
Individual contribution over the control	–	+0.5	+1.4	–
Expected oxygen index if effects were cumulative	–	–	–	20.0

Many organic polymers, and in particular the ABS resins, may be rendered less combustible by the addition of a fire retarding additive consisting essentially of a bentonite clay having an exchangeable monoatomic, elemental cation such as the alkali metals, alkaline earth metals and hydrogen, and a halogenated organic compound which when dispersed in an organic polymer and thermally decomposed will yield a hydrogen halide.

Economically, it is best that the bentonite clay be present in as large a concentration as possible since it is the least expensive ingredient. That upper concentration limit may be defined as that point at which physical properties of the polymers deteriorate beyond the acceptable limits for any given application.

PHOSPHORUS-CONTAINING ADDITIVE

Treatment of Styrene Beads with Intumescent Latex Paint

A.R. Ingram; U.S. Patent 4,035,315; July 12, 1977; assigned to ARCO Polymers, Inc. describes a procedure whereby fire resistant foams having 1.2 to 2.5 pcf density are prepared by coating 100 parts by weight of styrene polymer foam particles with at least 67 parts by weight of an intumescent paint and molding the coated prepuff into the desired end product foam.

A typical intumescent paint latex suitable for use in this process consists essentially of the following:

Ingredient	%, by Weight	Purpose
Water	40.1	Medium
Ammonium Polyphosphate	24.0	Carbonizing catalyst
Dipentaerythritol	6.9	Carbon former
Chlorinated paraffin	11.1	Blowing agent
Polyvinyl acetate	11.2	Binder

(continued)

Ingredient	%, by Weight	Purpose
Titanium dioxide	5.6	Pigment
Surfactants, thickeners, etc.	1.1	Miscellaneous
	100.0	

The coating may be applied to the expanded styrene polymer particles by any suitable method, such as stirring the particles into the paint, or by spraying the particles with the paint or by, for example, the method described in U.S. Patent 3,462,293 where the expanded beads are suspended in a stream of air at 100° to 130°F, contacted with the aqueous latex, and dried in the heated air stream to form a uniform particulate coating. The polymer particles may contain, prior to being coated, any of the usual additives, such as dyes, pigments, self-extinguishing agents, antistatic agents, plasticizers, etc.

Example 1: To 100 parts of polystyrene prepuff beads having a density of 1.0 pcf was added 400 parts of an intumescent latex paint consisting of 40.1% of water, 24.0% of ammonium polyphosphate, 7.0% of dipentaerythritol, 11.1% of paraffin hydrocarbon (70% chlorinated), 11.2% of polyvinyl acetate, 5.6% titanium dioxide and 1.0% of added surfactants and thickening agents. The mixture was stirred with a wooden spatula until the paint appeared uniformly dispersed over the polymer bead surfaces. The beads were then allowed to air dry for 18 hours.

The resultant beads had a foam density of about 2.2 pcf. They were not ignitable, but did shrink and crumble in the path of the flame. These foam particles were then placed in a conduction press, moistened, and pressed at 120°C for one minute. The resulting air-dried foam had about 20% fusion, few voids, a density of 4.8 pcf and was nonignitable.

Example 2: To 100 parts of polystyrene prepuff beads having a density of 1.0 pcf was added 133 parts of an intumescent paint consisting of 28.6% of water, 28.6% of ammonium polyphosphate, 8.2% of dipentaerythritol, 9.0% of melamine, and 25.6% of polyvinylidene chloride. The mixture was stirred to ensure complete coating of the beads with the paint and then dried. The coated prepuff was then molded in a compression mold using 20 to 25 psi steam and a back-pressure of 15 to 20 psi to give 8 inch diameter by 1 inch thick discs having a foam density of 2.2 pcf and a foam fusion of 65%. The dried foam moldings were difficult to ignite, extinguished immediately after removal of the external flame, and burned without smoke or flaming drips while in the external flame.

Example 3: An intumescent latex paint (150 parts) such as that used in Example 1 was spray-coated onto 100 parts of polystyrene prepuff beads and the coated beads molded while still wet in a molding press at 120°C for 1.5 minutes into plaques 5½ inches x 5½ inches x ½ inch. The initial prepuff had a foam density of 1.0 pcf and the density of the dried plaque was 2.4 pcf. The plaques had very low internal fusion and attempts to ignite the plaque when supported vertically by a clamp resulted in disintegration of the sample. When supported on a horizontal screen, the plaques did not burn, smoke, or drip when exposed to flame from a propane torch from above or below.

These polymer foams are highly desirable in building construction because of their light weight and good heat insulating property.

OTHER ADDITIVES

Hot Aqueous Alkali Silicate Solutions

H. Honer; U.S. Patent 3,994,836; November 30, 1976 describes a process for preparing molded articles of foamed polystyrene which are characterized by flame resistance properties. The process comprises foaming expandable polystyrene granulate or bead containing a blowing agent with an aqueous medium containing an alkali silicate, and optionally triglycidyl isocyanurate, a water-soluble amine, and a quaternary ammonium base.

The aqueous medium used contains a water-soluble alkali silicate having a content of at least 20% by weight of SiO_2 and at least 7% by weight of K_2O and 0 to 5% by weight of Na_2O and a viscosity of 35 to 150 cp at 20°C, the medium optionally further containing hardeners and/or fillers.

The triglycidyl isocyanurate used as additive contains more than 14% by weight of epoxide and is used in amounts of from 0.03 to 2.0% by weight, based on the solids content of the alkali silicate solution. The water-soluble amines used as additives are primary, secondary, aliphatic, alicyclic and aromatic amines. The amount of these amines used amounts to usually 0.1 to 5% by weight, based on the solids content of the alkali silicate solution. The amount should be sufficient to provide 0.6 to 1.1 equivalents of active hydrogen atoms to one equivalent of epoxide oxygen of the triglycidyl isocyanurate.

The quaternary ammonium bases used as additives are such containing at least one alkyl radical which is not substituted by a hydroxy group and further straight chain or branched chain alkyl, alkanol, alkoxy, aminoalkyl, cycloalkyl, morpholinyl or piperidyl radicals. It is preferred to use tetraalkyl ammonium hydroxides, having 2 to 12 carbon atoms in each alkyl group. The amount of the quarternary ammonium bases used amounts to 0.1 to 5% by weight, based on the content of SiO_2 of the alkali silicate solution.

The hardeners optionally used in the foaming medium are phosphates, carbonates and oxides of divalent and trivalent metals such as calcium hydrogen phosphate; calcium polyphosphate; phosphates of aluminum; etc.

The fillers to be used in the process are particulate minerals which are slightly or not combustible such as the silicate minerals asbestos, allophane, aluminite, anthopyllite, and alusite. The amount of fillers used is usually from 7 to 15% by weight, based on the weight of the finished molded article of polystyrene foam. The particle size of the fillers is about 1 μm to 1 mm, smaller or larger size fractions in a filler blend being possible.

The foamed polystyrene molded articles as prepared according to this process are resistant to temperatures up to at least 400°C. On the action of flames and higher temperature the articles form coke without structural decay but do not burn with free flames. Thus the articles are especially suited for insulating purposes, e.g., in building and construction with certain requirements of flame retarding or, respectively, self-extinguishing properties of the insulating material used.

Example 1: Expandable polystyrene beads containing pentane as the blowing agent were filled into a cube shape mold the walls of which consisted of fine wire network. The polymer filled 1/50 of the volume of the mold.

The mold containing the expandable polystyrene beads was placed into a vessel containing an aqueous solution at 70°C of 25% by weight of potassium silicate consisting of 24% by weight of SiO_2 and 10.7% by weight of K_2O and having a viscosity of 46 cp and a concentration of 35°Bé. Further the vessel contained a mixture of 5% by weight of $CaHPO_4$ and 5% by weight of sodium silicofluoride, based on the total solution. The vessel was equipped with a stirrer.

The vessel was closed and heated to about 125°C. After 20 minutes the vessel was cooled down and the mold which was filled with the cube-shaped foamed polystyrene article was isolated. After cooling, the product had a compressive strength of above 20 kg/cm² and did not disintegrate on heating at 950°C.

Example 2: Example 1 was repeated but the potassium silicate was replaced by potassium sodium silicate consisting of 2.7% by weight of Na_2O, 9.5% by weight of K_2O and 24% by weight of SiO_2 having a viscosity of 80 cp (at 20°C) and a concentration of 40°Bé which silicate was modified by reaction with 0.1% by weight of triglycidyl isocyanurate, 0.05% by weight of tetraethylene pentamine and 0.3% by weight of tetraethyl ammonium hydroxide, the latter percentage based on the solid content of the silicate. There was obtained a similar cube-shaped article being flame resistant at 1100°C.

By subsequent pressing of the still compressible article at a pressure of 5 kg/cm² followed by 30 minutes of heating at 70°C an article was obtained which had a 50% better compressive strength.

Example 3: 150 g of polystyrene particles foamed according to Example 1, 14 g of perlite dust (size about 1 to 30 mm), 40 g of calcium hydrogen phosphate, 20 g of sodium silicofluoride, 10 g of kaolin and 250 g of the modified potassium sodium silicate solution of Example 2 were mixed for a short time, filled into a perforated mold and subjected to a pressure of about 7 kg/cm². The block-shaped product obtained was dried at a temperature of 75°C for about 30 minutes. The compressive strength of the finished product amounted to over 40 kg/cm². The product was fire resistant up to 1050°C.

Hydrated Magnesium Oxychloride

D.F. Lawson and E.L. Kay; U.S. Patent 4,059,560; November 22, 1977; assigned to The Firestone Tire & Rubber Company describe a process whereby the evolution of smoke and the development of flame are retarded by the incorporation of hydrated magnesium oxychloride into organic polymers which contain aromatic components, particularly polymers and copolymers of styrene. The hydrated magnesium oxychloride has the added advantage that it has low tinting strength, i.e., does not greatly change the color or appearance of the polymers.

The polymers which may be smoke and flame suppressed in accordance with this process include any organic polymers containing a sufficient content of aromatic groups, say 10% or more, based on the weight of the polymers, so that in the absence of the inhibiting material, they will have a smoke density rating (Dmc) of at least 75. Such polymers will include polystyrene and copolymers of styrene with other ethylenically unsaturated compounds copolymerizable therewith such as butadiene, acrylonitrile and the like, e.g., copolymers and block copolymers such as ABS, SBS, BSBS, SBR, SAN, etc., which contain repeating aliphatic or aromatic groups joined with aromatic units.

Evaluations for the density of visible smoke were made using a commercial smoke density chamber modeled after one developed at the National Bureau of Standards by the Fire Research Group (see D. Gross, J.J. Loftus and A.F. Robertson, *ASTM Special Technical Publication 422,* pages 166-204, 1969).

Limiting oxygen indexes were obtained using the Michigan Chemical LOI apparatus. An Aminco-NBS smoke density chamber was used to obtain data on the rate of smoke generation as well as intensity of visible smoke.

Hydrated magnesium oxychlorides are generally prepared by treating magnesia with magnesium chloride in water. Several hydrated magnesium oxychloride compositions were prepared in the laboratory for purposes of illustrating this process. Others are available from commercial sources, with somewhat different contents of magnesium oxide, magnesium chloride and water.

Preparation A

Part A	
Magnesium oxide (MgO)	61.0 g (1.5 mol)
Water	400 ml
Part B	
Magnesium chloride ($MgCl_2 \cdot 6H_2O$)	101.5 g (0.5 mol)
Water	400 ml

A slurry of magnesium oxide was prepared from the ingredients of Part A, and a solution of magnesium chloride hydrate was prepared from the ingredients of Part B. The solution was slowly poured into the slurry with vigorous stirring at 25°C, and stirring was continued for 1 hour after addition. The resulting mixture was then filtered to remove water, and the solid product was air-dried for 16 hours, then dried in a forced-air oven at 100°C for 24 hours. The product was ground to a fine powder before mixing with polymer, as described in the example below.

The dried product weighed 89.5 g and it had a chlorine content of 9.6% by weight. Upon calcination, the product lost 39 to 40% of its weight, indicating an approximate formula of $8.7MgO \cdot MgCl_2 \cdot 16.4H_2O$.

Example:

	Parts by Weight
High impact polystyrene (Styron 475)	100
Calcium stearate	2
Flame and smoke retarding agent*	67

*Hydrated magnesium oxychloride prepared as in Preparation A

A series of compositions was made up in accordance with the above formulation, using the magnesium oxychloride as above prepared, and also, for comparison, other flame and smoke retarding agents in the several compositions are shown in the table below.

In each preparation, the polystyrene, the calcium stearate, and the flame and smoke retarding agent selected for that run were combined on a roll mill at 280° to 290°F (138° to 143°C), sheeted off, and molded into appropriate test specimens in a compression mold at 325°F (162°C). The specimens for smoke tests were of nominal 0.03 inch thickness, with other dimensions noted above. The specimens were then subjected to the smoke and flame generation tests. The results are tabulated below.

Type of Smoke and Flame Retardant	National Bureau of Standards Smoke Density Chamber Test			Oxygen Index Determination		Run No.
	Dmc	t.9Dm	tD16	LOI	LOI, % Increase Over Control	
Hydrated Magnesium* Oxychloride	210	1.88	0.94	28.0	47	1
Aluminum Hydroxide	224	1.69	0.75	23	22	2
Silica (325 Mesh)	228	1.34	0.49	21.0	11	3
Control	385	1.33	0.3	19.0	—	4

*Prepared as described in "Preparation A"

Note: Dmc = Specific optical density at maximum smoke intensity, corrected for fogging of lens seals. Dmc<25, light; 25-75, moderate; 100-400, dense; >400 very dense.

t.9Dm = Time (minutes to reach 90% of maximum optical density.

tD16 = Time (minutes) to D 16; corresponds to early visibility obscuration. tD16 <1, very fast; 1 to 3 fast; 4 to 6 moderate; 7 to 10 slow; >10, very slow smoker.

From the table, it will be seen that the maximum smoke density (Dmc) for the hydrated magnesium oxychloride (Run Number 1) was on a par with this figure for the other additives (Runs 2 and 3) in polystyrene. The time required for smoke generation was more protracted (t.9Dm and tD16) for the hydrated magnesium oxychloride as compared to the others.

In addition, the limiting oxygen index (LOI), which is an index of flame resistance, was increased far more in the presence of the hydrated magnesium oxychloride than was the case with the other additives. The color and appearance of the specimen of Run Number 1 containing hydrated magnesium oxychloride was very little changed in comparison with the unpigmented control specimen of Run Number 4.

POLYAMIDES AND ACRYLICS

POLYAMIDES

Cyclic Hydrogen Phosphonates

G.H. Birum; U.S. Patent 4,070,336; January 24, 1978; assigned to Monsanto Company describes a process for the preparation of cyclic hydrogen phosphonates, which comprises the reaction of phosphorochloridites with formic acid in accordance with the general reaction below:

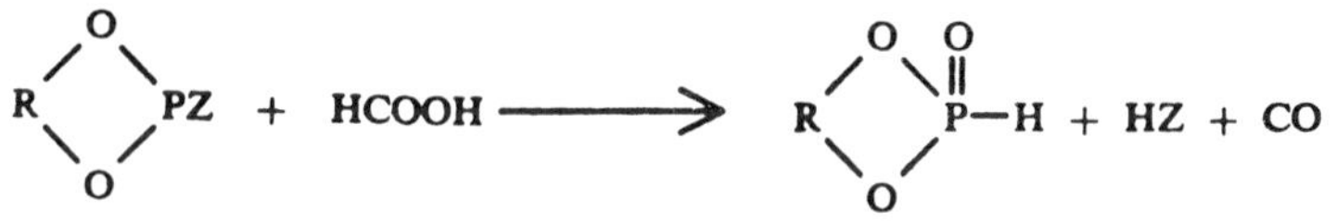

where R is an alkylene or haloalkylene group of a 1,2-glycol having from 2 to 6 carbon atoms, or a 1,3-glycol having from 3 to 8 carbon atoms, and Z is Cl or Br. Ketone and sulfenyl derivatives are also prepared.

The general formula for the cyclic phosphorus derivatives is:

$$X\overset{\overset{\displaystyle O}{\|}}{P}Y$$

where Y is selected from the group consisting of

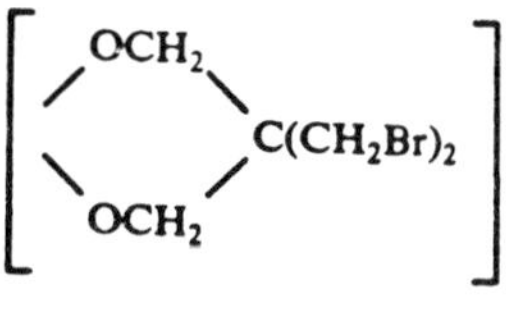

and

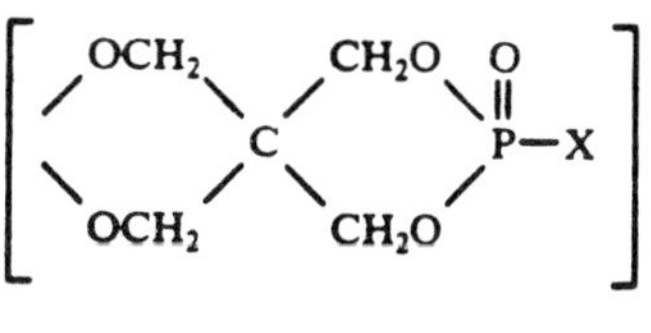

and where X is selected from the group consisting of —S(R_c) where R_c is phenyl with 0 to 5 halogens, or phenyl with 0 to 2 nitro groups, or methyl with 0 to 3 chlorine atoms and

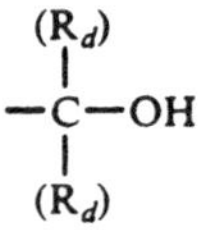

where R_d is an alkyl radical of 1 to 4 carbon atoms, or phenyl.

The above compounds are derivatives of cyclic hydrogen phosphonates. They are obtained by the reaction of a crude or purified hydrogen phosphonate with sulfenyl chlorides, or ketones, respectively.

Typical sulfenyl chlorides include chlorophenylsulfenyl chloride, nitrophenylsulfenyl chloride, bromophenylsulfenyl chloride, 2,4-dinitrophenylsulfenyl chloride, chloromethylsulfenyl chloride, and trichloromethylsulfenyl chloride.

Examples of ketones include acetone, methyl ethyl ketone, diethyl ketone, methyl propyl ketone, dibutyl ketone, and acetophenone.

The cyclic phosphorus compounds are useful as flame retardant materials which provide phosphorus as a component to reduce flammability. Improvement in flame retardant properties results when hydrogen phosphonates are reacted with sulfenyl chlorides and ketones. Such resultant products do not burn readily, and instead inhibit flammability of an organic polymer.

These compounds can be added directly to the molten polymer or the components of a foam composition before polymerization, e.g., before spinning fibers or forming films or other shaped objects including foamed plastics. Typical polymers are polyesters, polyamides, polyurethanes, polyolefins, nitrile polymers such as polyacrylonitrile, vinyl polymers such as vinyl chloride, styrene polymers and copolymers such as acrylonitrile-butadiene-styrene compositions.

The method of testing flame retardant properties is ASTM Designation D 2863-70. In the oxygen index (OI) testing procedure the relative flammability of a plastic material is determined by measuring the minimum concentration of oxygen in a slowly rising mixture of oxygen and nitrogen that will just support combustion.

Since these flame retardant compounds are quite effective they are normally combined in flame retarding proportions with the organic polymer at relatively low concentrations, e.g., about 1 to 20 weight percent, preferably about 3 to 15% based on additive plus the polymeric substrate, such as by milling, or impregnation, e.g., from a water or alcohol dispersion or solution or by dissolving or dispersing in the molten polymer before extrusion such as in the form of fibers or sheets. It should be noted that such ingredients as dyes, pigments, stabilizers, antioxidants, antistatic agents and the like can be incorporated. The following examples illustrate the process.

Example 1: A cyclic phosphorochloridite is first prepared as follows. A reaction vessel is charged with 272 grams (2.0 mols) of pentaerythritol, 567 grams of phosphorus trichloride, 0.1 gram of pyridine hydrochloride, and 272 grams of ortho-dichlorobenzene as a solvent. This mixture is warmed to 100°C in 2 hours and kept at 100° to 105°C for 1 hour, giving a clear, colorless solution having a ^{31}P NMR signal at -149 ppm for the cyclic phosphorochloridite desired, 3,9-dichloro-2,4,8,10-tetraoxa-3,9-diphosphaspiro[5.5] undecane.

This intermediate is converted to the hydrogenphosphonate by dropwise addition of 184 grams (4 mols) of 97 to 100% formic acid over a 65-minute period. The temperature is maintained at about 25° to 42°C during the addition. A stream of gaseous nitrogen is passed through the stirred reaction mixture until most of the by-product HCl and CO have been swept out. The reaction mixture is filtered to remove solids which are then washed with benzene, followed by acetonitrile and ether. The white solid hydrogenphosphonate product, ^{31}P NMR-6.4-(d, J_{PH} 696 cps), is 3,9-H-3,9-dioxa-2,4,8,10-tetraoxa-3,9-diphosphaspiro[5.5]-undecane, having the formula:

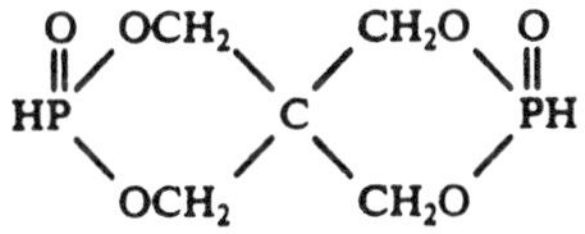

Analysis—Calculated for $C_5H_{10}O_6P_2$ (percent): C, 26.33; H, 4.42; P, 27.16. Found: C, 26.17; H, 4.76; P, 26.77. The use of PBr_3 leads to the same product.

Example 2: The acetone derivative of the product of Example 1 is prepared in a one-step reaction from the phosphorochloridite precursor as follows: A solution of 3,9-dichloro-2,4,8,10-tetraoxa-3,9-diphosphaspiro[5.5] undecane is prepared as in Example 1. Acetone, 348 grams (6.0 mols), is added, and then 202.4 grams (4.4 mols) of formic acid is added dropwise in 1 hour and 40 min with cooling at 27° to 40°C. The reaction mixture is warmed at 40° to 52°C for ½ hour, filtered, and the resulting solid is washed with acetonitrile and acetone, giving 499 grams of white solid, MP 207° to 209°C (from glacial acetic acid and acetone), ^{31}P NMR -22.7 ppm, having the structure,

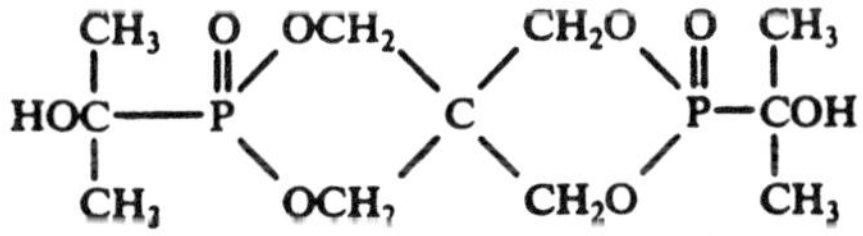

Analysis—Calculated for $C_{11}H_{22}O_8P_2$: C, 38.37; H, 6.44; P, 17.99. Found: C, 38.20; H, 6.44; P, 17.90.

Similar derivatives of methyl ethyl ketone, diethyl ketone, methyl propyl ketone, acetophenone, dibutyl ketone or of diphenyl ketone are obtained when these ketones are used in place of acetone in the above process. These ketone derivatives are also obtained when the hydrogenphosphonate is first prepared from a phosphorochloridite and formic acid and then the ketone added.

Example 3: A solution of 2.0 mols of 2-chloro-5,5-dimethyl-1,3,2-dioxaphosphorinane in 200 grams of 1,2-dichloroethane is stirred as 2.0 mols of formic acid is added dropwise at 20° to 30°C. The product is then distilled twice, giving 217 grams of colorless liquid, BP 98° to 100°C/0.15 mm, ^{31}P NMR-3.7 ppm (d, J_{PH} 660 cps), which is the cyclic hydrogenphosphonate.

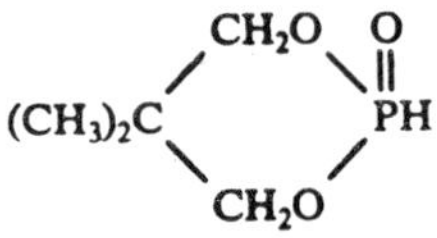

Analysis—Calculated for $C_5H_{11}O_3P$: C, 40.01; H, 7.39; P, 20.63. Found: C, 39.93; H, 7.34; P, 20.60.

Example 4: Phosphorus trichloride, 274.8 grams (2.0 mols), is added dropwise to a stirred solution of 524 grams (2.0 mols) of 2,2-di(bromomethyl)-1,3-propanediol and 0.2 grams of pyridine hydrochloride in 600 grams of benzene during 2.5 hours with enough warming to keep the temperature above 20°C. After several hours of stirring at room temperature, a ^{31}P NMR spectrum of the reaction mixture shows a peak for 90% of the phosphorus at -146.6 ppm for 2-chloro-5,5-bis(bromomethyl)-1,3,2-dioxaphosphorinane.

Stirring of the reaction mixture is continued as 106 grams (2.3 mols) of formic acid is added at room temperature during 2 hours, followed by warming at 45°C for 1.5 hours. A solid product separates on cooling. This is recrystallized from benzene, giving 333 grams of white solid, MP 91° to 96°C, ^{31}P NMR -5.2 ppm (d, J_{PH} 695 cps), which is the cyclic hydrogenphosphonate, 5,5-bis(bromomethyl)-2-H-2-oxo-1,3,2-dioxaphosphorinane, having the structure,

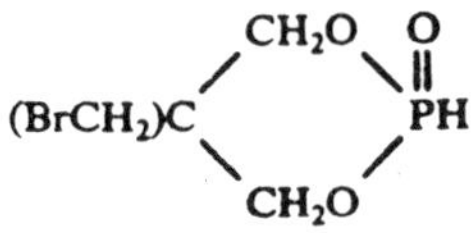

Similarly cyclic phosphorus compounds are obtained by the substitution of the above phosphorochloridite by other phosphorochloridites such as 2-chloro-5,5-dimethyl-1,3,2-dioxaphosphorinane, and 2-chloro-5-ethyl-5-methyl-1,3,2-dioxaphosphorinane, and 2-bromo-5,5-bis(chloromethyl-1,3,2-dioxaphosphorinane, 2-chloro-5-phenyl-1,3,2-dioxaphosphorinane, and 2-chloro-5-(4-fluoro-3-bromophenyl)-1,3,2-dioxaphosphorinane.

Example 5: Flame retardancy tests are conducted using typical compounds of this process, specifically compounds of the above examples. These compounds do not burn readily, subjected to heat and a flame, and also improve the flame retardant properties of polyamides, specifically nylon 66, at concentrations of 3-15% by weight, based upon the total mixture, such as by milling, or impregnation, or by dissolving or dispersing in the polymer in molten form before extrusion such as in the form of fibers or sheets. Test data for certain compounds in the Oxygen Index Test are shown on the following page.

Flame Retardant Data

Compound of Example	Oxygen Index*
1	22.6 (5%)
3	24.9 (10%)
4	26.3 (10%)

* % additive in total mixture

The other sulfenyl and ketone derivatives such as the product of Example 2 also have flame retardant properties, for polyethylene terephthalate and also polyamides, such as nylon 66.

In related work, *G.H. Birum; U.S. Patent 4,073,767; February 14, 1978; assigned to Monsanto Company* describes various derivatives of cyclic hydrogenphosphonate obtained by the reaction of a hydrogen phosphonate with azomethines, sulfonyl isocyanates, isocyanates and isothiocyanates respectively.

Representative examples of azomethines are N-(p-chlorobenzylidene)ethylamine, and N,N'-terephthalylidene-di-p-toluidine, and benzylidene ethylamine, and p-chlorobenzylidenemethylamine, and methyl benzylidenemethylamine. The general reaction for the preparation of the azomethine derivatives is:

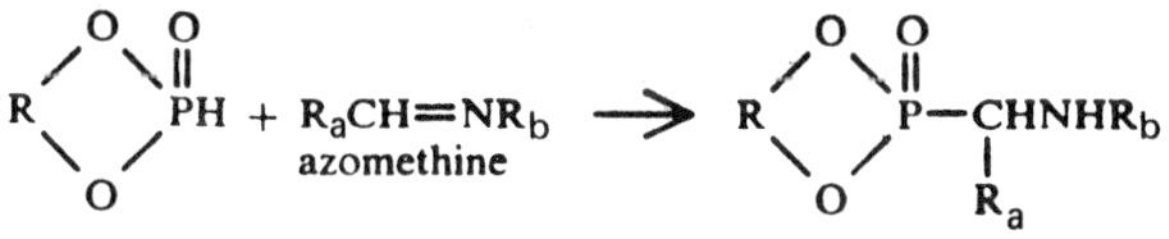

where R_a is a C_{1-4} alkyl and R_b is a phenyl radical of 6-10 carbons.

Examples of isocyanates that can be used are: p-toluenesulfonylisocyanate, methyl sulfonylisocyanate, ethyl sulfonylisocyanate, phenylsulfonylisocyanate, chlorophenylsulfonylisocyanate, etc.

Examples of isothiocyanates that can be used are: allylisothiocyanate, ethyl isothiocyanate, 4-fluorophenyl isothiocyanate, 2-naphthyl isothiocyanate, methyl isothiocyanate, and 4-chloro-2-naphthyl isothiocyanate. The general reactions for the preparation of the sulfonyl isocyanate derivatives, isocyanate derivatives and isothiocyanates are shown below:

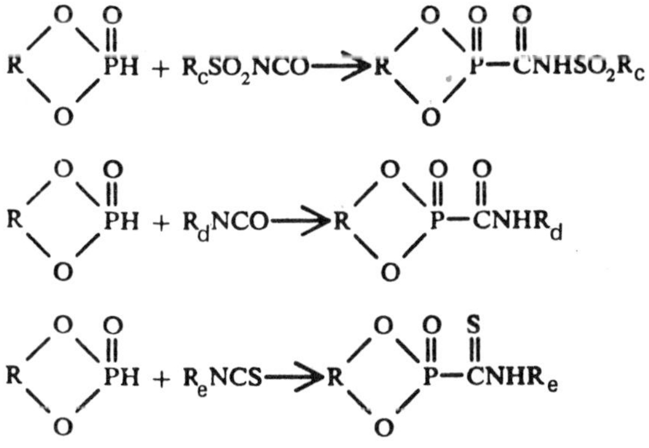

where R is an alkylene, or haloalkylene (halo is chlorine or bromine), group of a 1,2-glycol having from 2 to 8 carbon atoms, or of a 1,3 glycol having from 3 to 8 carbon atoms.

Cyclic Organophosphorus Amides

A process described by *G.H. Birum; U.S. Patent 3,980,618; September 14, 1976; assigned to Monsanto Company* relates to compounds containing urea and organophosphorus amide moieties in cyclic structures as shown in the following equation:

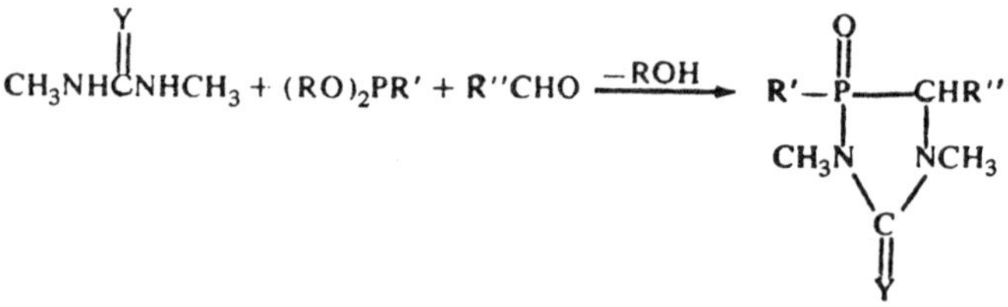

R is aryl or chloroalkyl of 2 to 10 carbon atoms
R' is aryl of 6 to 10 carbon atoms
R'' is hydrogen, alkyl, alkenyl, furanyl or aryl of 1 to 15 carbon atoms and substituted aryl forms where the substituent is fluorine, chlorine, bromine, cyano, hydroxyl, alkyloxy, alkylthio or mixtures of such substituents, and
Y is O or S

Example 1: 3-(2,4-Dichlorophenyl)-1,4-Dimethyl-2-Phenyl-1,4,2-Diazaphospholidin-5-one-2-Oxide —

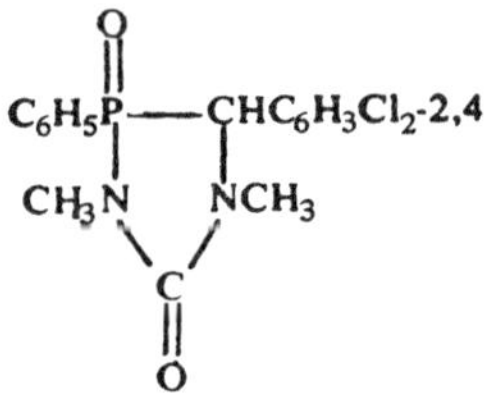

A solution of 0.3 mol each of diphenyl phenylphosphonite, 2,4-dichlorobenzaldehyde, and 1,3-dimethylurea in 150 grams of toluene is stirred under N_2 and warmed at 100° to 110°C for 2 hour. The resulting solution has ^{31}P NMR peaks at -30.6 and -26.3 ppm (~2:1 areas). The solvent and most of the by-product phenol are removed by stripping to 120°C/1 mm, and the residue is diluted with ether which causes separation of white solid (I), MP 204° to 206°C (from CH_3CN). Additional solid (II), MP 205° to 207.5°C, is isolated from the filtrate. There is a marked depression of the mixed MP of (I) and (II) and appreciable differences in their NMR spectra. Analysis (II)—Calculated for $C_{16}H_{15}N_2Cl_2O_2P$: C, 52.05; H, 4.10; Cl, 19.21, N, 7.59; P, 8.39. Found: C, 52.33; H, 4.14; Cl, 18.82; N, 7.50; P, 8.31.

Example 2: 1,4-Dimethyl-2,3-Diphenyl-1,4,2-Diazaphospholidin-5-one-2-Oxide —

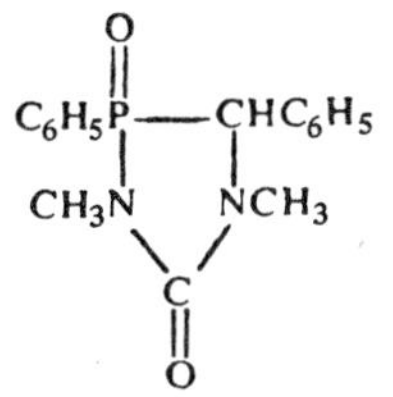

A solution of 0.3 mol each of bis(chloropropyl)-phenylphosphonite, 1,3-dimethylurea, and benzaldehyde in 100 grams of toluene is warmed at 105° to 110° for 1.5 hours, giving a light yellow reaction mixture in which about half of the phosphorus detectable by NMR had a chemical shift at -25.9 ppm. Stripping to 107°C/0.5 mm followed by dilution with ether and filtration gives 14.8 grams of white solid: MP 132° to 138°C (141.5° to 143.5°C from tetrahydrofuran-ether). Analysis—Calculated for $C_{16}H_{17}N_2O_2P$: C, 63.98; H, 5.70; N, 9.33; P, 10.31. Found: C, 63.70; H, 5.75; N, 9.06; P, 10.40.

A utility of the phosphorus amides of this process is for fire retardancy of paper. The test procedure is as follows: strips (6 x 2 inches) of paper (Whatman filter paper No. 3) are dried in an oven at 130°C for 10 minutes and then weighed. The weighed strips of paper are placed in glass bottles containing about 50 cc of a solution of ethyl alcohol having 10 to 20% of the phosphorus amide. The bottles are tilted if necessary to cover the paper strips completely with impregnating solution.

After one minute the paper is removed and pressed between two paper tissues with a 100-gram weight. The paper strips are dried in air for two minutes and then in an oven at 130°C for 10 minutes and weighed. The paper treated in this way usually contains from 5 to 15% of the additive amide chemical. Standard burning tests show that the samples having above about 10% of the additive phosphorus amides are significantly more resistant to burning than untreated control samples. Fire retardant properties are obtained with:

3-(2,4-dichlorophenyl)-1,4-dimethyl-2-phenyl-1,4,2-diazaphospholidin-5-one-2-oxide
1,4-dimethyl-2,3-diphenyl-1,4,2-diazaphospholidin-5-one-2-oxide
3-(4-cyanophenyl)-1,4-dimethyl-2-phenyl-1,4,2-diazaphospholidin-5-one-2-oxide
3-(2,4-dichlorophenyl)-1,4-dimethyl-2-phenyl-1,4,2-diazaphospholidin-5-one-2-oxide
1,4-dimethyl-2-phenyl-1,4,2-diazaphospholidin-5-thione-2-oxide
1,4-dimethyl-3-(4-bromo-2-fluorophenyl)-2-(2-naphthyl)-1,4,2-diazaphospholidin-5-one-2-oxide

The cyclic organophosphorus amides have biological activity and are also useful as fire retardants and as lead scavengers in gasoline.

Compositions of Polyvalent Metal Compounds and Hydroxylcarboxylic Acid

P.S. Minhas and B. Sukornick; U.S. Patent 4,061,810; December 6, 1977; assigned to Allied Chemical Corporation describe combinations of certain hydroxycarboxylic acids with certain metal oxides and hydroxides that provide enhanced flame retardancy to carpets. The flame retardancy is durable to the usual carpet cleaning procedures.

The additive flame retardant composition with which the carpet is treated comprises:

(a) about 10 to 90 weight percent of a polyvalent metal compound selected from the group consisting of oxides and hydroxides of tin, antimony, aluminum, and zinc; and
(b) about 10 to 90 weight percent of a hydroxycarboxylic acid selected from the group consisting of malic acid, citric acid, tartaric acid, gallic acid, and 2,4-dihydroxybenzoic acid.

The combination of the polyvalent metal oxide or hydroxide and the hydroxycarboxylic acid provides synergistic improvement in flame retardance of the carpet.

The polyamides which are useful in the improved flame retardant carpets of the process include polycaprolactam, the polyamides which are derived from the condensation of a dicarboxylic acid with a diamine, such as polyhexamethylene adipamide and polyhexamethylene sebacamide, and copolymers of these.

In treating the carpet from about 1 to 15 weight percent of the treating composition is applied to the carpet. The composition is dissolved in a solvent, preferably water, to make a solution of about 0.5 to 12 weight percent and more. The carpet is then soaked by the solution, which may contain other additives commonly used in finishing baths to improve properties such as penetration. The carpet is then squeezed with any suitable apparatus such as pad rollers to remove excess solution. The squeezing apparatus, such as the rollers, is adjusted to give from about 25 to 300 weight percent wet pick-up. The carpet is then dried and cured in a dryer or oven, preferably at temperatures of 100° to 150°C. The solution may be applied to the carpet in numerous ways. For example, the carpet may be immersed in the solution or the solution may be sprayed upon the carpet or applied to the carpet by means of pad rollers. The following examples illustrate the process.

Example 1: About 132 grams of gallic acid monohydrate is dissolved in 1,000 grams of hot distilled water. The solution is allowed to cool to about 40°C, and 87.6 grams of an aqueous dispersion of colloidal antimony oxide is added and uniformly distributed in the solution, the weight ratio of gallic acid monohydrate to antimony oxide solids being about 5.5:1.0. The resulting composition is made up to 3,500 grams with distilled water and used to produce a flame retardant carpet as described in the following example. For convenience, this composition may be called Composition A.

Example 2: Composition A of Example 1 is used to produce a flame retardant carpet having excellent durability to carpet cleaning procedures. A typical polyamide carpet may be produced as follows: a reactor equipped with a heater and stirrer is charged with a mixture of 1,520 parts of ε-caprolactam and 80 parts of aminocaproic acid. The mixture is then flushed with nitrogen and stirred and heated to 255°C over a 1 hour period at atmospheric pressure to produce a polymerization reaction. The heating and stirring is continued at atmospheric pressure under a nitrogen sweep for an additional four hours in order to complete the polymerization. Nitrogen is then admitted to the reactor and a small pressure is maintained while the polymer is extruded from the reactor in the form of a polymer ribbon. The polymer ribbon is subsequently cooled, pelletized, washed and then dried. The polymer is a white solid having a relative viscosity of about 50 to 60 as determined at a concentration of 11 grams of polymer in 100 ml of 90% formic acid at 25°C (ASTM D-789-62T).

The polymer is melt extruded under pressure of 1,500 psig to a 70 orifice spinnerette to produce a fiber having about 3,600 denier. The fiber is collected, drawn at about 3.2 times the extruded length and textured with a steam jet to produce yarn suitable for use in carpet. The yarn is two-plied by twisting two ends together with a 1.5 S twist. The yarn is tufted into a level loop 22 ounce/yard^2 carpet at about 8 stitch rate. A relatively pliable nonwoven polypropylene fabric is used as the primary backing. Tufting is carried out on a conventional tufting machine operated to give a pile having a height of 5/32 to 7/32 inch. An untreated portion of this carpet is used as a control and a second portion is padded with Composition A of Example 1 to give a wet pick-up sufficient to deposit 6.6% of the gallic acid monohydrate and 1.2% antimony oxide solids on the fibers.

The treated carpet is dried at 125°C in an air circulating oven until dry, then cured at 125°C for about 10 minutes. The treated and untreated carpets are then given a jute secondary backing using a conventional latex binder containing no flame retardant. For convenience, the treated carpet is called Carpet A and the untreated carpet is called Carpet B. The following table compares the carpets with respect to flame retardancy (before and after laundering) as measured by the Department of Commerce flame retardancy test DOC FF 1-70, together with the AATCC laundering procedure 124-1973. The test shows the highest temperature applied to the carpet before the carpet fails to extinguish within 90 seconds after the pill has finished burning. Measurements were made on the carpet initially and after 1-5 launderings.

	- - - Highest Temperature Before Failure, °C- - - -		
Carpet System	**Initial**	**After 1 Laundering**	**After 5 Launderings**
Carpet A	205	203	206
Carpet B	141	130	115

These results indicate that the fire retardancy of Carpet A is significantly greater than that of Carpet B and that the fire retardancy of Carpet A is durable to the cleaning procedure.

Fireproofing Shampoo Composition for Polyamide and Other Fibers

A process described by *G.A. Pearson; U.S. Patent 3,990,977; November 9, 1976;* relates to a shampoo composition particularly adapted for the cleaning of surfaces such as carpets, rugs, etc., and the fireproofing of such surfaces.

The composition for use in the method of this process comprises an aqueous composition containing an amine, phosphoric acid, one or more fireproofing agents and a detergent. A composition for use in this process is as follows:

Component	Weight Percent
Amine	40-60
Phosphoric acid	30-50
Fireproofing agents	35-60
Detergent	1-30
Water	250-550

The amine component used in the composition may be an alkyl amine, aromatic amine, aralkyl amine or alkaryl amine. Especially preferred amines are the polyolamines, particularly the water-soluble polyalkanol amines, such as those of the formula:

$$HO(CH_2)_n{-}N\left\langle\begin{matrix}R\\R\end{matrix}\right.$$

where R is hydrogen , a lower alkyl group, or a hydroxy lower alkyl group, and n is an integer of 1 to 5. The phosphoric acid is commercially available in 85% strength. The fireproofing agents may comprise diammonium phosphate, monoammonium phosphate, ammonium salts of sulfamic acids such as ammonium sulfamate, etc. A preferred composition contains about 2 parts of diammonium phosphate and 1 part ammonium sulfamate.

The class of detergents used are the alkali metal derivatives of aryl sulfonic acids, usually an alkali metal salt such as sodium benzenesulfonic acid. The aryl group may be further substituted by one or more alkyl substituents and mixtures of these detergents may also be used.

The aqueous composition is prepared by initially mixing the water and the amine, preferably an alkanolamine, until good solution is obtained. Thereafter, the other components are added to obtain the final mixture or solution of the process. The resulting solution will have a pH in the range of 5.5 to 6.5 and preferably about 6.0. The solution has good storage properties and may be stored in final mixed form for extended periods until ready for use. It may be diluted to any desired consistency by the addition of water.

A fireproofing shampoo which includes urea is prepared as follows. Substantially equal parts by weight of urea and phosphoric acid are heated to boil off water at atmospheric pressure, 90° to 115°C. After the water is removed, heating is continued until ammonia evolution begins (125° to 130°C) which can be detected by its odor. The mixture is then cooled, about twice the volume of water is added followed by about ½ the amount of the amine as used in the main composition (e.g., 15-20 wt %). To this mixture is then added ½ to 1 wt % of detergent. The resulting mixture has a pH of 5.5 to 6.5 and is also useful as a combination fireproofing agent and cleaner.

The substrates to which this process is applicable may be any surface such as a plastic, foam rubber, paper, wood and textiles such as carpets or rugs constructed by any known method. If the surface is a textile, it may be of the woven type consisting largely of fibrous material or it may be a tufted material in which the pile fibers are embedded in a nontextile backing, for example a rubber or resin backing. The fibers may be any of those known in construction of carpets and rugs including wool, cellulose, synthetic fibers such as polyamides, nylon, polyesters, polyolefins and polymers or copolymers of acrylonitrile as well as blends of fibers.

In use, the composition may be applied by spraying, as from an aerosol can or other sprayer or with any of the well-known types of applicators. When applied to a carpet, it is applied in excess to the carpet or rug in sufficient amounts to saturate the surface while it remains in place on the floor or other surface. After saturating the rug or carpet, it may then be worked in by conventional shampooing methods, such as by a rotary shampoo machine. At this point, a wet sponge or the like may be used to soak up excess material which will contain dirt foamed from the fibers. After completion, the mixture is then allowed to dry. The resulting carpet or rug will contain a transparent flexible film on the fibers which is fire retardant.

A composition was prepared by heating 90 parts by weight urea and 90 parts by weight 85% phosphoric acid at 115°C until all water was removed. Thereafter heating was continued to 127°C until NH_3 odor was detected. The mixture was then cooled and 300 parts water and 30 parts triethanolamine were added together with 1.0 wt % of the detergent Nacconol. The pH of the mixture was 6.0. When applied to a nylon carpet and allowed to dry the mixture was found to have cleaned dirt from the fibers and to have placed a fire retardant film on it.

This process has been described herein with reference to certain preferred embodiments. However, as obvious variations thereon will become apparent to those skilled in the art, the process is not to be considered as limited thereto.

Mold Release and Lubricant Additive Composition

R.E. Phillips, Jr.; U.S. Patent 3,988,271; October 26, 1976; assigned to Monsanto Company describes a polyamide molding composition having good flame retardant and mold release characteristics that comprises a polyamide and from 500 to 5,000 parts per million each of a C_{12-20} aliphatic alcohol and an aluminum salt of a C_{12-20} fatty acid such that the ratio of the parts of salt to the parts of alcohol is at least 1.0. The alcohol is one having from 12 to 20 carbon atoms such as lauryl alcohol, cetyl alcohol, stearyl alcohol, myristyl alcohol, and eicosyl alcohol.

The fatty acid from which the salt is derived can be for example lauric acid, myristic acid, palmitic acid, stearic acid or eicosoic acid. Of these alternatives the diacid salts are preferred and particularly aluminum dipalmitate, aluminum distearate and aluminum dieicosoate. The polyamide may be any of the polyamides known as nylons.

This process is illustrated by the following examples which demonstrate the effect of varying additive amounts on the flammability of a commercially available nylon 66 polymer as assessed by the UL-94 test applied to a strip of the polymer one thirty-second of an inch in thickness. All amounts given here are in parts per million. The additives were surface coated on the sample to be tested. Samples rated HB had unsatisfactory flammability characteristics while those having V2 rating are considered to meet the required level of performance. Examples 1 through 10 are for purposes of comparison only.

EXAMPLE	TOTAL LUBRICANT LEVEL (PPM)	STEARYL ALCOHOL (PPM) (1)	ALUM. DI-STEAR-ATE (PPM) (2)	RATIO OF (2) to (1)	UL-94 RATING
1 (C)	0	0	—	—	V2
2 (C)	1000	0	1000	—	V2
3 (C)	2000	2000	0	—	HB
4 (C)	3000	3000	0	—	HB
5 (C)	3000	2000	1000	0.5	HB
6 (C)	4000	4000	0	—	HB
7 (C)	4000	3000	1000	0.33	HB
8 (C)	5000	5000	0	—	HB
9 (C)	5000	4000	1000	0.25	HB
10 (C)	5000	3000	2000	0.67	HB
11	2000	1000	1000	1.0	V2
12	3000	1000	2000	2.0	V2
13	4000	2000	2000	1.0	V2
14	4000	1000	3000	3.0	V2
15	5000	2500	2500	1.0	V2
16	5000	2000	3000	1.5	V2
17	5000	1000	4000	4.0	V2

Compositions containing no stearyl alcohol were found to have poor mold release characteristics and all containing stearyl alcohol performed well in this respect.

The polyamide compositions can incorporate other additives such as fillers, whether in fibrous or granular form, antioxidants, stabilizers, coloring materials, antistatic additives, etc. without materially affecting the applicability of the process. These compositions find use in the automotive, electrical, appliance and consumer goods field.

Derivatives of Phosphoric Acid

Y. Imanaka and H. Uchiyama; U.S. Patent 4,008,345; February 15, 1977; assigned to Teijin Limited, Japan describe a process for fireproofing of shaped articles of aromatic polyamides which comprises contacting a shaped article of an aromatic polyamide with an aqueous solution containing a phosphorus compound selected from halogen- and sulfur-free, phosphorus-containing inorganic acids, and their ammonium, amine and urea salts in a concentration of as low as 2 to 30%, thereby causing the aqueous solution to be absorbed in the shaped article, drying the shaped article at a temperature of not more than about 150°C, and then post-heat-treating it at a temperature of about 300° to 450°C, preferably for a time period of not more than about 60 minutes.

Examples of aromatic polyamides that can be used are aromatic polyamides, aromatic polyamideimides, aromatic polyamide esters, aromatic polyamide benzimidazoles, and aromatic polyamide sulfones. Of halogen- and sulfur-free, phosphorus-containing inorganic acids and their ammonium salts, ortho-phosphoric acid, phosphorous acid, primary ammonium ortho-phosphate, secondary ammonium ortho-phosphate, primary ammonium phosphite, secondary ammonium phosphite and urea ortho-phosphate are preferred.

The aromatic polyamide shaped articles so treated find a wide range of applications, such as filter bags, ironing stand coverings, fire-protective clothing, electrical insulating materials, curtains, wall materials, indoor finishing of decorative materials, or finishing materials for furniture. The following examples illustrate the process.

Each of the aromatic polyamide shaped articles in Table 1 was treated under the conditions shown (except as noted below) using various treating agents and then post-heat-treated under varied conditions. The results are shown in Table 2. Tables 1 and 2 are shown on the following pages.

In Example 3, the treating agent was dissolved in a concentration of 5.0% by weight in an N-methylpyrrolidone containing 20% by weight of poly(m-phenylene isophthalamide). The resulting dope was coated on a glass plate using a doctor knife, and dried at 140°C for 2 hours. The coated glass plate was immersed in water to remove the solvent thoroughly. Then, it was dried at 120°C for 2 hr, and a resulting film having a thickness of 30 microns was post-heat-treated under the conditions shown in Table 2. In Examples 10 and 11, a 50 cm long steel hot plate was used as a heat source instead of the hot air dryer.

Bispentabromophenoxyethane

A.L. Anderson; U.S. Patent 4,064,105; December 20, 1977; assigned to Velsicol Chemical Corporation describes plastic compositions containing a three component system consisting of (1) nylon, (2) 1,2-bis(2,3,4,5,6-pentabromophenoxy)-ethane of the formula:

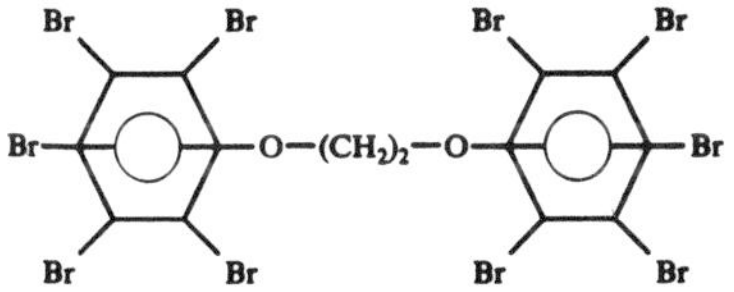

and (3) an enhancing agent, preferably an oxide of antimony, arsenic or bismuth. Similar bisphenoxy compounds and examples have been described above in the section on polyesters.

ACRYLICS

Acrylate Type Liquid Polymerizable Compositions

H.A. Whittum and M.J. Hjelm; U.S. Patent 4,019,922; April 26, 1977; assigned to Advance Coatings Company describe a liquid polymerizable material which may be used for the coating of wood or other substrate, which material, when cured, is fire resistant and protects the substrate. The material cures without application of heat and is formed from methyl methacrylate monomer, a hydroxy alkylacrylate or diacetone acrylamide monomer, a chlorinated hydrocarbon, an antimony compound, a zinc compound, a filler, and a catalyst-promoter system.

An example of a practical use is shown in Figure 7.1. A wall panel **10** is shown with a magnified cutaway of the layers with which it is fabricated. The base layer or substrate **11** is ¾" construction-grade plywood. To this substance is

TABLE 1: PREPARATION OF THE PRODUCTS

		Treatment with a treating agent				Post heat-treatment	
Examples	Shaped article of aromatic polyamide	Kind	Concentration (%)	Drying temperature (°C)	Take-up (%)	Temperature (°C)	Time (minutes)
1	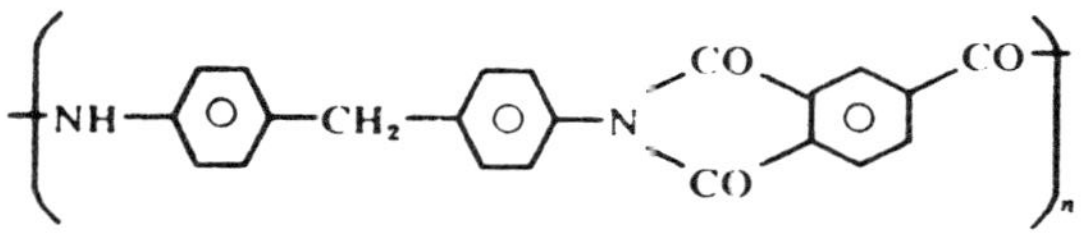 fabric	H_3PO_4	10	110	3.5	350	30
2	 filament	H_3PO_3	15	110	8.6	"	"
3	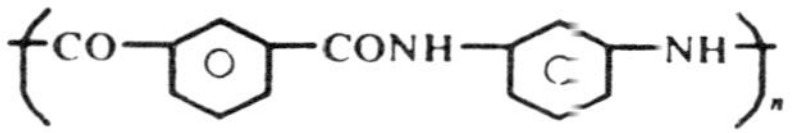 film	"	—	—	5.0	"	10
4	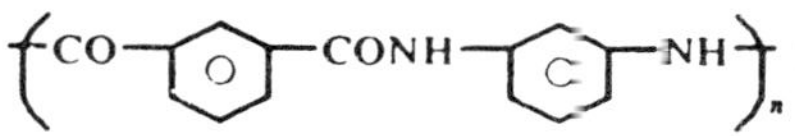 fabric	"	5	110	2.3	370	5
5	"	$(NH_4)H_2PO_4$	10	"	7.0	350	30
6	"	$H_3PO_4NH_2CONH_2$	"	"	6.1	"	"
7	"	$(H_3C)_2NH_2H_2PO_3$	"	"	5.7	"	"

Source: U.S. Patent 4,008,345

(continued)

TABLE 1: (continued)

		Treatment with a treating agent				Post heat-treatment	
Examples	Shaped article of aromatic polyamide	Kind	Concentration (%)	Drying temperature (°C)	Take-up (%)	Temperature (°C)	Time (minutes)
8	 fabric	$H_6P_4O_{13}$	10	110	7.8	370	5
9	"	H_3PO_2	"	"	4.3	"	15
10	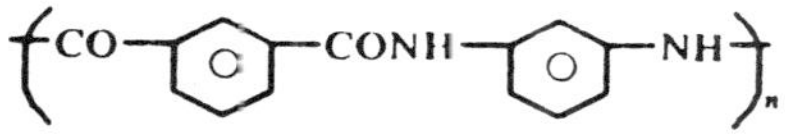 filament	H_3PO_4	"	90	2.6	370	1/3
11	"	H_3PO_3	"	"	3.2	"	1/6
12	 fabric	$NH_4H_2PO_3$	"	110	5.8	350	30
13	"	$(NH_4)_3PO_4$	"	"	5.3	"	"
14	 filament	H_3PO_3	15	110	7.3	"	"
15	 filament	H_3PO_4	"	110	8.5	"	"

Source: U.S. Patent 4,008,345

TABLE 2: PROPERTIES OF THE PRODUCTS

Examples	Fire-proofness (A) LOI value	Fire-proofness (B) Grade	Acid resistance (%)	Strength retention (%)	Evolution of noxious gases: Kind	Evolution of noxious gases: Amount (ppm)	Evolution of noxious gases: Allowability	Fire-proofness retention (%)
1	45	1	81	71		Not detected	Yes	100
2	52	''	93	68	PH_3	below 0.1	''	100
3	51	''	97	78	''	''	''	100
4	43	''	92	73	PH_3	below 0.1	''	100
5	49	''	78	55		Not detected	''	100
6	46	''	79	61		''	''	100
7	46	2	75	63		''	''	100
8	49	1	81	40		Not detected	Yes	85*
9	43	2	76	78	PH_3	0.2	''	100
10	45	2	59	65		Not detected	''	94
11	43	1	67	67	PH_3	0.1	''	100
12	48	1	85	65	PH_3	below 0.1	''	100
13	48	1	76	58		Not detected	''	100
14	49	1	91	71	PH_3	below 0.1	''	100
15	51	1	83	63		Not detected	''	95

*When the tested sample was again tested by the same procedure, the sample showed a fireproofness retention of at least 90% based on the fireproofness of the tested material.

Source: U.S. Patent 4,008,345

FIGURE 7.1: STRUCTURE PANEL FOR FIRE RESISTANT RESIN

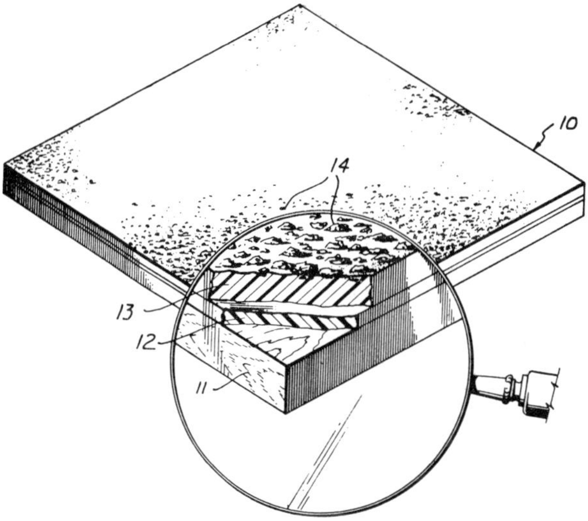

Source: U.S. Patent 4,019,922

applied a thin tiecoat **12** of a polyester resin. This coat seals the plywood and provides a uniform surface for further layers. While the tiecoat is still tacky, the fire retardant coat **13** is applied. This coat is applied with a thickness of about one-sixteenth inch. In addition, a cosmetic coat **14**, which in the preferred embodiment consists of a sprinkling of small stones, can be applied to enhance the appearance of the exposed surface. The following examples illustrate the process.

Example 1: The first step was the preparation of the mixture of monomers and flame retardant chlorinated organic material. The Formula W18-222-1 was as follows:

	%	g
Hydroxy propylacrylate monomer (CX-2383, Dow Chemical Company)	17.95	718
Methyl methacrylate monomer (Union Carbide Corp.)	17.95	718
Chlorez 700AX (Dover Chemical Co.) chlorinated paraffin (70% chlorine)	63.50	2540
Super Airout (Isochem Resins Co.) (an air release agent comprising a silane modified product of a silicone reacted with a tallate)	0.10	4
Wetting Agent	0.50	20
	100.00	4000

The ingredients were added, in the order shown, to a one-gallon friction-top can with mechanical agitation.

The second step was the establishment of a suitable curing system which will give an initial cure at room temperature without application of heat. The following systems were tried:

	W18-222-		
	2	3	4
	100 g.	100 g.	100 g.
2,4-Pentanedione (Union Carbide Corp.)	0.4	—	—
Lupersol PDO (50% t-butyl peroctoate) (Lucidol Division, Pennwalt Corp.)	2	—	—
Luperco ANS-50 (50% benzoyl peroxide) (Lucidol Division, Pennwalt Corp.)	—	0.70	0.70
Cobalt catalox 12% (Ferrol Chemical Division, Ferro Corporation)(12% cobalt catalytic paint dryer)	0.4	—	—
N,N-Dimethyl-p-toluidine (Naftone, Inc.)	—	0.14	—
N,N-Dimethylaniline (American Cyanamid Company)	—	—	0.14

In each experiment, the ingredients were added, in the order shown, to a 6-oz plastic-coated paper hot-drink cup with hand mixing with a wooden spoon after each addition. The timing for gel time was begun at the time of mixing in of the cobalt or amine promoter. Each cup was allowed to stand at room temperature (about 72°F) without external heating. Experiment W18-222-2 began to gel in 30 minutes; W18-222-3 in 5 to 8 minutes; W18-222-4 in about 20 minutes. Of the three systems tested, the dimethyl-p-toluidinebenzoyl peroxide (W18-222-3) was considered most suitable.

An antimony oxide masterbatch, W18-222-5, was prepared by the addition of 400 grams of Thermogard S (M&T Chemicals, Inc.) to 400 grams of monomer mixture W18-222-1 with mechanical agitation. The Thermogard-monomer mix was passed once through a laboratory hand homogenizer.

The next step was the preparation of a cast cured panel for evaluation of flame retardant properties. The formula for this preparation (W18-222-6) was as follows:

	g
W18-222-1 monomer mix	600
W18-222-5 antimony oxide masterbatch	76
Industrial ground gypsum (United States Gypsum Co.)	825
Iron oxide	15
Luperco ANS-50 (50% benzoyl peroxide) (Lucidol Division, Pennwalt Corp.)	4.2
N,N-Dimethyl-p-toluidine (Naftone, Inc.)	0.60

The ingredients were added, in the order shown, to a one-quart friction-top can with mechanical agitation. Immediately following the mixing in of the dimethyl-p-toluidine, the mix was poured into a steel mold 22¾ inches long, 3⁹⁄₁₆ inches wide and 1 inch deep, lined with 0.5 mil Mylar film. The mix was smoothed out to form a layer one-fourth inch deep. Gelation occurred without application of heat in 15 to 20 minutes at 80°F. The panel appeared to be well cured after overnight standing, as judged by qualitative surface hardness. After 3 days, the Barcol hardness was 10-20 as determined with the Barcol Impressor.

The Flame Spread Rating (FSR) of the cast panel was determined by a modification of the method described in Special Report No. 6090 entitled "Use of a Small Flame Tunnel in the Laboratory Evaluation of Flame Spread Rating," dated May 5, 1964, prepared by H.L. Vandersall of Monsanto Company.

The FSR for the W18-222-6 test panel was 17.5. No afterflaming was observed, and the afterglowing (length of time after all flaming ceases but while red-glowing embers are still visible) was 10 seconds. The char length was 12½ inches. The foam thickness at the point of flame impingement on the panel was about 12 millimeters (original panel thickness about 6 mm). The results of the FSR tests show that the composition described is capable of being fabricated into a panel with excellent flame retardant properties.

Example 2: The effect of filler addition on the mixture of Example 1 was examined. A base resin, similar to that of Example 1, was prepared as follows.

Hydroxy propylacrylate monomer	18.0
Methyl methacrylate monomer	27.0
Chlorinated paraffin (70% chlorine)	55.0

was mixed with filler to form:

Base resin	100.
Antimony oxide	5.5
Zinc oxide	20.0
Hydrated alumina	150.0
50% benzoyl peroxide paste (ANS-50)	1.0
N,N-dimethyl-p-toluidine	0.2

The result was a mixture that gelled at 75°F in 5 to 8 minutes, had a Barcol of 40 and a FSR of 25 in a 2-foot tunnel.

Halogenated Bis-Acrylate Comonomers

J. Habermeier; U.S. Patent 4,057,558; November 8, 1977; assigned to Ciba-Geigy Corporation describes a process whereby halogenated bis-acrylates are obtained when a N,N'-di(hydroxyalkyl)benzimidazolone which is halogenated in the nucleus is esterified with acrylic acid or methacrylic acid or an acrylate or methacrylate.

The halogenated bis-acrylates are valuable comonomers since they contain halogen bonded in a stable manner, can readily be copolymerized with polymerizable compounds and are suitable for the manufacture of flame resistant polymers.

The polymerizable mixtures according to the process can be used for surface protection, in molding compositions, as casting resins and the like. The polymerizable mixtures which are suitable for the manufacture of coatings and molding compositions can additionally also contain flexibilizers, fillers, and preferably pigments, for example titanium dioxide.

Example 1: Bis-acrylate obtained from 1,3-di(2'-hydroxyethyl)-4,5,6,7-tetrabromobenzimidazolone – A mixture of 1,075.7 g (2.0 mols) of 1,3-di(2'-hydroxyethyl)-4,5,6,7-tetrabromobenzimidazolone, 720.6 grams (10.0 mols) of acrylic acid, 2.5 liters of toluene, 35 ml of 50% strength sulfuric acid, 3.0 ml of triphenyl phosphite and 0.5 gram of phenothiazine are heated to 100° to 102°C, while stirring, in a 6-liter glass apparatus provided with a stirrer, a thermometer and a water separator with a reflux condenser. The water/toluene azeotrope starts to distil off at a boiling point of 92° to 96°C. After 3 hours under these conditions, 89.3 ml of water have separated off (theory: 89.5 ml) and the reaction has ended. The clear, homogeneous, pale yellow-brown reaction solution is immediately filtered hot and 7 grams of hydroquinone and 0.5 grams of copper naphthenate are added to the filtrate and the mixture is cooled to room temperature.

The product which has crystallized out is filtered off and, while still moist, recrystallized from 4.5 liters of acetone. The product is then dried in vacuo at 20° to 30°C. This gives 517 grams of a colorless pure substance in the form of needle-shaped crystals with a melting point of 137.9°C (Mettler FP 51, 1°C/min). A further 103 grams of the product, which is also very pure, can be obtained from the mother liquor (MP 136°C). After recrystallization, the total yield is, accordingly, 620.0 grams (48% of theory).

This bis-acrylate corresponds to the formula:

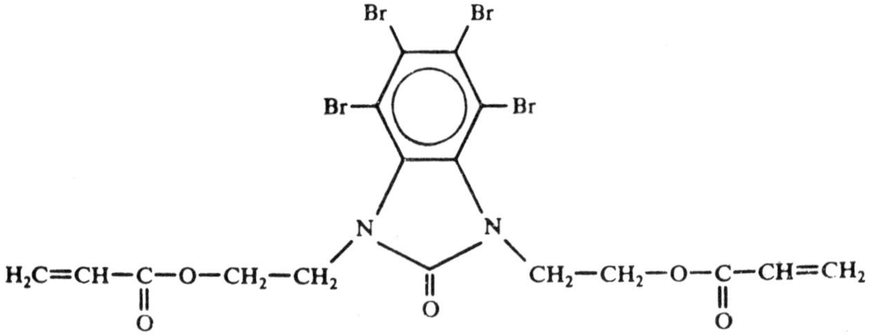

Example 2: This example illustrates the use of the bis-acrylate according to Example 1 as a comonomer for styrene and for inherent flame retardation of polystyrene.

60 grams of the bromine-containing bis-acrylate prepared according to Example 1 are added, at 60°C, to 120 grams of freshly distilled styrene and the two compounds are mixed by stirring. After a short time, the bis-acrylate used as the comonomer has completely dissolved in styrene (clear colorless solution); 3.5 grams of dried dibenzoyl peroxide are then added and mixed in. The colorless, clear mixture is poured into an aluminum mold (4 mm wall thickness) which has been prewarmed to 60°C and in order to effect polymerization the mold is placed in a warming cabinet at 55° to 60°C for 36 hours. The mold is then removed and allowed to cool to room temperature and the transparent colorless sheets of the styrene copolymer are taken out. When the various tests are carried out, the sheets display the properties given below and, for comparison,

these are contrasted with the characteristic values for commercially available polystyrene (data from the *Kunststoff-Handbuch,* Volume V, "Polystyrol" by Vieweg/Daumiller, page 390).

Properties	Polystyrene modified according to Example 1	Commercially available polystyrene
Combustibility (according to UL* 94)	57 seconds burning time	burns away completely
Limiting Oxygen Index (according to ASTM** D 2863-70)	23.4	17 - 18
Heat distortion point, according to Martens (DIN*** 53,461)	120° C	75 - 88° C
Flexural strength (according to DIN 53,452)	10.2 kp/mm^2	5 - 8.0 kp/mm^2
Deflection (according to DIN 53,352)	3.8 mm	—
Absorption of water (4 days at 20° C; DIN 53,472)	0.18%	0.1 - 0.2%

*UL = Underwriter's Laboratories. The combustibility test UL 94 prescribes a vertical arrangement of the test piece.
**ASTM = American Standard Test Method. The "Limiting Oxygen Index" indicates the minimum amount of oxygen in a nitrogen/oxygen mixture at which the test piece still just burns.
***DIN = Deutsche Industrie-Norm (German Industrial Standard)

The comparison shows that, compared with unmodified polystyrene, the polystyrene rendered flame resistant with a bis-acrylate displays better mechanical properties, such as heat distortion point or flexural strength.

Polymethyl Methacrylate Based Molding Compositions

A process described by *G.L. Deets; U.S. Patent 4,065,428; December 27, 1977; assigned to Monsanto Company* relates to polyblend molding compositions which exhibit smoke evolution levels of less than 300 D_m (Flaming) and UL-94 ratings of at least V-1. The polyblends comprise:

(a) from 45 to 84% by weight based on the total weight of the polyblend of a polymethylmethacrylate (PMMA) matrix polymer, which optionally contains up to 40% by weight based on the weight of the matrix polymer of a comonomer, preferably a halogenated comonomer;

(b) from 10 to 40% by weight based on the total weight of the polyblend of a polychloroprene rubber component which is grafted with from 10 to 100 parts of MMA per hundred parts of rubber; provided that the total amount of halogen in the polyblend due to the polychloroprene rubber component and the halogenated comonomer is at least 9.0%, preferably 10.0%, by weight;

(c) from 6 to 15% by weight based on the total weight of the polyblend of a metal oxide, such as Sb_2O_3.

Example 1: This example illustrates the preparation of the polymethylmethacrylate (PMMA) matrix used in the process. The PMMA used in the working examples is prepared using the following charge:

100 parts MMA monomer
0.6 part n-dodecyl mercaptan
0.1 part $K_2S_2O_8$ (initiator)
0.04 part $NaHSO_3$ (initiator)
0.00033 part $Fe_2(SO_4)_3$
1 part Dowfax 2A (emulsifier)
1 part Duponol ME (emulsifier)
Water to provide 35% of polymeric solids

The water, emulsifiers (as 10% solutions), and $Fe_2(SO_4)_3$ are added initially to the reaction vessel and heated to 60°C under nitrogen. The monomer, n-dodecyl mercaptan and initiator solutions are added continuously over a 4-hr period. The bath is then held at 70°C for 2 hours at the end of the addition period. Conversion of monomer to polymer is 98%.

Example 2: This example illustrates the grafting of polychloroprene with MMA monomers. A polychloroprene latex (Latex 842A from Du Pont), having an average particle size of about 0.12 micron, is grafted with methylmethacrylate using the following charge:

100 parts polychloroprene in latex form
1 part Dowfax 2A (emulsifier)
1 part Duponol ME (emulsifier)
100 parts MMA monomer
1 part $K_2S_2O_8$ (initiator)
0.5 part n-dodecyl mercaptan
Water to provide 35% of polymeric solids

The water and emulsifiers (as 10% aqueous solutions) are added to the polychloroprene latex which is then heated to 85°C under a nitrogen blanket. The monomer, n-dodecyl mercaptan and $K_2S_2O_8$ (4% aqueous solution) are added continuously to the latex over a four-hour period. At the end of the addition period the batch is held at 85°C for two hours. The resulting conversion of monomer to polymer is in the range of from 94 to 98%. This includes the polymer which is grafted onto the rubber and that which polymerized to form a matrix polymer. The amount of grafting of PMMA onto the polychloroprene rubber is about 50% by weight.

Example 3: Example 2 is repeated here except that a polybutadiene latex is used in place of the polychloroprene latex used in Example 2.

Example 4: Example 2 is repeated here except that a polybutyl acrylate latex is used in place of the polychloroprene latex used in Example 2.

Examples 5 to 9: Various polyblends are prepared using the polymeric products from Examples 1, 3 and 4 above. The PMMA matrix polymer from Example 1, the grafted rubbers from Examples 3 and 4 and antimony oxide and/or decabromodiphenyl ether are blended by melt processing and then formed into test samples by compression molding. The samples are then tested for flame retardancy and smoke evolution, according to the procedures outlined in Underwriters Laboratory UL-94 Test, September 17, 1973 and "Method for Measuring Smoke from Burning Materials," Symposium on Fire Test Methods—Restraint and Smoke 1966, ASTM STP 422, ASTM, 1967, page 166.

Summary of Examples 5 to 9*

Example	% Sb_2O_3	% DBDPE**	% Rubber***	UL-94	D_m(Flaming)
5 (Control)	—	—	—	Fail	40
6 "	—	—	15 PBD	Fail	180
7 "	—	—	15 PBA	Fail	60
8 "	7	15	15 PBD	V-1	380
9 "	9	19	15 PBA	V-1	380

* Example 5 is the PMMA from Example 1 without rubber or fire retardant additives.
**DBDPE is decabromodiphenyl ether.
***Percent by weight of rubber in polyblend. The weight given is that of the rubber component before grafting.
PBD is polybutadiene of Example 3.
PBA is polybutyl acrylate of Example 4.

Examples 5 to 7 indicate that PMMA with and without rubber modification burns quite readily. The data further indicate that these materials are fairly clean burning exhibiting low levels of smoke. Examples 8 and 9 indicate that when the rubber modified PMMA is compounded with antimony oxide and decabromodiphenyl ether, then the samples achieve a degree of fire retardancy (V-1 rating) with the smoke levels increasing markedly. The total halogen content in these examples is 12.5 and 15.8% by weight, respectively.

The polyblends of this process are useful for preparing a wide variety of molded objects such as radio and television cabinets, appliance housings, parts and components for vehicles, furniture and other related items.

In related work, *G.L. Deets; U.S. Patent 4,080,404; March 21, 1978; assigned to Monsanto Company* describes polyblend molding compositions which exhibit smoke evolution levels of less than 360 D_m (Flaming) and UL 94 ratings of at least V-1. The polyblends comprise (A) poly(styrene/acrylonitrile/methylmethacrylate) matrix polymer; (B) polychloroprene rubber which is grafted with monomers of the type used to prepare the matrix; (C) metal oxide; and (D) smoke suppressants.

The composition of the polyblend comprises

(A) from 10 to 75% by weight based on the total weight of the polyblend of a matrix polymer which comprises the interpolymerization product of;

  (1) from 20 to 40% by weight of a styrene type monomer;

  (2) from 20 to 40% by weight of a nitrile type monomer; and

  (3) from 40 to 60% by weight of methylmethacrylate; where the weight percents of (1), (2) and (3) are based on the total weight of the matrix polymer with the proviso that the total weight of the styrene and nitrile type monomers is equal to or less than 60% by weight and the amount of methylmethacrylate monomer is at least 40% by weight;

(B) from 20 to 40% by weight based on the total weight of the polyblend of a polychloroprene rubber component which is grafted with from 10 to 100 parts of monomers per hundred parts of rubber; where the monomers are of the type used to prepare the matrix polymer;

(C) from 3 to 15% by weight based on the total weight of the polyblend of a metal oxide;

(D) from 2 to 25% by weight based on the total weight of the polyblend of a smoke suppressant selected from the group consisting of dawsonite, $[NaAlCO_3(OH)_2]$, magnesium carbonate, ferric oxalate, zinc borate, zinc acetate and combinations of zinc oxide and magnesium oxide, where the ratio of zinc oxide to magnesium oxide is in the range of from 30/70 to 70/30;

(E) from 0 to 10% by weight based on the total weight of the polyblend of a halogen fire retardant additive; provided that the total amount of halogen in the polyblend is at least 9% by weight.

Vinyl Polymer of Crosslinked Structure

A process described by *K. Kosaka, H. Miyamoto, O. Nakagawa, S. Maekawa, T. Miyamatsu and Y. Hiromoto; U.S. Patent 3,997,515; December 14, 1976; assigned to Mitsubishi Rayon Co., Ltd., Japan* is based on the finding that a shaped material based on a vinyl polymer such as fiber, film, or the like, can manifest an excellent flame retardancy when it is modified so as to have a crosslinked structure and to bear COOM (M: a mono- to trivalent metal) groups in its vinyl units by introducing the crosslinked structure during or subsequent to the shaping step into a base polymer containing in its vinyl units 40 mol percent or more, based on the polymer, of COOH groups, and then allowing the resulting shaped material to undergo salt formation or crosslinking reaction with a mono- to trivalent metal.

The vinyl polymers used are those having the recurring units of the following general formula:

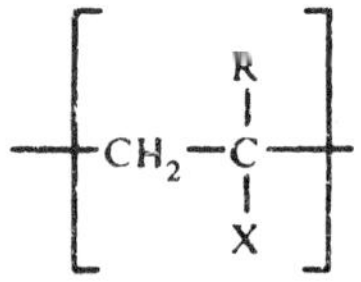

where R is hydrogen or methyl group, and X is CN, COOH, $CONH_2$, or COOR' (R' is methyl or ethyl). Such polymers include homopolymers and copolymers of the vinyl monomers containing in the vinyl units COOH groups or such groups convertible to COOH groups as CN, $CONH_2$, and COOR' (R is an alkyl group).

The methods for forming a crosslinked structure in the vinyl base polymer utilize one of the following reactions:

(1) reaction of CN groups in a polymer comprising acrylonitrile, methacrylonitrile, etc.

(2) reaction of $CONH_2$ groups in a polymer comprising acrylamide, methacrylamide, etc.
(3) reaction of COOH groups in a polymer comprising acrylic acid, methacrylic acid, etc., or reaction of COOH groups formed by conversion of a part of CN groups or $CONH_2$ groups in a polymer comprising acrylonitrile, acrylamide, etc.
(4) reaction of OH groups in a saponified polymer comprising vinyl acetate or in a polymer comprising a hydroxyalkyl acrylate, etc.

For example, in the method utilizing the reaction (1), the base polymer is treated with a solution containing hydrazine in the form of hydrate or mineral acid salt, or with a solution containing hydroxylamine in the form of hydrochloride or sulfate; in the method utilizing the reaction (2), the base polymer is treated with an aldehyde such as formaldehyde or benzaldehyde in the presence of an acid catalyst; in the method utilizing the reaction (3), the base polymer is reacted with a diisocyanate such as tolylene diisocyanate, diphenylmethane diisocyanate, xylylene diisocyanate, or an alkyl diisocyanate; and in the method utilizing the reaction (4), the base polymer is reacted with the above-noted diisocyanates, melamine, N-methylolmelamine, N-methylbenzoguanamine, alkyl ethers of melamines, or diepoxides such as ethylene glycol diglycidyl ether and 2,2-bis[p-(β-γ-epoxy)propoxyphenyl]-propane.

The crosslinking reaction should be carried out prior to the step of salt formation or crosslinking by use of a metal ion. In effecting the salt formation or crosslinking with a mono- to trivalent ion, use is made of a monovalent metal salt, such as potassium or sodium slat, of a weak acid such as carbonate or acetate; or an oxide, halogenide, alcoholate, inorganic or organic acid salt.

An approximately 10% aqueous solution of these metal compounds is used to effect the salt formation or crosslinking at a temperature of 60° to 100°C, the reaction conditions being variable according to the type of metal compound, reaction temperature, reaction time, type of solvent, COOH group content, and degree of crosslinked structure. To achieve the flame retardancy aimed at by the process, the shaped polymer should contain preferably 1 to 30%, by weight (in terms of metal) based on polymer of COOM (M: a mono- to trivalent metal) groups which have been formed from COOH groups in the shaped polymer by replacing hydrogen with mono to trivalent metal by way of salt formation or crosslinking. A metal content exceeding 40% by weight is undesirable because of deterioration in physical properties of the shaped material.

The flame retardant shaped materials of this process can be fibrous materials in the form of flocks, yarns, woven fabrics, or knitted fabrics and are suitable for use in wearing apparel such as nightgowns and nightshirts; interior furnishings such as curtains and carpets; and mattress waddings; particularly in interior furnishings for railway carriages and aircraft where a high degree of flame retardancy is required, and in protective garments for fire fighters, welders, iron and steel furnace workers, racing car drivers, pilots, etc.

These materials can also be films, nonwoven fabrics, or other shaped articles and are suitable for use in fabricated articles requiring flame retardancy, such as construction panels, pipes, bars, sheeting, wallpapers, filter cloths, etc. The following example illustrates this process.

Example: A tow of an acrylic fiber comprising a copolymer of 93 mol percent of acrylonitrile and 7 mol percent of vinyl acetate was immersed in an aqueous solution containing 200 g/l of hydrazine hydrate (the fiber to liquor ratio = 1/20) and allowed to react at 90° to 100°C for 4 hours. After completion of the reaction, the tow was washed with water, dehydrated, dried in the air, and heat-treated at 150°C for 30 minutes to obtain an orange-yellow fiber with a weight increase of 5.3%. It was found that in the treated fiber (referred to as Fiber A) 12 mol percent of the CN groups contained in the initial fiber had been reacted with hydrazine, resulting in crosslinkage.

The Fiber A was subjected to hydrolysis by immersion in a 0.5 N aqueous solution of sodium hydroxide (the fiber to liquor ratio = 1/30) at 90° to 100°C for 2 hours. After completion of the reaction, the fiber was immersed in a large excess of 1 N hydrochloric acid, and then dried in a hot-air dryer. After having been thus dried, the fiber (referred to as Fiber B) showed a weight increase of 20%. It was found that in the fiber 74 mol percent of the CN groups contained in the initial fiber had been converted to COOH groups. The Fiber B was reacted with a 10% aqueous solution of each of the seven metal compounds shown in the table (the fiber to liquor ratio = 1/20) at 70° to 80°C for 1 hour.

After completion of the reaction, the fiber was washed with water, dehydrated, and dried to obtain 7 types of fibers (referred to as Fibers C, D, E, F, G, H and I). Each fiber was evaluated for its metal content and performance. The results obtained were as shown. As is seen from the table, polyvalent metals showed an outstanding effect.

Fiber	Metal Compound	Metal Content (% by wt)	Light Stability, 60 Hours Exposure	Flame Retardancy Burning Test	LOI	Combustion Behavior
A	–	–	no brittleness	burned out on first ignition	19.0	burned with flaming and carbonized
B	–	–	no brittleness	burned a little and extinguished	24.0	burned with flaming and carbonized*
C	$Ca(OCOCH_3)_2$	7.6 22.9**	no brittleness	hardly burned	34.5 35.0	instantly carbonized without flaming***
D	$Mg(OCOCH_3)_2$	5.8 13.0**	no brittleness	hardly burned	31.5 33.5	instantly carbonized without flaming***
E	$Mn(OCOCH_3)_2$	5.9	no brittleness	hardly burned	32.0	instantly carbonized without flaming***
F	$Zn(OCOCH_3)_2$	6.1	no brittleness	hardly burned	36.5	instantly carbonized without flaming***
G	$Cu(OCOCH_3)_2$	14.6	no brittleness	hardly burned	34.0	instantly carbonized without flaming***
H	$Fe(OCOCH_3)_2$	2.6	no brittleness	hardly burned	31.5	instantly carbonized without flaming***
I	$Al(OCOCH_3)_3$	6.7	no brittleness	hardly burned	34.5	instantly carbonized without flaming***
untreated	–	–	no brittleness	burned out on first ignition	19.5	burned with flaming while melting and shrinking†

*Leaving behind traces of initial form.
**Reaction at 100°C for 2 hours.
***Leaving behind a mass in the original form.
†Left behind a hard mass.

Fiber-Forming Acrylonitrile Copolymer

A process described by *A. Maranci; U.S. Patent 4,081,498; March 28, 1978; assigned to American Cyanamid Company* relates to a highly lustrous, flame retardant acrylic fiber of low apparent soiling tendencies which comprises a first acrylonitrile polymer containing at least 50 weight percent acrylonitrile, an amount of a halogen-containing vinyl monomer sufficient to impart flame retardancy to the fiber, and any balance of a halogen-free monomer. Heterogeneously dispersed within the first polymer there is a second acrylonitrile polymer free from any halogen content and containing at least 70 weight percent acrylonitrile and any balance of one or more halogen-free vinyl monomers, the second polymer being incompatible with the first polymer and being present in an amount of about 2 to 15 weight percent based on the total weight of the first and second polymers.

The content of halogen-containing monomer may vary from 5 to 30 weight percent and is generally selected on the basis of the degree of flame retardance desired. Suitable flame retardant halogen-containing monomers are exemplified by vinyl chloride, vinyl bromide, vinylidene chloride, vinylidene bromide, etc. Suitable halogen-free monomers are exemplified by monoolefinic monomers such as the acrylate and methacrylate esters, the corresponding alkyl derivatives of acrylamide and methacrylamide, methacrylonitrile, methyl vinyl ketone, etc.

The first and second polymers described above must be separately soluble in the aqueous inorganic solvent to be employed in fiber spinning. Solubility should be at least about 8 weight percent of polymer in 92 weight percent of the solvent. The various useful solvents are the various aqueous solutions that are conventional and include concentrated aqueous solutions of such salts as zinc chloride, sodium thiocyanate, calcium thiocyanate, lithium bromide and the like, as well as various salt mixtures.

In carrying out the process the first acrylonitrile polymer is dissolved in an aqueous inorganic solvent in a polymer concentration of at least 8 weight percent. A separate solution of the second polymer is then prepared using an aqueous solvent of the same inorganic salt. The second polymer, which need not be a fiber-forming polymer, is employed in solution in an amount which preferably provides a solution viscosity which closely matches that of the solution of the first polymer so as to aid in obtaining an intimate mixture. The concentration of the second polymer in weight percent in solution is not critical so long as the viscosity relationship is maintained and the addition of the appropriate quantity of the solution of the second polymer does not reduce the total polymer content of the mixed solution below about 8 weight percent.

Fibers obtained from such copolymers are particularly suitable for use in the fabrication of carpets, upholstery fabrics, and other home furnishings as a result of their flame retardancy. The following examples illustrate the process.

Example 1: A spinning solution was prepared containing 10% of a fiber-forming polymer composed of 81.1% acrylonitrile, 9.2% methyl methacrylate, and 9.7% vinylidene chloride in 90% of an aqueous solution of 46% sodium thiocyanate. The solution had a viscosity of 34 poises at 28°C and was extruded through a spinnerette having 10 orifices, each of 200 microns diameter, into an aqueous 12% sodium thiocyanate solution maintained at -2°C, to form filaments. The filaments were continuously withdrawn from the bath, stretched at a stretch ratio of 2, washed with water, and drawn a second time in water at 99°C so as to provide a cumulative stretch ratio of 12 and a denier of 9.6.

The stretched filaments were dried in a free-to-relax state at 127°C dry bulb and 60°C wet bulb and then further relaxed in saturated steam at 130°C. The filaments obtained had a high degree of luster and a high degree of light transmission, indicating that the filaments had an undesirably high level of apparent soiling tendencies.

Example 2: A portion of the wet-stretched filaments of comparative Example 1 prior to drying were first exposed to saturated steam at 110°C in a free-to-relax state for 10 minutes and thereafter dried at 100°C dry bulb and 36°C wet bulb. The filaments obtained were highly delustered but had a very low light transmission, thus indicating that the apparent soiling tendencies were low but the luster value of 9.2 was unsatisfactory.

Example 3: A solution of 11.2% of a polymer having a composition of 89.3% acrylonitrile and 10.7% methyl methacrylate, was prepared in 88.8% of an aqueous solution of 40% sodium thiocyanate. Five parts of this solution were mixed with 95 parts of the solution used in Example 1. The mixed solution was turbid, thus indicating incompatibility between the two solutions. The mixed solution was spinnable, however, and stretched wet gel filaments were made following the procedure of Example 1.

The wet-stretched filaments were then exposed to saturated steam and dried as in Example 2. The filaments obtained had a high degree of luster, 19.8%, and a low degree of light transmission, indicating low apparent soiling tendencies.

Example 4: The procedure of Example 3 was followed in every material detail except that 10 parts of the solution of Example 3 were mixed with 90 parts of the solution of Example 1. The resulting fiber had a low degree of light transmission, thus indicating a desirably low apparent soiling tendency, and a high degree of luster, 17.5%.

In addition to the light transmission and luster properties reported in the examples above, the fiber obtained in each of the examples had a desirable level of flame retardancy when tested according to standard procedures.

Complex Antimony Compounds Insoluble in Water

A process described by *E. Radlmann, H. Schaffner, G. Lorenz and G. Nischk; U.S. Patent 4,010,104; March 1, 1977; assigned to Bayer AG, Germany* relates to solutions of complex antimony compounds substantially insoluble in water which are produced by reacting antimony oxides with α-hydroxy carboxylic acids and further reaction with monoisocyanates. The solutions may be added to solutions of acrylonitrile polymers and copolymers thereby improving the noninflammability of shaped articles produced. The antimony oxides used are Sb_2O_3 and Sb_2O_5, preferably Sb_2O_3. Examples of suitable α-hydroxy carboxylic acids include tartaric acid, malic acid and lactic acid.

Suitable monoisocyanates are compounds corresponding to the general formula R—N=C=O, where R is a linear or branched alkyl radical with from 1 to 20 carbon atoms, or the radical:

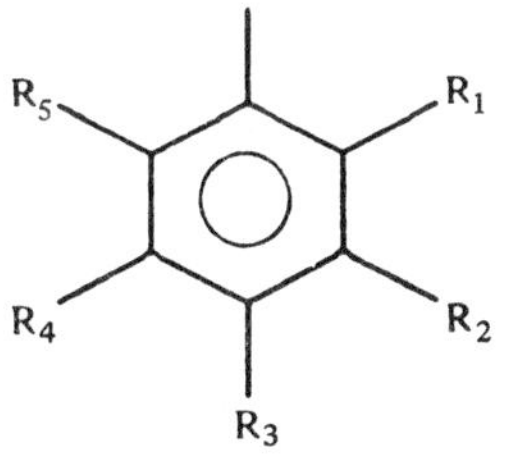

in which R_1, R_2, R_3, R_4 and R_5, which may be identical or different, each represents hydrogen, an alkyl radical with from 1 to 5 carbon atoms, or fluorine, chlorine or bromine.

Examples of polar organic solvents include formamide, monomethyl formamide and dimethyl formamide. The following general procedure describes this process for making the complex compounds.

Antimony(III) or (V) oxide is heated under nitrogen at 100° to 200°C with an α-hydroxy carboxylic acid, for example tartaric acid, in a molar ratio of from 1:2 to 1:4 (ratio of Sb to α-hydroxy carboxylic acid) in the presence of from 30 to 100% by weight of a polar organic solvent, accompanied by the elimination of water, until a solution is formed. The solution is then cooled to temperatures below 100°C and diluted with more polar solvent and the stoichiometric quantity (based on the free carboxyl groups) of a monoisocyanate is added. Carbonamide groups are formed, their formation being accompanied by the evolution of carbon dioxide.

The solutions obtained are colorless to pale yellow in color and may be used as polymer additives as such, i.e., without any need for further purification. If the solvent is removed by distillation, a substantially colorless solid residue is left. The great advantage of this procedure is the fact that it is possible to use antimony oxides, preferably antimony trioxide, as the starting antimony compound. There is no need to start with antimony halides and, hence, to remove the halogen by precipitation with ammonia. Accordingly, the process provides nonhydrolyzing, i.e., noncorrosive additives, which are insoluble in water and, hence, cannot be washed out during the usual aftertreatments. The following examples illustrate this process.

Example 1: Solution A — 150 parts by weight of tartaric acid are dissolved at approximately 100°C in 600 parts by weight of dimethyl formamide. Following the addition of 72.9 parts by weight of antimony trioxide, the temperature is increased with stirring to 150°C while nitrogen is passed over. Under these conditions, a clear solution is formed after about 55 minutes, during which water is distilled off. After cooling to 20°C another 600 parts by weight of dimethyl formamide are added and phenyl isocyanate is added dropwise to the almost colorless solution at 20° to 40°C until there is no further evolution of carbon dioxide. 136.5 parts by weight have to be added to achieve this result. The reaction mixture is then stirred at 40°C for 2 hours, during which a slight deposit may form, although it is completely dissolved again on mixing with a copolymer of acrylonitrile and vinyl or vinylidene chloride. The molar ratio of tartaric acid to antimony is 2:1. Whereas antimony/tartaric acid complexes are soluble in water, the product precipitates when water is added to it.

Example 2: Solution B — 750 parts by weight of tartaric acid and 364 parts by weight of antimony trioxide, together with 400 parts by weight of dimethyl formamide, are heated with stirring to 150°C over a period of about 30 minutes, during which nitrogen is passed over and water is distilled off, until a clear solution is formed. The resulting solution is then diluted with another 1,500 parts by weight of dimethyl formamide, cooled to 25°C and a solution of 290 parts by weight of methyl isocyanate in 500 parts by weight of dimethyl formamide is added dropwise in such a way that the temperature does not exceed 40°C. The evolution of CO_2 stops after a little more isocyanate is added.

The reaction mixture is then stirred for 1.5 hours at 40°C. A small deposit may crystallize out of the colorless solution in the event of prolonged standing, although it disappears completely on mixing with an acrylonitrile-vinyl chloride copolymer solution in dimethyl formamide. The addition of water to the solution of the complex antimony compound causes the compound to precipitate.

Example 3: Solution C — 364.3 parts by weight of antimony trioxide and 400 parts by weight of dimethyl acetamide, together with 825 parts by weight of tartaric acid, are heated with stirring to 155°C over a period of 40 minutes, during which nitrogen is passed over and water is distilled off, until a clear melt is obtained. The melt is then diluted with another 2,000 parts by weight of dimethyl acetamide, cooled to 20°C and 1,128 parts by weight of 3,4-dichlorophenyl isocyanate are added in portions, so that the temperature does not exceed 40°C. The evolution of CO_2 stops after the abovementioned quantity of isocyanate has been added. The almost colorless solution is then stirred for 3 hours at 40°C. Any deposit which may form dissolves completely on admixture with an acrylonitrile/vinylidene chloride copolymer solution in dimethyl acetamide.

Example 4: Solution D — 670 parts by weight of malic acid, 364.3 parts by weight of antimony trioxide and 500 parts by weight of dimethyl formamide are heated with stirring to 150°C. Water distils off until a clear solution is formed. The solution is then diluted with 1,800 parts by weight of dimethyl formamide, cooled to 20°C and n-butyl isocyanate is added in portions at 20° to 40°C until the evolution of CO_2 stops. 499 parts by weight of isocyanate are consumed. The solution formed is only the palest yellow in color and is stirred for 2 hours at 40°C. Any deposit which may form is completely redissolved when the solution is added to a copolymer solution in dimethyl formamide. The complex antimony compound is insoluble in water.

The complex antimony Solutions A, B, C and D described above are added in various quantities to a 35% dimethyl formamide solution of an acrylonitrile/vinylidene chloride copolymer (composition: 58.6% of acrylonitrile, 38.5% of vinylidene chloride and 2.9% of sodium methallyl sulfonate) with a K-value (according to Fikentscher) of 74. The polymer solutions with the additions are then dry spun into filaments. The filaments thus produced are stretched in hot water and washed in the usual way.

The results of burning tests (vertical burning test as specified in DIN 53906) carried out on pieces of fiber yarn knitting are set out in the following table.

Solution Added	Antimony Content [%] added	Antimony Content [%] Found in piece of knitting	Vertical burning test According to DIN 53866 Ignition time secs.	Burning time * secs	Length burnt [cm] max 35 cm
			3	48	32.5
			3	29	32.5
—	0	0	3	27	20.0
Comparison test without			15	25	32.5
any addition			15	27	32.5
			15	26	32.5
			3	12	11.0
			3	15	11.5
A	0.5	0.4	3	17	9.0
			15	22	12.0
			15	15	14.5
			15	18	15.5
			3	—	4.5
			3	—	5.0
A	1.0	0.9	3	2	4.0
			15	—	10.0
			15	2	11.5
			15	5	13.0
			3	—	2.0
			3	—	1.0
A	1.5	1.5	3	—	1.5
			15	—	3.5
			15	1	3.0
			15	—	3.0
			3	1	1.0
			3	—	2.0
B	1.5	1.4	3	—	1.0
			15	1	3.0
			15	—	3.5
			15	—	3.0
			3	—	1.0
			3	—	1.0
C	1.8	1.6	3	—	1.5
			15	—	1.0
			15	—	1.5
			15	—	2.0
			3	1	2.0
			3	2	1.0
D	1.5	1.5	3	—	1.0
			15	—	3.0
			15	—	3.0
			15	1	3.0

*Following removal of the ignition flame.

Polyphosphonates

J.S. Clovis and F.R. Sullivan; U.S. Patent 4,041,107; August 9, 1977; assigned to Rohm and Haas Company describe a thermally stable flame resistant phosphonate polymer of the structure

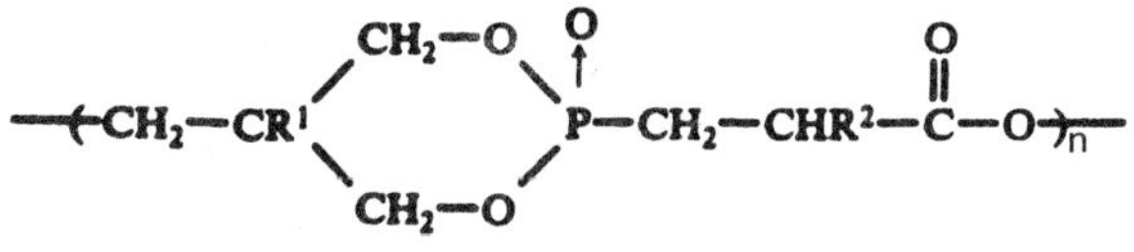

where R^1 is methyl or ethyl, R^2 is hydrogen or methyl and n is 3 to 100. The polymer is formed by the reaction of acrylic or methacrylic acid with an alkyl bicyclic phosphite of the structure

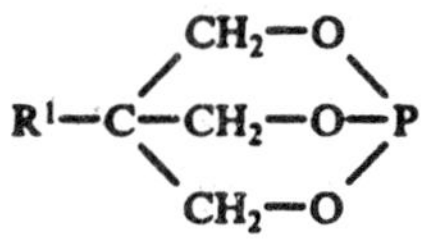

where R^1 is as above. The phosphonate polymer is useful in forming flame resistant acrylic compositions such as molding powders and sheets from mixtures comprising the polymer and methyl methacrylate.

The following examples illustrate the process, parts and percentages being by weight unless otherwise specifically noted.

Example 1: This example illustrates the preparation of the alkyl bicyclic phosphite. A slurry of 134 grams of trimethylolpropane (1.0 M) in 175 grams of ethylene dichloride is brought to reflux at a pressure of about 100 mm to azeotrope off any water present. The reaction mixture is then cooled to 15°C. To the cooled and stirred solution which is kept under nitrogen is added dropwise 140 grams of PCl_3 (1.02 M) over a 3-hour period. The reaction mixture is stirred overnight at room temperature after which time the solution is gradually heated to 70°C and refluxed under vacuum (about 50 mm) until the acid number is less than two. An IR spectrum of the solution is taken to insure the complete reaction of the $-CH_2OH$ groups. The distilled yield of phosphite is about 90%, which agrees well with the 92 to 98% solution yields obtained from bromine number determinations.

Example 2: This example illustrates the preparation of the flame resistant phosphonate polymer of this process. To the solution described above is added about 60 grams of ethylene dichloride (EDC) and 2.34 grams of acetic anhydride (0.5 weight percent on final reaction solution). The solution is brought to reflux for one-half hour and is maintained under a nitrogen atmosphere. Seventy-two grams (1.0 M) of freshly distilled acrylic acid containing 2.34 grams of acetic anhydride is added over molecular sieves to the refluxing solution during a 4-hour period. The addition funnel is washed down with 20 to 25 ml of EDC. The reaction solution is maintained under reflux for an additional 2 hours. The final solution acid number is between 25 and 35.

The product is recovered by stripping off the volatiles under vacuum in the presence of 1% benzyl chloride (by weight on solids) at 170° to 180°C. Alternatively, the product may be recovered by precipitation into ether. The product is a colorless to yellow solid with a softening point of 80° to 90°C. The yield is about 70% (precipitation) to about 95% (stripping).

Characterization of the product by elemental analysis, infrared spectra and unsaturation by bromine number indicates a product consistent with a polymer characterized by the structure, shown on the following page, where n varies from 3 to 10 depending upon the method of recovery. The product is characterized by excellent thermal stability.

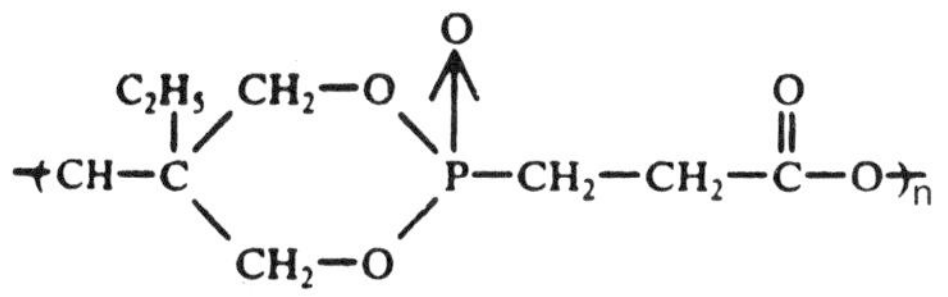

Example 3: This example illustrates the preparation of acrylic sheet material of this process. Fifteen grams of the product obtained from Example 2 is dissolved in 80 grams of methyl methacrylate and 5 grams of glacial methacrylic acid to which has been added 0.03 grams of a 25% solution of acetyl peroxide in dimethyl phthalate. The resulting solution is polymerized between glass plates. The resulting sheet has a self extinguishing rating by ASTM D635-56T.

Example 4: This example illustrates the preparation of the acrylic molding powder of this process. Fifteen grams of the product obtained from Example 3 is dry blended with 85 grams of a copolymer of methyl methacrylate/methacrylic acid/ethyl acrylate; 92.1/5.9/2.0 $\eta_{EDC} = 0.8$. The resulting blend is extruded into pellets which are injection molded into bars which are SE II by test UL-94.

Acrylic Fiber Containing a Regulated Water Content

H. Suzuki, S. Sawanishi and T. Kusunose; U.S. Patent 4,091,066; May 23, 1978; assigned to Japan Exlan Company Limited, Japan describe a process for producing a flame retardant acrylic fiber by wet-spinning a halogen-containing acrylic spinning solution containing antimony oxide particles dispersed in it. This acrylic fiber, having a high degree of flame retardancy and improved fiber properties, such as transparency, luster and brightness of colors of dyed products, is produced in an industrially advantageous manner by regulating the water content in the hydrogel fiber obtained by hot stretching after wet-spinning within a range of 50 to 130% based on the fiber-forming polymer.

Thermal Stabilization of Acrylic Fibers

K.H. Gump and D.E. Stuetz; U.S. Patent 4,004,053; January 18, 1977; assigned to Celanese Corporation describes a process for the thermal stabilization of an acrylic fibrous material or film. The fibrous precursor or film is impregnated with an organic antioxidant by contact with a solution at a moderate temperature, dried to remove the solvent, and heated in an oxygen-containing atmosphere at a more highly elevated temperature until a stabilized fibrous material or film is formed. The organic antioxidant serves to moderate the oxidative portion of the stabilization reaction and enables the cyclization portion to proceed in the absence of substantial polymer chain degradation. The resulting stabilized fibrous material or film is flexible and nonburning, and may be utilized as a fire resistant fiber, fabric or film, or optionally carbonized or carbonized and graphitized to form a carbonaceous fibrous material or film.

The acrylic polymer utilized as the starting material is formed primarily of recurring acrylonitrile units. For instance, the acrylic polymer should generally contain not less than about 85 mol percent of acrylonitrile units and not more than about 15 mol percent of units derived from a monovinyl compound which is copolymerizable with acrylonitrile such as styrene, methyl acrylate, methyl methacrylate, etc.

The preferred acrylic precursor is an acrylonitrile homopolymer. Preferred acrylonitrile copolymers contain at least about 95 mol percent of acrylonitrile units and up to about 5 mol percent of one or more monovinyl units copolymerized with it.

The acrylic precursor is provided as a continuous length of fibrous material and may be in a variety of physical configurations. For instance, the acrylic fibrous materials may be present in the form of continuous lengths of multifilament yarns, tows, tapes, strands, cables, or similar fibrous assemblages.

Prior to heating the acrylic fibrous material or film in an oxygen-containing atmosphere to accomplish the desired stabilization, the precursor is impregnated with a minor quantity of an organic antioxidant capable of moderating or inhibiting the oxidative crosslinking reaction of adjoining acrylic molecules, e.g., when heated in an oxygen-containing atmosphere at 200°C, or above. The organic antioxidant optionally may be nitrogen-free phenolic antioxidant, an aminophenol antioxidant, a nonphenolic aromatic amine antioxidant, a quinone antioxidant, a sulfur-containing antioxidant, or mixtures. The organic antioxidants selected preferably have a melting point of at least 200°C.

Following impregnation the acrylic fibrous material or film is next dried so that the solvent is substantially removed, and a minor quantity of the organic antioxidant provided in intimate association with it. The drying step may be conducted in any convenient manner commonly at a temperature of about 10° to 100°C. The impregnated acrylic precursor may be simply exposed to ambient conditions until solvent adhering to it is substantially evaporated. For instance, drying may be conducted by exposure to a gaseous atmosphere at a temperature of about 10° to 40°C. The drying step can, of course, be expedited by exposure to a circulating gaseous atmosphere at a more highly elevated temperature, e.g., 80° to 100°C, or even in the same zone where the stabilization reaction is carried out.

The resulting impregnated and dried acrylic material is heated to an oxygen-containing atmosphere at a temperature of about 200° to 360°C until a stabilized fibrous product or film is formed which retains its original configuration substantially intact and which is nonburning when subjected to an ordinary match flame. The following examples illustrate the process.

Example 1: A continuous length of an 800 fil dry spun acrylonitrile homopolymer continuous filament yarn having a total denier of 1,200 is selected as the starting material. The yarn is initially dry spun from a solution of the same in N,N-dimethylformamide solvent into an evaporative atmosphere of nitrogen. The yarn is spun as a 40 fil bundle, and plied to form the 800 fil yarn which exhibits a twist of about 0.5 tpi. The yarn is next drawn at a draw ratio of about 5:1 to a single filament tenacity of about 4 grams per denier by stretching while passing over a hot shoe at a temperature of about 160°C for a residence time of about 0.5 second.

A solution of tetrachlorohydroquinone is formed in methyl alcohol at room temperature (i.e., 25°C). Tetrachlorohydroquinone is present in the solution in a concentration 10% by weight. The acrylonitrile homopolymer yarn is immersed in the solution which is provided at 25°C by continuous passage through a vessel containing the solution for a total residence time of 10 minutes. While present in the solution, the yarn becomes impregnated with tetrachlorohydroquinone.

The yarn is next removed from the solution and is dried at a temperature of 90°C by placement in a circulating air oven so that the methyl alcohol solvent adhering to it is substantially evolved. The resulting impregnated and dried yarn contains tetrachlorohydroquinone in a concentration of about 6% by weight.

The yarn is next stabilized on a continuous basis by passage for 7 minutes through a tube furnace provided with a circulating heated air atmosphere having a maximum temperature of 340°C while axially suspended in the furnace.

The resulting stabilized yarn is shiny black in appearance, flexible, has a textile-like hand, retains its original fibrous configuration substantially intact, is non-burning when subjected to any ordinary match flame, retains strength after glowing in a match flame, and has an oxygen content of about 7.5% by weight as determined by the Unterzaucher analysis.

In a control run, an identical sample of the acrylonitrile homopolymer yarn is passed through the tube furnace in an identical manner with the exception that it has not been previously impregnated with tetrachlorohydroquinone. The control yarn is incapable of withstanding the 340°C stabilization temperature and breaks.

Encapsulation of Electrical Elements with UV Curable Resins

A. Otsuki and M. Nishino; U.S. Patent 4,073,835; February 14, 1978; assigned to Toyo Ink Manufacturing Co., Ltd., Japan describes a process for the formation of a protective coating on an electrical element. This comprises housing an electrical element such as a capacitor in a mold made of an ultraviolet light-transmissible material and introducing a fire retardant liquid resin composition into the space between the mold and the electrical element to enclose the element with the composition. The liquid resin is then cured in two steps, firstly by irradiating the composition with ultraviolet light for a partial cure, and secondly by heating the partially cured resin to complete the cure, thus encapsulating the electrical element.

Referring to Figure 7.2a an electrical element **1A** with lead wires or terminals **2** such as an electrical resistor or capacitor, is housed in a cup-like mold **3A** coaxially with aid of supports **2** for the electrical element **1**, the mold being made of an ultraviolet light-transmissible material. A liquid resin composition **4** is introduced into the space between the element **1A** and the mold, an air preventing layer **5** is formed on the portion of the resin composition **4** at the open top of the mold **3A**, the mold is irradiated with ultraviolet light for suitable times to cure the outer side **4a** of the liquid resin composition **4**. Then the resin composition **4** so partially cured is heated to complete the cure thus obtaining a protective coating for the element **1A**. The degree of cure of the liquid resin composition layer at **4a** effected only by the irradiation with ultraviolet rays varies decreasingly from the outside to the inside of the resin composition layer.

Referring then to Figure 7.2b an electrical element **1B** with lead wires or terminals **2** is housed in a mold **3B** having an open top and consisting of two separable parts, followed by taking the same procedure as previously mentioned with reference to Figure 7.2a to perfectly cure the liquid resin composition whereby a protective coating for the element **1B** is obtained.

FIGURE 7.2: ENCAPSULATED ELECTRICAL ELEMENT

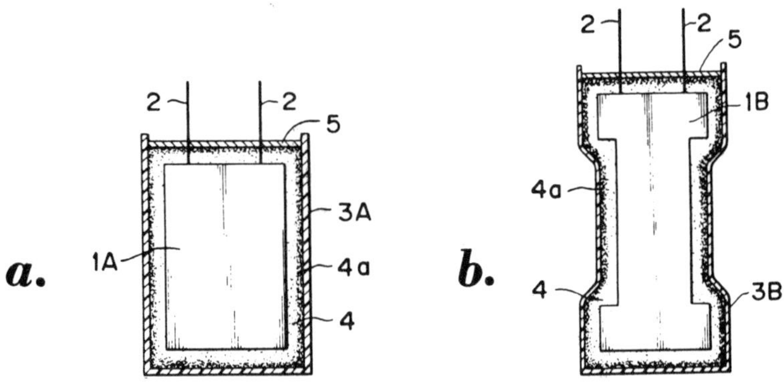

Source: U.S. Patent 4,073,835

The liquid resin compositions include as the essential components at least one ethylenically unsaturated monomer and/or prepolymer, a photopolymerization initiator and thermopolymerization initiator and may further include various additives such as fire retardants, fillers and pigments.

The monomers having a double bond include styrene and its derivatives; acrylic esters such as methyl acrylate, ethyl acrylate, butyl acrylate, 2-ethylhexyl acrylate and 2-hydroxyethyl acrylate; methacrylic esters such as methyl methacrylate, ethyl methacrylate, butyl methacrylate and 2-ethylhexyl methacrylate; etc.

The prepolymers having double bonds include adducts of epoxy resins with acrylic acid or methacrylic acid. The photopolymerization initiators which may be used include benzoin type compounds such as benzoin methyl ether, benzoin ethyl ether, benzoin isobutyl ether, etc.

The thermopolymerization initiators include benzoyl peroxide, tert-butyl perbenzoate, methyl ethyl ketone peroxide and azobisisobutyronitrile. The following examples illustrate the process.

Example 1: An electrical resistor (8 mm in diameter and 20 mm in length) was held in a polymethylmethacrylate-made cup-like mold (1 mm in thickness, 12 mm in inner diameter and 24 mm in length) so that it was approximately 2 mm apart from the inner wall of the mold and the inner side of the bottom as shown in Figure 7.2a, and a liquid resin composition of the following composition was introduced into the mold to cover the resistor with the resin composition. The mold with the resistor and liquid resin composition was irradiated uniformly at the wall for 30 seconds, at the top for 5 seconds and at the bottom for 5 seconds in a stream of nitrogen with ultraviolet rays by a 2 Kw high voltage mercury lamp while being kept 10 cm apart from the mold, to incompletely cure the liquid resin composition. The incompletely cured resin composition with the resistor enclosed was not deformed even after withdrawal from the mold, and it was then heated at 100°C for 2 hours in a hot air oven to complete the cure, thus forming around the resistor a protective coating which had satisfactory surface gloss and did not crack even when immersed in warmed water.

Liquid Resin Composition

	Parts by Weight
Acrylic acid-modified epoxy resin*	33
1,6-hexanediol diacrylate	20
Aluminum hydroxide	30
Benzoin ethyl ether	4
Benzoyl peroxide	1

*Prepared by addition reacting an epoxy resin (Epikote 828) with acrylic acid in the molar ratio of from 1 to 2 using a conventional method.

The above liquid resin composition was prepared by kneading the ingredients other than the benzoin ethyl ether and benzoyl peroxide on a three-roll mill and then homogeneously mixing the thus milled ingredients with the nonmilled initiators.

Example 2: A box-like mold (15 x 30 x 40 mm) made of a 1 mm-thick polymethylmethacrylate sheet was allowed to house an electrical condenser or capacitor so that the capacitor was held 3 mm, on the average, apart from the inner wall and bottom, and a liquid resin composition was poured into the mold to enclose the capacitor.

After vacuum degassing the mold and pouring 0.1 gram of liquid paraffin to form a layer on the portion of the liquid resin composition at the open top of the mold, the mold with the capacitor and liquid resin housed was irradiated at the wall, top and bottom respectively for 5 seconds with ultraviolet rays by the use of a 2 Kw high voltage mercury lamp while being kept 10 cm apart from the mold, whereby the liquid resin composition in the mold was crosslinked and cured at the surface with the balance remaining uncured.

Since the partially cured resin composition enclosing the capacitor was thus impossible to release without the deformation from the mold, it was then placed, as it remained in the mold, in an oven and heated therein at 100°C for two hours to complete the cure of the partially cured composition, after which the completely cured resin composition enclosing the capacitor was released from the mold thus obtaining the capacitor enclosed with the completely cured composition which was a protective coating having excellent gloss and other properties.

Liquid Resin Composition

	Parts by Weight
Acrylic acid-modified epoxy resin*	40
Ethylene glycol diacrylate	13
Aluminum hydroxide	40
Titanium dioxide	4
Benzoin ethyl ether	2
Benzoyl peroxide	1

*Prepared by addition reacting an epoxy resin (FRRA-4211) with acrylic acid in the molar ratio of from 1 to 2 using a conventional method.

The liquid resin composition was prepared from the above ingredients in the same manner as in Example 1.

Example 3: The procedure of Example 1 was followed except that curing was effected only by the irradiation of ultraviolet rays for the same period of time as Example 1, without subsequent curing effected by heat. The resin composition so cured appeared to have been completely cured; however, it exhibited cracks on the surface in a warm water resistance test. The warm water resistance test is done as follows: a sample is tested for warm water resistance by immersing the sample in water at 80°C for 50 hours and then making visual estimation of the state of the sample surface.

The abovementioned drawback of the partially cured resin composition was not eliminated even after further irradiation of ultraviolet rays for an additional time ten times as long as the initial curing time.

Example 4: The procedure of Example 1 was followed except that curing was effected by heating at 100°C for two hours in the oven, without ultraviolet rays irradiation; in this case, the resin composition so heated and the mold were melt bonded to each other thereby rendering it impossible to release from the mold the partially cured resin composition with the capacitor enclosed.

The protective coatings enclosing, or forming on, electrical elements according to this process are required to be, per se, fire retardant; to this end, it is preferable to contain in the liquid resin composition 10-70% by weight of aluminum hydroxide, a combination of a halide and antimony oxide, or each of phosphoric acid esters such as cresyldiphenyl phosphate and triaryl phosphate. Among those, aluminum hydroxide is the most preferable since it is excellent in ultraviolet ray transmissibility, thereby not hindering the liquid resin composition from being cured.

Bisphenoxy Compounds

A.L. Anderson; U.S. Patent 3,971,761; July 27, 1976; assigned to Michigan Chemical Corporation describes plastic compositions containing polyacrylates and bisphenoxy compounds having the formula:

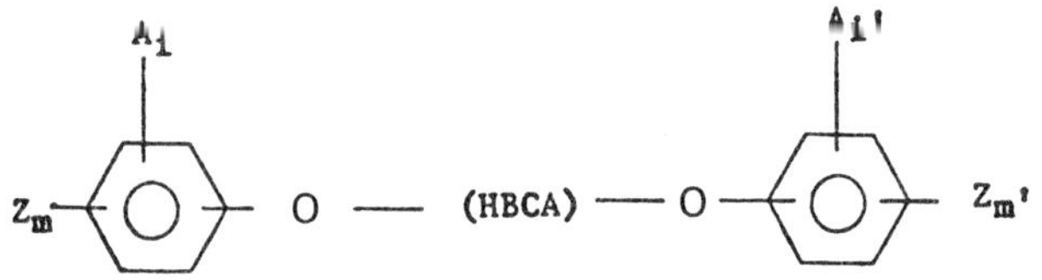

where Z is bromine or chlorine, m and m' are integers having a value of 1 to 5, i and i' are integers having a value of 0 to 2, HBCA is a halo-branched chain alkylene group having from 1 to 6 carbon atoms and A is cyano, nitro, lower alkoxy, lower alkyl, fluorine, dialkylamino, phenyl, halo-phenyl, benzyl or halobenzyl. These compounds have been described above in the section on polyesters (see page 94).

POLYCARBONATES, PVC AND OTHER THERMOPLASTICS

POLYCARBONATES

Metal Salts of Sulfur

J.L. Webb; U.S. Patent 4,028,297; June 7, 1977; assigned to General Electric Company describes a mixture of an aromatic polycarbonate and a flame retardant additive selected from an alkali or alkaline earth metal salt of an inorganic sulfur compound. The flame retardant polycarbonates can be molded or formed into fire retardant articles, such as films, sheets, fibers, laminates, or reinforced plastics by conventional techniques.

Any nonhalogenated aromatic polycarbonate can be employed including those derived from aromatic bis-hydroxy compounds, free of halogen substituents, and in particular, those manufactured from dihydric phenols, for example, resorcinol, hydroquinone or dihydroxydiphenyl and from bis(hydroxyphenyl)-alkanes. A preferred polycarbonate of a dihydric phenol is represented by the formula:

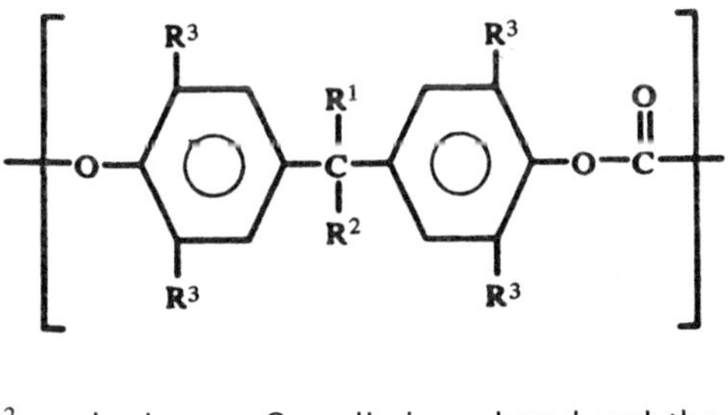

wherein R^1 and R^2 are hydrogen, C_{1-4} alkyl or phenyl and the R^3 groups are selected from hydrogen or C_{1-4} alkyl.

Generally useful aromatic polycarbonates have an intrinsic viscosity of at least 0.3 and preferably about 0.6 dl/g as measured in methylene chloride, or similar solvent systems at 25° to 30°C. The upper intrinsic viscosity limit is not critical, however, and will generally be about 1.5 dl/g. Especially useful polycarbonate resins generally have an intrinsic viscosity within the range of from about 0.35 to about 0.7 dl/g.

The flame retardant inorganic sulfur additive can be any alkali or alkaline earth metal salt of the general formulas: M_xSO_3 (sulfite), $M_xS_2O_3$ (thiosulfate), $M_xS_2O_4$ (dithionate) and $M_xS_2O_5$ (pyrosulfite), where M is lithium, sodium, potassium, rubidium, and cesium when x is a number equal to 2, and M is beryllium, magnesium, calcium, strontium, or barium when x is a number equal to 1. The following examples illustrate the process.

General Procedure: A series of aromatic polycarbonate compositions were prepared from commercially available materials comprising: (a) the polycarbonate of bisphenol-A, i.e., bis(4-hydroxyphenyl)-2,2-propane, a thermoplastic polycarbonate commercial product having the following properties: intrinsic viscosity $[\eta]$ of 0.55 dl/g as measured in dioxane at 25°C and an M_n equal to about 12,300, and a P_n equal to about 48.5 units; and (b) an inorganic sulfite, thiosulfate, dithionite or pyrosulfite compound, by dry blending and precompounding in a hot melt extruder at about 550°F. The extrudate was pelletized. For comparative purposes, compositions containing aromatic polycarbonate free of any flame retardant additive were also prepared in pellet form.

The aromatic polycarbonate and inorganic sulfur compound compositions and control compositions were separately molded into appropriate test shapes, i.e., bars, rods, etc., for flammability testing according to the standards of Underwriters' Laboratories Item 94, (UL 94).

Individual UL 94 test specimens $^1/_{16}$" x ½" x 5" were vertically clamped and twice exposed for 10 seconds during each exposure to a blue gas flame of 20 mm height, the top of the gas burner being positioned about 10 mm from the lower end of the test specimen. The specimens were positioned 12 inches above a horizontal layer of absorbant surgical cotton. For classifying material in a fire class, five test pieces were tested. The following criteria were employed for classification:

	V-2	V-1	V-0
Average flame out time (FOT)	≤25	≤25	≤5
Maximum after burning time (ABT) after any one flame application. The material may drip off burning. (For V-0 the material must not drip off burning.)	30	30	10

Additionally, the polycarbonate compositions were evaluated accordingly: (1) ASTM D256 notched Izod impact strength measured in ft-lb/in employing test bars 2½ inches long, ½ inch wide and ⅛ inch thick, notched to a depth of 100 mils, having a notched tip radius of 10 mils, and (2) visual determination of color, translucence, transparence, opaqueness, etc.

Examples 1 through 7: In the table below the data of these examples illustrate that the inorganic sulfur flame retardant aromatic polycarbonate compositions meet, whereas aromatic polycarbonates containing no flame retardant additives or aromatic polycarbonates containing other inorganic sulfur compounds do not meet, the Underwriters' Laboratories UL 94 test standards. In the examples, on a weight basis, one part of flame retardant additive was combined with 100 parts of aromatic polycarbonate.

Example	Additive	FOT*	UL 94 Rating
1	None	40.0	burning
2	Na_2SO_4	24.0	burning
3	$Na_2S_2O_6$	35.0	burning
4	Na_2SO_3	5.45	V-2
5	$Na_2S_2O_3$	5.33	V-2
6	$Na_2S_2O_4$	4.47	V-2
7	$Na_2S_2O_5$	5.95	V-2

*Average flame-out time measured in seconds.

Examples 8 through 13: In the table below the data of these examples illustrate that the flame retardant additive efficacy varies in accordance with the alkali or alkaline earth metal cation associated with the sulfite, thiosulfate, dithionite or pyrosulfite flame retardant employed. In the examples, on a weight basis, one part of flame retardant was combined with 100 parts of aromatic polycarbonate.

Example	Additive	FOT*	Σ Drip**	UL 94 Rating
8	None	40	16	burning
9	Na_2SO_3	5.45	7	V-2
10	K_2SO_3	21.30	12	V-2
11	$CaSO_3$	7.13	9	V-2
12	$Na_2S_2O_5$	5.95	13	V-2
13	$K_2S_2O_5$	12.10	8	V-2

*Average flame-out time measured in seconds.
**Total drips.

Incorporation of Organic Fibers in Polycarbonates

D. Margotte and H. Vernaleken; U.S. Patent 4,061,691; December 6, 1977; assigned to Bayer AG, Germany describe molding compositions having an oxygen index of greater than 0.30 based on high molecular weight aromatic polycarbonates containing 0.5 to 20% of organic fibers. The incorporation of only 2% of organic fibers in bisphenol-A-polycarbonate results in comparably good flame resistance, as determined by the oxygen index according to ASTM/D 2863/70, as is otherwise only achievable by halogen contents of 6 to 10% by weight.

Organic fibers used are polymeric materials spun into filaments according to known processes or prepared by the fibrillation process. The fibers are capable of withstanding the heat exposure at not less than 180°C during the process of their incorporation into the thermoplastic polycarbonate and during thermoplastic processing of the molding composition.

Polymeric substances of this nature include highly crystalline or highly heat distortion-resistant saturated, thermoplastic polyesters, such as polyethylene terephthalate, polybutylene terephthalate and poly-1,4-dimethylenecyclohexane terephthalate, polyamides preferably those based on caprolactam, aliphatic and aromatic dicarboxylic acids and diamines, such as adipic acid and hexamethylenediamine, terephthalic acid or isophthalic acid and phenylenediamine or diaminodiphenyl ether, polybenzamides, and also polyacrylonitrile. Among the highly heat distortion-resistant materials, polyimides, polybenzimidazoles, polybenzoxazinediones and polyhydantoins. For reasons of practical manipulation, fiber lengths of between 500 and 50,000 μ have proved successful, fibers in a length range of between 1,000 and 40,000 μ being used preferentially.

The fiber material can be processed together with the polycarbonate granules in a single-screw or twin-screw extruder, or the fibers are directly incorporated into the polycarbonate melt according to known processes, and the mixture extruded. A further process of incorporation consists of thorough mixing in an internal kneader, with subsequent ribbon granulation. Incorporation of the fibers in the polycarbonate solutions, with subsequent evaporation of the solvent in a vented extruder, is also possible.

The molding compositions are employed wherever the use of materials of low flammability appears advisable and desirable for safety reasons. Thus, they are employed for holders of light bulbs and neon tubes, backs of television sets, and terminal bars and for a large number of individual components of electrical instruments. Some examples are given below.

Example 1: Manufacture of the Polycarbonate – Approximately 454 parts of 4,4'-dihydroxydiphenyl-2,2-propane and 9.5 parts of p-tert-butylphenol are suspended in 1.5 liters of water. The oxygen is removed from the reaction mixture in a 3-necked flask equipped with stirrer and gas inlet tube by passing nitrogen through the reaction mixture for 15 minutes, while stirring. 355 parts of 45% strength sodium hydroxide solution and 1,000 parts of methylene chloride are then added. The mixture is cooled to 25°C. 237 parts of phosgene are added over a period of 120 minutes while maintaining this temperature by cooling. An additional amount of 75 parts of a 45% strength sodium hydroxide solution is added after 15 to 30 minutes or after the absorption of phosgene has commenced. 1.6 parts of triethylamine are added to the resulting solution and the mixture is stirred for a further 15 minutes.

A highly viscous solution is obtained, the viscosity of which is regulated by adding methylene chloride. The aqueous phase is separated off. The organic phase is washed with water until free of salt and of alkali. The polycarbonate is isolated from the washed solution and dried. The polycarbonate has a relative viscosity of 1.29 to 1.30, measured in an 0.5% strength solution of methylene chloride at 20°C. This corresponds approximately to a molecular weight of 32,000. The polycarbonate thus obtained is extruded and granulated.

In the examples that follow the relative viscosities (η_{rel}) of the polycarbonates are measured in a 0.5% strength solution of methylene chloride at 20°C.

Example 2: 99 grams of a polycarbonate based on 4,4'-dihydroxydiphenyl-2,2-propane (bisphenol A), prepared according to Example 1, (η_{rel} = 1.30) together with 1 gram of an aromatic polyamide of the following structure having a fiber length of 1,000 μ, are extruded from a twin-screw extruder at 310°C, and granulated.

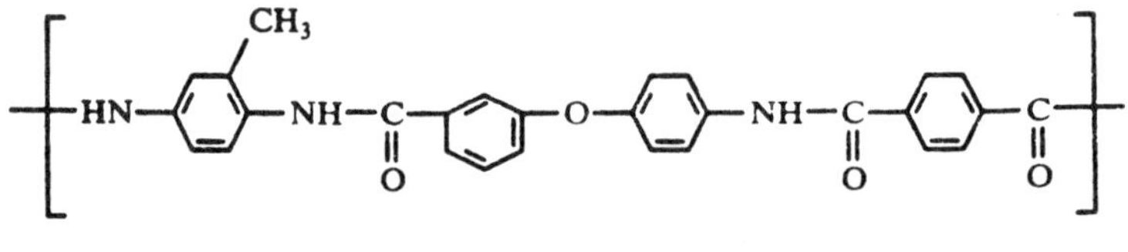

Polyamides of this structure are distinguished by high tensile strength and E-modulus. The fibers used have the following properties: 67 to 75 kp/mm^2 tensile strength; 1.2 g/cm^3 density; approximately 1.5% elongation at break; and 6,000 to 6,500 kp/mm^2 E-modulus. The oxygen index and other properties of the polycarbonates of low flammability which are obtained are summarized in the table below.

Example 3: 95 grams of a polycarbonate prepared according to Example 1, having a relative viscosity of η_{rel} = 1.28 and 5 grams of a polyamide fiber described in Example 2, are extruded in a twin-screw extruder at 290°C and granulated. The oxygen index and other properties can be seen from the table below.

Example 4: 98 grams of a polycarbonate prepared according to Example 1, of relative viscosity η_{rel} = 1.30 and 5 grams of an aromatic polyamide of the following structure having a fiber length of 50,000 μ are extruded from a twin-screw extruder at 300°C and subsequently granulated.

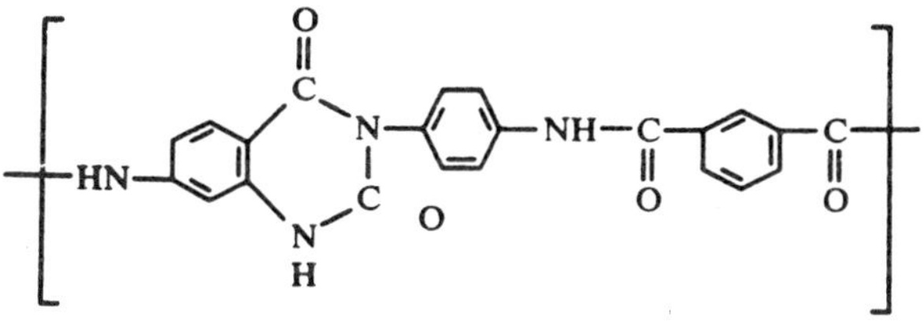

The polyamide used is synthesized from isophthalyl chloride and 3-aminophenyl-7-amino-2,4-quinazolinedione and is used as curled fiber having a titer of 3.3 dtex and an E-modulus of 5,000 kp/mm^2. The oxygen index and other properties can be seen from the table below.

Example 5: 99 grams of a polycarbonate according to Example 1, having a relative viscosity of η_{rel} = 1.26 and 1 gram of a polyester fiber based on terephthalic acid dichloride and cyclohexanedimethanol having the following structure and a fiber length of 50,000 μ are extruded from a twin-screw extruder at 280°C and granulated.

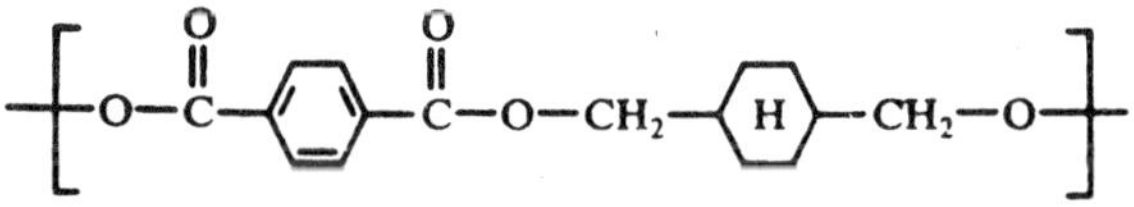

The polyester fiber used is characterized by elongation at break 40%, ultimate tensile strength 3 pont/dtex, titer 3 dtex. The oxygen index and other properties can be seen from the table below.

Polycarbonate from	Oxygen Index ASTM-D 2,863/70	Impact Strength cm.kp/cm²	E-Modulus kp/cm²
Example 1	0.24	not fractured	23,000
Example 2	0.33	not fractured	23,000
Example 3	0.35	80	24,100
Example 4	0.31	not fractured	23,700
Example 5	0.30	not fractured	22,900

Mixture of Siloxanes with Organic Metal Salts

C.A. Bialous, J.B. Luce and V. Mark; U.S. Patent 3,971,756; July 27, 1976; assigned to General Electric Company describe a flame retardant polycarbonate composition of an aromatic polymer in a mixture with an organic alkali metal salt or an organic alkaline earth metal salt, where the composition is mixed with a siloxane. The siloxanes which may be employed include compounds of the formula:

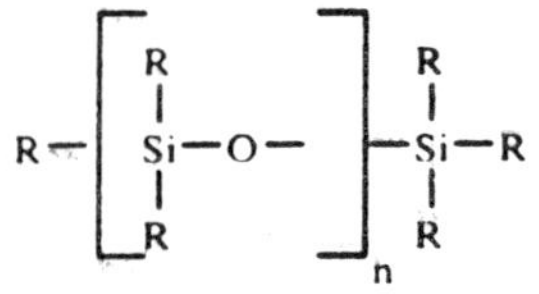

where the Rs are independently selected from the group consisting of hydrogen, hydroxy, alkyl of from 1 to 6 carbons, alkoxy of from 1 to 6 carbons, alkenyl from 2 to 8 carbons, alkynyl from 2 to 10 carbons, aryl from 6 to 18 carbons, aryloxy from 6 to 10 carbons.

Examples of these compounds include octaphenylcyclotetrasiloxane; 2,4,6,8-tetramethyl-2,4,6,8-tetravinylcyclotetrasiloxane; α,ω-divinylpolydimethylsiloxane; 1,1,3,3-tetramethyl-1,3-divinylsiloxane; branched methyl phenyl silicone fluid; linear dimethylsiloxane (1,000 cp viscosity); trimethylsilyl terminated methylhydrogensiloxy diphenylsiloxy copolymers; α,ω-dimethylpoly(dimethylmethylvinylsiloxane); trimethylsilyl terminated polymethylhydrogen siloxane, and octachlorooctaphenylcyclotetrasiloxane.

The amount of the siloxane to be used can be from about 0.01 to about 5.0 weight percent based on the total weight of the resin. Some of the organic alkali metal and alkaline metal salts employed include sodium 2,4,5-trichlorobenzenesulfonate; sodium benzenesulfonate; disodium naphthalene-2,6-disulfonate; sodium p-iodobenzene-sulfonate; sodium 4,4'-dibromobiphenyl-3-sulfonate; sodium 2,3,4,5,6-pentachloro-betastyrenesulfonate; sodium 4,4'-dichlorodiphenylsulfide-3-sulfonate; disodium tetrachlorodiphenyletherdisulfonate, and disodium 4,4'-dichlorobenzophenone-3,3'-disulfonate.

The aromatic carbonate polymers employed are homopolymers and copolymers that are prepared by reacting a dihydric phenol with a carbonate precursor. The dihydric phenols that can be employed are bisphenols such as bis(4-hydroxyphenyl)methane, 2,2-bis(4-hydroxyphenyl)propane, 2,2-bis(4-hydroxy-3-methylphenyl)propane, 4,4-bis(4-hydroxyphenyl)heptane, and 2,2-bis(4-hydroxy-3,5-dichlorophenyl)propane. The following examples illustrate the use of the flame retardant additives.

Example 1: A polycarbonate composition is prepared by extruding a homopolymer of 2,2-bis(4-hydroxyphenyl)propane (bisphenol-A) prepared by reacting essentially equimolar amounts of bisphenol-A and phosgene in an organic medium with triethylamine, sodium hydroxide and phenol, under standard conditions. The resulting polymer is then fed to an extruder. The extruder is operated at about 265°C and the extrudate is comminuted into pellets. The pellets are then injection molded at about 315°C into test bars of about 5 inches by $\frac{1}{2}$ inch by about $\frac{1}{16}$ to $\frac{1}{18}$ inch thick. The test bars (five for each additive listed in the table) are subjected to the test procedure set forth in Underwriters' Laboratories, Inc. Bulletin UL-94, Burning Test for Classifying Materials.

Example 2: To the polycarbonate of Example 1 is added 0.5 part of sodium 2,4,5-trichlorobenzenesulfonate. The mixture is extruded, molded and tested in the same manner as described in Example 1.

Examples 3 through 5: To the polycarbonate Example 1 is added 0.5 part of sodium 2,4,5-trichlorobenzenesulfonate and 0.05 part of octaphenylcyclotetrasiloxane. The mixture is extruded, molded and tested in the same manner as described in Example 1. Compositions of Examples 4 and 5 were prepared and tested by the methods of examples 1 through 3, utilizing the salt additives and siloxanes below.

Ex. No.	Salt Additive	Parts	Siloxane	Parts	Average Flame-Out Time (sec)	No. of Drips per Five Bars	Rating
1	–	–	–	–	13.4	10	Burning
2	Sodium 2,4,5-trichloro-benzenesulfonate	0.5	–	–	3.4	8	UL-94 V-2
3	Sodium 2,4,5-trichloro-benzenesulfonate	0.5	A	1.0	2.6	0	UL-94 V-0
4	Sodium benzenesulfonate	1.0	B	0.6	4.4	0	UL-94 V-0
5	Disodium naphthalene-2,6-disulfonate	1.0	C	1.0	4.8	0	UL-94 V-0

Note: A is octaphenylcyclotetrasiloxane.
B is branched methyl phenyl silicone fluid.
C is poly(methylhydrogensiloxane).

As can be seen from the data in the table, when the polycarbonate compositions contain the siloxanes in addition to the flame retardant additives, the polycarbonate does not drip and is rated UL-94 V-O.

Polyphenylene Sulfides

S. Adelmann, D. Margotte, J. Merten and H. Vernaleken; U.S. Patent 4,046,836; September 6, 1977; assigned to Bayer AG, Germany describe thermoplastic polycarbonate molding materials having improved flame resistance and hydrolysis resistance. These materials comprise blends of aromatic high molecular weight polycarbonates and about 0.5 to 50 weight percent of polyphenylene sulfides.

These additions of polyphenylene sulfide to pure aromatic polycarbonates raise the fire resistance of aromatic polycarbonates without loss of the good mechanical properties and of the processing properties. The oxygen index (according to ASTM-D 2863-70) which is a measure of the inflammability in a mixture of oxygen and nitrogen, is also markedly raised through the mixed polyphenylene sulfide. Thus, for example, a test specimen containing 5% by weight of polyphenylene sulfide attains an oxygen index of 0.30 while pure aromatic polycarbonate based on bisphenol-A has an oxygen index of 0.24.

Aromatic polycarbonates used are homopolycarbonates and copolycarbonates which are based on one or more of the following bisphenols: hydroquinone, resorcinol, dihydroxydiphenyls, bis(hydroxyphenyl)alkanes, bis(hydroxyphenyl)-

cycloalkanes, bis(hydroxyphenyl)sulfides, bis(hydroxyphenyl)ethers, bis(hydroxyphenyl)ketones, bis(hydroxyphenyl)sulfoxides, bis(hydroxyphenyl)sulfones and α,α'-bis(hydroxyphenyl)diisopropylbenzenes. The aromatic polycarbonates should as a rule have molecular weights $\overline{M}_w$ of about 10,000 to 200,000, preferably of about 20,000 to 80,000.

Polyphenylene sulfides used are reaction products of polyhalogen compounds with alkali metal sulfides (according to U.S. Patent 3,354,129) or of polyhalogeno-aromatics with sulfur and a metal sulfide (according to U.S. Patent 2,513,188) or polymers of metal salts of halogenothiophenolates obtained under nitrogen at a polymerization temperature of 250° to 310°C (according to British Patent 962,941).

The homogeneous incorporation of the polyphenylene sulfides into polycarbonate may be effected by using a twin-screw extruder at about 250° to 290°C. The melt may be extruded to form a ribbon and subsequently granulated. The polycarbonate molding materials can, if appropriate, additionally contain stabilizers, dyestuffs or fillers or glass fibers in the usual amounts and the good properties of the polycarbonate-polyphenylene sulfide mixtures can thereby be modified in the desired way. The homogeneity of the mixture and the good compatibility of the components of the mixture manifests itself in the excellent mechanical properties.

The flame resistant and hydrolysis resistant polycarbonate molding materials in the form of powders or granules, can be converted by injection molding and extrusion, to very diverse types of moldings. Molding materials of the above can be used for the manufacture of hydrolysis resistant and impact resistant moldings of low inflammability, such as are used, in general instrument construction, precision engineering, electrical engineering and telecommunication, and also for the manufacture of semi-finished goods such as, for example, sheets, slabs, bars, profiles and the like. The examples illustrate the use of polyphenylene sulfide. Polycarbonates A, B, and C were prepared by the following general procedure.

About 454 parts of 4,4'-dihydroxydiphenyl-2,2-propane and 9.5 parts of p-tert-butylphenol are suspended in 1.5 liters of water. The oxygen is removed from the reaction mixture in a 3-neck flask equipped with a stirrer and gas inlet tube, by passing nitrogen through the reaction mixture for 15 minutes while stirring. 355 parts of 45% strength sodium hydroxide solution and 1,000 parts of methylene chloride are then added. The mixture is cooled to 25°C. While maintaining this temperature by cooling, 237 parts of phosgene are added over a period of 120 minutes. An additional amount of 75 parts of a 45% strength sodium hydroxide solution is added after 15 to 30 minutes or after the absorption of phosgene has started. 1.6 parts of triethylamine are added to the resulting solution and the mixture is stirred for a further 15 minutes.

A highly viscous solution is obtained, the viscosity of which is regulated by adding methylene chloride. The aqueous phase is washed with water until free from salt and alkali. The polycarbonate is isolated from the washed solution and dried. The polycarbonate has a relative viscosity of 1.29 to 1.30, measured in an 0.5% strength solution in methylene chloride at 20°C. This corresponds approximately to a molecular weight of 32,000. The polycarbonate thus obtained is extruded and granulated.

A is an aromatic polycarbonate based on 2,2-bis(4-hydroxyphenyl)propane (bisphenol-A), having a relative viscosity of η_{rel} = 1.29, M_{LS} = 28,000. The viscosities are measured at 25°C in methylene chloride at a concentration of 5 g/l (M_{LS} = molecular weight determined by light scattering).

B is an aromatic polycarbonate based on 70 mol percent of bisphenol-A and 30 mol percent of 2,2-bis(3,5-dimethyl-4-hydroxyphenyl)propane (tetramethylbisphenol-A), having a relative viscosity of η_{rel} = 1.28, M_{LS} = 31,000.

C is an aromatic polycarbonate based on 95 mol percent of bisphenol-A and 5 mol percent of 2,2-bis(3,5-dibromo-4-hydroxyphenyl)propane (tetrabromobisphenol-A), having a relative viscosity of η_{rel} = 1.31, M_{LS} = 39,000.

The polyphenylene sulfide can be prepared according to U.S. Patent 3,354,129. For this purpose, 60 grams of $Na_2S \cdot 9H_2O$ in 100 ml of N-methylpyrrolidone are heated to 190°C in an autoclave flushed with nitrogen. This initial heating removes the water of crystallization of the sodium sulfide. 36.7 grams of p-dichlorobenzene are added to the solution. The solution is then kept in a closed autoclave for 44 hours at 231°C, 20 hours at 225°C and 24 hours at 260°C. A polymer which is dried after washing with water and acetone and can be converted to a hard film at 290°C is obtained. The melting point of the polyphenylene sulfide is between 275° and 285°C and the yield is about 80%.

D is a nonreinforced polyphenylene sulfide, glass transition temperature 94°C, melting point 289°C, density 1.34 g/cm^3 (DIN 53 479), impact strength 8 kj/m^2 (DIN 53 453), notched impact strength 1 kj/m^2 (DIN 53 453).

E is a polyphenylene sulfide reinforced with 40% by weight of glass fibers, density 1.66 g/cm^3 (DIN 53 479), impact strength 15.7 kj/m^2 (DIN 53 453), notched impact strength 6.6 kj/m^2 (DIN 53 453).

Examples 1 through 4: The properties of the pure polycarbonates A, B and C without polyphenyl sulfide are described in the table below, as are those of the polyphenylene sulfide D.

Example 5: 2% by weight of polyphenylene sulfide D are mixed to pure polycarbonate A in a twin-screw extruder at 250°C. The mixture is extruded and granulated. Moldings of this mixture are injection-molded at 270°C. The properties are described in the table below.

Example 6: 10% by weight of polyphenylene sulfide D are mixed to pure polycarbonate B in a twin screw extruder at 270°C. The mixture is extruded and granulated. Moldings of this mixture are injection-molded at 280°C. The properties are described in the table below.

Example 7: 20% by weight of polyphenylene sulfide D are mixed to pure polycarbonate A in a twin-screw extruder at 250°C. The mixture is extruded and granulated. Moldings of this mixture are injection-molded at 270°C. The properties are described in the table below.

Example 8: 5% by weight of polyphenylene sulfide D are mixed to pure polycarbonate C in a twin-screw extruder at 290°C. The mixture is extruded and granulated. Moldings of this mixture are injection-molded at 280°C. The properties are described in the table below.

Example 9: 40% by weight of polyphenylene sulfide E (that is to say 24% by weight of pure polyphenylene sulfide and 16% by weight of glass fibers) are admixed to pure polycarbonate A in a twin-screw extruder at 250°C. The mixture is extruded and granulated. Moldings of this mixture are injection-molded at 280°C. The properties are described in the table below.

	DIN	Example Number 1	2	3	4	5	6	7	8	9
Polycarbonate, %	–	100	100	100	–	98	90	80	95	60
Polyphenylene sulfide, %	–	–	–	–	100	2	10	20	5	24
Elongation at break, %	53,455*	>120	100	110	<9	105	98	107	115	4
E-modulus, MPa	53,457**	2,300	2,400	2,300	4,390	2,310	2,420	2,440	2,500	3,700
Impact strength, kJ/m²	53,453	nb	nb	nb	8	nb	nb	nb	nb	35
Weld line strength, kJ/m²	–	88	80	90	7	80	78	74	77	7
Notched impact strength, kJ/m²	53,453	>35	10	25	1	34	14	25	24	8
Vicat B, °C	53,460***	153	160	154	227	150	148	147	152	150
O_2 index	ASTM-D 2863-70	0.24	0.24	0.34	–	0.27	0.33	0.35	0.35	0.33
Burning test according to UL 94										
1/16"	–	V-2	V-2	V-2	–	V-2	V-2	V-2	V-0	V-1
1/8"	–	V-2	V-2	V-0	–	V-1	V-1	V-0	V-0	V-1
¼"	–	V-1	V-1	V-0	–	V-1	V-0	V-0	V-0	V-0
Impact strength, kJ/m²†	53,453	3 x 92	nb	nb	–	nb	nb	2 x 50	nb	20
after 100 hr	53,453	7 x nb	–	–	–	–	–	8 x nb	–	–
after 200 hr	53,453	25	60	40	–	5 x 75	4 x 71	6 x 46	5 x 60	20
		–	–	–	–	5 x nb	6 x nb	4 x nb	5 x nb	–
after 300 hr	53,453	20	50	20	–	6 x 50	6 x 42	65	5 x 50	20
		–	–	–	–	4 x nb	4 x nb	–	5 x nb	–
after 400 hr	53,453	15	50	5	–	30	35	50	30	19
after 500 hr	53,453	8	10	1	–	18	20	25	20	16
after 700 hr	53,453	8	10	1	–	15	17	25	10	14

Note: nb means not broken.

*DIN 53,455 is directly comparable to ASTM D-638.

**DIN 53,457 is comparable to ASTM D-638 tensile test, ASTM D-790 flexural test and ASTM D-695 compression test.

***DIN 53,460 is similar to ASTM D-1525.

†The impact strength after storage in boiling water is taken as a measure of the hydrolysis resistance.

Molding Compositions Containing Reinforcing Agents and Red Phosphorus

P. Horn and C. von Rumohr; U.S. Patent 4,014,849; March 29, 1977; assigned to BASF AG, Germany describe self-extinguishing reinforced molding compositions based on aromatic polycarbonates and containing from 0.5 to 15% by weight, based on the weight of polycarbonate, of red phosphorus and a reinforcing agent such as glass fibers or glass spheres or a filler such as asbestos. The molding compositions are suitable for processing to shaped articles.

Commercial-grade red phosphorus may be used, it contains stabilizing additives comprising metal oxides and metal salts in amounts of up to 3% by weight. The red phosphorus may be incorporated in a pure form or as a mixture with the filler or reinforcing agent or by way of a masterbatch such as is described, e.g., in *Kunststoffe* 49, pp. 543-546 (1959). The masterbatches may be prepared using dispersing or passivating agents which are thermally stable up to 320°C, are miscible with the polycarbonate and are inert to the carbonate melt.

Furthermore, the dispersing agent, after incorporation of the red phosphorus, should be readily removable from the polycarbonate melt, for example by vacuum distillation. It is also advantageous to use a dispersing agent which is solid at room temperature and has a passivating effect on the red phosphorus so as to avoid fires which could be caused by electrostatic charges occurring during blending of the polycarbonate with the red phosphorus at room temperature or during incorporation of the red phosphorus into the melt.

In addition to red phosphorus, the self-extinguishing reinforced polycarbonate molding compositions contain reinforcing agents or fillers, usually in amounts of from 5 to 50% and preferably from 15 to 35%, by weight of the polycarbonate. Examples of suitable fillers are mineral substances such as chalk, kaolin and Aerosil and preferably asbestos.

Incorporation of the red phosphorus or phosphorus-containing mixture is effected by means of a single-screw or twin-screw extruder, an injection molding machine, a roller mixer, a ribbon blender or a twin-shaft kneader. The reinforcing agents or fillers may be added to the polycarbonate separately or together with the red phosphorus.

When use is made of a masterbatch or a slurry of red phosphorus in a precondensate of diphenylcarbonate and 2,2,-[4,4'-dihydroxydiphenyl] propane according to German Published Application 1,495,730 it is advantageous to provide the mixer, such as the extruder, ribbon blender or twin-shaft kneader, with devolatilizing means so that the volatiles may be removed through a port mounted on a portion of the housing at which the mixture passing therethrough is molten.

In the self-extinguishing reinforced polycarbonate molding compositions the red phosphorus is present in the form of a uniformly distributed heterogeneous phase. The average size of the phosphorus particles should be between 0.001 and 0.5 mm and preferably between 0.01 and 0.15 mm in diameter. The molding compositions of the process are suitable, for example, for the manufacture of supports for electrical parts and of shaped articles used in the electrical industry, for example, cover plates and coil formers.

The fire resistant properties were tested on rod-shaped injection moldings having a length of 12.7 cm, a width of 1.27 cm and a thickness of 0.16 cm, by a method similar to that described in *Self-Extinguishing Test* in Modern Plastics, October 1970, pp. 92 to 98, particularly pp. 95-96.

Example 1: A granular polycarbonate based on phosgene and 2,2-[4,4'-dihydroxydiphenyl] propane and having a relative viscosity of 1.305 (measured on a 0.5% solution in methylene chloride at 25°C) was thoroughly mixed with red phosphorus and glass fibers in a tumbling drum, melting and homogenized in an extruder and extruded in the form of a rope which was then granulated on cooling.

Specimens having a length of 12.7 cm, a width of 1.27 cm and a thickness of 0.16 cm were made from the dry granules and subjected to the self-extinguishing test. This test was repeated after storage of the test specimens for 7 days at 70°C and 4 hours at 23°C over calcium chloride as desiccant.

Comparative Examples 2 through 5: Test specimens were made as described in Example 1 except that the addition of glass fibers and/or red phosphorus was omitted.

The components of the polycarbonate molding compositions and the results of the self-extinguishing test are listed in the table below. The examples show that only the reinforced polycarbonate molding compositions give specimens having a fire resistance which can be classified as SE-0.

			Results of Self-Extinguishing Test: Immediately After Manufacture of Specimens			After Storage for 7 Days at 70°C and 4 Hours at 23°C		
Ex. No.	Red Phosphorus (% w/w)**	Glass Fibers (% w/w)**	Classification (SE)	Burning Time* (sec)	Dripping Observed	Classification (SE)	Burning Time* (sec)	Dripping Observed
1	5	30	0	1.0	no	0	1.1	no
2	–	–	2	7.0	yes	2	5.4	yes
3	5	–	2	0.8	yes	2	0.7	yes
4	–	10	2	7.7	yes	2	6.1	yes
5	–	30	1-2	4.6	yes	2	3.8	yes

*The burning time is the total burning time following two ignition periods each of 10 sec duration.
**The percentages are by weight of the polycarbonate.

Halogenated Dihydric Phenols

A.L. Baron and P. Sivaramakrishnan; U.S. Patent 4,043,980; August 23, 1977; assigned to Mobay Chemical Corporation describe polycarbonates having improved critical thickness values based on an aromatic diphenol, a halogenated dihydric phenol and an aromatic diphenol thioether. By incorporating an effective amount of the aromatic thiodiphenol based on the total diphenol content into the polymer the critical thickness and flame retardant characteristics of the polycarbonate are substantially improved. The aromatic diphenols used are those represented by the structural formula

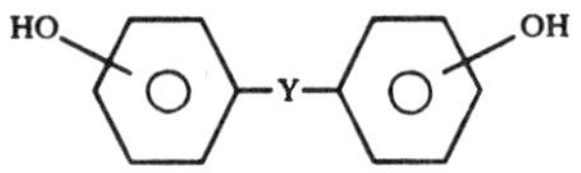

where Y is a single bond, an alkylene or alkylidene radical with 1 to 7 carbons, a cycloalkylene or cycloalkylidene radical with 5 to 12 carbons, —O—, —CO—, —SO—, or $—SO_2—$, preferably methylene or isopropylidene. Suitable aromatic diphenols are, for example, (4,4'-dihydroxy-diphenyl)methane, 2,2'-dihydroxy-diphenyl)propane, 1,1-(4,4'-dihydroxydiphenyl)cyclohexane, 1,1-(4,4'-dihydroxy-3,3'-dimethyldiphenyl)cyclohexane and 1,1-(2,2'-dihydroxy-4,4'-dimethyldiphenyl)-butane.

The halogenated phenolic diols are any suitable bis-hydroxyaryl components such as, for example, the halogen containing bisphenols such as 2,2-(3,5,3',5'-tetrachloro-4,4'-dihydroxydiphenyl)propane, 2,2-(3,5,3',5'-tetrabromo-4,4'-dihydroxy-diphenyl)propane, 2,2-(3,3-dichloro-4,4'-dihydroxydiphenyl)propane and 2,2-(3,5-dichloro-4,4'-dihydroxydiphenyl)propane. These are represented by the structural formula below.

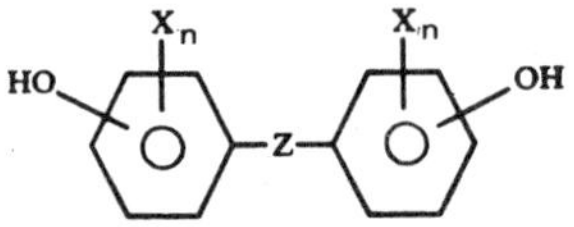

In the formula Z is a single bond, an alkylene or alkylidene radical with 1 to 7 carbons, a cycloalkylene or cycloalkylidene radical with 5 to 12 carbons, —O—, —CO—, —SO—, or $—SO_2—$, preferably methylene, isopropylidene or $—SO_2—$, and X is a halogen, preferably chlorine or bromine, most preferably bromine and n is 1 to 4.

These halogenated diols are incorporated into the polycarbonate at levels sufficient to impart flame retardant characteristics. For example, a halogen content of about 3 to 10% by weight of the polymer is normally sufficient. The thiodiphenols used are those represented by the structural formula

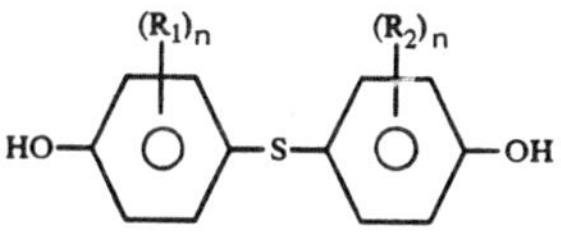

where R_1 and R_2 are hydrogen or lower alkyl selected from the group consisting of methyl, ethyl, propyl and butyl, and n is equal to from 0 to 2.

The terpolycarbonate resins not only exhibit the flame retardant synergism of sulfur and halogen but also demonstrate critical thickness values similar to the critical thickness values demonstrated by polycarbonates based solely on an aromatic diol. Such an improvement in critical thickness is surprising because it is known that the critical thickness of polycarbonate copolymers based on an aromatic diol and a halogenated dihydric phenol is significantly lower than the critical thickness of polycarbonates based solely on an aromatic diol. This difference in critical thickness is attributable to the detrimental effect of the halogenated dihydric phenol.

Thus, the combination of the thiodiphenol with the aromatic diol and the halogenated dihydric phenol overcomes this detrimental effect of the halogenated dihydric phenol and results in polycarbonate terpolymer resins having critical thickness values similar to the critical thickness values of polycarbonates based only on an aromatic diol.

Example 1: An aqueous solution was prepared by simultaneously charging to a suitable premix vessel with agitation 67.6 kg of water, 13.29 kg of bisphenol-A, 0.34 kg of 4,4'-thiodiphenol, 9.7 kg of 50% aqueous sodium hydroxide solution and 0.30 kg of tert-butyl phenol. 33.75 kg/hour of the above solution was continuously phosgenated with 2.87 kg of phosgene/hour in 50.38 kg of 1:1 methylene chloride:chlorobenzene solvent. 2.5 kg/hour of 25% aqueous sodium hydroxide solution was continuously added to the reaction mixture to provide the proper basicity for the interfacial polycondensation.

The phosgenated solution as obtained above was then reacted with 0.56 kg/hour of tetrabromobisphenol-A dissolved in 4.2 kg of 1:1, methylene chloride:chlorobenzene solvent. The precondensate as above obtained was then mixed with a solution/hour of 25 grams of triethylamine and 2.8 kg of 25% aqueous sodium hydroxide solution and was further reacted in a stirred kettle cascade over the course of an average dwell time of 30 minutes. The organic phase was separated from the aqueous phase and the organic phase washed with 1% aqueous sodium hydroxide solution, separated and then the organic phase was washed with 1% aqueous phosphoric acid solution and separated. The organic phase was simultaneously washed and separated three times with water. The polymer was recovered from the organic phase by concentrating the polymer by solvent evaporation and subsequently passing the polymer through a devolatilizing extruder. The polymer as above prepared had the following composition: 89.63 wt % (93.14 mol %) bisphenol-A; 2.25 wt % (2.45 mol %) 4,4'-thiodiphenol; and 10.13 wt % (4.41 mol %) tetrabromobisphenol-A.

The polymer was pelletized and tested for physical, optical and rheological properties. The polymer was found to be highly transparent. The test results are reported in the table below with the test results of the following examples.

Examples 2 through 5: Example 1 was repeated except that the stoichiometric amounts of bisphenol-A, tetrabromobisphenol-A and thiodiphenol were varied. The table shows both the weight and mol percent of the constituents. Each of these polymers were found to be highly transparent.

Example 6: This example is a comparison between thiodiphenol and sulfonyl diphenol demonstrating the constrasting critical thickness values.

Example 7: This example is a control having no thiodiphenol therein.

Ex. No.	Monomer Composition of Terpolymer wt %, (mol %) BPA**	TDP***	SDP†	TBBPA††	Izod Impact* (notched) 1/8"	1/4"	Critical Thickness (mils)	Melt Index (g/10 min)	O.I. (%)	UL-94 1/16" Thick Samples, Average Class (sec)
1	89.63 (93.14)	2.25 (2.45)	–	10.3 (4.41)	14.46	2.47	136	3.1	28.5	V-0 (3.3)
2	85.36 (90.50)	4.51 (5.00)	–	10.13 (4.5)	14.22	2.30	146	2.6	28.2	V-0 (2.2)
3	71.5 (75.09)	18.8 (20.64)	–	9.7 (4.27)	12.24	3.01	197	1.19	28.5	V-0 (3.6)
4	61.5 (64.27)	28.8 (31.48)	–	9.7 (4.25)	12.29	5.60	232	1.8	31.3	V-0 (1.7)
5	51.5 (53.53)	38.9 (42.29)	–	9.6 (4.18)	12.95	11.28	>255	2.5	32.2	V-0 (2.4)
6	84.8 (90.50)	–	5.14 (5.00)	10.06 (4.5)	9.53	2.13	132	2.5	29.1	V-0
7	90 (95.55)	–	–	10 (4.45)	14.53	2.2	130	2.9	28.0	V-2

*Expressed in ft lb/in.
**Bisphenol-A.
***4,4'-Thiodiphenol.
†Sulfonyldiphenol.
††Tetrabromobisphenol-A.

Metal Salts of Phenol Ester Sulfonic Acids

V. Mark; U.S. Patents 4,073,768; February 14, 1978 and 4,093,590; June 6, 1978; both assigned to General Electric Company describes a nonopaque flame retardant polycarbonate composition comprising a mixture of an aromatic carbonate polymer and an additive which may be the metal salts of either monomeric or polymeric phenol ester sulfonic acids. The aromatic carbonate polymer and additive have a refractive index in the range of 1.54 to 1.65.

The types of sulfonic acids employed may be either the monomeric form or the polymeric form or mixtures of these. When considering the monomeric form, the metal salt of the unsubstituted and substituted monomeric phenol ester sulfonic acid can best be represented by the following formula:

(1) $$(R)_z[(O_qA)_u(X)_y(O_rB)_v]_m$$

where A and B are independently selected from the following formula:

(2) $$R'(SO_3M)_nR''_p$$

where R' is an aromatic radical of 1 to 4 aromatic nuclei, M is a metal which may be selected from the periodic table of either an alkali metal or an alkaline earth metal, R" is an electron withdrawing radical, n is an integer of 0 to 4 and p is an integer of 0 to 10, provided, however, that the sum of n must be at least 1.

In formula (1) above, (R) is an organic radical of 1 to 20 carbons and is either alkyl, aralkyl, alkenyl, aralkenyl, aryl, arylene, alkylene, aralkylene, aralkenylene, alkylidene, aralkylidene, alkenylidene or aralkenylidene and the group comprising (A) and (B) as well as a trivalent and tetravalent aromatic nucleus. The radical (R) can also contain halogen substituents which would be an electron withdrawing radical and sulfonic acid salt substituents. In formula (1), (X) is a di-, tri- or tetravalent radical selected from the following:

$$-\underset{\underset{O}{\|}}{C}-,\ -\underset{\underset{S}{\|}}{C}-,\ -\underset{\underset{O}{\|}}{S}-,\ -\overset{\overset{O}{\|}}{\underset{\underset{O}{\|}}{S}}-,\ -\overset{|}{\underset{\underset{O}{\|}}{P}}-,\ -\overset{|}{\underset{\underset{S}{\|}}{P}}-,\ \text{and}\ -\overset{|}{\underset{|}{Si}}-$$

y is an integer of 1 to 2, m, u, q and r are integers of 1 to 4 and z is an integer of 0 to 3, v is an integer of from 0 to 4, and O is oxygen.

In the above formula (1), the presence of the (SO_3M) radical on the phenol ester is the feature that offers the excellent flame retardant properties when mixed with an aromatic polycarbonate.

While there are many compounds that meet the requirements of formula (1) and which offer excellent flame retardant characteristics to an aromatic polycarbonate, the preferred additive employed in the monomeric form is the sodium salt of the sulfonic acid of dichlorophenyl 2,4,5-trichlorobenzenesulfonate. This has the following formula:

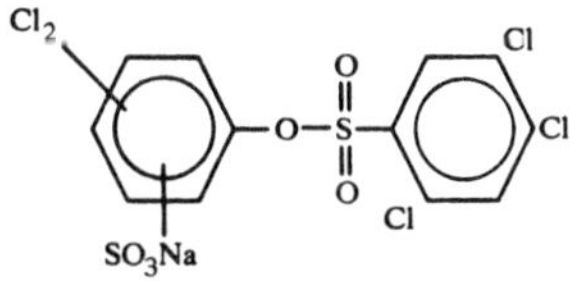

When the polymeric form of the unsubstituted and substituted aromatic sulfonic acid is employed, it can best be represented by the following formula:

(3) $$\langle(R)_z[(A)(X)_y(B)_{1\text{-}3}]\rangle_a\langle(R)_z[(A)(X)_y(B)_{1\text{-}3}]\rangle_b$$

where A and B are independently selected from the following formula:

(4) $$(O)_qR'(SO_3M)_nR''_p$$

In the formula (O), R', M, R", n, p and q have the same meaning as cited previously. In addition, (R), (X) and z have the same meaning as cited previously. Also, the same requirements for n are also applicable. In formula (3), a is an integer of from 0 to 2,000 and b is an integer of from 0 to 2,000, provided, however, that the sum of a and b must be at least 4 and can be as high as 2,000. As shown, the selections of a and b can be random or equal or one can be zero (0).

The polymeric structure can be either a homopolymer, a copolymer, a random copolymer, a block copolymer or a random-block copolymer, or mixtures of these polymeric forms. In addition, the ratio of sulfonated phenolic rings to unsubstituted phenolic rings can vary from 1 to 1 to as high as that which is barely necessary to render the polycarbonate flame retardant and this may be 1 to 1,000. In the examples that follow all parts and percentages are on a weight basis unless otherwise specified.

Example 1: 100 parts of an aromatic polycarbonate, prepared by reacting 2,2-bis(4-hydroxyphenyl)propane and phosgene in the presence of an acid acceptor and a molecular weight regulator and having an intrinsic viscosity of 0.57 is mixed with 0.01 part of a finely ground dehydrated additive listed below in the first table by tumbling the ingredients together in a laboratory tumbler. The resulting mixture is then fed to an extruder. The extruder is operated at about 265°C and the extrudate is comminuted into pellets.

The pellets are then injection molded at about 315°C into test bars of about 5" x ½" x 1/16" to ⅛" thick into test squares of about 2" x 2" x ⅛" thick. The test bars (5 for each additive listed in the first table) are subject to the test procedure set forth in Underwriters' Laboratories, Inc. Bulletin UL-94, "Burning Test for Classifying Materials". In accordance with this test procedure, materials so investigated are rated either V-O, V-I or V-II based on the results of 5 specimens. The test squares are tested for light transmission in a Gardner XL 10-CDM instrument. The data shows the amount of incident light transmitted by the test squares using air as 100% transmission. The results of an additive is as follows with a control being the aromatic polycarbonate as prepared above without the additive of the type mentioned.

Additive (0.01 parts per hundred)	Light Transmission (%)	Flame Out Time Seconds	No. of Drips Per five test bars	UL-94 Rating
CONTROL	86	24	13	Burning
Potassium salt of 4-chloro-3-sulfophenyl benzoate	86	8.6	3	V-II

Example 2: This example demonstrates the effect of the additives at limits of 0.10 part per hundred parts of the polycarbonate. In preparing the test specimens for this example, 100.00 parts of the polycarbonate of Example 1 is mixed with 0.10 part of the additives listed below (one additive per sample) employing the same procedure. Test specimens are then molded using the same procedure employed in Example 1. The test specimens are subjected to the same test procedure of Example 1 with the results as shown below.

Additive (0.10 parts per hundred)	Light Trans-mission (%)	Flame Out Time Seconds	No. of Drips Per five test bars	UL-94 Rating
CONTROL	86	26	13	Burning
Sodium [phenyl 2,4,5-trichloro-benzenesulfonate]-4'-sulfonate	83	3.6	0	V-0
Potassium triphenyl phosphate-4-sulfonate	80	7.0	0	V-I
Dipotassium bis-(4-sulfophenyl) benzenephosphonate	83	5.6	0	V-I

Example 3: This example demonstrates the effect of the additives at limits of 0.20 part per hundred parts of the polycarbonate. In preparing the test specimens for this example, 100.00 parts of the polycarbonate of Example 1 is mixed with 0.20 part of the additives listed below in the third table employing the same procedure. Test specimens are then molded using the same procedure employed in Example 1. The test specimens are subjected to the same test procedure of Example 1 with the following results:

Additive (0.20 parts per hundred)	Light Trans-mission (%)	Flame Out Time Seconds	No. of Drips Per five test bars	UL-94 Rating
CONTROL	86	26	13	Burning
Potassium diphenylsulfate-4-sulfonate	78	6.2	0	V-I
Polysodium (bis-phenol-A)(tetra-bromobisphenol-A) copolycarbonate-polysulfonate	84	2.8	0	V-O

Example 4: This example is set forth to show the effect of a known commercially available flame retardant additive. Example 1 is repeated except that in place of the additives employed, only one part decabromodiphenyl ether is used. The results obtained upon evaluating five test bars are the same as obtained for the control shown in the table under Example 1 above.

Metal Salts of Sulfonic Acids of Aromatic Sulfides

In related work on a nonopaque flame retardant polycarbonate composition, *V. Mark; U.S. Patent 4,067,846; January 10, 1978; assigned to General Electric Company* describes the use of another additive, that of metal salts of sulfonic acids of aromatic sulfide. The compositions of these also have a refractive index of 1.54 to 1.65. The metal salt of the sulfonic acid of an aromatic sulfide can best be represented by the following formula: $R'(S)_{1-2}R''(SO_3M)_{1-6}(X)_{0-11}$ where X is an electron withdrawing radical, M is a metal which may be selected from the Periodic Table groups of alkali metals and alkaline earth metals and R' and R'' may be either an aryl radical of 1 to 2 aromatic rings or an aliphatic radical of 1 to 6 carbons and they may be the same or different. However, R' and R'' together must contain at least one aromatic ring. Also, as shown in the above formula, the additive may or may not have an electron withdrawing radical. For optimum results, it is preferred that the additive contain an electron withdrawing radical and that both the electron withdrawing radical and the (SO_3M) radical be on the same aromatic ring. The preferred additive employed is sodium 4,4'-dichlorodiphenylsulfide-3-sulfonate. This has the following formula:

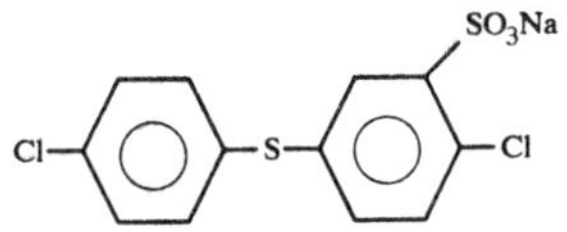

The examples and tests performed are similar to the above work.

Metal Salts of Aromatic Sulfonesulfonic Acid

In further related work, *V. Mark; U.S. Patent 4,092,291; May 30, 1978; assigned to General Electric Company* describes the additive in the flame retardant polycarbonate composition as being the metal salts of either monomeric or polymeric aromatic sulfonesulfonic acids.

Metal Salts of Oxocarbon Acids

V. Mark; U.S. Patent 4,069,201; January 17, 1978; assigned to General Electric Company describes a flame retardant polycarbonate composition comprising a mixture of an aromatic carbonate polymer and a flame retardant additive which may be the metal salts of unsubstituted or halogenated oxocarbon acids, or mixtures of these. The best known examples of oxocarbon acids are the following:

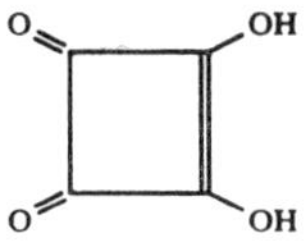

Squaric Acid

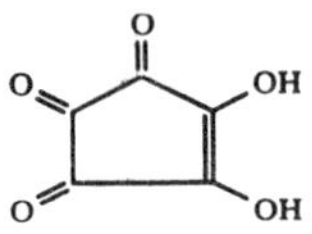

Croconic Acid

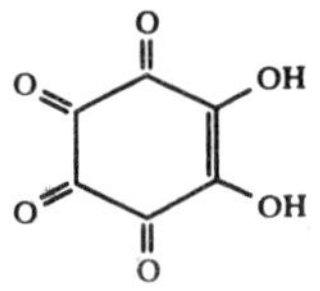

Rhodizonic Acid

These oxocarbon acids may be substituted with aromatic rings. Examples of these substituted oxocarbon acids are the following:

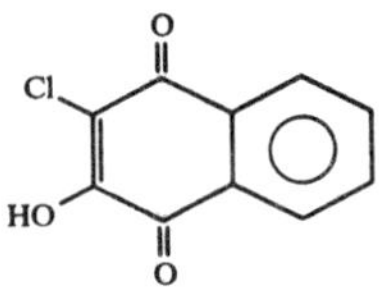

3-chloro-1,4-naphthoquinone-2-ol

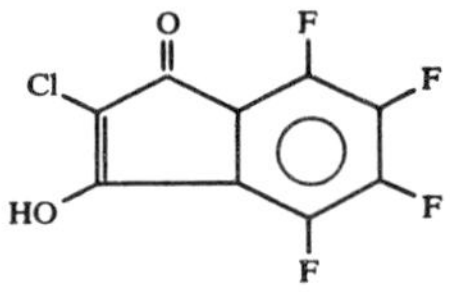

2-chloro-4,5,6,7-tetrafluoroindenone-3-ol

Halooxocarbon acids contain the basic acidic feature of the oxocarbon acids, that is, the —CO—C=C—OH moiety, but in some or in all of the other OH and CO functions the oxygen is replaced by halogen.

The metal salts of the oxocarbon and halooxocarbon acids have the following formula:

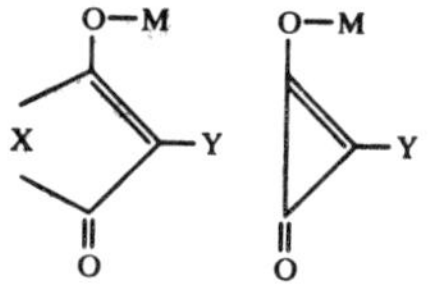

where M is a metal which may be selected from the periodic table of either an alkali or an alkaline earth metal, Y is a halogen radical selected from chlorine, fluorine and bromine, or an —OM radical where M is the same as cited above, and X is a divalent radical selected from the following groups:

or

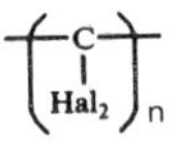

or (ortho-arylene) where m and n are integers from 1 to 2 and Hal is a halogen selected from chlorine, fluorine and bromine radicals and arylene comprises the phenylene, naphthylene and anthracenylene radicals. The following examples illustrate the process.

Example 1: 99 parts of an aromatic polycarbonate, prepared by reacting 2,2-bis-(4-hydroxyphenyl)propane and phosgene in the presence of an acid acceptor and a molecular weight regulator and having an intrinsic viscosity of 0.57 is mixed with one part of a finely ground dehydrated additive listed in the table below by tumbling the ingredients together in a laboratory tumbler. The resulting mixture is then fed to an extruder, which extruder is operated at about 265°C and the extrudate is comminuted into pellets.

The pellets are then injection molded at about 315°C into test bars of about 5" x ½" x 1/16" to ⅛" thick. The test bars (5 for each additive listed in the table) are subject to the test procedure set forth in Underwriters Laboratories, Inc. Bulletin UL-94, "Burning Test for Classifying Materials." The results of the different additives are as follows with a control being the aromatic polycarbonate as prepared above without the additive of the type mentioned.

Additive (1.0 weight percent)	Flame Out Time Seconds (Avg.)	No. of Drips Per Test Bar	Rating
CONTROL	31.6	4+	Burns
Disodium squarate	5.3	2.5	SE-II
Barium rhodizonate	5.3	3.2	SE-II
Disodium chloranilate	2.4	0	SE-O
Calcium chloranilate	2.8	2.4	SE-II
Disodium croconate	5.5	2.8	SE-II

Example 2: This example demonstrates the effect of the flame retardant additives at the lower limits of 0.01 weight percent based on the weight of the polymer composition. In preparing the test bars for this example, 99.99 parts of the polycarbonate of Example 1 is mixed with 0.01 weight percent of the additives listed in the table below employing the same procedure. Test bars are then

molded using the same procedure employed in Example 1. The test bars are subjected to the test procedure of Example 1 with the following results:

Additive (0.01 weight percent)	Flame Out Time Seconds (Avg.)	No. of Drips Per Test Bar	Rating
CONTROL	31.6	4+	Burns
Disodium chloranilate	9.6	4.2	SE-II
Barium chloranilate	10.6	3.2	SE-II

Example 3: This example shows the effect of a known commercially available flame retardant additive.

(a) Example 1 is repeated except that in place of the additives employed, only one part 1,2,5,6,9,10-hexabromocyclododecane is used. The results obtained upon evaluating five test bars are the same as obtained for the Control shown in the table of Example 1.

(b) Part (a) above is repeated but using 5 weight percent of the above additive, namely 1,2,5,6,9,10-hexabromocyclododecane. The results obtained are the same as obtained in part (a) above.

(c) Part (a) above is repeated by using 10 weight percent of the above additive, namely 1,2,5,6,9,10-hexabromocyclododecane. At this level of additive, test bars are rated SE-II. However, the polycarbonate is badly degraded as evidenced by severe dark streaking of the molded test bars, which degradation does not occur with the additives.

Example 4: Example 3 is repeated except that hexabromobiphenyl is employed. The results obtained are essentially the same as those of Example 3.

Example 5: Example 3 is repeated except that the additive employed is a combination of antimony oxide and a material which is a mixture of polychlorinated biphenyl. The proportion of the ingredients of the additive employed in this example is based on 3 parts of chlorine per one part of antimony. The results obtained at 1 weight percent and 5 weight percent amounts are the same as in Example 3.

However, at the higher amount, namely 10 weight percent, flame retardancy effect is noted, but with, again, severe degradation of the polycarbonate, as evidenced by the substantial reduction in the intrinsic viscosity of the molded test bars. As molded, the intrinsic viscosity of the test bars with 1 weight percent of the above additive is about 0.50. The intrinsic viscosity of the molded test bars containing 10 weight percent of the flame retardant additive of this example is 0.253. This shows the severe degradation of the polycarbonate when employing this type of well known flame retardant.

Metal Salts of Aromatic Sulfonic Acids

In related work, *V. Mark; U.S. Patent 4,075,164; February 21, 1978; and A. Factor, D.B.G. Jaquiss and V. Mark; U.S. Patent 4,093,589; June 6, 1978; both assigned to General Electric Company* describe a flame retardant polycarbonate composition comprising a mixture of an aromatic carbonate polymer and a flame retardant additive which may be the metal salts of either mixtures of monomeric and polymeric substituted aromatic sulfonic acids, or polymeric substituted aromatic sulfonic acids. When considering the monomeric form, the metal salt of the substituted monomeric aromatic sulfonic acid can best be represented by the following formula:

(1) $$A_{0-1}R_{0-1}B$$

where A and B can, in turn, be independently selected from the following formula:

(2) $$R'_x(SO_3M)_yR''$$

where R' is an electron withdrawing radical, M is a metal which may be selected from the Periodic Table groups of alkali metals and alkaline earth metals, R'' is an aryl radical of 1 to 4 aromatic rings, x is an integer of 0 to 17 and y is an integer of 0 to 10. However, x and y in formula (2) must equal at least 1.

In the formula (1) R is an organic radical of 1 to 20 carbons and is either alkyl, aralkyl, alkenyl, aralkenyl, aryl, arylene, alkylene, aralkylene, alkenylene, aralkenylene, alkylidene, aralkylidene, alkenylidene or aralkenylidene. The radical R can also contain halogen substituents, an electron withdrawing radical. However, this would be in addition to the above requirement that in formula (2) the value of x must be at least 1.

In formula (1), the combination of the electron withdrawing radical and the (SO_3M) radical offers the excellent flame retardant properties when mixed with an aromatic polycarbonate. Preferably, for optimum results, the electron withdrawing radical and the (SO_3M) radical should be on the same aromatic ring, namely either in A or B of formula (1). As further indicated in formula (1) in its simplest form, the flame retardant additive can consist merely of B which in turn would be simply formula (2). This is shown where A and R may be 0. The simplest formula may be as follows:

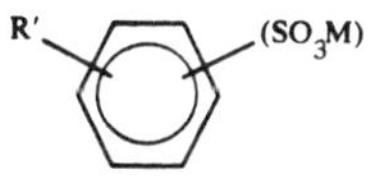

The polymeric form of the substituted aromatic sulfonic acid can best be represented by the following formula:

(3) $$A_mR_{1\,m+n}B_n$$

where A and B are independently selected from the following formula:

(4) $$R'_x(SO_3M)_yR''$$

where R', M, R'', x and y have the same meaning as recited previously. In addition, the same requirements for x and y apply. However, A and B units can be randomly selected. In addition R_1 is an organic radical of 1 to 20 carbons and is either arylene, alkylene, aralkylene, alkenylene, aralkenylene, alkylidene, aralkylidene, alkenylidene and aralkenylidene. In the polymeric form, R_1 cannot be monovalene as in the case of the monomeric form of the additive employed here and described by formula (1). In formula (3), the sum of m and n must be at least 4 and can be as high as 2,000. As shown, the selection of m and n can be random or equal or one can be 0. In formula (3), R_1 can also contain halogen substituents which would be an electron withdrawing radical. However, this would be in addition to the requirement that in formula (3), the value of x must be at least 1. The examples and tests performed are similar to those of the previous works.

Metal Salts of Sulfonic Acids of Aromatic Carboxylic Acids and Esters

In further related work *V. Mark; U.S. Patent 4,064,101; December 20, 1977; assigned to General Electric Company* describes a flame retardant polycarbonate employing a flame retardant additive which may be the metal salts of sulfonic acids of either mixtures of monomeric and polymeric aromatic carboxylic acids and esters or polymeric aromatic carboxylic acids and esters. When considering the monomeric form, the metal salt of the unsubstituted sulfonic acids of aromatic carboxylic acids and esters have the following formula:

$$X_pR_v(SO_3M)_m(COOY)_nR'_q$$

where X is an electron withdrawing radical, M is a metal which may be selected from the periodic table of either an alkali metal or an alkali earth metal, R is an aryl radical of 1 to 4 aromatic rings, R' is an organic radical of 1 to 4 carbons and is either alkyl, alkenyl, alkylene, alkenylene, alkylidene or alkenylidene, Y is a metal described under M or an organic radical of 1 to 20 carbons and is either alkyl, cycloalkyl, aralkyl, alkenyl, cycloalkenyl, aralkenyl or aryl, m and n are integers of 1 to 6 and p and q are integers from 0 to 10 and v is an integer of from 1 to 10. The preferred additive employed in the monomeric form is sodium dimethyl 2,4,6-trichloro-5-sulfoisophthalate. This has the following formula:

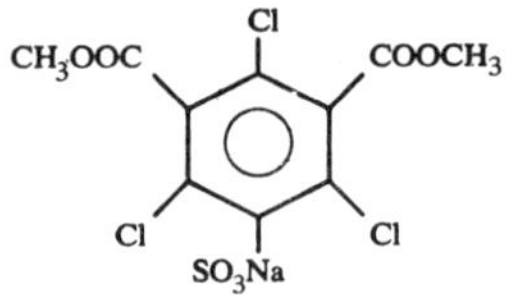

When the polymeric form of the sulfonic acids of the unsubstituted and substituted aromatic carboxylic acid esters is employed, it can best be represented by the following formula:

$$[Hal]_p[R(COO)_{2-4}R']_m[R''(COO)_{2-4}R''']_n[SO_3M]_x$$

where Hal is a halogen selected from the group comprising fluorine, chlorine and bromine, R, R', R'' and R''' are either aromatic nuclei of 1 to 4 aromatic rings or aliphatic and cycloaliphatic radicals of from 2 to 20 carbons, provided that at least one of the Rs must be aromatic, M is a metal which may be selected from the Periodic Table groups of alkali metals and alkaline earth metals, m plus n is an integer from 50 to 300, x is an integer from 1 to 600 and p is an integer from 0 to 600, where the SO_3M group is attached to an aromatic ring.

Metal Salts of Halogenated Nonaromatic Carboxylic Acids

In an additional and further related work, *V. Mark; U.S. Patent 4,066,618; January 3, 1978; assigned to General Electric Company* describes a flame retardant polycarbonate composition comprising a mixture of an aromatic carbonate polymer and a flame retardant additive which may be the metal salt of a halogenated nonaromatic carboxylic acid. The halogen function may be either bromine, chlorine or fluorine. However, when fluorine is used as the only halogen substituent, the carboxylic acid must be a dicarboxylic acid. The carboxylic acid function contains one to two carboxylic functions and from 1 to about 20 carbons. The preferred metal salts of a halogenated nonaromatic carboxylic acid are disodium hexafluoroglutarate, NaOOC—CF_2—CF_2—CF_2—COONa and disodium 1,4,5,6,7,7-hexachlorobicyclo[2.2.1]-5-norbornene-1,2-dicarboxylate, also known as sodium chlorendate,

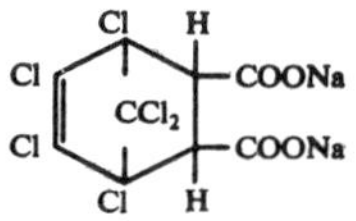

PVC

Organic Ester Plasticizers with High Oxygen Content

J.S. Heckles; U.S. Patent 4,085,081; April 18, 1978; assigned to Armstrong Cork Company describes smoke retardant compositions which can be used alone or together with other additives in polyhalocarbon resins to reduce emission of smoke and promote enhanced flame retardancy upon the burning of such compositions.

The composition can be obtained by the incorporation of a high oxygen content organic ester plasticizer into a halocarbon polymeric composition comprising a vinyl chloride resin and stabilizer. High oxygen content plasticizers for use as additives in vinyl chloride and vinylidene chloride rigid and plasticized polymer compositions comprise various alkyl and alkoxy organic esters. The term "high oxygen content" is related to di-2-ethylhexyl phthalate (DOP). This material has an oxygen content of 16%. The high oxygen content plasticizers of this process are those having an oxygen content which is greater than that of DOP, which is 16%.

The high oxygen ester-type oxalate plasticizers have the following general formula ROC(O)C(O)OR' where R and R' can be the same or different and are C_{1-8} alkyl or C_{1-8} alkoxy substituted C_{1-8} alkyl. The method used to prepare the oxalate plasticizers can generally be described as the ester interchange reaction between an alcohol and diethyl oxalate.

In a second composition, the high oxygen ester-type acetate plasticizers are used in the low smoke polyhalocarbon formulations. They have the general formula $RC(CH_2O)_3(CH_2CH{=}CH_2)_{3-x}[C(O)CH_3]_{3-y}$ where R is defined above and x and y are 1 or 2 with the proviso that $x \neq y$.

The general preparation of these alkeneoxy esters is by the acetylation of the respective alcohol with acetic anhydride with a small amount of base catalyst. They are useful as low smoke plasticizers in polyhalocarbon formulations in an amount based on 100 parts of resin of 1 to 80 parts, preferably 5 to 50 parts.

A further composition uses the high oxygen ester-type levulinates as polyhalocarbon plasticizers. These low-smoke plasticizers have the following formula $[CH_3C(O)CH_2CH_2C(O)O]_xR''$ where x is as previously defined and R" is C_{1-8} alkyl, C_{6-12} cycloalkyl, C_{1-8} alkoxy-substituted alkyl, C_{2-8} alkylene or C_{2-8} oxyalkylene. The majority of levulinic acid esters used in the composition were prepared by direct esterification of alcohol-acid with dibutyltin dioctoate catalyst with a xylene azeotrope. The following examples illustrate the process.

Example 1: Dibutyl Oxalate (Bu-Oxalate) – 81 grams (0.55 mol) diethyl oxalate, 89 grams (1.2 mol) n-butyl alcohol and 1.2 grams dibutyltin dioctoate were heated 6 hours at 100° to 164°C under a slight nitrogen pressure while collecting ethanol, 47 grams, at the top of a Vigreaux column and returning butyl alcohol to the reaction. Dibutyl oxalate was collected in fraction 2, BP 96° to 100°C/1 mm. By gas chromatography, fraction 2 contained 89% dibutyl oxalate.

Example 2: Trimethylolpropane Monoallyl Ether Diacetate – 57 grams (0.3 mol) TMP monoallyl ester, 92 grams (0.9 mol) acetic anhydride and 12 grams pyridine were mixed in a stoppered flask and held 3½ hours. The reaction produced 90% acetate product by gas chromatography. The reaction product was washed with diluted HCl, saturated $NaHCO_3$, water, dried with $MgSO_4$ and distilled. The acetate product was collected in fraction 2, BP 99° to 101°C/1 mm. Yield 64 grams, 95% (gas chromatography).

Example 3: 2-(2-Butoxyethoxy)Ethyl Levulinate – 363 grams (2.24 mols) 2-(2-butoxyethoxy)ethanol (butyl Carbitol), 186 grams (1.6 mols) levulinic acid and 1.5 grams dibutyltin dioctoate were heated with a xylene azeotrope for 18 hours to Acid No. 2.4 while collecting 33.5 grams water distillate. The ester product was washed with saturated $NaHCO_3$, water, dried with anhydrous $MgSO_4$ and distilled. The ester product was collected in fraction 3, BP 141° to 153°C/1 mm. Yield 282 grams, 68%, 99% ester by gas chromatography.

Examples 4 through 10: Test samples of polyvinyl chloride were made up in a typical formula using heat and light stabilizers. The masterbatch formulation is set forth below (additives are employed in parts by weight per 100 parts of resin):

PVC, Escambia 2160, copolymer of specific viscosity 0.02 by ASTM D-1243-60-B	100
Admex 710 (epoxidized soybean oil)	1.0
Synpron 744 (zinc octoate/barium phenate/arylphosphite)	4.0
High oxygen content organic ester plasticizer	Varied

Processability, compatibility and thermal stability were observed while determining torque in meter-grams with a Brabender Plastograph at 300°F. Each batch was additionally pressed at 300° to 320°F to a 30 to 40 mil thickness. Long-term compatibility of these plasticizers was determined by visual observation of the Brabender mixed and pressed samples, noting whether the plasticizer exuded or bloomed from the sample. Observations on compatibility were made after storing the samples at ambient conditions for six months to one year. In the NBS Smoke Density Chamber Test (Gross, D. et al, ASTM Special Technical Publication 422, p. 166-204, 1969) D_m (corr) is the maximum specific optical density corrected for soot deposits on chamber cell windows, established for flaming (fl) and nonflaming (s) modes.

NBS Smoke Density of Oxalate Ester
Plasticized PVC ROC(O)C(O)OR'*

Ex. No.	R & R'	Brabender Viscosity (m/g, 295°F)	NBS Smoke Density, D_m (corr) fl	s	Average
4	Butyl	600	327	80	204
5	Ethylhexyl	925	349	328	339
6	Cyclohexyl	750	323	326	329
7	2-Ethoxyethyl	425	226	156	191

*The plasticizer was used at 35 phr; Admex 710 at 0.8 phr and Synpron 744 at 2.0 phr. All samples exhibited good compatibility after compounding and forming. The sheets were clear and flexible.

NBS Smoke Density of Levulinic Acid Ester Plasticized PVC
$[CH_3C(O)CH_2CH_2C(O)O]\,R_x$ where x is 1 or 2

Ex. No.	R Mono- and Dilevulinates**	x	Brabender Viscosity (m/g, 300°F)	NBS Smoke Density, D_m (corr) fl	s	Average	Sample* Comment
8	2-Butoxyethyl	1	525	260	181	222	c, flex.
9	2-(2-Ethoxyethoxy)ethyl	1	525	287	192	240	c, flex.
10	2-(2-Butoxyethoxy)ethyl	1	575	258	320	289	c, flex.

*c means compatible, and flex. means flexible.
**Esters used at 38 phr.

Ferric Oxide and Hydrated Metallic Compounds as Additives

A process described by *D.M. Florence; U.S. Patent 3,993,607; November 23, 1976; assigned to Armstrong Cork Company* relates to a resin composition containing a PVC homopolymer or copolymer, where the copolymer contains a major portion of vinyl chloride. The combination of ferric oxide with hydrated metallic compounds results in polyvinyl chloride homopolymer and copolymer-containing compositions having both low smoke and low flame spread characteristics.

Generally, the incorporation of ferric oxide when compounding resinous compositions containing polyvinyl chloride homopolymer and copolymer resins, in particular plasticized polyvinyl chloride resins and polyvinyl chloride modified rubbers, significantly reduces the propensity of these compositions to generate smoke on burning and further evaluation of compounded polyvinyl chloride compositions and resinous compositions containing mixtures of polyvinyl chloride with other resinous materials confirmed the above.

In order to significantly reduce the propensity of the resinous compositions to generate smoke on burning, it is preferable to incorporate, on compounding the compositions, from 5 to 75 parts by weight ferric oxide, based on 100 parts by weight of resin, a marked reduction in smoke density being achieved at levels of Fe_2O_3 of about 5% and above.

The flame spreads are also materially reduced by incorporating, while compounding, in addition to the ferric oxide, from 5 to 200 parts by weight of a hydrated metallic compound based on the total weight of the resinous components in the compositions. In particular, hydrated compounds of magnesium carbonate containing five or six waters of hydration have been found to be particularly effective in reducing the flame spread of such resinous compositions while still achieving the significant reduction in smoke from the incorporation of the ferric oxide. Other hydrated metallic compounds such as aluminum trihydrate also significantly reduce the flame spread of these compositions, although the use of the hydrated magnesium carbonates gives compositions having reduced flame spreads over a wider temperature range. The following examples illustrate the process.

Examples 1 through 14: A masterbatch was made up of the following ingredients:

Ingredients	Parts by Weight
Paracril OZO-50*	200.0
EPO (epoxidized soybean oil)	8.0
Admex 710 (epoxidized soybean oil)	16.0
Thermax-carbon black (medium thermal)	40.0
Santicizer 160 (butylbenzyl phthalate)	20.0
Di-2-ethyl hexyl phthalate	24.0
Antimony trioxide (Sb_2O_3)	13.3

*A 50-50 mixture of a polyvinyl chloride homopolymer and an acrylonitrile-butadiene rubber having approximately 30% acrylonitrile and 70% butadiene.

The polyvinyl chloride modified rubber was first added to a Banbury mixer and the remaining ingredients, including ferric oxide and hydrated magnesium carbonate in the amounts indicated in the table below, were added incrementally with thorough mixing up to a temperature of about 280° to 300°F. After thorough mixing, the mixture was sheeted out on a two roll rubber mill and cooled.

The masterbatch thus formed was compounded by mixing 482 parts by weight of the masterbatch with 1.0 parts by weight of diphenyl guanidine, 2.5 parts by weight of zinc diethyldithiocarbamate, 8 parts by weight of Spider sulfur, a fine grade sulfur and 44 parts by weight of Celogen-OT [p,p'-oxybis(benzene sulfonylhydrazide)] in a Banbury mixer at 150°F maximum temperature until thorough mixing of the ingredients was achieved. This mix was then sheeted out on a two roll rubber mill and cooled. It was then remilled and warmed and strips (at 110°F) were removed and fed directly to an extruder. The compounded polyvinyl chloride modified rubber was extruded as a sheet of about 0.26 inch thickness. The sheet was cooled, cut into samples about 10" x 5" and the samples were expanded and cured in an oven to give cellular modified rubber sheets about 0.56" thick. The cure cycle utilized in the oven was five minutes at 220°F, a 20 minute rise to 300°F and a final cure at 300°F for five minutes.

The cured cellular sheets were trimmed to a 0.50" thickness and cut to proper sample size for smoke and flame spread testing. Samples for the smoke test were three inches square and those tested for flame spread rating were 3¾" in width and 30" in length.

The table below shows the amounts of ferric oxide (iron red oxide NR 4284 from Charles Pfizer and Company) and hydrated magnesium carbonate added in compounding the polyvinyl chloride rubber samples tested. Flame spread ratings and smoke densities as well as sample densities are given for each of the samples.

In the table FSR is a measurement of the flame spread rating achieved in a 30/30 Tunnel Test, a standard test reported in the April 1967 *Journal of Cellular Plastics* in an article by M.M. Levy entitled "A Simplified Method for Determining Flame Spread."

NBS smoke was determined according to "Method for Measuring Smoke from Burning Materials" by D. Cross et al, *ASTM Special Technical Publication No. 422* (1967). The measurements were made on Model 4-5800, Aminco-NBS Smoke Density Chamber, American Instrument Company (average value from three tests).

Example 10 was prepared from a masterbatch in which the amount of carbon black was reduced from 40 to 30 parts in the masterbatch to give a filler level equivalent to the control (Example 1) and to show that the reduction in smoke is not due to the filler level (Fe_2O_3 plus carbon black) but to the level of Fe_2O_3 present.

Ex. No.	Fe_2O_3	$(MgCO_3)4Mg(OH)_2 \cdot 5H_2O$	30/30 Tunnel Test FSR	NBS Smoke Test D_m (corr)
	...... percent by weight			
1	–	–	113	584
2	0.25	–	111	606
3	0.50	–	109	583
4	2.0	–	109	547
5	5.0	–	104	439
6	10.0	–	91	367
7	15.0	–	87	344
8	30.0	–	83	356
9	60.0	–	77	215
10	10.0	–	96	355
11	5.0	20.0	85	400
12	5.0	10.0	99	388
13	10.0	20.0	84	415
14	10.0	10.0	86	372

Examples 1 through 14 clearly illustrate that the smoke producing propensity of polyvinyl chloride resin-containing compositions can be dramatically reduced by adding effective amounts of Fe_2O_3 to the compositions. Furthermore, the flame spread ratings for these compositions can be lowered at the same time by adding effective amounts of a hydrated metallic compound.

Plasticizers for Polyvinyl Chloride

A process described by *F.-J. Dany, J. Wortmann, P. Münch and H. Vollmer; U.S. Patent 4,032,498; June 28, 1977; assigned to Hoechst AG, Germany* relates to flame retardant PVC plasticizers of the general formula:

(1)

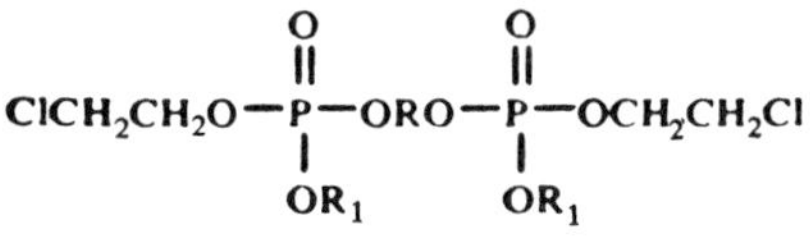

In the above formula R stands for an alkylene radical or a halogen-substituted alkylene radical having from 2 to 6 carbons, a $-CH_2CH_2-O-CH_2CH_2-$ radical, a phenylene or an alkyl or halogenoalkyl-substituted diphenylenemethane radical and R_1 stands for a branched or unbranched alkyl radical having from 6 to 18 carbons. These plasticizers may be used in a mixture with known plasticizers being free from, or having not fully satisfactory, flame retardant properties.

The plasticizers of this process are highly viscous, colorless and undistillable liquids. The process for making flame retardant plasticizers is shown in the following examples.

Example 1: A plasticizer of the following formula was made:

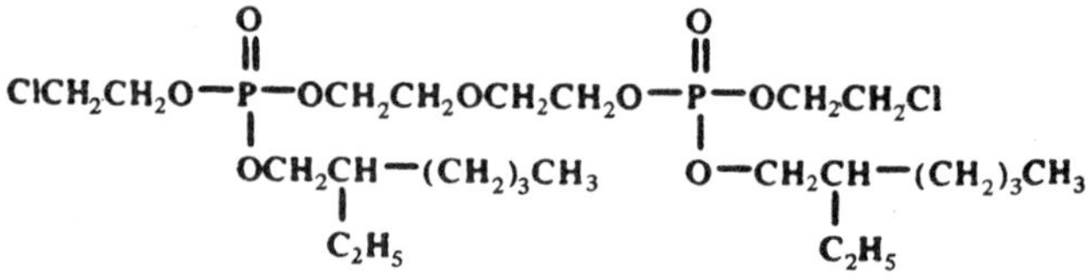

A solution of 1,373.4 grams (10.0 mols) of PCl_3 and 3.0 liters of methylene chloride was introduced into a reactor provided with reflux condenser and agitator and 620.7 grams (10.0 mols) of ethylene glycol was added dropwise with agitation at room temperature. The reaction was complete after about one hour. After that time, a further 5.0 mols of diethylene glycol was added to the reaction mixture so that the reaction temperature remained within the limits of 20° to 25°C.

Following this, chlorine was introduced at 0° to 5°C into the solution obtained until the solution assumed a yellow-green coloration which indicated the end of the reaction. Chlorine gas in excess and resulting chlorohydrocarbon were expelled by means of nitrogen. The introduction of nitrogen was continued and 651 grams (5.0 mols) of 2-ethylhexanol was added at room temperature. After all had been added, the temperature was increased for a period of about 1 hour to 40°C, the solution was distillatively freed from the solvent and a highly viscous colorless liquid was obtained as the residue.

It was mixed at 90°C first with 1 gram of Na_2HPO_4 and then with ethylene oxide until continuous reflux after about 15 to 30 minutes indicated that ethylene oxide ceased to be absorbed. 1.5 mols of ethylene oxide underwent reaction. Ethylene oxide in excess and resulting ethylene-chlorohydrin were distilled off under vacuum. The distillation residue was the phosphoric acid ester having the formula indicated above. It had an acid number of less than 1 mg KOH/g substance and was obtained in a yield of 97% of the theoretical. The ester was analyzed and the following results were obtained:

	Percent	
	Found	Calculated
P	9.9	10.5
Cl	12.3	12.2

Example 2: In a similar manner, a plasticizer of the formula shown below was made.

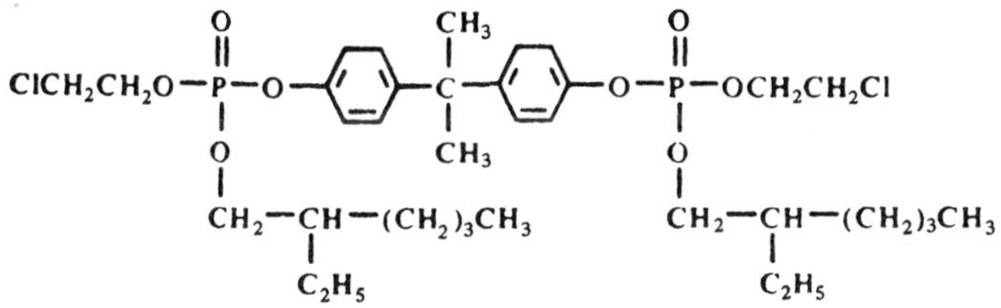

Example 3: The products of this process were tested as to their flame retardant efficiency in polyvinyl chloride, their influence on the notched impact strength of polyvinyl chloride and their influence on the migration of the plasticizer in the plastics. To this end, various polyvinyl chloride specimens, identified as specimens A through E were prepared. Specimen A was free from flame retardant agent, specimens B and C each contained a known flame retardant agent and specimens D and E each contained a flame retardant agent made in accordance with this process. Specimens B through E having the flame retardant agents incorporated in them were composed as follows:

	Parts by Weight
Suspension-made polyvinyl chloride, K-value of 70	70
Blend of 75 wt % of di-n-octyltin mercaptide and 25 wt % of glycerol mono fatty acid ester	1.05
p,p'-Isopropylidene diphenol	0.105
Dioctyl phthalate	22.5
Epoxidized soybean oil	2.5
Flame retardant agent	5.0

As regards composition, specimen A differed from specimens B through E by the fact that it was free from flame retardant agent but contained 27.0 parts by weight of dioctyl phthalate and 3.0 parts by weight of epoxidized soybean oil. The individual powder blends were plasticized on a roller for 10 minutes at 160°C and made into sheets having the thickness necessary for the tests described.

The plasticizer migration in specimens A through E was tested as described in DIN test (German Industrial Standard) No. 59 407, process A. To this end, the specimens were made into round molded plates 1 mm thick and 50 mm in diameter which were embedded in active carbon and stored for 24 hours at 100°C. After further storage for 24 hours at room temperature, the loss in weight in percent, of the individual plates was determined.

The flexibility and freeze resistance of the specimens were identified by determining the notched impact strength at 0° and -10°C, respectively, in accordance with DIN test (German Industrial Standard) No. 53 453. High notched impact strength values mean high flexibility.

The burn-up behavior of the specimens was determined by the small burner test described in DIN test (German Industrial Standard) No. 53 438, which permitted the specimens to be categorized (burning classes K1, K2 or K3) and the individual burning classes to be further identified by burning time and burn-up length. The test results obtained are indicated in the following table.

Specimen	1*	2**	3***	4†	5††	6†††
A	3.3	6.1	2.7	K3/0.2 mm	–	–
B	2.1	3.3	1.9	K3/0.2 mm	–	–
C	3.0	4.0	2.9	K1/0.2 mm	48	140
D	2.2	9.3	2.9	K1/0.2 mm	1	60
E	2.1	3.1	1.9	K1/0.2 mm	1	61

*Migration of plasticizer (DIN 53 407), loss in weight in percent.
**Notched impact strength (DIN 53 453) identified at 0°C in kg cm/cm^2.
***Notched impact strength (DIN 53 453) identified at -10°C in kg cm/cm^2.
†Burning class, sheet specimens 0.2 mm thick.
††Burning period in seconds.
†††Burn-up length in millimeters.

In the above table, the specimen symbols have the following meanings:

A — Polyvinyl chloride free from flame retardant agent
B — Polyvinyl chloride containing tricresyl phosphate
C — Polyvinyl chloride containing liquid chlorinated paraffin (chloroparaffin) with 56 wt % of chlorine therein
D — Polyvinyl chloride containing the product of Example 1
E — Polyvinyl chloride containing the product of Example 2

The results indicated in the above table show that polyvinyl chloride specimens D and E having the flame retardant agents of this process incorporated compare favorably with comparative specimens B and C, and that the initial properties of the specimens remain unaffected by the addition of the products of this process.

Flexible PVC Containing Manganese Borate as Smoke Retardant

D.F. Lawson and E.L. Kay; U.S. Patent 4,035,324; July 12, 1977; assigned to The Firestone Tire & Rubber Company describe a process whereby the addition of manganese borate (MnB_4O_7) to flexible vinyl chloride resin compositions greatly reduces the evolution of smoke when the compositions undergo combustion processes. Experimentation indicates that the smoke inhibiting effect of manganese borate in rigid, unplasticized polymer compositions is less pronounced than in plasticized flexible filled or unfilled compositions. The use of MnB_4O_7 is therefore useful chiefly in plasticized compositions, which may be defined as compositions containing 25 to 150 parts by weight of a plasticizer per 100 parts by weight of the polymer. Any of the plasticizers commonly used to plasticize polyvinyl chloride and other chlorine-containing polymers may be used.

Smoke Suppressant Compositions of Amine Molybdates and Copper Oxalate

A.W. McRowe and W.J. Kroenke; U.S. Patent 4,053,453; October 11, 1977; assigned to The B.F. Goodrich Company describe a process whereby copper oxalate and amine molybdates provide an effective combination in vinyl halide polymer compositions to suppress smoke formation when vinyl halide polymers burn, without attendant adverse effects on the properties of the polymer compositions. The copper oxalate and amine molybdates are normally used in powder form in amounts greater than 1 to about 20 total weight parts per 100 weight parts of vinyl chloride polymers. From about 1 to 10 total weight parts is a very practical range providing a good balance of desirable properties in the compounds. The useful ratio of copper oxalate to amine molybdate is from about 1:10 to 10:1 on a weight basis. The copper oxalate and amine molybdate are readily mixed with the vinyl chloride polymers in powder form as is any other powder ingredient mixed with vinyl chloride polymers.

Cupric oxalate is readily prepared by mixing stoichiometric amounts of hot concentrated solutions of cupric sulfate and oxalic acid. The precipitate is filtered and may be dried in vacuum or at 100°C and atmospheric pressure.

Amine molybdates are produced by reacting an amine with MoO_3, molybdic acid or a molybdenum salt such as ammonium molybdate, ammonium dimolybdate, ammonium heptamolybdate (also known as ammonium paramolybdate), ammonium octamolybdate, sodium molybdate, etc.

This reaction normally is conducted in the presence of water and preferably in the presence of an acid in order to maximize the amine molybdate yield in the aqueous reaction medium. Amines suitable for preparing the organoamine molybdates used include melamine, piperazine and alkyl amines where the alkyl contains 1 to 8 carbons.

In a typical preparation of melamine molybdate, 10 grams of melamine was slurried with 150 ml of water and heated to 60° to 90°C. 13.4 grams of 37% HCl was added to the slurry to dissolve the melamine. A hot solution of 28 grams of ammonium heptamolybdate dissolved in 50 ml of hot water was added to the melamine solution and the reaction mixture heated to reflux for 2 hours. After cooling at room temperature the white precipitate was filtered, washed 4 times with water and vacuum dried. The total yield of melamine molybdate of a molar ratio of 2 mols of molybdenum to 1 mol of melamine was 33.02 grams.

Melamine molybdate with a molar ratio of 1 mol of molybdenum to 1 mol of melamine was similarly prepared with 10 grams of melamine dissolved in 250 ml of water and a solution formed by adding 6.7 grams of 37% HCl solution. A solution of 14 grams of ammonium heptamolybdate was added to the melamine solution and after refluxing, the precipitate was recovered and dried for a total yield of 21.08 grams.

Example 1: To demonstrate the synergistic reduction in smoke formation obtained with the combination of copper oxalate and melamine molybdate, varying amounts of these two materials were used in the following standard compound: 100 weight parts of polyvinyl chloride having a specific viscosity of 0.38 to 0.42 in nitrobenzene at 30°C was dry mixed with 2 weight parts of dibutyltin bisisooctyl thioglycolate, 4 weight parts of a processing aid of a copolymer of a major proportion of methyl methacrylate and lesser amounts of ethyl acrylate, 3.0 weight parts of titanium dioxide, 1.0 weight part of calcium stearate and 1.0 weight part of ethylene bis(stearamide).

The powder mixture was placed on a heated 10 inch 2-roll mill and formed into sheets which were cut into predetermined size samples. To these samples there were then added on a mill the amounts of copper oxalate and melamine molybdate in a mol ratio of 2 mols of Mo to 1 mol of melamine indicated in the table below. 6 x 6 x 0.025 inch test pieces were pressed at 320°F for 3 minutes to 25± mil sheets. 3 x 3 inch samples were tested in the NBS smoke chamber in the flaming mode. The test results in the tables are averages of several runs and the data is NBS smoke chamber flaming mode data. The copper oxalate/melamine molybdate combination is particularly useful and valuable in synergistically providing reduced smoke formation in burning vinyl chloride polymer compositions. This combination does not form objectionable color reaction products with standard tin-sulfur stabilizers.

	Weight Parts		Sample			% Smoke
Run	Copper Oxalate	Melamine Molybdate	Weight grams	thickness mils	D_m/g*	Reduction
1	0	0	5.18	27.1	55.94	—
2	0	2.5	5.84	29.5	28.31	49.39
3	0	5.0	5.63	28.3	27.05	51.65
4	0	10.0	5.57	27.7	25.72	54.02
5	2.5	0	4.95	25.2	29.30	47.60
6	5	0	5.81	29.7	26.04	53.43
7	10	0	6.08	30.2	19.138	65.78
8	1.25	1.25	5.53	27.3	19.03	65.97
9	2.5	2.5	5.64	28.3	13.96	75.05
10	2.5	2.5	5.63	30.7	12.91	76.92
11	1	4.0	6.11	30.7	15.48	72.31
12	1.80	3.20	5.98	29.5	12.51	77.62
13	3.20	1.80	6.54	32.5	13.86	75.22
14	5.0	5.0	5.99	29.0	12.38	77.87
15	7.0	3.0	6.16	29.7	11.75	78.99

*D_m/g = Maximum smoke density per gram of sample

Example 2: To demonstrate the utility of other amine molybdates, a series of compounds were prepared using 2.5 weight parts of copper oxalate and 2.5 weight parts of the amine molybdate in a polyvinyl chloride compound containing 100 weight parts of the PVC of Example 1, 2 weight parts of polyethylene and 2 weight parts of dibutyltinbisisooctyl thioglycolate. Controls containing: (a) 5 weight parts of copper oxalate, (b) 5 weight parts of amine molybdate, and (c) neither were also prepared for comparison purposes. The results obtained, in terms of maximum smoke density and percent smoke reduction are described in the data table below.

Run No.	Amine Molybdate (phr)		Copper Oxalate (phr)	Maximum Smoke Density - D_m/g	% Smoke Reduction
1	Control	0	0	52.87	—
2	—	0	5	20.27	62
3	Ethylamine molybdate	5	0	34.11	35
4	Ethylamine molybdate	2.5	2.5	13.83	74
5	Piperazine molybdate	5	0	24.77	53
6	Piperazine molybdate	2.5	2.5	15.24	71
7	Control	0	0	54.78	—
8	—	0	5	17.76	68
9	Guanidine molybdate	5	0	29.98	45
10	Guanidine molybdate	2.5	2.5	12.16	78
11	Aniline molybdate	5	0	26.16	52
12	Aniline molybdate	2.5	2.5	15.59	72
13	2,2,4-trimethyldecahydroquinoline	5	0	30.01	45

(continued)

Run No.	Amine Molybdate (phr)		Copper Oxalate (phr)	Maximum Smoke Density - D_m/g	% Smoke Reduction
14	2,2,4-trimethyldeca-hydroquinoline	2.5	2.5	16.07	71
15	Pyridine molybdate	5	0	30.04	45
16	Pyridine molybdate	2.5	2.5	14.75	73
17	Hexamethylmelamine molybdate	5	0	22.38	59
18	Hexaethylmelamine molybdate	2.5	2.5	13.34	76

Synergistic Mixture of Melamine Molybdate and Nickel Derivatives

In related work, *W.J. Kroenke; U.S. Patent 4,053,454; October 11, 1977; assigned to The B.F. Goodrich Company* describes smoke retardant vinyl chloride and vinylidene chloride polymer compositions obtained by including a synergistic mixture of: (a) at least one melamine molybdate or substituted melamine molybdate, and (b) one compound selected from the group consisting of nickel oxide, nickel(II) oxalate, nickel(II) acetylacetonate, and their hydrates. Substantial smoke retardation is also obtained by including the above compounds individually in the vinyl chloride or vinylidene chloride polymer compositions.

Examples 1 through 3: The following formulation was used:

Material	Parts by Weight
Polyvinyl chloride*	100
Polyethylene powder	2
Processing aid, dibutyltinbisisooctyl thioglycolate	2
Additive (a)**	Variable
Additive (b)***	Variable

*Homopolymer having an inherent viscosity of about 0.98 to 1.04; ASTM Classification GP-5-15443.

**Melamine molybdate having a 2/1 molybdenum/melamine molar ratio. The control sample contained no additive.

***Compound selected from the group consisting of NiO, nickel(II) oxalate and nickel(II) acetylacetonate. The control sample contained no additive.

Each experimental sample was prepared by milling the formulation materials on a two-roll mill for about 5 minutes at a roll surface temperature of about 320°F. The milled samples were pressed into 6" x 6" x 0.025" sheets. Pressing was done at about 320° to 330°F using 40,000 pounds of force applied to a 4-inch ram. The samples were given a 2 to 5 minute preheat prior to pressing for 8 minutes under full load.

The molded samples were cut into 2⅞" x 2⅞" x 0.025" sections. Testing was performed using the flaming mode of the NBS smoke chamber test (ASTM STP 422, pp. 166-204). Test results are given in the table below.

The test data in the table demonstrate that the additive mixtures are synergistic and substantially retard smoke formation during burning of rigid polyvinyl chloride in the NBS smoke chamber test. The results also demonstrate the smoke retardant effects of individual additives in the same test.

Ex. No.	Additive a (phr)	Additive b (phr)	Maximum Smoke Density per Gram of Sample (D_m/g)	Smoke Reduction (%)
Control	–	–	61.02	–
1	Melamine molybdate (5)	–	28.82	53
	–	NiO (5)	36.20	41
	Melamine molybdate (2.5)	NiO (2.5)	22.54	63
2	Melamine molybdate (5)	–	28.82	53
	–	Ni(II) acetylacetonate (5)	39.71	35
	Melamine molybdate (2.5)	Ni(II) acetylacetonate (2.5)	18.26	70
3	Melamine molybdate (5)	–	28.82	53
	–	Nickel(II) oxalate (5)	41.01	33
	Melamine molybdate (2.5)	Nickel(II) oxalate (2.5)	21.46	65

Melamine Molybdate and Metal Oxides

In a variation of the two component system, *W.J. Kroenke; U.S. Patent 4,055,537; October 25, 1977; assigned to The B.F. Goodrich Company* describes the use of a synergistic mixture of: (a) at least one melamine molybdate or substituted melamine molybdate, and (b) at least one compound selected from the group consisting of Bi_2O_3, CdO, $MnCO_3$, SnO_2, Ta_2O_5, TiO_2, H_2WO_4, ZnO, Zn_2SiO_4, zinc borate, silica and their hydrates.

The additive melamine molybdate and substituted melamine molybdates (hereinafter called amine molybdates), may be polycrystalline or amorphous fine powders, preferably with an average particle size from about 0.01 to about 800 microns. The additives are used in smoke retardant amounts of 0.1 to 20 parts by weight, per 100 parts by weight of polymer. The amount of each additive may constitute 1 to 99% by weight, of the additive compound mixture. Supporting media such as SiO_2, Al_2O_3 and the like may be used for the smoke retardant additives and in many cases are preferred.

The additives were incorporated into polymers and tested as described in U.S. Patent 4,053,454. They were effective in smoke reduction individually and the effect was intensified synergistically when both types of additives were used in the same formulation.

Melamine Molybdate and Iron Derivatives

In another two component system, *W.J. Kroenke; U.S. Patent 4,055,538; October 25, 1977; assigned to The B.F. Goodrich Company* describes the use of iron derivatives with melamine molybdate. The following examples illustrate this work.

Examples 1 through 5: Experimental samples were prepared with the same formulation and procedure used in Examples 1 through 3 of U.S. Patent 4,053,454, above, except that additive (b) was selected from the group consisting of red iron oxide, yellow iron oxide, $FeSO_4$, $Fe_2(SO_4)_3$ hydrate, and iron(III) acetylacetonate. The control sample contained no additive. The experimental samples were prepared as before. Results are shown in the table on the following page.

Ex. No.	Additive a (phr)	Additive b (phr)	Maximum Smoke Density per Gram of Sample (D_m/g)	Smoke Reduction (%)
Control	–	–	54.57	–
1	Melamine molybdate (5)	–	27.46	50
	–	Red iron oxide (5)	30.70	44
	Melamine molybdate (2.5)	Red iron oxide (2.5)	22.60	59
2	Melamine molybdate (5)	–	27.46	50
	–	Yellow iron oxide (5)	29.98	45
	Melamine molybdate (2.5)	Yellow iron oxide (2.5)	24.79	55
3	Melamine molybdate (5)	–	27.46	50
	–	$FeSO_4$ (5)	36.44	33
	Melamine molybdate (2.5)	$FeSO_4$ (2.5)	24.24	56
4	Melamine molybdate (5)	–	27.46	50
	–	$Fe_2(SO_4)_3$ hydrate (5)	29.96	45
	Melamine molybdate (2.5)	$Fe_2(SO_4)_3$ hydrate (2.5)	21.06	61
5	Melamine molybdate (5)	–	27.46	50
	–	Iron(III) acetylacetonate (5)	36.21	34
	Melamine molybdate (2.5)	Iron(III) acetylacetonate (2.5)	25.53	53

Amine Molybdate

In further related work, *W.J. Kroenke; U.S. Patent 4,053,455; October 11, 1977; assigned to The B.F. Goodrich Company* describes the use of amine molybdates as the only additive for vinyl chloride and vinylidene chloride polymers as an effective smoke retardant. The additive amine molybdates used may be polycrystalline or amorphous fine powders, preferably with an average particle size from about 0.1 to about 200 microns. The molybdates are used in smoke retardant amounts, typically from about 1 to about 10 parts by weight, per 100 parts by weight of polymer. Supporting media such as SiO_2, Al_2O_3 and the like may be used for the smoke retardant additives and in many cases are preferred, since additive surface area is increased greatly for smoke reduction purposes.

OTHER THERMOPLASTICS

Resin Compositions for Coaxial Cables

K. Suzuki, T. Nanao and H. Miyazaki; U.S. Patent 4,014,770; March 29; 1977; assigned to Kanegafuchi Kagaku Kogyo KK, Japan describe a foam and flame breakable resin composition comprising a noninflammable agent, an auxiliary noninflammable agent, a foaming agent, an ash coagulating agent and a resin system. This consists essentially of a blend of a copolymer of ethylene and vinyl acetate having a high melt viscosity with the vinyl acetate content being between 40 to 80 weight percent and the molecular weight being more than 200,000, and polyvinyl chloride in a ratio of copolymer to polyvinyl chloride of 2:8 to 7:3. The composition may be further subjected to vulcanization treatment.

The noninflammable agents used include halogen compounds such as chlorinated paraffin and phosphorus compounds, such as ammonium salt of polyphosphoric acid. As the auxiliary noninflammable agent, those compounds which react with halogen and enhance the noninflammability can be used. Among them, diantimony trioxide and metal borate are used. As the ash coagulating agent, such compounds as boron compounds, lead compounds, etc, are used.

As the foaming agent in this process, although it is possible to use an organic or inorganic foaming agent, a necessary condition is to use a foaming agent which decomposes only at high temperature in order that no foaming of the resin compound takes place during manufacture of the insulated wire and cable. Examples of foaming agents include hydrazine sulfate, inorganic foaming agents such as aluminum hydroxide, magnesium carbonate, calcium carbonate, and other inorganic compounds which have a water of crystallization which is liberated at temperature higher than 200°C.

In order to increase the flame breakability of the obtained resin composition, it is effective to add a peroxide to the compound or to irradiate the compound with an electron beam to derive the polymer contained in the resin compound into a three dimensional structure to a certain degree. For the vulcanization of the resin compound by irradiation, the irradiation may be done by an electron beam of 2 to 10 Mrad.

Examples 1 through 7: After preparing 7 different compositions in different proportions, as listed in the table below, by mixing at 160°C for 10 minutes and using mixing roll, test pieces having dimensions of 10 mm width, 3 mm thickness and 150 mm length, were prepared by pressing at 155°C for 7 minutes. The tendency to drop by fusion and the foaming property were tested in the following manner. Each test piece was contacted with a flame of about 12 cm length at an angle of 20°C for 15 seconds. After contact for 15 seconds, the flame was kept away for 15 seconds and again was contacted for 15 seconds. This procedure was repeated until the test piece dropped by fusion or was cut off by burning and the number of times the contact was repeated was recorded. At the same time, the foaming tendency and the residual flame on the test piece were observed. The results are also shown in the table below.

	Example Number						
Composition*	**1**	**2**	**3**	**4**	**5**	**6**	**7****
PVC ($\bar{P}$ is 1,300)							
Kanevinyl S1003	100	80	60	40	20	60	60
EVA (VA is 60 wt %)							
MW ~1,000,000	–	20	40	60	80	40	40
Dioctyl phthalate	60	40	20	–	–	20	20
Tribasic lead sulfate	5	5	5	5	5	5	5
Stearic acid	0.5	0.5	0.5	0.5	0.5	0.5	0.5
Clay 33	30	30	30	30	30	30	30
Diantimony trioxide	5	5	5	5	5	5	5
Foaming agent (hydrazine sulfate)	5	5	5	5	5	5	5
Dicumyl peroxide	–	–	–	–	–	1	–
Total	205.5	185.5	165.5	145.5	145.5	166.5	165.5
Flame resistance***	1.0	2.0	3.0	3.5	3.0	5≤	5≤
Foaming tendency	none	poor	gentle	good	good	very good	very good
Residual flame on test piece	none	none	none	none	present	none	none

*All parts and percentages are by weight.

**Irradiated by electron beam of 5 Mrad (all parts and percentages are by weight).

***Number of repeated contacts with flame.

Example 8: After mixing 60 parts PVC ($\overline{P}$ = 1300), 40 parts of EVA (commercial name Evathlene containing 65 weight percent vinyl acetate and having about 500,000 MW); 25 parts of plasticizer of polyester type; 5 parts of stabilizer of tribasic lead sulfate; 0.5 part of lubricant stearic acid; 10 parts of noninflammable agent chlorinated paraffin; 5 parts of auxiliary noninflammable agent diantimony trioxide; 20 parts ash coagulating agent zinc borate and 5 parts of foaming agent magnesium carbonate, by mixing roll for 15 minutes at 150°C, 2 parts of vulcanizing agent 2,5-dimethyl-2,5-(t-butyl peroxy)-hexyne was added to the compound and the compound was again mixed for 8 minutes by mixing roll and then prepared into a pellet. All parts and percentages are by weight.

Using this pellet, a coaxial cable of a construction such as shown in Figure 8.1 was made. The thickness of the insulating layer of polyethylene **4** was 1.3 mm; and the sheathing layer **1** was 0.9 mm. The cable was made by an extrusion process.

FIGURE 8.1: COAXIAL CABLE

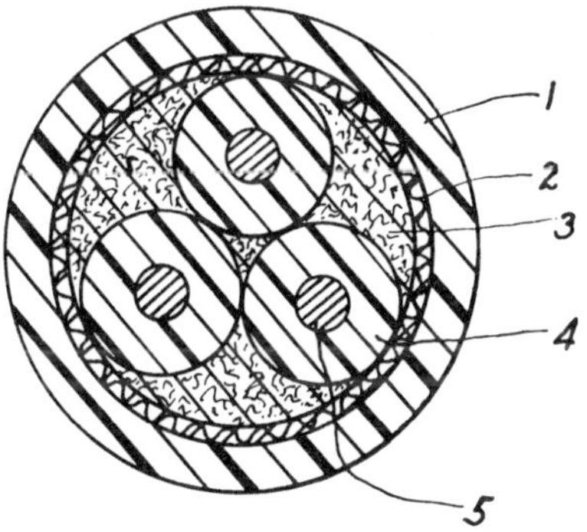

Source: U.S. Patent 4,014,770

Turning to Figure 8.1, there are depicted in cross section, three conductors of 1 mm diameter, each containing 7 twisted wires, arranged parallel to each other and to the axis, and each covered by polyethylene insulators **4**. The conductors are numbered **5**. Filled between the insulator and the outer sheath **1**, are jute intermediate covering **3** and cloth layer **2** in the manner depicted.

Each test piece of the cable of about 50 cm of length was fixed perpendicularly and contacted with a flame of a bunsen burner where the flame was adjusted so as to have a reducing flame of about 4 cm and an oxidizing flame of about 7 cm. Upon contact, the top of the reducing flame was held just touching the cable with an inclination of 20°C for 15 seconds. After contact for 15 seconds, the flame was held away from the sample for 15 seconds and again contacted for 15 seconds. After repeating the flame contacting procedure 5 times, the duration of contact of a residual flame on the cable was observed. In this test, when the duration of contact of a residual flame on the cable was longer than 15 seconds, the flame of the Bunsen burner was brought to the cable to contact with it just after the disappearance of the flame on the cable. The observed results showed that the duration of a residual flame on the cable was zero.

Comparative Example: After mixing 100 parts of PVC ($\overline{P}$ = 1300); 60 parts of plasticizer DOP (dioctyl phthalate); 5 parts stabilizer tribasic lead sulfate; 0.2 part of lubricator stearic acid and 40 parts of filler calcium carbonate, by a mixing roll for 10 minutes at 150°C, a pellet was prepared. Using the pellet, an insulated cable was prepared as in Example 8 and the duration of residual flame on the cable was observed. As a result of this test, it was proven that the duration of residual flame on the cable in this example was as long as 150 seconds.

Honeycomb-Laminate Composite Structure

W.J. Gilwee, Jr. and J.A. Parker; U.S. Patent 4,061,812; December 6, 1977; assigned to NASA describe a honeycomb-laminate composite structure comprising a cellular core of a polyquinoxaline foam in a honeycomb structure, and a layer of a noncombustible fibrous material impregnated with a polyimide resin laminated on the cellular core.

Suitable honeycomb structures which can be employed include those made of polyamide and a combination of a polyimide and glass. A preferred honeycomb material is one of a polyamide because of its low weight, e.g., about 1.5 lb/ft^3 (24 kg/m^3) and ready commercial availability. Honeycomb structures can be easily produced using techniques as disclosed in the *Encyclopedia of Polymer Science & Technology,* vol. 7, p. 513. Examples of suitable materials which can be used to form the polyquinoxaline foam in the cells of the honeycomb structure include ammonium 4-nitroaniline-2-sulfonate, p-nitroaniline bisulfate, and the reaction product of p-benzoquinone dioxime and phosphoric acid.

The compound capable of forming the polyquinoxaline foam can be placed in the cells of the honeycomb structure as a finely divided powder. A suitable particle size for the powder can be that which will pass a 65 mesh (U.S. standard screen) and be retained on a 270 mesh screen, preferably pass a 100 screen and be retained on a 150 mesh screen. Particle sizes within this range can be easily achieved by grinding and sieving to obtain particles within the above described range.

Once the cells of the honeycomb structure have been filled with the compound capable of forming the polyquinoxaline foam, the compound is thermally polymerized to form the polymeric material and also to form the foam. Suitable heating temperatures which can be employed in order to achieve the thermal polymerization can range from about 270° to about 300°C. Upon completion, the honeycomb structure is filled with a black foam material having a low density, in general ranging from about 1.5 to about 5 lb/ft^3.

Once the honeycomb structure described above has been filled with the polyquinoxaline foam, generated as described above, the honeycomb structure containing the foam can then be machined, if desired, to remove any surface irregularities, cut to an appropriate size and subsequently one or more layers or plies of a noncombustible fibrous material impregnated with a polyimide resin can be laminated on it.

Suitable noncombustible fibrous materials which can be employed include glass fibers, carbon fibers, graphite fibers, boron fibers, silicon carbide fibers, asbestos fibers, steel fibers, and other metallic fibers. Glass fibers are preferred from the standpoint of cost, weight, and ease of processing. The noncombustible fibrous material can be employed in the form of a woven or nonwoven material.

The fibrous material is impregnated with a polyimide resin. Suitable examples of polyimide resins which can be employed include bismaleimide resins, for example, a bismaleimido resin produced by the polymerization of 4,4'-bismaleimido-diphenylmethane monomer as disclosed in W.J. Gilwee et al, "High Temperature Resistant Composites for Bismaleimide Resin," *Proceedings of the 18th National SAMPE Symposium, SAMPE,* Los Angeles, California, April 1973.

The honeycomb core-laminate composite structures are useful as lightweight panels or structural members in aircraft, watercraft and housing and commercial structural applications. Figure 8.2a shows an exploded view of the honeycomb core-laminate composite structure of the process, in which **1** is the honeycomb structure formed of hexagonal cells **2**, **3** shows the polyquinoxaline foam present in the hexagonally shaped cells of the honeycomb structure **1** and **4** represents a layer of a noncombustible fibrous material impregnated with a polyimide resin laminated on the foam-filled honeycomb structure.

Figures 8.2b and 8.2c show an exploded view of an article containing a honeycomb-laminate composite structure and an article for comparison containing a different honeycomb-laminate composite as described in the example, respectively. Figure 8.2d shows the thermal insulation characterics of the articles shown in Figure 8.2b as described in the example below.

FIGURE 8.2: HONEYCOMB-LAMINATE COMPOSITE STRUCTURE

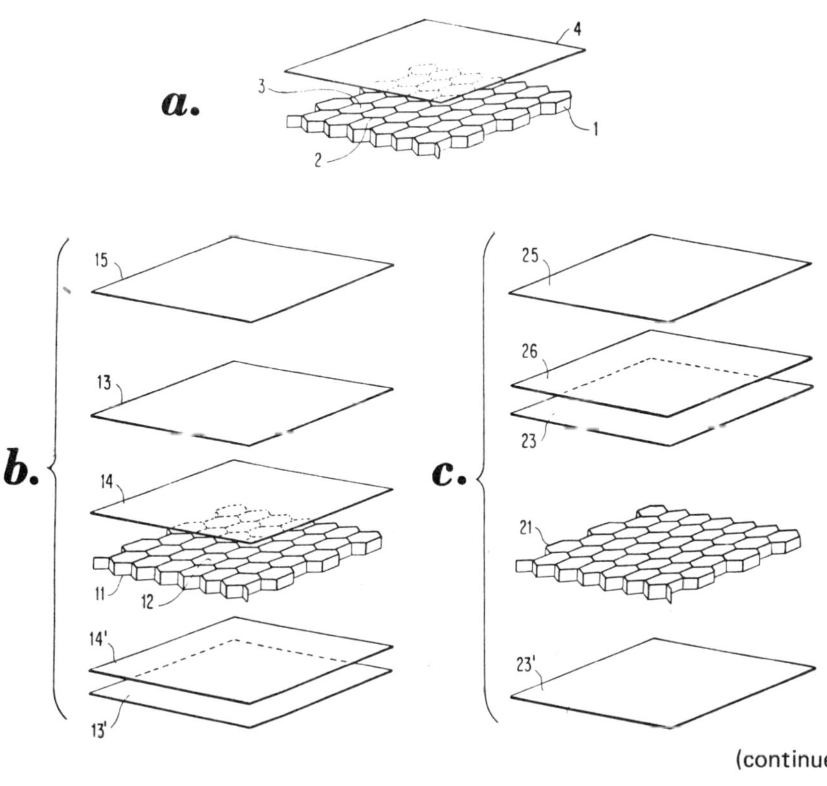

(continued)

FIGURE 8.2: (continued)

d.

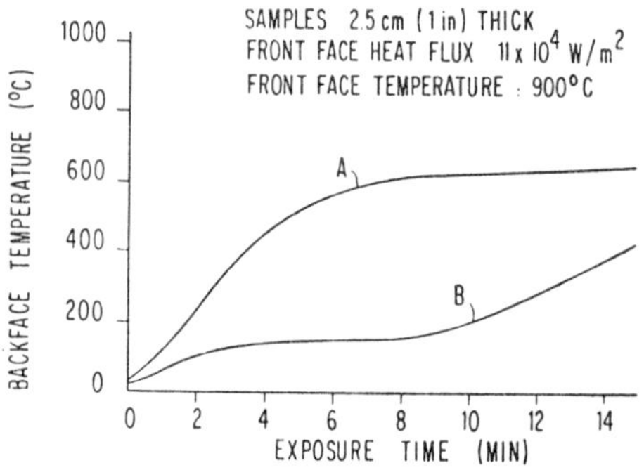

Source: U.S. Patent 4,061,812

Example: Ammonium 4-nitroaniline-2-sulfonate was prepared according to the procedure described in the *Journal of Paint Technology,* vol. 44, No. 571, p. 51, August 1972, and was then converted into a coating composition using the following formulation:

Component	Parts by Weight
Ammonium 4-nitroaniline-2-sulfonate	23
Nitrocellulose*	7
Methyl ethyl ketone (18%)	34
Toluene	18
Ethyl acetate	18

*5 to 6 sec viscosity ASTM D 301-56; 11.8 to 12.2 wt % nitrogen; 30 wt % solids in methanol.

This coating was coated on a polyamide honeycomb structure comprising cells of a hexagonal configuration generally having a size of about 0.32 cm and allowed to dry in the air for about 2 to 24 hours.

In coating the coating composition described above onto the honeycomb structure, a wire screen was placed on either side of the honeycomb structure panel and three layers of a glass cloth, style 181 were placed next to each wire screen. This assembly was then restrained between heated plattens and heated to 280°C whereby the ammonium 4-nitroaniline-2-sulfonate expanded, foamed and filled the cellular voids in the honeycomb structure. The foaming reaction was completed in about 1 to 5 minutes, and the evolution of gas, which was allowed to escape, ceased within about 10 minutes. The density of the polymeric foam formed in the honeycomb structure was between about 1.0 to 2.0 lb/ft^3 (16 to 32 kg/m^3).

Once the honeycomb structure described above and filled with the polymeric foam had been produced, the layers of glass cloth on either side of the wire screen and the wire screen on either side were removed and a prepreg comprising a glass cloth of woven glass fibers (coating style 181, A1100 silane finish) was laminated on it. The prepreg was produced by coating the woven glass with a varnish of 4,4'-bismaleimide-diphenylmethane. This varnish was prepared by dissolving the 4,4'-bismaleimide-diphenylmethane in 1-methyl-2-pyrrolidinone at a 40% by weight monomer solids concentration. The viscosity of this coating varnish at 23°C was 50 cp. After coating the varnish as described above onto the glass cloth, the glass cloth was dried for 15 minutes at 79°C and partially polymerized (staged) for 20 minutes at 150°C. The prepreg was found to have a volatile content of 0.02% by weight, a resin flow of 15% and a resin content of 40% by weight. After staging, the prepreg was found to have substantially no retained solvent. The resin flow is a value calculated on the amount of resin which flows under pressure and is obtained on dividing the weight of the laminate (plies cured together) with flash minus the weight without the flash (i.e., the amount of polymer resin which flows out from the assembly of prepregs) by the weight of the laminate with the flash minus the weight of the glass.

After the prepreg was prepared as described above, the prepreg was cut into a suitable size, stacked, and pressed between aluminum plates with a Teflon (0.127 millimeter) film release sheet. The cure of the prepreg was performed in a flat platten press and the assembly was placed in a cold press and heated at a rate of 3°C/min at 689 kN/m^2 (100 psi) to 240°C and held at 240°C for 2 hours. The prepreg thickness comprised 1 to 3 plies.

This prepreg was then adhered to one surface of the honeycomb core-structure produced above using a film or liquid adhesive, e.g., a film adhesive such as FM-34 or a liquid adhesive such as Kerimid 500, which was placed between the honeycomb and the prepreg face sheet. A suitable amount of the film adhesive ranged between 0.02 and 0.06 lb/ft^2 (0.1 to 0.3 kg/m^2), preferably 0.025 to 0.035 lb/ft^2 (0.12 to 0.17 kg/m^2) and a suitable amount of the liquid adhesive was 0.2 to 0.5 g/in^2 (0.03 to 0.08 g/cm^2), preferably 0.3 to 0.4 g/in^2 (0.05 to 0.06 g/cm^2). The adhesive layer was then bonded at 180° to 200°C, preferably 180° to 185°C, and under a pressure of 1 to 50 lb/in^2(6.9 to 344 kN/m^2), preferably 40 to 50 lb/in^2 (275 to 344 kN/m^2).

Comparison – As described in Figure 8.2b, a polyamide honeycomb of hexagonal cells **11** and with a cell size of 0.312 cm was coated with a composition of the formulation and with the properties shown in the tables below.

Components, wt %	
Methyl ethyl ketone	32.8
Nitrocellulose*	6.9
p-nitroaniline	27.5
Ethyl alcohol	13.1
Sulfuric acid	19.7
Properties of Coating Composition	
Nonvolatile (wt %)	54
Density (g/cm^3)	1.09
Viscosity (Brookfield, 25°C cp)	1300
Fineness of grind (Hegeman Scale, ASTMD-1210)	4+
p-nitroaniline + sulfuric acid/nitrocellulose ratio (by weight)	6.9
p-nitroaniline/sulfuric acid ratio (by weight)	1.4

*As described in the Example.

The amount (dry basis) of the coating composition used was approximately 0.20 to 0.23 g/cm^2 (honeycomb face area). After coating, the coating was dried at 100°C for 1 hour. The coated honeycomb was heated at a temperature of 316° to 371°C to form the polyquinoxaline foam **12**, and any excess amount of foam was removed from the faces of the honeycomb.

The bismaleimide used in producing the prepreg to be laminated on each side of the honeycomb core was a hot melt, addition type polyimide, Kerimid 353. The prepregs **13, 13'** were prepared from this resin and 1 to 4 pieces of 181E-glass cloth were used to form flat laminates. Softening of the polyimide resin occurred at 70°C and the resin became tacky at 50°C.

Laminates are prepared by placing them in a rubber bagging and cured in an autoclave at a heating rate of 1° to 2°C/min to 180°C under a vacuum of 29 to 30 in Hg (95 to 102 kN/m^2) and with an applied pressure of 690 kN/m^2. The laminates were cured for 1 hour at 180°C and adhered to the polyquinoxaline-filled honeycomb structure with a polyimide resin **14, 14'** (Kerimid 501), using contact pressure in a heated press at 170°C for 2 hours. After production of the honeycomb prelaminate composite, a phenolphthalein polycarbonate film **15** was adhered to the surface of the composite with a chlorendic anhydride-epoxy adhesive and bonded by applying a contact pressure in a mold at 150°C.

For comparison, a conventional honeycomb-laminate composite structure was produced using an unfilled polyamide, hexagonal-cell honeycomb structure as described above, **21**. An epoxy resin-preimpregnated prepreg of glass fibers (Type 120E) previously bonded to a glass fiber (181E glass/polyvinyl fluoride laminate with polymethylmethacrylate), **23, 23'**, was bonded to each side of the honeycomb core structure. A polyvinyl fluoride film layer aesthetically decorated, **26**, was bonded onto the surface of one of the prepregs **23** and a polyvinyl fluoride film **25** was further adhered thereon with a polymethylmethacrylate adhesive. Curing was accomplished at 110°C with 50 cm Hg minimum vacuum pressure.

Exploded views of the composite structures of this process and for comparison produced as described above are shown schematically in Figures 8.2b and 8.2c, respectively. Typical flammability and smoke characteristics of the composites produced as described above are shown in the table below.

Material	Amount (wt %)	Char Yield (N_2, 700°C) (%)	Polymer Decomposition Temperature N_2 (°C)	Limiting Oxygen Index $O_2/(N_2 + O_2)$ (ASTM D-2863)
Process				
Phenolphthalein polycarbonate	5.0	58	425	38
Glass–SiO_2	30.0	100	–	>70
Bismaleimide	30.0	45	430	46
Polyamide–poly(m-phenylene isophthalamide)	20.5	48	420	44
Polyquinoxaline	14.5	90	530	48

(continued)

Material	Amount (wt %)	Char Yield (N_2, 700°C) (%)	Polymer Decomposition Temperature N_2 (°C)	Limiting Oxygen Index $O_2/(N_2 + O_2)$ (ASTM D-2863)
Comparison				
Polyvinyl fluoride	7.6	10	400	20
Glass–SiO_2	419	100	–	>70
Epoxy–4,4'-isopropylidene-diphenol (bisphenol A)	30.0	23	180	21
Polyamide–poly(m-phenylene isophthalamide)	20.5	48	420	48

*Molecular weights of the polymers used are from 1,000 to 100,000.

Properties	Composite Structure: Comparison		Composite Structure: Process	
Density (ASTM D-71) (kg/m³)		96		96-113
Tensile strength flatwise (ASTM C307, 24°C) (kN/m²)		610		517-690
Fire endurance (time to reach back face temperature of 204°C with front face heat flux of 11 x 10^4W/m²) (min)		2		10
Smoke density (NBS, Ds 4 min) specific optical density		8.7		45
Component smoke density (maximum specific optical density (NBS smoke chamber)	polyvinyl fluoride	7	phenol polycarbonate	5
	polyamide core	2	polyamide core	2
	epoxy resin	90	bismaleimide resin	9
	glass	–		–

From the above, it can be seen that at equivalent density the fire containment or fire endurance capability of the composite structure has been improved by a factor of five in comparison with the comparison composite structure. This improvement is believed to be due to the thermal protection efficiency of the high-char-yield polyquinoxaline foam filling the honeycomb of the composite structure of this process. It can also be seen from the above that the composite structure of this process can result in a significant reduction in the specific optical smoke density when tested in the NBS smoke chamber. Examination of the smoke density values for the constituent components of these composites clearly shows that the principal contributor to smoking is the epoxy-laminating resin in the comparative composite structure. The bismaleimide resin replacing the epoxy effectively reduces the total smoke produced from the composite structure of this process.

Figure 8.2d shows the fire endurance of containment capability of the comparative composite structure (curve A) and process composite structure (curve B) under conditions of the NASA-Ames T-3 test conducted as disclosed in "The Performance of Lightweight Plastic Foams Developed for Fire Safety," Proceedings of the NASA Conference on Materials for Improved Fire Safety, May 1970. In Figure 8.2d, the backface temperature increase is shown as a function of the time in minutes when the composite structure is subjected to a front face heat flux of 11 x 10^4 W/m^2. It can be seen that the backface temperature of the conventional composite structure reaches 200°C in 2 minutes whereas it takes as long as 9 minutes to reach a comparable backface temperature in the composite structure of this process. In addition, the composite structure exhibited less distortion and better structural integrity after testing than the comparative composite structure.

Polyphenylene Ether and EPDM Modified Alkenyl Aromatic Resin Compositions

G.D. Cooper and V. Abolins; U.S. Patent 3,974,235; August 10, 1976; assigned to General Electric Company describe a thermoplastic molding composition. This comprises 20 to 65% by weight of a polyphenylene ether resin; 35 to 80% by weight of an alkenyl aromatic resin that is modified with a rubbery interpolymer of a mixture of monoolefins and a polyene by polymerizing an alkenyl aromatic monomer in the presence of the rubbery interpolymer; and a flame retardant amount of a flame retardant additive. These rubbery interpolymers include those prepared from ethylene, an alpha-olefin and a polyene. Preferred types comprise 10 to 90 mol percent of ethylene, 10 to 90 mol percent of an alpha-olefin containing 3 to 16 carbons and 0.1 to 10 mol percent of a polyene that is a nonconjugated cyclic or open-chain diene that has from 4 to 20 carbons. The alkenyl aromatic resins are well known and are derived from monomers of the formula:

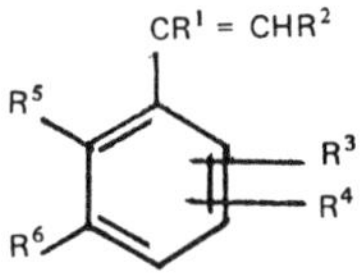

where R^1 and R^2 are selected from the group consisting of lower alkyl or alkenyl groups of from 1 to 6 carbons and hydrogen; R^3 and R^4 are selected from the group consisting of chloro, bromo, hydrogen and lower alkyl of from 1 to 6 carbons; R^5 and R^6 are selected from the group consisting of hydrogen and lower alkyl and alkenyl groups of from 1 to 6 carbons or R^5 and R^6 may be concatenated together with hydrocarbyl groups to form a naphthyl group. These compounds are free of any substituent that has a tertiary carbon atom. The polyphenylene ethers are of the formula:

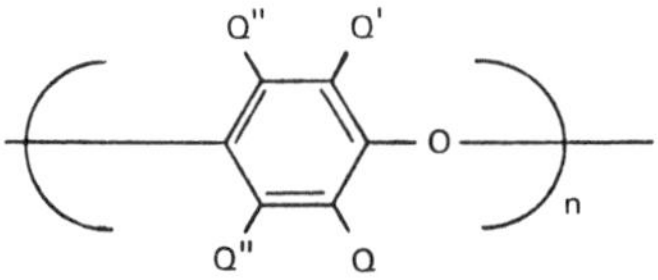

where Q is selected from the group consisting of hydrogen, hydrocarbon radicals, halohydrocarbon radicals having at least 2 carbons between the halogen atom and the phenol nucleus, hydrocarbonoxy radicals and halohydrocarbonoxy radicals having at least 2 carbons between the halogen atoms and the phenol nucleus, Q' and Q" are the same as Q and in addition, halogen with the proviso that Q and Q' are all free of a tertiary carbon atom; and n is an integer of at least 50.

It is preferred that the flame retardant additive comprise a halogenated organic compound (brominated or chlorinated); a halogen-containing organic compound in a mixture with antimony oxide; elemental phosphorus or a phosphorus compound; a halogen-containing compound in a mixture with a phosphorus compound or compounds containing phosphorus-nitrogen bonds or a mixture of two or more of the foregoing. The resinous compositions of this process find applications in molded articles. The following examples illustrate the process.

Example 1: A blend having the following composition was prepared:

	Grams
Poly(2,6-dimethyl-1,4-phenylene)ether (PPO, General Electric Co., I.V. about 0.52 dl/g measured in chloroform at 30°C)	35
Polystyrene modified with a terpolymer of ethylene, propylene and 5-ethylidene-2-norbornene, having a Mooney viscosity of 140 (ML-8)*	65

*Prepared according to Example 8 of U.S. Patent 3,538,192.

This blend is prepared with a twin screw extruder at 585°F and the extruded pellets were molded into standard test pieces with a Newbury injection molding machine, barrel temperature 450°F.

Example 2: A blend having the following composition was prepared:

	Grams
Poly(2,6-dimethyl-1,4-phenylene)ether*	700.0
Polystyrene resin**	1,300.0
Triphenyl phosphate	140.0
Tridecyl phosphate	10.0
Polyethylene (Microthene M-710)	30.0
Titanium dioxide	100.0

*Same as described in Example 1.
**Modified with a rubbery interpolymer of a mixture of mono-olefins and a polyene (Taflite 925-01, Mitsui Toatsu Chemical Co.)

The blend was prepared with a twin screw extruder at 585°F and the extruded pellets were molded into standard test pieces such as tensile bars with a Newbury injection molding machine, barrel temperature 450°F.

Example 3: The following composition was prepared according to the procedure set forth in Example 2:

	Grams
Poly(2,6-dimethyl-1,4-phenylene)ether (as used in Example 2)	1,000.0
Polystyrene resin (as used in Example 2)	1,000.0
Triphenyl phosphate	60.0
Tridecyl phosphate	10.0
Zinc sulfide	3.0
Zinc oxide	3.0
Polyethylene (Microthene M-710)	30.0
Titanium dioxide	100.0

Control Specimens – A first control specimen was prepared according to the method of Example 2 except that a polybutadiene modified, high-impact polystyrene (Foster Grant 834) was substituted for the polystyrene that was modified with a rubbery interpolymer of a mixture of monoolefins and a polyene.

A second control specimen was prepared according to the method of Example 3 except that a polybutadiene modified, high-impact polystyrene (Foster Grant 834) was substituted for the polystyrene that was modified with a rubber interpolymer of a mixture of monoolefins and a polyene.

A test specimen prepared in Example 2 and the first control specimen were heat aged in an air oven at 105°C after initial physical properties were measured. At periodic intervals the specimens were removed and tested on an Instron machine at a speed of one inch per minute. The tensile bars of Example 3 and the second control specimen were processed similarly except that the oven temperature was 115°C. The results were as follows:

Specimen	Alkenyl Resin	Weight Ratio (PPO:AR)	HDT (°C)	TS (psi)	Elongation (%)	Time to Embrittlement, days at 105°C	at 115°C
Example 2	rubbery interpolymer modified polystyrene	35:65	202	7,000	36	100	–
First control	polybutadiene modified polystyrene	35:65	208	7,200	73	20	–
Example 3	rubbery interpolymer modified polystyrene	50:50	250	8,300	22	–	90-100
Second control	polybutadiene modified polystyrene	50:50	252	8,500	58	–	25-28

It can be seen that the compositions at the 50:50 level of Example 3 show a four-fold increase in the time to embrittlement as compared to the control specimen. The improvement was even greater for the 35:65 blends. The test bars according to this process also showed much less darkening on exposure to heat-aging than did the prior control specimens.

Resin Compositions of Polysulfones and Polyphosphonates

N. Okada and T. Iwabuchi; U.S. Patent 4,065,517; December 27, 1977; assigned to Nissan Chemical Industries Ltd., Japan describe a polysulfone composition comprising 1 to 20 wt % of polyphosphonate having the formula:

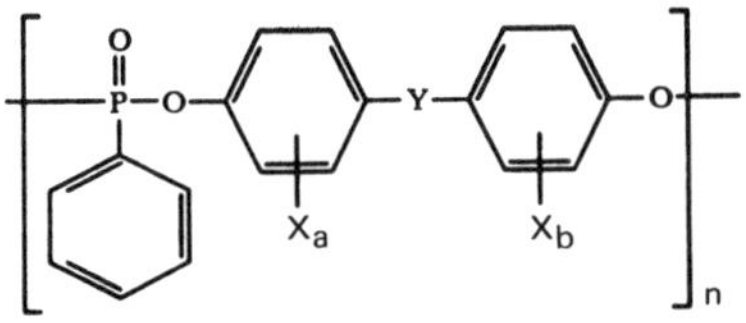

where X represents Br or Cl; Y represents a C_{1-8} alkylidene group or sulfone group; a and b are respectively integers of 1 to 4 and n is an integer of 2 to 100. These polysulfones have high heat resistance and excellent mechanical and electrical properties and have been used for various construction materials as the engineering plastics. In the examples which illustrate the process the evaluation of flameproof property was conducted by the oxygen index method stated in ASTM D-2863-70. The tensile strength was measured by ASTM D-638. The dielectric loss tangent and the dielectric constant were measured by ASTM D-150.

Example 1: (a) Preparation of Polyphosphonate – In a 500 ml four-necked flask equipped with a stirrer, a thermometer and a dropping funnel, 250 ml of chloroform, 54.4 grams of 2,2-bis(3,5-dibromo-4-hydroxyphenyl)propane and 24.2 grams of triethylamine were charged.

A mixture of 50 ml of chloroform and 19.5 grams of phenylphosphonic dichloride was added dropwise to the mixture in the flask with stirring under cooling with ice. The reaction was conducted at 50°C for 2 hours and then the reaction mixture was cooled and was added dropwise to 3 liters of methanol to form a precipitate. The precipitate was washed with methanol and dried to obtain the polyphosphonate (A) having the following formula at a yield of 91%.

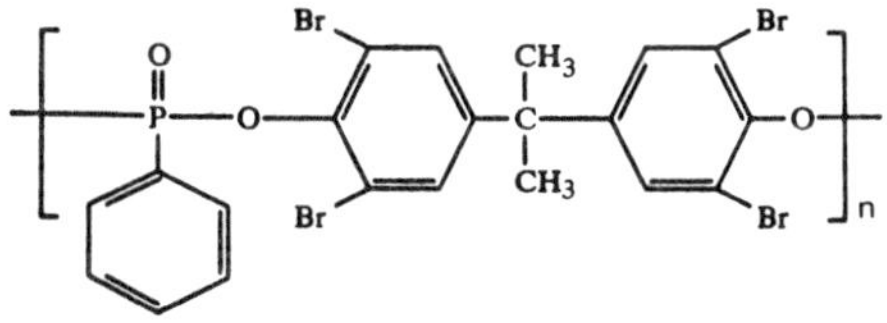

According to the capillary method, the melting point of the polyphosphonate (A) was 228° to 232°C. The viscosity $[\eta]$ of the polyphosphonate in a mixture of 3:2 of phenol to 1,1,2,2,-tetrachloroethane at 30°C was 0.276.

(b) Preparation of Polysulfone Composition – The polysulfone having the formula below where n = 65, was crushed and dried at 120°C for 6 hours under a reduced pressure.

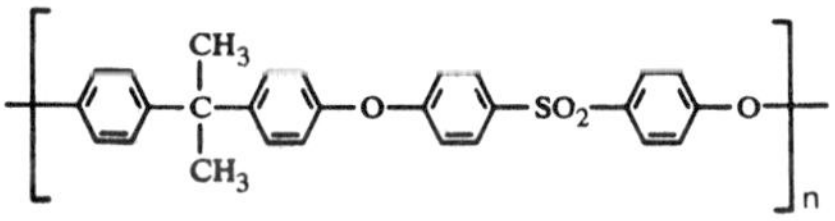

The polyphosphonate (A) produced in the step (a) was mixed with the polysulfone at each ratio stated in the table below. Each mixture was kneaded at 270° to 280°C for 5 minutes with a roll mill to obtain polysulfone compositions (1) and (2).

(c) Tests of Flameproof Property and Physical Properties – The polysulfone compositions (1) and (2) and the polysulfone without polyphosphonate were respectively used to prepare each sheet by a press-molding machine. The sheet was cooled and each test sample was cut from the molded sheet. According to the abovementioned test methods using the test samples, the oxygen index, the tensile strength, the elongation, the dielectric loss tangent and the dielectric constant were measured. The results are shown in the table below.

The polysulfone composition containing polyphosphonate (A) had a higher oxygen index and higher flameproof property than those of the polysulfone. The tensile strength, the dielectric loss tangent and the dielectric constant were substantially the same and no coloring and no transparency change were found.

Example 2: (a) Preparation of Polyphosphonate – In accordance with the process of Example 1 except using phenylphosphonic dichloride and 3,3',5,5'-tetrabromo-4,4'-dihydroxydiphenylsulfone, the reaction was conducted to obtain the polyphosphonate (B) having the formula below at a yield of 80%. The polyphosphonate had a melting point of 150° to 155°C and [η] of 0.05 dl/g.

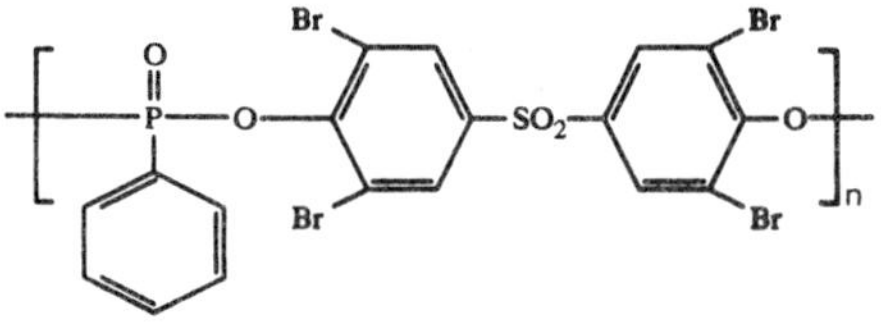

(b) Preparation of Polysulfone Composition – In accordance with the process of Example 1 except using the polyphosphonate (B), the polysulfone compositions (3) and (4) having the ratio stated in the table below were prepared.

(c) Tests of Flameproof Property and Physical Properties – In accordance with the process of Example 1 (c) except using the polysulfone compositions (3) and (4), each test sample was prepared and the flameproof property and the physical properties were tested. The results are shown in the table below. The polysulfone compositions having high strength and improved flameproof property were obtained by the addition of the polyphosphonate (B).

Composition No.	—	(1)	(2)	(3)	(4)
Polyphosphonate					
Type	—	(A)	(A)	(B)	(B)
Amount (wt.%)	0	8	15	8	15
Oxygen index	30.0	40.0	45.0	35.5	41.0
Tensile strength	748	753	686	730	658
(Kg/cm^2)	748	753	686	730	658
Elongation in yield (%)	5.9	6.0	4.8 —	—	
Dielectric loss tangent	0.0008	0.0006	0.0013	—	—
25° C 10^3 cp					
Dielectric constant					
25° C 10^3 cp	2.97	3,01	3.01	—	—

Resinous Polymer of Benzene Phosphorus Oxydichloride, 4,4'-Thiodiphenol and $POCl_3$

K.S. Kim and S. Altscher; U.S. Patent 4,039,512; August 2, 1977; assigned to Stauffer Chemical Company describe flame retardant polymers of benzene phosphorus oxydichloride, 4,4'-thiodiphenol and $POCl_3$. They are formed by reacting these monomers in an inert organic solvent at ambient temperature in the presence of an amine acid acceptor. The copolymers of this process can be used as materials for the production of nonflammable fibers, molding compositions and the like. The following examples illustrate the process.

Example 1: This example describes a procedure for forming a copolymer of benzene phosphorus oxydichloride, 4,4'-thiodiphenol and $POCl_3$ (hereinafter called BPOD/TDP/$POCl_3$ copolymer). A three-necked flask with a mechanical stirrer, addition funnel and condenser were used. The following reagents were charged into the flask: 89.8 grams (0.412 mol) 4,4'-thiodiphenol; 88.8 grams (0.88 mol) triethylamine; and 600 ml methylene chloride.

The following reagents were placed in the addition funnel: 78.0 grams (0.40 mol) benzene phosphorus oxydichloride; 1.22 grams (0.008 mol) $POCl_3$; and 250 ml methylene chloride.

The benzene phosphorus oxydichloride was slowly added to the flask over a period of 20 to 30 minutes with rapid stirring at room temperature. The stirring was continued for one hour after the addition had been completed, and the resulting solution was washed with an equal volume of 1 N hydrochloric acid and an equal volume of distilled water. The solution was added to methanol to yield the desired product as a precipitate. The precipitate was removed from solution and was vacuum dried.

The relative viscosity of the product was 1.92 when measured as a 1% by weight solution in a 60/40 phenol/tetrachloroethane mixture. The yield of product was 90%. It had an intrinsic viscosity of 0.65 and a melting point of 180°C.

Example 2: The same procedure as used in Example 1 was repeated to form a copolymer of benzene phosphorus oxydichloride, 4,4'-sulfonyldiphenol and $POCl_3$, as described in U.S. Patent 2,716,101 by H.W. Coover Jr., et al. This polymer is abbreviated BPOD/SDP/$POCl_3$ copolymer. The reagents in the flask are: 65.38 grams (0.2575 mol) 4,4'-sulfonyl diphenol; 55.5 grams (0.55 mol) triethylamine; and 400 ml methylene chloride.

The reagents in the addition funnel are: 48.75 grams (0.25 mol) benzene phosphorus oxydichloride; 0.77 gram (0.005 mol) $POCl_3$; and 100 ml methylene chloride. The yield of product was 94%, it had an intrinsic viscosity of 0.42 and a melting point of 185° to 190°C.

Example 3: Examples 1 and 2 were repeated using a higher amount of $POCl_3$ so as to yield copolymers having a 2 mol % content of $POCl_3$ based on the molar sum of benzene phosphorus oxydichloride, the diphenol and $POCl_3$. The amount in Examples 1 and 2 was 1%.

The copolymer products in this example were tested for flammability using the Limiting Oxygen Index (LOI) ASTM D-2863 test. This method directly relates to the minimum percentage concentration of oxygen in an oxygen-nitrogen mixture which permits a sample to burn. A high LOI is indicative of a higher degree of flame retardancy. The following LOI values were observed and demonstrate the greater flame retardancy for the product of this process.

Sample	LOI	MP (°C)	T_g (°C)
BPOD/TDP/$POCl_3$ copolymer	60	180	88
BPOD/SDP/$POCl_3$ copolymer	50	190	151

The polymers of this process have an intrinsic viscosity of from about 0.30 to about 1.1 when measured in a 60/40 mixture (by weight) of phenol and tetrachloroethane at 30°C. A preferred intrinsic viscosity is greater than about 0.35 and less than about 0.7.

Poly(Phenylene Oxide) Plastic Compositions

A.L. Anderson; U.S. Patent 3,971,762; July 27, 1976; assigned to Michigan Chemical Corporation describes poly(phenylene oxide) plastic compositions which are flame retardant. This can be accomplished by the incorporation of bisphenoxy compounds in PPO to subsequently provide compositions which exhibit outstanding chemical, physical and mechanical properties. The bisphenoxy compounds used in the composition have the formula:

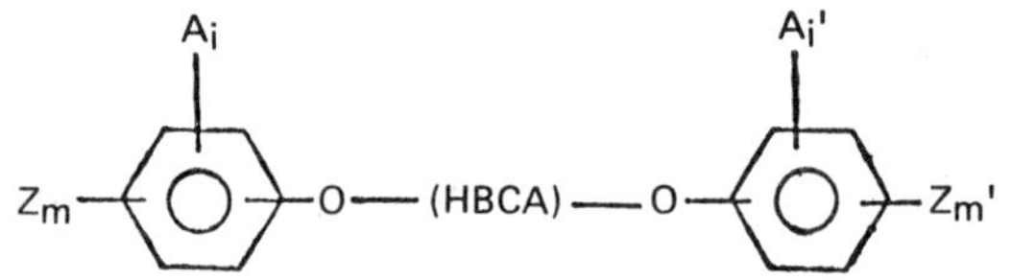

In the formula, Z is bromine or chlorine; m and m' are integers each independently having a value of 1 to 5; i and i' are integers each independently having a value of 0 to 2; HBCA is a halo-branched chain alkylene group having from 1 to 6 carbons (e.g., $CH_2C(CH_2Cl)H$; $CH_2C(CH_2Cl)_2$; $CH_2CH_2C(CH_2Br)H$; and $CH_2C(CH_2Cl)HCH_2$; and A is from the group cyano (—CN), nitro ($—NO_2$), lower alkoxy (e.g., $—OCH_3$, OC_2H_5), lower alkyl (e.g., CH_3, C_2H_5, C_3H_7, C_4H_9), fluorine, dialkylamino, e.g., $—N(CH_3)_2$ or $—N(C_2H_5)_2$, phenyl ($—C_6H_5$), halo-phenyl, benzyl ($—CH_2C_6H_5$), and halo-benzyl. In the formula, the sum of i + m or i' + m' is not greater than 5. Illustrations of some of the bisphenoxy compounds are shown below.

Compound No.	Z	m	m'	A	i	i'	HBCA
1	Br	3	3	–	0	0	$CH_2C(CH_2Cl)H$
2	Br	3	3	–	0	0	$CH_2C(CH_2Cl)_2$
3	Br	5	5	–	0	0	$CH_2C(CH_2Cl)H$
4	Br	2	2	$-CN$	1	1	$CH_2CH_2C(CH_2Br)H$
5	Br	2	2	$-NO_2$	1	1	$CH_2C(CHCl_2)H$
6	Cl	2	2	$-OCH_3$	1	1	$CH_2C(CH_2Cl)H$
7	Br	3	3	$-OCH_3$	1	1	$(CH_2)_3C(CH_2Cl)_2$
8	Br	2	2	$-CH_3$	1	1	$CH_2C(CH_2Cl)H$
9	Br	2	2	$-F$	1	1	$CH_2C(CH_2Cl)_2$
10	Br	2	2	$-N(CH_3)_2$	1	1	$CH_2C(CHBr_2)H$
11	Br	2	2	$-C_6H_5$	1	1	$(CH_2)_2C(CBr_3)H$
12	Cl	2	2	$-C_6H_3Br_2$	1	1	$CH_2C(CH_2Cl)H$
13	Br	2	2	$-CH_2C_6H_5$	1	1	$CH_2C(CCl_3)_2$
14	Br	2	2	$-CH_2C_6H_3Br_2$	1	1	$CH_2C(CH_2Cl)H$
15	Cl	3	3	$-C_6H_3Cl_2$	1	1	$CH_2C(CH_2Cl)H$
16	Br	3	3	$-F$	1	1	$CH_2C(CCl_3)_2$
17	Cl	5	5	–	0	0	$CH_2C(CH_2Cl)H$
18	Br	4	4	–	0	0	$CH_2C(CH_2Cl)H$
19	Br	3	3	$-C_6H_2Br_3$	1	1	$(CH_2)_3C(CH_2Cl)_2$
20	Br	3	3	–	0	0	$CH_2C(CH_2Cl)HCH_2$
21	Br	2	2	–	0	0	$CH_2C(CCl_3)H$
22	Br	4	4	–	0	0	$CH_2C(CHBr_2)H$
23	Br	3	3	$-F$	2	2	$CH_2C(CH_2Cl)H$
24	Br	1	1	$-C_4H_9$	1	1	$CH_2C(CH_2Cl)H$
25	Br	1	1	$-OC_4H_9$	1	1	$(CH_2)_3C(CH_2Cl)H$

The amount of bis-phenoxy compound employed in the plastic compositions is from about 1 to 25% by weight, based on the total weight of the composition.

To enhance the flame retardancy of the poly(phenylene oxide) plastic composition, certain metal compounds are used. These enhancing agents are from the group antimony, arsenic, bismuth, tin and zinc-containing compounds. Examples of the enhancing agents include Sb_2O_3, $SbCl_3$, $SbBr_3$, SbI_3, SbOCl, As_2O_3, As_2O_5, $ZnBO_4$, $BaB_2O_4 \cdot H_2O$, $2ZnO \cdot 3B_2O_3 \cdot 3.5H_2O$ and stannous oxide hydrate. The preferred enhancing agent is antimony trioxide. The amount employed is from about 1 to about 15% by weight, based on the total weight of plastic composition. The following example illustrates several plastic compositions.

Example: A poly(phenylene oxide), PPO, plastic material is utilized as the base resin in order to prepare 26 formulations (plastic compositions). With the exception of formulation No. 1, the particular bis-phenoxy compound (and the antimony trioxide enhancing agent where indicated) is incorporated into the plastic by adding both to a Brabender mixer. The mixer is equipped with a pair of roller type blades positioned within a head provided with heat transfer means.

The resultant mixture is heated to about 320°C; at this temperature, it is in a molten state. The percentages by weight of each component utilized in the respective formulations are listed in the table below. Each formulation is discharged from the mixer and upon cooling solidifies and is ground into chips. The chips are subjected to compression molding in a Wabash press by placing the chips between two platens, the bottom of which contains four equal size depressions 3" x 5" x ⅛" deep. The top platen is then placed over the bottom platen and heat transfer means supplied to it in order to melt the chips and thus provide solid samples (after cooling) for testing.

Portions of the samples of each respective formulation (No. 1 through 26) prepared according to the above described procedure are then subjected to two different standard flammability tests, i.e., UL 94 and ASTM D-2863-70. The results of these flammabllity tests are shown in the table below.

Formulation No.	Bisphenoxy No.	Compound (%)	Enhancing Agent Sb_2O_3 (%)	O.I. (%)	UL 94
1	–	0	0	25.5	SB
2	1	20	0	27.5	SE-2
3	1	20	10	34.5	SE-0
4	3	20	0	28.5	SE-1
5	3	20	10	38.0	SE-0
6	4	20	0	26.5	SE-2
7	4	20	10	31.0	SE-0
8	5	20	0	27.5	SE-2
9	5	20	10	34.0	SE-0
10	6	20	0	27.0	SE-2
11	6	20	10	30.5	SE-0
12	8	20	0	28.0	SE-1
13	8	20	10	34.5	SE-0
14	12	20	0	27.5	SE-2

(continued)

Formulation No.	Bisphenoxy No.	Compound (%)	Enhancing Agent Sb_2O_3 (%)	O.I. (%)	UL 94
15	12	20	10	36.0	SE-0
16	14	20	0	29.5	SE-0
17	14	20	10	39.0	SE-0
18	16	20	0	28.5	SE-1
19	16	20	10	37.5	SE-0
20	17	20	0	29.0	SE-0
21	17	20	10	36.0	SE-0
22	20	20	0	28.5	SE-1
23	20	20	10	37.5	SE-0
24	23	20	0	27.0	SE-2
25	23	20	10	34.0	SE-0
26	25	20	10	33.5	SE-0

Referring to the above table, the bis-phenoxy compound number relates to the structural formulas heretofore set forth in the previous table; a difference of 2% in the Oxygen Index values is considered significant, and the UL 94 values are on a graduated scale wherein the highest degree to lowest degree of flame retardancy is respectively SE-0, SE-1, SE-2, SB and Burns.

The results shown in the table above demonstrate the unique effectiveness of these bis-phenoxy compounds as flame retardants for poly(phenylene oxide). Specifically, formulation No. 1 (the control) had a O.I. of 25.5 and UL 94 value of SB. In the even numbered formulations, the use of the particular bis-phenoxy compound results in a significant increase in fire retardancy as measured by O.I.; however, the individual UL rating of SE-0 for most of these formulations is, in this case, more indicative of increased flame retardancy, where compared to the SB rating of the control.

The use of an enhancing agent such as Sb_2O_3 to promote a cooperative effect between such agent and the bis-phenoxy compound is fully demonstrated via the results obtained from testing the odd numbered formulations. The highest UL 94 ratings and significantly higher O.I. values are obtained by the use of an enhancing agent.

Polymeric Alloys

A process described by *J. Jagur-Grodzinski, D. Vofsi and I. Cabasso; U.S. Patent 4,008,191; February 15, 1977; assigned to Yeda Research and Development Co., Ltd., Israel* relates to polymeric materials and to a process for preparing them, which comprises blending suitable polymers into polymeric alloys. The process further relates to fire-retarding and flame-resistant materials comprising such polymeric alloys and to membrane materials and membranes, adapted to provide processes of separation based on membrane technology. The polymeric alloys comprise commercially available polymers, blended with polymeric phosphonates of the general formulas:

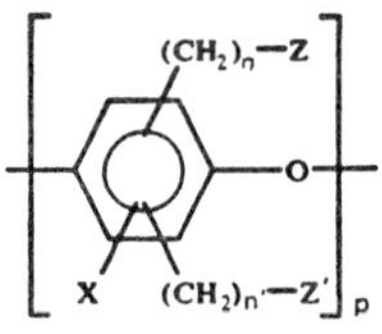

or

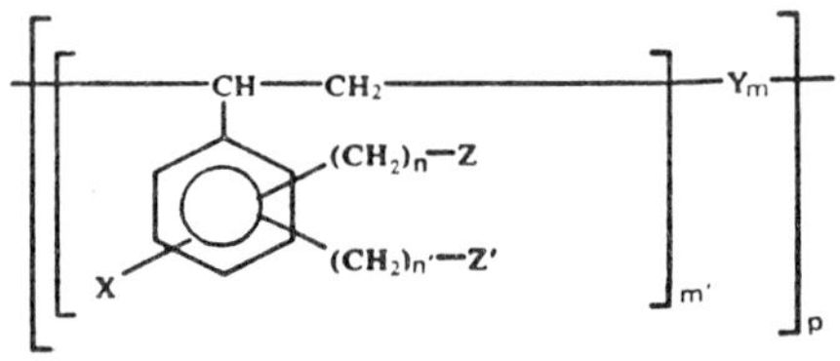

In the formulas m, m', n, and n' are zero or any integer, p is an integer higher than 10, X designates a hydrogen atom, halogen atom or a phosphorus containing group, Y designates a CH_2CHCl or CH_2CCl_2 group, Z and Z' designates H or a phosphonate group, PO(OR')(OR") (at least one being a phosphonate group), where R and R' are alkyl or aryl groups.

Polymeric alloys based on sulfonated polystyrene and copolymers of acrylonitrile have been used as ion exchange membranes and polymeric alloys based on polyvinylpyrrolidone and polyisocyanates have been tested as membranes in desalination processes. The following examples illustrate the process.

Example 1: 10 grams of styrene-vinylidene chloride copolymer η = 0.34 in toluene at 25°C (Cl = 26.65%) are dissolved in 50 ml of dry carbon disulfide and 50 ml of chlorodimethyl ether. 10 grams of $AlCl_3$ are slowly added to the vigorously stirred solution. Reaction is continued for 7 hours at room temperature. Wet dioxane is added until purple coloration of the solution disappears.

A polymeric material is precipitated with methanol, redissolved in methyl ethyl ketone, filtered and again precipitated with methanol. 5 grams of so prepared chloromethylated copolymer (Cl = 35.9%) are dissolved in 25 ml of dry bis(2-ethoxyethyl) ether (diethyl Carbitol) and the solution is added dropwise to 30 milliliters of vigorously stirred triethylphosphite at 150° to 160°C. The reaction mixture is refluxed for 4 hours. After cooling to room temperature, polymer is precipitated with pentane.

Poly(vinylidene chloride - benzyldiethyl phosphonate) (Cl = 15.1%, P = 9.2%) referred to as PVBP is obtained in 90% yield. The polymer is nonflammable and highly hygroscopic. When immersed in water, 1 part absorbs up to 2.5 parts of water at room temperature. PVBP is soluble in aromatic hydrocarbons, ketones, dioxane and hot methanol. Its formula is verified by PMR spectroscopy.

Example 2: 5.5 grams of PVBP prepared as described in Example 1, and 8.5 grams of acetyl cellulose (Eastman-Kodak 394-45) are dissolved in 100 grams of dioxane. A 100 μ thick film is cast with a doctor blade on a glass plate. Dioxane is evaporated at room temperature for 24 hours. Remaining dioxane is removed in vacuum at 30°C for 3 hours. A transparent homogenous membrane is stripped off the glass plate after it is immersed in water. It is self-extinguishing (see table below) and its tensile strength is 6.3 kg/mm^2.

Tensile Strength and Flammability of Polymeric Alloys

. . . . Component 1 (%).		Component 2 (%)		Tensile Strength (kg/mm^2)	Flammability* ASTM D 1433-58
Cellulose acetate**	100	–	–	8.0	burn, 3.5 cm/sec
Cellulose acetate	80	$PPOBrP_2$	20	7.3	self-extinguishing (2-2.5 sec, 3-6 cm)
Cellulose acetate	75	$PPOBrP_2$	25	6.4	self-extinguishing (0-2 sec, 0-4 cm)
Cellulose acetate	70	$PPOBrP_2$	30	6.3	self-extinguishing (0-1.5 sec, 0-2 cm)
Cellulose acetate	60	$PPOBrP_2$	40	6.0	nonflammable
Cellulose acetate	50	$PPOBrP_2$	50	5.8	nonflammable
Cellulose acetate	50	PPOP	50	5.9	nonflammable
Cellulose acetate	60	PVBP	40	6.3	self-extinguishing (2-3 sec, 5-6 cm)

(continued)

.... Component 1 (%)		Component 2 (%)		Tensile Strength (kg/mm^2)	Flammability* ASTM D 1433-58
Cellulose acetate**	50	PBP	50	5.7	self-extinguishing (1-2 sec, 2-4 cm)

*Determined for 20 μ thick films. **Eastman 394-45.

Note—$PPOBrP_2$ is a poly(2-methyl-6-methylenedimethylphosphonate-3-bromo-1,4-phenylene oxide); PBP is polybenzyldiethylphosphonate; PPOP is brominated and phosphonated PPO [Poly(2,6-dimethyl-1,4-phenylene oxide)].

Example 3: Polymeric films consisting of a polymeric alloy are prepared from PPOP and acetyl cellulose in the same way as described in Example 2 for PVBP, the ratio of PPOP/acetyl cellulose being, however, 1:1. A transparent, mechanically strong (tensile strength 5.9 kg/mm^2) film is obtained. It is nonflammable and acts as a highly permeable membrane for aromatic hydrocarbons (cf table above).

In related work, *I. Cabasso, J. Jagur-Grodzinski and D. Vofsi; U.S. Patent 4,073,754; February 14, 1978* describe additional polymeric alloys useful as fire-retardant and flame-resistant materials and as membranes useful in the separation of solvents exhibiting differential permeabilities therethrough.

Polyimide Prepolymer Resin and Red Phosphorus

M. Ducloux; U.S. Patent 4,067,897; January 10, 1978; assigned to Rhone-Poulenc Industries, France describes heat-curable compositions based on a prepolymer obtained by reacting: (a) a N,N'-bis-imide of an unsaturated dicarboxylic acid of the formula

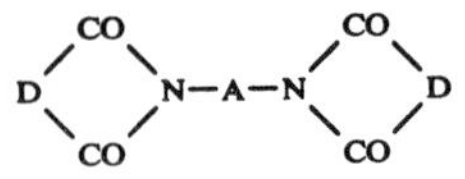

in which D represents a divalent radical containing a carbon-carbon double bond and A is a divalent radical which possesses at least 2 carbons; and (b) a di-primary diamine of the general formula $NH_2—B—NH_2$, in which B represents a divalent radical which does not have more than 30 carbons, in a ratio of number of mols of bis-imide/number of mols of diamine of between 1.2 and 5.

The composition contains up to 15% by weight, relative to the weight of the prepolymer, of red phosphorus in the form of a powder having a mean particle size of less than 200 μ. The polyimide resins possessing flameproof properties can be converted into shaped articles or can be used as coatings and adhesives. The following examples illustrate the process.

Example 1: 18 grams of a mixture homogenized on rollers and consisting of 5% by weight of red phosporus powder having a mean particle size of 20 to 30 μ and 95% of a prepolymer obtained by the reaction of 2.5 mols of 4,4'-bis-maleimido-diphenylmethane and 1 mol of 4,4'-diamino-diphenylmethane, and having a softening point of 148°C, are introduced into a circular mold of 75 mm diameter.

The above prepolymer is prepared by heating the two reactants at 150°C for 45 minutes. The mold is introduced between the platens of a press heated to 250°C.

A pressure of 200 bars is applied for 1 hour. The molded article is withdrawn and reheated for 24 hours at 250°C. Small plates of size 50 x 6 x 3 mm are cut from the molded article and the limiting oxygen index is measured by the LOI test in accordance with Standard Specification ASTM D 2863. More small plates were prepared under otherwise the same conditions but without adding the red phosphorus. The following results are obtained:

	LOI Index
Polyimide reference sample	29
Polyimide + red phosphorus	45

As can be seen, the flameproof properties of the polyimide alone, which were already good (an LOI index of 29) are improved very considerably by the addition of 5% of red phosphorus.

It was found furthermore that in the course of the preparation of the small plates, of their heat treatment and of the LOI test, there was no evolution of a toxic gas such as phosphine. On using various methods such as trapping gases by means of an aqueous solution of mercury chloride, or using a Draegger tube, it was not possible to detect any trace thereof.

Example 2: The following composition is mixed and homogenized: 110 grams of N-methylpyrrolidone, 90 grams of the same prepolymer as that used in Example 1 and 4.5 grams of red phosphorus powder (particle size 20 to 30 μ).

This rather viscous composition is applied by brushing to the two faces of a glass fabric weighing about 300 g/m^2. The coated article is dried for 5 minutes in a ventilated oven at 130°C. A second layer is applied. This is dried and the residual solvent is then evaporated by heating for 30 minutes at 130°C under a pressure reduced to 40 cm of mercury.

The amount of prepolymer and of red phosphorus thus deposited represents about 38% of the total weight of the coated article. 14 samples of size 14 x 14 cm are stacked and compressed by means of a press, the platens of which are at 160°C, under a pressure of 3 bars. The temperature and the pressure are increased simultaneously so as to bring the pressure to 40 bars at the end of 10 minutes and the temperature to 200°C at the end of 30 minutes.

The material is kept for 30 minutes at 200°C under 40 bars. It is then allowed to cool while maintaining the pressure. A laminate is obtained. This is then reheated for 48 hours at 250°C.

Samples of size 127 x 12.7 mm, and of a thickness of about 3 mm, are cut from the laminate and subjected to the UL (Underwriters Laboratories) test 94 described in document BNMP 9, 750/1 of the (French) Bureau for the Standardization of Plastics. This laminate gave a classification 94 VE-0.

An otherwise similar laminate in which, however, the impregnation composition did not contain red phosphorus was prepared under the same conditions. This laminate gave a classification of 94 VE-1 in the UL test.

Mixtures of Unsaturated Polymers and Red Phosphorus

J. Cerny and G. Vivant; U.S. Patent 4,092,460; May 30, 1978; assigned to Rhone-Poulenc Industries, France describe compositions for flameproofing plastics without the release of toxic products during use in which the compositions consist of a pulverulent mixture containing from 50 to 95% by weight of red phosphorus and from 5 to 50% by weight of polymers having unsaturated malenoid, fumaroid or allyl bonds.

Compounds having malenoid, fumaroid or allyl unsaturated bonds in their chain are essentially represented by prepolymers or polymers of the unsaturated polyester or polyimide type, or copolymer based on maleic anhydride or on esters of maleic or fumaric acid. The prepolymers of the polyimide type are the same ones used in the above work.

NATURAL POLYMERS

COTTON AND POLYESTER-COTTON BLENDS

Brominated Phosphoramidates

J.J. Duffy, R.D. Carlson, and J.C. Watson; U.S. Patent 4,034,141; July 5, 1977; assigned to Hooker Chemicals & Plastics Corporation describe a method of imparting fire retardant properties to a combustible material, which comprises treating the material with a fire retardant amount of a compound of the following structure.

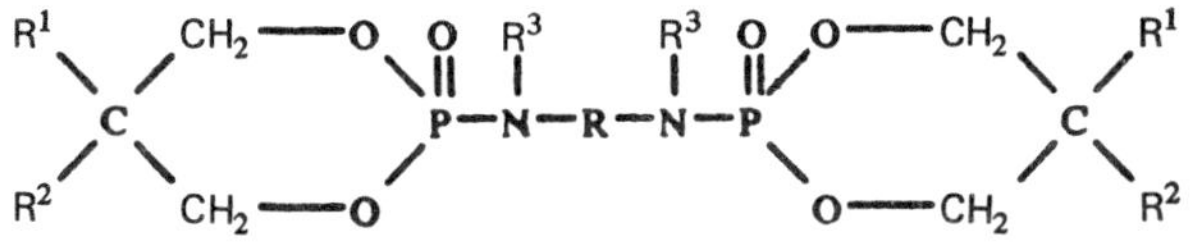

R^1 and R^2 are independently selected from the group consisting of hydrogen, chloro, bromo and chloro or bromo substituted alkyl of from 1 to 8 carbon atoms. R^3 is selected from the group consisting of hydrogen, alkyl of from 1 to 8 carbon atoms, hydroxy substituted alkyl of from 1 to 8 carbon atoms and chloro or bromo substituted alkyl of from 1 to 8 carbon atoms. R is selected from the group consisting of alkylene of from 2 to 8 carbon atoms, phenylene, biphenylene and dicyclohexylene; provided that at least one of R^1 and R^2 contains a bromine atom.

These compounds are particularly effective in imparting flame retardant character to cellulosics-containing material such as cotton and rayon as well as blends of cellulosics. The following examples illustrate the process.

Example 1: A 500 ml flask was charged with 100 g (0.292 mol) of dibromoneopentyl phosphorochloridate and 150 ml THF. The mixture was heated to 45°C, and 17.5 g (0.292 mol) of ethylene diamine was added dropwise. The temperature was maintained at 45° to 55°C with an ice bath. The mixture was then allowed to cool to room temperature, and filtered to remove the hydrochloride.

The filtrate was then stripped to final conditions at 70°C and 20 mm Hg for 2 hours, yielding 73 g (75% of theory) of product. NMR analysis confirmed the structure as:

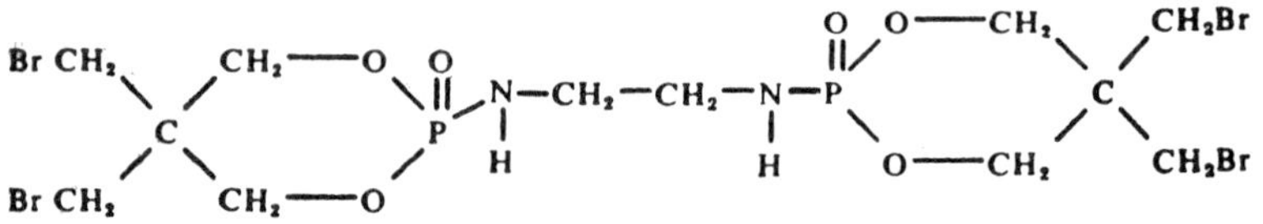

Example 2: A 1-liter flask was charged with 100 g (0.149 mol) of the product of Example 1 and 9 g (0.298 mol) paraformaldehyde in 500 ml methanol. The pH was adjusted to 8 to 10 with sodium methoxide, and the mixture was refluxed for 4 hours. The solvent was stripped to final conditions of 0.5 mm Hg and 70°C for 1 hour, yielding 100 g of residue product (92% of theory) and having the structure listed below.

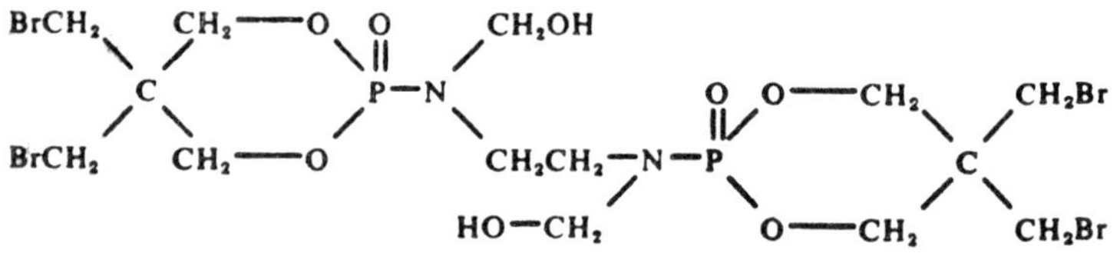

Example 3: General Example for Application to Cloth – Samples of fabric were cut to approximately 15 x 24 inch size and immersed in a pad solution of the formulation shown below. Samples were squeezed in a 2 roll laboratory padder with the pad pressure adjusted to give approximately 90 to 100% wet pickup. Samples were then dried by being hung in a circulating air oven for the time and temperature indicated in the following table. Samples were then cured at the time and temperatures indicated in the circulating air oven.

After cooling to room temperature, the samples were washed in a Sears washing machine with the following added to the wash water, sodium perborate (20 g), soda ash (30 g), and Orvus detergent (30 g) (Tide), which is a phosphate-based detergent, for one cycle. The samples were then dried in a tumble dryer, being overnight in the laboratory and then weighed to determine initial resin add-on. Oxygen index was then determined.

To determine durability, the samples were washed 10 cycles in a Sears home washer with 50 g of Orvus detergent added per wash cycle. At the end of 10 washings, the samples were dried in a tumble dryer and hung overnight prior to testing. Oxygen indices of the initial fabric were as follows:

Cotton	19
50/50 Polyester/Cotton	19
65/35 Polyester/Cotton	19

A solution was formulated of the following components and applied to cloth samples as described above.

Formulation	Parts by Weight
Product of Example 2	25
Tetrakis(hydroxymethyl) phosphonium chloride (30% by wt)	20
Triethanol amine	20
NaOH (50%)	3
H_2O	32

The flame retardant character was determined as shown in the table below.

	Sample A	Sample B	Sample C
Fabric (cotton or polyester/cotton blend)	cotton	50/50	65/35
Wet pickup, %	93	96	93
Drying time, min	2.5	2.5	2.5
Drying temperature, °F	200	200	200
Cure time, min	3.5	3.5	3.5
Cure temperature, °F	300	300	300
Resin add-on, %	14.2	16.3	20.0
Initial oxygen index	29	26	25
Oxygen index after ten home washings	27	25	24

Example 4: Following the same procedure of Example 3, the following composition was formulated.

Formulation	Parts by Weight
Product of Example 2	40
Trimethylol melamine	30
Water	27.5
$Zn(NO_3)_2 \cdot 6H_2O$	2.5

The results of flame retardant tests are shown in the table below.

	Sample A	Sample B	Sample C
Fabric (cotton or polyester/cotton)	cotton	50/50	65/35
Wet pickup, %	100	105	95
Drying time, min	2.5	2.5	2.5
Drying temperature, °F	200	200	200
Cure time, min	3.5	3.5	3.5
Cure temperature, °F	300	300	300
Resin add-on, %	11.7	16.6	12.0
Initial oxygen index	23	24	23
Oxygen index after ten home washings	24	23	23

Brominated Carbamoyl Derivatives

D.J. Scharf; U.S. Patent 3,969,230; July 13, 1976; assigned to Hooker Chemicals & Plastics Corporation describes brominated carbamoyl derivatives of the structure:

$$(X)(X^1)(X^2)C–CH_2–(R^2)_n–O–C(O)N(R)(R^1)$$

R and R^1 are independently selected from the group consisting of hydrogen, alkyl, alkanol, cycloalkyl of from 5 to 7 carbon atoms, phenyl, alkyl substituted by phenyl, phenyl substituted by alkyl, hydroxy terminated oxyalkylene, hydroxy substituted hydroxy terminated oxyalkylene and halogen substituted hydroxy terminated oxyalkylene. R^2 is independently selected from the group consisting of alkylene, oxyalkylene, hydroxy substituted oxyalkylene, halogen substituted oxyalkylene and alkylene or oxyalkylene substituted by the group –O–C(O)N(R)(R^1). X, X^1 and X^2 are independently selected from the group consisting of halogen and halogen substituted alkyl, where the halogen is chlorine or bromine; the alkyl, alkanol and alkylene each independently has from 1 to 4 carbon atoms; the oxyalkylene having from 2 to 4 carbon atoms; n is 0 to 20 (preferably 0 to 6), provided one of X, X^1 or X^2 is a bromo or a bromo substituted alkyl, and that each carbamoyl group has a replaceable hydrogen in the R or R^1 group.

In general, to prepare the compounds, react an alcohol (1 mol) with phosgene (2 mol) in the liquid phase. The second step is the formation of the carbamoyl group by reacting the product of the first step with ammonia. The hydrogen of the carbamate group may then be replaced or substituted by the appropriate R or R^1 groups. For example, the initial alcohol which could be a halogenated alcohol may be reacted with an epihalohydrin, such as epichlorohydrin or epibromohydrin. Alternatively, the alcohol may be reacted with various alkylene oxides, such as ethylene oxide, propylene oxide or butylene oxide to give the desired oxyalkylene product which can then be reacted with phosgene.

Regardless of how these compounds are prepared, N-alkanol derivatives are the most preferred compounds. They are prepared by reacting the carbamate with aqueous formalin at a pH of 8 to 9 and a temperature of about 45°C.

Listed below are preferred alcohols that may be employed in the preparation of the compounds:

$(BrCH_2)_3C–CH_2OH$
$(BrCH_2)_2(ClCH_2)C–CH_2OH$
$Br_3C–CH_2OH$
$(BrCH_2)_3C–CH_2–[OCH_2–CH(CH_2Br)]OH$
$(BrCH_2)_3C–CH_2–O–[CH_2–CH(CH_2Br)–O–]_nH$
$(BrCH_2)_3C–CH_2–O–[CH_2–CH(CH_2Cl)–O–]_nH$

Although the brominated materials may be applied alone, i.e., without other flame retarding compositions, a preferred sequence of applying the flame retardant composition is to employ additional flame retardant compositions such as phosphorus containing materials in combination with the brominated compounds.

In general, among the preferred phosphorus flame retardants are Pyrovatex $[(CH_3O–)_2P(O)–C_2H_4–C(O)NHCH_2OH]$, tetrakis(α-hydroxyorgano)phosphonium salts, such as the inorganic salts or salts of organic materials such as the acids, acetic, formic, oxalic and the like, or the product of neutralization of the tetrakisphosphonium compound with alkali metal hydroxide or a tertiary amine. One preferred material is tetrakis(hydroxymethyl)phosphonium hydroxide.

The brominated carbamoyl derivatives are useful compounds for imparting flame retarding properties to combustible substrates such as synthetic resins, wood, paper, cellulose containing fibers and fabrics, etc. In particular, the compounds impart flame retardant property to cellulosic containing fabrics by providing reactive sites with the fabric. The cellulose materials, which can be treated to impart flame retardant properties, include cotton, rayon, jute, ramie, and mixtures thereof, as well as blends of cellulosics, such as cotton or rayon with synthetic blends, such as nylon, polyesters, acrylics and with proteinaceous fibers, such as wool, etc. Further details of the process are given below where all percentages are by weight and all temperatures degrees centigrade unless otherwise indicated.

Example 1: Preparation of Carbamoyl Tribromoneopentane – Into a 500 ml, 3-neck, round-bottom flask equipped with a gas inlet tube and an exhaust tube connected to a caustic trap was added a solution of 108 g (0.3 mol) tribromoneopentyl alcohol in 250 ml tetrahydrofuran (THF). To this solution was added 35 g (0.35 mol) of phosgene meted from a cylinder. The reaction flask was cooled in an ice-water bath followed by the slow addition of triethylamine whereupon a precipitate of triethylamine·HCl was observed.

After stirring at 5°C for 2 hours, a large excess of ammonia gas was bubbled into the reaction flask (slight exotherm noted). After standing at room temperature overnight, the reaction mixture was filtered and the filtrate stripped in vacuo (55°C, 20 mm) to an oil which solidified on standing. The crude yield was 120 g (99% of theory).

Recrystallization of a small quantity of crude from EtOH/H_2O gave a white crystalline solid, MP 109° to 111°C, literature 110° to 110.5°C. Combustion analysis calculated for $C_6H_{10}Br_3NO_2$: C, 19.60%; H, 2.72%; N, 3.81%; Br, 65.20%. Found: C, 20.14%; H, 2.84%, N, 3.67%; Br, 64.81%.

Example 2: Preparation of N,N-Bishydroxymethyl Carbamoyltribromoneopentane – Into a 250 ml flask was placed 100 ml of 37% aqueous formalin. pH was adjusted to 8 to 9 with 10% aqueous NaOH and 20 g of carbamoyltribromoneopentane was added over 45 minutes as a pulverized powder. After the first 5 g was added, the reaction solution was heated to and maintained at 45°C throughout the addition and for an additional hour. Periodic readjustment of pH was required. The water was stripped in vacuo on a rotary evaporator to constant weight. NMR analysis (DMSO-d_6) was consistent with the desired product.

Example 3: Application of N,N-Bishydroxymethyl Carbamoyltribromoneopentane to 100% Cotton and 50/50 Polyester/Cotton – N,N-Bishydroxymethyl carbamoyltribromoneopentane prepared according to Example 2 was diluted with methanol to a 40% solution. 5% catalyst AC (2-amino-2-methylpropanol·HCl) was added and the solution padded onto separate fabrics of 100% cotton and 50/50 polyester/cotton. The treated fabric samples were dried at 200°F/1.5 minutes, cured at 320°F/2.5 minutes and scoured. After line drying, the 100% cotton had 26.9% resin add-on and the 50/50 polyester/cotton had 20.3% resin add-on. Flame retardancy was indicated by oxygen index measurements.

Oxygen Index – The O.I. for 100% cotton was 24 and for 50/50 PE/cotton was 23. The O.I. of untreated 100% cotton was 20.1 and for untreated 50/50 PE/cotton was 17.6.

Example 4: One Step Application of THPOH/N,N-Bishydroxymethyl Carbamoyltribromoneopentane to 100% Cotton and 50/50 Polyester/Cotton – A 50% THPOH aqueous solution was formulated with an equal volume of a 30% (20 g CH_3OH, 15 g H_2O) solution of N,N-bishydroxymethyl carbamoyltribromoneopentane prepared according to Example 2. The resultant solution (40% solids) was padded onto separate fabric samples of 100% cotton and 50/50 polyester/cotton. Each sample was dried at 200°F for 2 minutes, cured at 300°F for 2 minutes, cured in an ammonia atmosphere for 2 minutes and scoured. The 100% cotton had 8.2% resin add-on and the 50/50 polyester/cotton had 11.3% resin add-on. Flame retardancy was indicated by oxygen index measurement.

Oxygen Index – The O.I. for 100% cotton was 25 and for 50/50 PE/cotton was 25. The O.I. of untreated 100% cotton was 20.1 and for untreated 50/50 PE/cotton was 17.6.

In related work, *D.J. Scharf; U.S. Patent 4,058,466; November 15, 1977; assigned to Hooker Chemicals & Plastics Corporation* describes other brominated carbamoyl derivatives and their application to cotton and polyester/cotton blends.

Polycyclic Phosphorous Triamides

T.J. Atkins; U.S. Patent 3,996,276; December 7, 1976; assigned to E.I. DuPont de Nemours and Company describes polycyclic phosphorous triamides of the formula:

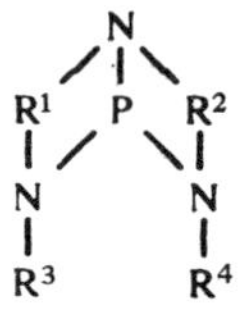

R^1 and R^2, alike or different, are alkylene of 2 to 6 carbons. R^3 and R^4, alike or different, are alkyl of 1 to 8 carbons, cycloalkyl of 5 to 8 carbons, or aralkyl where the aryl group is of 6 to 12 carbons and the alkyl is of 1 to 8 carbons or R^3 and R^4 are joined together to form an alkylene group of 2 to 6 carbons which may be interrupted by:

(1) a

$$\begin{array}{c} -N- \\ | \\ Q \end{array}$$

group where Q is hydrogen or alkyl of 1 to 18 carbons, or

(2) one or two –O– linkages; there are at least 2 carbons between each two hetero atoms in the outer ring system and, when the triamide is tricyclic, at least one of the chains between the nitrogens linked to phosphorus contains at least three atoms.

The products are prepared by reacting the appropriate polyamine with a hexa(lower alkyl)phosphorous triamide (also known as a tris[di(lower alkyl)amino] phosphine). The process can be represented by the following equation:

$$R^3\text{–NH–}R^1\text{–NH–}R^2\text{–NH–}R^4 + P(NR_2)_3 \longrightarrow \begin{array}{c} R^1\text{–N–}R^2 \\ |\quad\;|\quad\;| \\ R^3\text{–N –P –N –}R^4 \end{array} + 3R_2NH$$

The chemical reaction involved is a transamidation or amine exchange. These products are useful as flame retardants for cellulosics such as cotton.

The following examples illustrate the process. All preparations were carried out in an atmosphere of nitrogen. Mass-spectral analyses were relied on to confirm the empirical formulas of the products.

Example 1: 10-Oxa-1,4,7-triaza-13-phosphatricyclo[5.5.1.$0^{4,13}$] tridecane is prepared as follows:

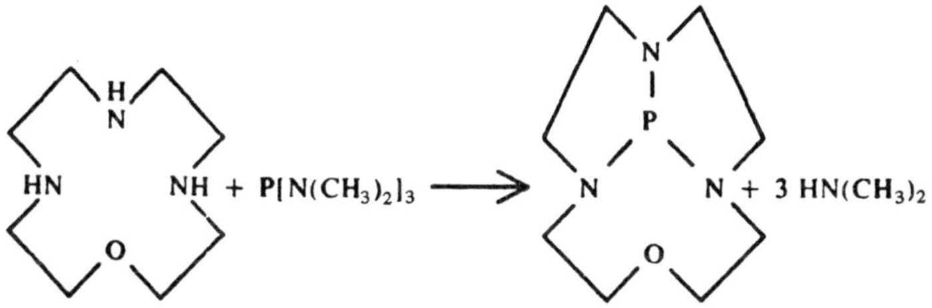

(A) A solution of 2.50 g of 1-oxa-4,7,10-triazacyclododecane and 2.35 g of hexamethylphosphorous triamide in 25 ml of toluene is refluxed for 40 hours. It is then concentrated to dryness in a rotary evaporator to give 2.97 g (100%) of 10-oxa-1,4,7-triaza-13-phosphatricyclo[5.5.1.$0^{4,13}$] tridecane as a brittle, white solid. Sublimation at 80°C (0.007 mm) gives 2.08 g of white crystals having a melting point of 68° to 70°C. An infrared absorption spectrum of this material in mineral oil shows absorptions at 3.5, 6.82, 7.38, 7.59, 8.00, 8.24, 8.50, 8.83, 9.69, 9.89, 10.55, 11.18, 11.77, and 12.56 μ.

(B) 10-Oxa-1,4,7-triaza-13-phosphatricyclo[5.5.1.$0^{4,13}$] tridecane is obtained in the absence of a solvent by heating 10.0 g of 1-oxa-4,7,10-triazatricyclododecane and 9.40 g of hexamethylphosphorous triamide at 75°C for 2 hours, by which time evolution of dimethylamine is complete. The product is sublimed at 80°C (0.40 mm) and identified by comparison of its infrared absorption spectrum with that of the product of part A. Mass-spectral analysis of this sample shows an M^+ ion at 201; measured mass, 201.1094; calculated 201.1031; which confirms the empirical formula of the product of part A.

If 1,4-dioxa-7,10,13-triazacyclopentadecane were used in place of 1-oxa-4,7,10-triazacyclododecane in essentially the procedure of Example 1, the product would be

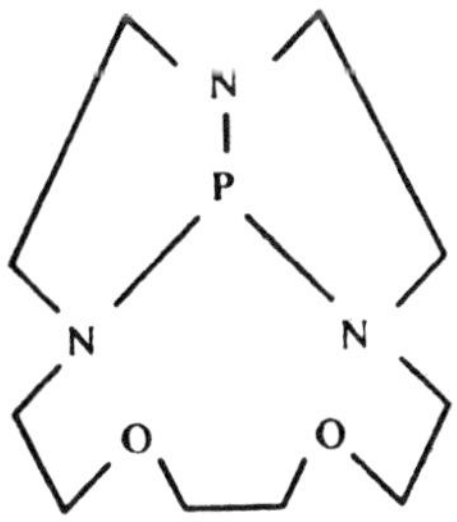

Example 2: 1,4,7,10-Tetraaza-13-phosphatricyclo[5.5.1.$0^{4,13}$] tridecane is prepared as follows:

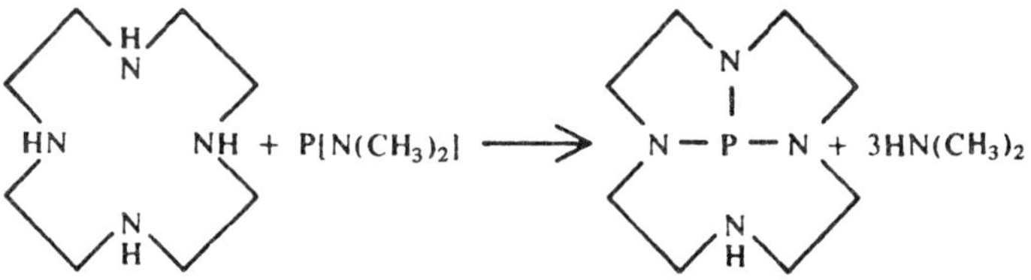

(A) A solution of 1.75 g of 1,4,7,10-tetraazacyclododecane and 1.63 g of hexamethylphosphorous triamide in 50 ml of toluene is refluxed for 12 hours. Titration of the off-gases with 1 N HCl indicates that 97% of the theoretical amount of dimethylamine is evolved in this time. The toluene is removed under reduced pressure to give 2.0 g of 1,4,7,10-tetraaza-13-phosphatricyclo[5.5.1.$0^{4,13}$] tridecane as a white solid which melts at 109° to 111°C with sintering from 90°C (possibly because of the presence of a trace of toluene). Sublimation of this product at 75°C and 0.55 mm gives large colorless crystals which melt at 111° to 113°C. The mass spectrum shows an M^+ ion at 200; measured mass, 200.1194; calculated, 200.1190.

(B) 1,4,7,10-Tetraaza-13-phosphatricyclo[5.5.1.$0^{4,13}$] tridecane is prepared in the absence of a solvent by heating 5.0 g of 1,4,7,10-tetraazacyclododecane and 4.7 g of hexamethylphosphorous triamide together at about 75°C for between 2 and 3 hours and subliming the crude product at 75°C at 0.4 mm. The yield is 5.3 g (92%) of product melting at 106° to 108°C. The infrared absorption spectrum of the product in mineral oil has absorptions at 3.05 (very weak), 4.31, 7.51, 8.00, 8.20, 8.36, 8.92, 9.50, 10.22, 10.5, 11.5, 13.4 and 14.6 μ. Measured mass (mass spec): 200.1226.

Example 3: Solutions of the products of Examples 1 and 2 in dimethylformamide (10 weight/volume %) are prepared. Cotton swabs are soaked in these solutions for 10 minutes and for 1 hour and dried overnight. Both the treated swabs and an untreated control are tested for flammability by holding them to a flame. The untreated control burns completely and glows after the flame extinguishes. All the treated swabs are self-extinguishing when removed from the flame, the swabs themselves are charred.

Example 4: Strips of cotton cloth are soaked overnight in 10 weight/volume % solutions of the products of Examples 1 and 2 in dimethylformamide and then dried. The treated fabrics, together with an untreated control, are tested for flammability by being held vertically and touched with a flame at their bottoms. The untreated control burns profusely. The fabric treated with the product of Example 1 self-extinguishes in less than 1 second and burns less than 5% of its length. The fabric treated with the product of Example 2 self-extinguishes in about 1 second and also burns about 5% of its length.

Tetracyclic Tetraamino Phosphoranes

J.E. Richman; U.S. Patent 3,987,128; October 19, 1976; assigned to E.I. DuPont de Nemours and Company describes a process whereby tetracyclic tetraamino phosphoranes of the following general formula:

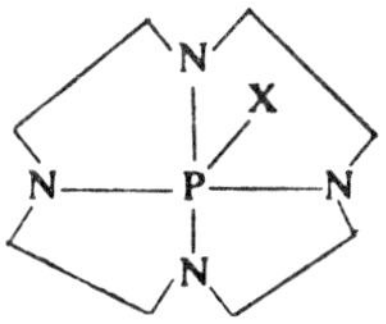

where X is F, C_6H_5, or C_{1-6} alkyl, and tricyclic polyamino phosphine oxides of the formula:

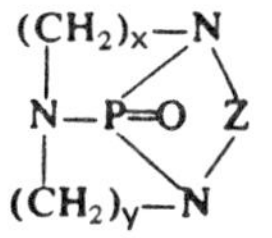

where x and y are each 2 or 3; Z is $-(CH_2)_Z-$ where z is 2 to 10 and x + y + z is 7 to 14; or Z is $-(CH_2)_V-NH-(CH_2)_W-$ where v and w are each 2 or 3; are prepared by reacting a silylated cyclic polyamine of the formula:

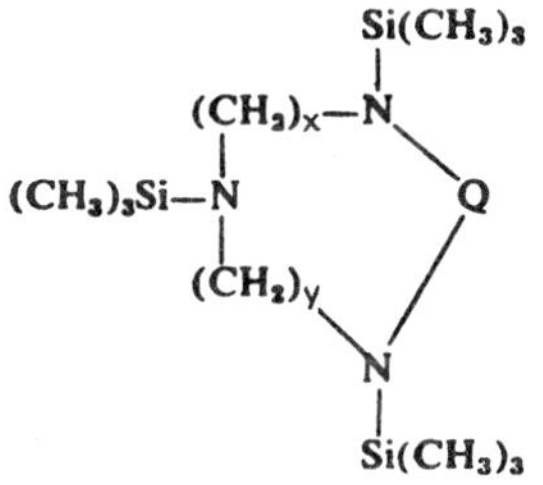

with a fluorophosphorane of the formula, XPF_4 or POF_3. Q may be $-(CH_2)_Z-$ or

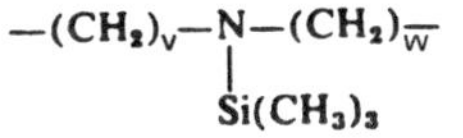

provided that when Q is $-(CH_2)_Z-$, the fluorophosphorane is POF_3. The corresponding alkoxy or amino

[X = OR, NR_2 or $\overbrace{N(CH_2)_{2-6}}$, where R is C_{1-6} alkyl]

tetracyclic tetraamino phosphoranes are obtained by reaction of an alkali metal alkoxide or amide with the tetracyclic fluorophosphorane. All of the polycyclic polyamino phosphoranes and phosphine oxides are useful as flame retardants for cotton.

The following examples illustrate the tetracyclic tetraamino phosphoranes and tricyclic polyamino phosphine oxides, their preparation, and their use as flame retardants. All percentages are by weight, and all temperatures are in degrees Centigrade.

Example 1: (A) 1,4,7,10-Tetrakis(trimethylsilyl)-1,4,7,10-tetraazacyclododecane was prepared as shown on the following page.

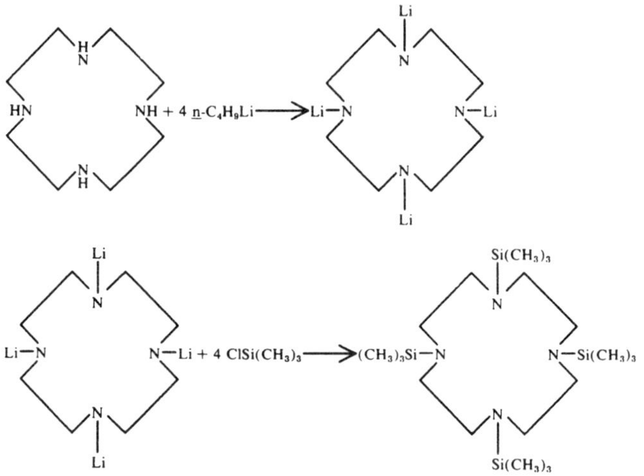

A slurry of 0.69 g of 1,4,7,10-tetraazacyclododecane in 20 ml of anhydrous tetrahydrofuran was cooled in a dry-ice/acetone bath and stirred under nitrogen as 10 ml of a 1.6 M n-butyllithium/hexane solution was quickly added. The milky slurry was stirred for 15 minutes at -78°C, and then was warmed and stirred at 0°C for 30 minutes. The mixture was again cooled in a dry-ice/acetone bath and 2.1 ml of trimethylchlorosilane was added. This mixture was stirred at room temperature overnight, and then filtered under nitrogen and concentrated to 2.3 g of white solid while avoiding undue exposure to air. This solid was slurried in petroleum ether, filtered, and vacuum dried. The analytical sample was sublimed at 110°C/0.1 mm giving a white solid melting at 140° to 150°C. Analysis–Calculated for $C_{20}H_{52}N_4Si_4$: C, 52.11%; H, 11.37%; N, 12.15%. Found: C, 51.78%, 51.86%; H, 11.32%, 11.20%; N, 12.28%, 12.34%.

(B) 13-Fluoro-1,4,7,10-tetraaza-13-phosphatetracyclo$[5.5.1.0^{4,13}.0^{10,13}]$ tridecane was prepared as follows:

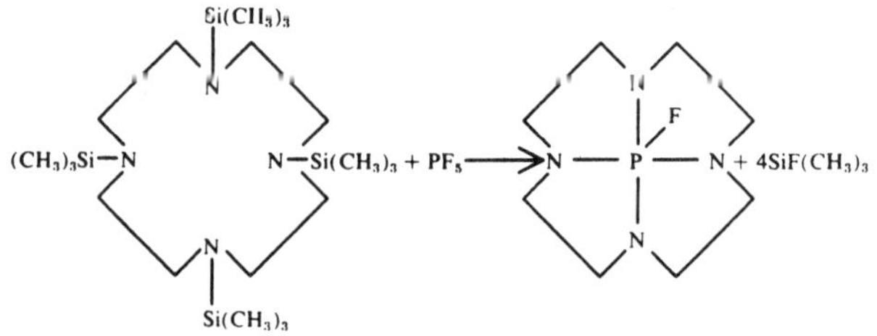

A solution of 1.0 g of 1,4,7,10-tetrakis(trimethylsilyl)-1,4,7,10-tetraazacyclododecane (sublimed) in 10 ml of o-dichlorobenzene was frozen and evacuated. This mixture was melted and stirred at 25°C as 50 ml of phosphorus pentafluoride gas was added. Then, nitrogen gas was added to atmospheric pressure. After 30 minutes, the solution was heated at 100°C for 1 hour, then at 145°C for 4 hours. The solvent was removed under vacuum and the residue distilled at 110°C and 0.1 mm pressure thereby giving clear colorless crystals of 13-fluoro-1,4,7,10-

tetraaza-13-phosphatetracyclo[5.5.1.$0^{4,13}$.$0^{10,13}$] tridecane, MP 87° to 95°C. X-ray crystal structure determination confirmed the structure.

Example 2: The following example shows the flame retardant properties of the products of this process. In a 1.75 ml vial, a 100 mg sample of the product of Example 1(B), 13-fluoro-1,4,7,10-tetraaza-13-phosphatetracyclo[5.5.1.$0^{4,13}$.$0^{10,13}$]-tridecane, was dissolved in 1.0 g of dimethylformamide. Samples (0.5 g) of 100% cotton cloth cut in strips (~0.75 x 4 inches) were soaked for 1 hour in this vial and in a vial containing only the solvent. After drying 20 hours in the air, these strips were vertically suspended and lighted at the bottom with a match. The blank sample completely burned in 20 to 25 seconds. The flame rapidly climbed and the ash was a white powder after total consumption of the glowing residue.

In the case of the product of Example 1(B), the cloth was totally inflamed in 30 seconds. Progress of the flame climbing was clearly slower than for the blank and there was considerable smoking. The ash was a black carbonaceous residue which did not glow. Similarly, samples of the product of Example 1(B) (0.02 g each) were dissolved in 1.0 g of acetone and soaked into cotton strips. After 1 hour of drying, these strips were similarly tested along with a blank. The burning rate was similar (15 seconds) in all samples. However, in the samples containing the product of Example 1(B) the ash did not glow and left a black carbon residue. The blank was totally consumed leaving a white ash.

Fiber Blends Containing Tetrakis(Hydroxymethyl)Phosphonium Chloride

B.R. Baird and L.A. Sherbeck; U.S. Patent 4,014,829; March 29, 1977; assigned to E.I. DuPont de Nemours and Company describe a flame-resistant blend of textile fibers comprising (a) from about 60 to 80% by weight of a poly(meta-phenylene isophthalamide) (MPD-I) fiber containing distributed substantially throughout its whole interior a finely divided crosslinked reaction product of at least 5% by weight (unreacted basis) of a tetrakis hydroxymethyl phosphonium compound and a resin containing active hydrogen and (b) from about 20 to 40% by weight of a normally flammable textile fiber aesthetically suitable for wearing apparel. The amount of the phosphorus-containing fiber in the blend will normally be the minimum amount that is sufficient to render fabrics of the fiber blend flame resistant.

The normally flammable textile fibers suitable for wearing apparel are polyethylene terephthalate fibers in the amount of about 20 to 40% of the blend with MPD-I fiber; cotton fibers in the amount of about 20 to 30% of the blend with MPD-I fiber; and acrylic fibers in the range of about 20 to 25% of the blend with MPD-I fiber.

The preferred tetrakis hydroxymethyl phosphonium compound is tetrakis hydroxymethyl phosphonium chloride (THPC), but may also be tetrakis hydroxymethyl phosphonium oxide (THPO) or a reactive derivative thereof.

The resin component is both water soluble and reactive at elevated temperature with the phosphorus compounds to form a crosslinked, insoluble product; yet reactive only at a very slow rate at ambient temperatures with the phosphorus compound to permit maximum lifetime of the treating solution. Preferred resin components are condensates, e.g., melamine-formaldehyde (such as Aerotex UM), phenolformaldehyde, and hexamethylol-melamine. These components may optionally be combined with other reactive materials such as guanidine phosphate. A catalyst such as ammonium chloride or magnesium chloride, which facilitates the reaction between the phosphorus compound and resin component at elevated temperature, may optionally also be included. The following examples illustrate the use of fiber blends.

Example 1: (A) Stock Tow Preparation — A filtered spinning solution is prepared consisting of 18.5%, based on the weight of the solution, of poly(meta-phenylene isophthalamide) in N,N-dimethylacetamide (DMAc) that contains 45% calcium chloride, based on the weight of the polymer. The polymer has an inherent viscosity of 1.60 as measured in a 0.5% solution in DMAc/4% LiCl at 25°C. The spinning solution is heated to 133° to 140°C and dry spun through multihole spinnerets. The extruded filaments are converged at a guide at the bottom of each cell where they are flooded with an aqueous solution containing 7 to 10% DMAc and 5 to 7% calcium chloride. Filaments from several adjacent cells are combined to give a large bundle of filaments, referred to as a tow, each filament being about 12 dpf as spun.

The wet tow is fed at 71 meters/minute to a 10-tank apparatus where the tow is extracted and drawn in stages, a total of about 4.2X draw, in aqueous baths, the temperatures of the aqueous baths ranging from 82°C in the first three baths by stages to 98°C in the last bath. Finish is applied to the wet tow, and the filaments are crimped in a steamed stuffer-box crimper. The crimped, never-dried filaments are stored wet in a sealed container.

(B) Preparation of the Phosphorus-Containing Fibers — The crimped, never-dried filaments of poly(meta-phenylene isophthalamide) are treated in stock dyeing procedure with a solution of 72% (based on the weight of the filaments) of THPC (tetrakis hydroxymethyl phosphonium chloride), 18% (based on the weight of the filaments) melamine-formaldehyde resin (Aerotex UM), and 0.4% (based on the weight of the filaments) of ammonium chloride as a catalyst. The never-dried tow is soaked for 60 minutes at 80°C in the solution of the flame retarder and the resin. The liquor is then drained from the tow and the resin in the tow is cured by drying at 135°C for 1.5 hours. The tow is then scoured, rinsed, dried, and cut to staple. The phosphorus-containing staple fiber product of poly(meta-phenylene isophthalamide) produced contains 6% THPC and is referred to as Fiber 1-B.

(C) Preparation of Polyester Fiber Blends and Fabrics — Double Knit Fabrics: a staple fiber blend of 75% of Fiber 1-B and 25% polyethylene terephthalate staple fibers is prepared, and the staple fiber blend is spun to a 20/1 cc (cotton count) yarn of 13.5Z twist and knitted into a 5.8 oz/yd² double knit fabric. In a vertical flame test (4 samples), the char length was 1.5 inches and the after-flame extinguishes itself in 3 seconds. In a control test, a blend of 25% polyethylene terephthalate staple fibers with unmodified poly(meta-phenylene isophthalamide) staple fibers containing no phosphorus is prepared and spun to a 20/1 cc yarn of 13.5Z twist. A 5.8 oz/yd² double knit fabric is knitted from this control yarn. This control fabric fails the vertical flame test (1 sample burns of 4 tested).

Jersey Knit Fabrics: Fiber 1-B is blended with polyethylene terephthalate staple fibers in various proportions, as indicated in table below. The staple fiber blends are spun to 20/1 cc yarns of 13.5Z twist and knitted into a 4.2 oz/yd² jersey knit fabric. The flame resistance of the fabrics so produced is determined in the vertical flame test. As shown in the table, fabrics made of yarns containing from 25 to 40% of the polyethylene terephthalate fibers (75 to 60% of Fiber 1-B), pass the vertical flame test. However, fabrics containing higher proportions of the polyethylene terephthalate staple fibers fail to pass the vertical flame test and are regarded as unsatisfactory.

Polyethylene Terephthalate Content	Number of Samples	Char Length (inches)	Duration of Afterflame (seconds)
Satisfactory fabrics, %			
25	2	1.5	2
30	6	2.0	2
35	6	2.0	4
40	8	2.0	4
Unsatisfactory fabrics, %			
45	4	one sample burns	
50	3	one sample burns	

Batiste Fabrics: a staple fiber blend of 60% of Fiber 1-B and 40% polyethylene terephthalate staple fibers is prepared, and the staple fiber blend is spun to a 30/1 cc yarn of 13.5Z twist and woven into a 3.0 oz/yd batiste fabric. In a vertical flame test (5 samples), the char length is 3.0 to 4.0 inches and the after-flame extinguishes itself in 3 seconds. In a control test, a blend of 55% of Fiber 1-B and 45% polyethylene terephthalate staple fibers is similarly spun to a 20/1 cc yarn and woven into a batiste fabric. However, this fabric fails the vertical flame test (3 samples burn of 5 tested).

Example 2: A staple fiber blend of 75% of Fiber 1-B and 25% cotton is prepared. The staple fiber blend is spun to a 20/1 cc yarn of 13.5Z twist and knitted into a 5.8 oz/yd^2 double knit fabric. In the vertical flame test (2 samples), the char length is 1.5 inches and the after-flame extinguishes itself in 3 seconds. A similar blend of 70% Fiber 1-B and 30% cotton, when spun into yarn and knitted into a double knit fabric, gives similar results. However, a staple fiber blend of 60% of Fiber 1-B and 40% cotton, when spun into yarn and knitted into a double knit fabric, fails the vertical flame test (both samples burning of 2 tested).

In a control test, a blend of 25% cotton with 75% unmodified poly(meta-phenylene isophthalamide) staple fibers containing no phosphorus is prepared and spun to a 20/1 cc yarn of 13.5Z twist. A 5.8 oz/yd^2 double knit fabric is knitted from this control yarn. This control fabric fails the vertical flame test (both samples burning of 2 tested).

Aminated Trichlorophosphazene

In a process described by *A.Y. Garner; U.S. Patent 4,053,560; October 11, 1977; assigned to Monsanto Research Corporation*, a $Cl_3P{=}N{-}N{=}PCl_3$ + NH_3 reaction product has been found to be an excellent flame retardant for material made from cellulose such as cotton, paper and sponge, polyester, wool and blends. The material can be treated with an aqueous solution containing a sufficient amount of the product and the material dried to make the material self-extinguishing. Then, the treated and dried material is cured at a sufficient temperature to bond the product to the material.

Preparation of $Cl_3P{=}N{-}N{=}PCl_3$ + NH_3 Reaction Product:

$$H_2N{-}NH_2{\cdot}HCl + 2PCl_5 \longrightarrow Cl_3P{=}N{-}N{=}PCl_3 \xrightarrow{NH_3} (H_2N)_3P{=}N{-}N{=}P(NH_2)_3$$

A slurry of 34.3 g (0.5 mol) of hydrazine monohydrochloride and 208.5 g (1.0 mol) of PCl_5 in 500 ml of benzene was stirred and heated at reflux for 17 hours. After the first half-hour of reflux, most of the solids had dissolved, and the surface of the yellow solution was covered with small bubbles like a foam. Hydrogen chloride gas was detected at the end of the drying tube on top of the condenser. The reaction mixture was cooled, and the dissolved hydrogen chloride and most of the solvent were removed under vacuum leaving a white solid which is $Cl_3P{=}N{-}N{=}PCl_3$.

The solid was slurried with 500 ml of chloroform, and 1,200 ml of liquid ammonia was added cautiously with stirring. The ammonia was contained by a Dry Ice condenser for at least 6 hours; then, it was allowed to evaporate overnight. The resultant white solid was allowed to settle. A portion of the supernatant liquid left no residue upon evaporation indicating the insolubility of the product in chloroform.

Two hundred and nineteen grams (3 mols) of diethylamine was added to the slurry which was heated to reflux and kept there for 23 hours for the purpose of converting the by-product ammonium chloride to soluble diethylamine hydrochloride and ammonia. The resulting slurry was filtered, and the white filter cake was washed with 750 ml of chloroform in 250 ml portions. Air drying followed by vacuum drying at room temperature yielded 121.1 g of product. The infrared spectrum of this material showed strong –NH absorption, a strong band at 8.1 μ and a weaker but broader band at 10.7 μ. An aqueous solution of the product was weakly alkaline. The material was insoluble in chloroform, acetone and cold methanol.

A thermogravimetric analysis revealed that nearly 41% of the material was lost between 25° and 300°C, but from 300° to 715°C only 3% additional weight loss occurred in helium. In air, the weight loss was similar, but from 650° to 900°C another 14% was lost.

Treatment of Fabrics: A solution containing the desired weight percent of the product in sufficient water to just saturate the cloth was poured on a weighed piece of cloth lying flat in a plastic bag. The solution was worked over the surface of the cloth, until it was uniformly wet. After standing for about 15 minutes, the cloth was placed in an oven at the desired temperature and cured. After curing and drying, the cloth was allowed to equilibrate before being weighed.

Cure Conditions: Cure was effected at 140°C using the one operation to also dry the wet sample. Formaldehyde and magnesium chloride catalyst were included in the treatment in some instances to see if this supplemental treatment made the material more durable to washing.

Flammability Tests: Samples of cloth 10" x 3½" were clamped in a metal stand and tested according to AATCC Test Method 34-1969 using a propane torch in place of the special gas mixture.

Tabulation of Data: The test data are tabulated in the following table. The table is divided into three main sections: Fabric Treatment, Flammability Tests and Miscellaneous Conditions. The following column headings are used.

Column	Heading	Explanation
1	Cloth Type	—
2	Reagent	$Cl_3P{=}N{-}N{=}PCl_3 + NH_3$ product
3	Auxiliary	Material used to bind to cloth such as formaldehyde and catalyst;
4	% Final Add-On	Weight percent of product added to the cloth after all processes including laundering if indicated in column 9
5	Distance Burned, in.	This represents the length of the sample that was burned out, charred or scorched from the ignited edge
6	Time, Sec.	Time from ignition to removal of flame even though self-extinguishment had already occurred
7	Self-extinguish (SE)	Answers question–Did the fire self-extinguish before burning the entire sample length? Y = yes; N = no
8	Cure Temp.	Temperature at which the wet cloth was dried and cured in a single operation
9	Post Treatment	Indicates treatment of sample after curing but before flammability test. L = laundered, detergent wash and dried; NL = not laundered

Fabric Treatment				Flammability Test			Miscellaneous Conditions	
Cloth Type	Reagent	Auxiliary	% Final Add - On	Distance Burned, in.	Time Sec	SE	Cure Temp °C	Post Treatment
65/35 polyester/cotton	*	—	19.0	5	5-30	Y	140	NL
65/35 polyester/cotton	*	—	12.15	5	5-30	Y	140	L
cotton	*	—	13.2	4	5-30	Y	140	NL
cotton	*	—	5.0	4	5-30	Y	140	L
cotton	*	CH_2O, $MgCl_2$	5.4	10	10	N	140	L
65/35 polyester/cotton	*	CH_2O, $MgCl_2$	6.67	6	6-30	Y	140	L
100 polyester[1]	*	—	27.9			Y	140	NL
100 polyester[2]	*	—	11.1			Y	140	L
100 polyester[2]	*	CH_2O, $MgCl_2$	35.0	melts		Y	140	NL
100 polyester[2]	*	CH_2O, $MgCl_2$	8.1	melts		Y	140	L
cotton	*	CH_2O, $MgCl_2$	10.6	5	30	Y	140	L
cellulose sponge[3]	*	CH_2O, $MgCl_2$	19.5			Y	140	L
wool	*	—		3	45	Y	140	L
100 polyester[4]	*	—	18.0	3		Y	140	NL
100 polyester[5]	*	—	4.7			N	140	L
cotton[6]	*	—	5.8	$3\frac{1}{4}$	30	Y	140	L
cotton[6]	*	—	5.8	$2\frac{3}{8}$	30	Y	140	L

* $Cl_3P{=}N{-}N{=}PCl_3 + NH_3$ product.
SE = Self-extinguishing.
Y = Yes.
NL = Not laundered.
N = No.
L = Laundered.

[1] Melts and shrinks in flame, does not burst into flame.
[2] No flaming when removed from burner.
[3] Charred and smoked but failed to ignite, some intumescence.
[4] Melted, but no dripping.
[5] Burned with tendency to extinguish.
[6] Same sample on different end.

Reaction Product of Cyanamide and Phosphorus Pentachloride

A.Y. Garner; U.S. Patents 4,040,780; August 9, 1977; and 4,000,191; Dec. 28, 1976; both assigned to Monsanto Research Corporation describes a process whereby flame retardant protection has been demonstrated on cotton and on polyester-cotton materials by the product obtained from the reaction of cyanamide and PCl_5 followed by reaction with anhydrous ammonia in excess of that required to react with the chlorine on the cyanamide and PCl_5 intermediate product.

One hundred and twenty-four and a half grams (0.60 mol) of PCl_5 were added rapidly to a slurry of 25.0 grams (0.60 mol) of cyanamide in 500 ml of chloroform with stirring. There was immediate reaction with the evolution of hydrogen chloride gas. After the initial reaction had subsided, the mixture was heated to reflux and kept there for 18 hours. The resultant, milky slurry was placed under vacuum to remove dissolved hydrogen chloride. An additional 500 ml of chloroform was added followed by the cautious addition of 500 ml of liquid ammonia. The ammonia was contained by a Dry Ice condenser. The excess ammonia was allowed to evaporate overnight.

Evaporation of a sample of the supernatant liquid left no residue. The slurry was mixed with 131.4 grams (1.8 mols) of diethylamine and refluxed for 7 hours to remove the ammonium chloride as ammonia and chloroform soluble diethylamine hydrochloride. After standing over the weekend, the white solid was filtered from the red solution and washed with 500 ml of chloroform. The material appeared to be hygroscopic, so it was subjected to vacuum to remove the remaining chloroform, weight 85.0 grams. The infrared spectrum of this product showed –NH absorption and had a moderately strong band at 2,155 cm^{-1} which suggests cyclization to give a carbodiimide structure. There was also strong absorption at 6.4 and 8.5 μ.

The product, a free-flowing, nonhygroscopic, white powder, was moderately soluble in water yielding a weakly alkaline solution. The material was insoluble in hot cyclohexene, xylene, toluene, THF, DMAC, DMF, pyridine, ethyl acetate, methylene chloride, trichloroethylene, ethanol, nitrobenzene and HMPA. The material shrinks at 93°C in a melting point tube, softens at 120°C forming a soft foam, expanding up the tube. The evolution of gas is noted. The crude product had 19.8% phosphorus and contained chlorine. Recrystallization from water raised the phosphorus content to 22% (theoretical 26%) and the chlorine was almost entirely removed. Analysis—Calculated for CH_6N_5P: C, 10.10%, H, 5.04%, N, 58.80%; P, 26.10%; Cl, 0.00%. Recrystallized Product: C, 12.10%; H, 4.95%; N, 48.80%; P, 21.89%; Cl, 0.30%.

Treatment of fabrics, curing conditions and flammability tests were performed in the same manner as the previous work. The following test data are tabulated below.

Column	Heading	Explanation
1	Cloth Type	–
2	Reagent	Cyanamide/PCl_5/NH_3 product
3	Auxiliary	Material used to bind to cloth such as formaldehyde and catalyst

Explanations 4 to 9 are the same as in the table on p 335.

Fabric Treatment				Flammability Test			Miscellaneous Conditions	
Cloth Type	Reagent	Auxiliary	% Final Add-On	Distance Burned, in.	Time, Sec	SE	Cure Temp. °C.	Post Treatment
65/35 polyester/cotton	*	—	20.8	5	30	Y	140	NL
65/35 polyester/cotton	*	—	12.6	10		N	140	L
cotton	*	—	16.8	3½	30	Y	140	NL
cotton	*	—	5.78	3	30	Y	140	L
65/35 polyester/cotton	*	CH_2O, $MgCl_2$	2.8	10	10	N	140	L
cotton	*	CH_2O, $MgCl_2$	3.4	10	10	N	140	L
65/35 polyester/cotton	*	—	5.67	10	15	N	140	L
50/50 polyester/cotton	*	—	10.8	10	20	N	140	L

*Cyanamide/PCl_5/NH_3.
SE = Self-extinguishing.
Y = Yes.
NL = Not laundered.
N = No.
L = Laundered.
Note: Untreated cotton and cotton/polyester cloths are not self-extinguishing in the flammability test.

Derivatives of Melamine

In related work, *A.Y. Garner; U.S. Patent 3,989,702; November 2, 1976; assigned to Monsanto Research Corporation* describes mixtures comprising compounds of the formula:

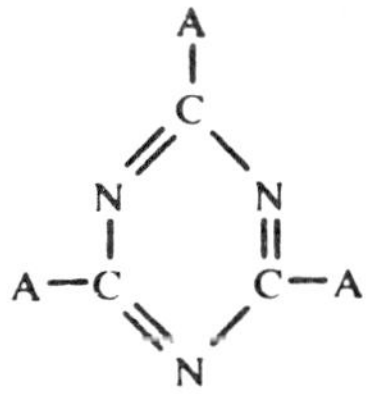

A is either NH_2 or $N{=}P(NH_2)_3$, provided at least one A is $N{=}P(NH_2)_3$. These mixtures have been found to be good flame retardants for material made from cotton or polyester cotton having about 30 to 70% by weight cotton.

The flameproofing reagents were prepared as follows. A mixture of 30.0 grams (0.24 mol) of melamine and 177.5 grams (0.85 mol) of PCl_5 with 200 ml of $CHCl_3$ was heated at gentle reflux with stirring until the evolution of HCl had ceased (44 hours). The top of the condenser was protected by a Drierite tube. The mixture was allowed to cool, and the solvent and trapped HCl were removed on a rotary evaporator leaving 190.7 grams of a white solid which smelled of $CHCl_3$. The solid was reactive to water but was not as reactive as PCl_5.

The solid was placed in a 3-liter flask with 500 ml of toluene. Eight hundred milliliters of liquid ammonia was added in portions. A Dry Ice condenser was used to contain the ammonia. This mixture was stirred overnight. The copious white solid that had formed was filtered to remove the toluene. Twelve hundred milliliters of ammonia was poured cautiously over the filter cake. The solid left on the filter and the solid left after evaporation of the ammonia from the filtrate had identical infrared spectra indicating the presence of both product and ammonium chloride and that the product was soluble in ammonia. The residue from evaporation of the ammonia filtrate was added to the filter cake giving a combined weight of product of 230 grams. The product was soluble in water. The mixture was also soluble in methanol and ethanol and formed a salt with acetic acid. There was no apparent solubility in acetone, acetonitrile, THF, benzene, toluene, chloroform or methylene chloride.

The 230 grams of solids product was slurried in 1,200 ml of chloroform and 315 ml (3 mols) of diethylamine. The mixture was heated slowly to reflux. (This method, see below, affords a chloroform-soluble amine hydrochloride from the ammonium chloride, thus enabling separation of the phosphazene.)

$$(C_2H_5)_2NH + NH_4Cl \longrightarrow (C_2H_5)_2NH_2Cl + NH_3$$

The ammonia was detected at the top of the condenser. The mixture was refluxed overnight. The cooled solution was filtered to remove the red chloroform solution of the amine hydrochloride. The filter cake product was washed with chloroform and air dried, weight 89.9 grams. The infrared spectrum of this product showed no ammonium chloride. An aqueous solution of the product was neutral; however, there was a slight indication of chloride ion using a silver nitrate test.

When heated in a capillary tube, gas was evolved at 113°C to form a foam; at 260°C, more expansion occurred and at 280° to 300°C, the material softened to a thick syrup with more gas evolution. A thermogravimetric analysis on this material in air from room temperature to 980°C left 18% residue. At 300°C, only 20% of the material had decomposed. Another 30% was lost between 300° and 500°C. There was slightly less weight loss in helium. An elemental analysis of the product was as follows: C, 10.17%; H, 5.61%; N, 41.04%; P, 19.06%; Cl, 1.70%.

Treatment of fabrics, curing conditions and flammability tests were performed in the same manner as in the previous work. The test data are shown below preceded by explanations of column headings where necessary.

Column	Heading	Explanation
1	Cloth Type	
2	Reagent	Melamine/PCl_5 + NH_3 product
3	Auxiliary	Material used to bind to cloth such as formaldehyde and catalyst; QB = quaternary base, TMM = trimethylolmelamine

Explanations 4 to 9 are the same as in the table on p 335.

Fabric Treatment				Flammability Test			Miscellaneous Conditions	
Cloth Type	Reagent	Auxiliary	% Final Add-On	Distance Burned, in.	Time, Sec	SE	Cure Temp °C	Post Treatment
cotton 65/35	—	—	0.00	10	1	N	—	NL

(continued)

Fabric Treatment				Flammability Test			Miscellaneous Conditions	
Cloth Type	Reagent	Auxiliary	% Final Add-On	Distance Burned, in.	Time, Sec	SE	Cure Temp °C	Post Treatment
polyester/cotton	—	—	0.00	10	1	N	—	NL
cotton	*	—	5.27	3½	3, 15	Y	120	NL
cotton	*	—	5.27	10	3	N	120	L
65/35 polyester/cotton	*	—	15.2	4	3	Y	140	NL
65/35 polyester/cotton	*	—	15.2	10	3	N	140	L
cotton	*	—	2.5	5½	12	Y	150	NL
cotton	*	—	2.5	10	15	N	150	L
65/35 polyester/cotton	*	—	9.3	6	15	Y	150	NL
65/35 polyester/cotton	*	—	9.3	10	15	N	150	L
65/35 polyester/cotton	*	CH_2O	4.45	10	8	N	120	L
cotton	*	CH_2O, Q.B.[1]	20.3	3	20	Y	130	NL
cotton	*	CH_2O, Q.B.	14.7	6	15	Y	130	L
65/35 polyester/cotton	*	CH_2O, Q.B.	30.7	3	20	Y	130	NL
65/35 polyester/cotton	*	CH_2O, Q.B.	0	10	15	N	130	L
cotton	*	CH_2O, NaOH	7.7	10	10	N	140	L
65/35 polyester/cotton	*	CH_2O, NaOH	6.45	8	8	N	140	L
cotton	*	Q.B.	5.7	10	5	N	140	L
65/35 polyester/cotton	*	Q.B.	6.25	10	5	N	140	L
cotton	*	TMM[2],Q.B.	6.4	4	5-30	Y	140	L
65/35 polyester/cotton	*	TMM, Q.B.	8.9	10	5	N	140	L
65/35 polyester/cotton	*	TMM, Q.B.	9.5	10	10	N	140	L
50/50 polyester/cotton	*	TMM, Q.B.	20.0	4½	3	Y	140	NL
50/50 polyester/cotton	*	TMM, Q.B.	7.6	10	10	N	140	L
65/35 polyester/cotton	*	CH_2O, Q.B.	3.85	10	8	N	140	L
65/35 polyester/cotton	*	TMM, Q.B.	10.7	10	15	N	140	L
65/35 polyester/cotton	*	CH_2O, $MgCl_2$	4.86	10	10	N	140	L
cotton	*	CH_2O, $MgCl_2$	7.1	10	10	N	140	L

[1]Quaternary base.
[2]Trimethylol melamine.
*Melamine/PCl_5 + NH_3.
SE = Self extinguishing.
Y = Yes.
NL = Not Laundered.
N = No
L = Laundered
Note: Untreated cotton and polyester/cotton cloth are not self-extinguishing in the test and data for these materials are given for comparison purposes.

Condensation Products of Hydroxymethyl Phosphonium Compounds and Bicyclic Dienes

H. Nachbur and A. Maeder; U.S. Patent 4,026,889; May 31, 1977; assigned to Ciba-Geigy Corporation describe a process for the manufacture of water-soluble condensation products of hydroxymethyl phosphonium compounds and bicyclic dienes. This is characterized in that (a) 1 mol of a tetrakis(hydroxymethyl)phosphonium compound is condensed with (b) 0.02 to 1 mol, preferably 0.1 to 0.5 mol, of a 2,4,6,8,9-pentaazo-bicyclo-[3.3.1]-nona-2,6-diene of the formula on the following page or its salt.

(1) 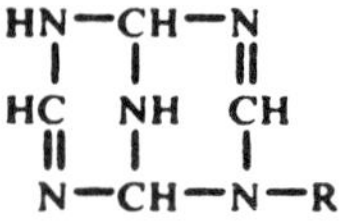

R represents hydrogen, hydroxyl or an optionally substituted aliphatic, cycloaliphatic, aromatic or heterocyclic radical; at 40° to 120°C, optionally in the presence of formaldehyde or a formaldehyde releasing agent.

An appropriate procedure is to heat the tetrakis(hydroxymethyl) phosphonium compound, which is in the form of an aqueous solution, together with the component (b) to the boil and to distill off the water. Among the tetrakis(hydroxymethyl) phosphonium salts used, the halides such as the bromide or the chloride are preferred. Tetrakis(hydroxymethyl) phosphonium chloride is referred to as THPC. The formaldehyde, which is optionally used conjointly in the manufacture of the phosphorus-containing condensation products is preferably in the form of an aqueous solution. Of possible formaldehyde-releasing agents, paraformaldehyde is preferable.

The condensation products are used for flameproofing organic fiber material, especially textiles. An appropriate procedure for this is to treat these materials with an aqueous preparation which contains at least (1) a condensation product and (2) a polyfunctional compound which differs from the condensation products according to (1) and to finish the materials treated in this way by the wet batch process, moist batch process, ammonia process, or the thermofixing process. The component (2) consists of polyfunctional nitrogen compounds. Possible epoxides are all epoxides which are liquid at room temperature and have at least two epoxide groups which are derived from polyhydric phenols. Polyfunctional nitrogen compounds are polyalkylenepolyamines or aminoplast-forming agents or aminoplast precondensates.

Aminoplast-forming agents are those nitrogen compounds which can be methylolated, and aminoplast precondensates are those addition products of formaldehyde to nitrogen compounds which can be methylolated. As aminoplast-forming agents or nitrogen compounds which can be methylolated, there may be used: 1,3,5-amino triazines such as N substituted melamines, for example, N butylmelamine, N-trihalogenomethylmelamines, triazones, and also ammeline, guanamines, for example, benzoguanamines or acetoguanamines, or also diguanamines.

The organic fiber materials to be provided with a flameproof finish are, for example, wood, paper, furs, hides or textiles. In particular, fiber materials of polyamides, cellulose, cellulose-polyester or polyester are flameproofed with fabrics of wool or polyester or mixed fabrics of polyester and cellulose, where the ratio of the polyester constituent to the cellulose constituent is 1:4 to 2:1, being preferred. Thus, for example, so-called 20/80, 26/74, 50/50 or 67/33 polyester-cellulose mixed fabrics can be used. The percentages and parts in the examples which follow are parts by weight or percentages by weight. The relationship of parts by volume to parts by weight is as of ml to g.

Example 1: 244 parts of a 78% strength aqueous solution of THPC (1 mol) and 203 parts (0.5 mol) of a 40% strength aqueous solution of a compound of the formula shown on the following page are combined.

(2) 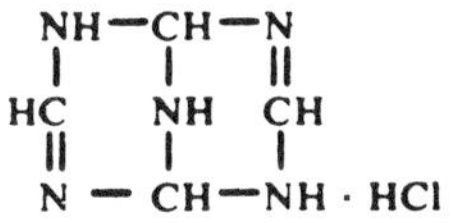

The solution is treated for 2 hours at 100°C internal temperature in a stirred vessel of 500 parts by volume capacity, which is equipped with a thermometer and reflux condenser. Thereafter, the solution is concentrated in vacuo at 60°C to 80% active substance content. 307 parts of a solution of low viscosity are obtained. The active substance content is 81.5%.

Example 2: 244 parts (1 mol) of a 78% strength aqueous THPC solution and 145 parts (0.5 mol) of a 66% strength aqueous solution of a compound of the formula:

(3) 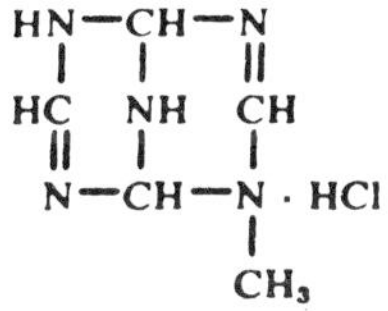

are condensed for 2 hours at 100° to 105°C in a stirred vessel of 500 parts by volume capacity equipped with a reflux condenser and thermometer. After cooling, 388 parts of a yellowish-colored aqueous solution of the condensation product are obtained. The phosphorus content of this solution is 8%.

Example 3: 244 parts (1 mol) of a 78% strength aqueous THPC solution, 21.2 parts of a 35.4% strength aqueous formaldehyde solution (0.25 mol) and 101.5 parts of a 40% strength aqueous solution of a compound of the formula (2) are condensed for 2 hours at 100° to 105°C in a stirred vessel of 500 parts by volume capacity which is equipped with a reflux condenser and thermometer. After cooling, 363 parts of a colorless aqueous solution of the condensation product are obtained. The phosphorus content is 8.55%.

Example 4: Mixed fabrics of polyester-cotton (PES/CO), 50:50 and 67:33, are padded with the liquors according to the following table, dried for 30 minutes at about 80°C and, subsequently, cured for 5 minutes at 150°C.

The fabric is then washed for 5 minutes at 60°C in a liquor which per liter contains 5 ml of hydrogen peroxide (35% strength), 3 g of sodium hydroxide solution (30% strength) and 1 g of a 25% strength aqueous solution of a condensation product of 1 mol of p-tert-nonylphenol and 9 mols of ethylene oxide. Afterwards, it is rinsed and dried. The degree of fixing indicates the amount of the product present on the fiber material after rinsing (relative to the amount originally taken up).

The fabrics are then washed up to 40 times for 45 minutes at 60°C in a domestic washing machine, using a liquor which contains 4 g/l of a domestic detergent (SNV 198,861 wash). The individual fabric samples are then tested for their flameproof character (DIN 53,906 vertical test; ignition time, 6 seconds). The results are summarized in the table shown on the following page.

	Treated with liquors PES:CO 67:33	PES:CO 50:50
Product according to Example 1, g/l	710	710
Dimethylolmelamine, g/l	96.5	96.5
pH value of the liquor (adjusted with NaOH)	5.8	5.8
g of phosphorus per kg of fabric	52	52
Liquor uptake, %	75	75
Degree of fixing, %	70	66
Flameproof character		
After rinsing		
Smouldering time (seconds)	0	0
Tear length (cm)	10.5	8.5
After 20 washes		
Smouldering time (seconds)	0	0
Tear length (cm)	9	7.5
After 40 washes		
Smouldering time (seconds)	0	0
Tear length (cm)	9.5	7

Untreated fabric burns in all the tests of flameproof character.

Condensation Products of Hydroxymethyl Phosphonium Compounds and Cyanamide

In related work, *H. Nachbur and A. Maeder; U.S. Patent 3,994,971; November 30, 1976; assigned to Ciba-Geigy Corporation* describe the manufacture of water-soluble condensation products of hydroxymethyl phosphonium compounds and cyanamide, characterized in that (a) 1 mol of a tetrakis(hydroxymethyl) phosphonium compound is condensed with (b) 0.1 to 0.3 mol of cyanamide at 40° to 120°C. The following examples illustrate this work.

Example 1: 244 parts of a 78% strength aqueous THPC solution (1 mol of THPC), 10.5 parts of cyanamide (0.25 mol) and 200 parts of a xylene isomer mixture are heated to the boil, with rapid stirring, in a stirred vessel of 500 parts by volume capacity which is equipped with a water separator and thermometer. The azeotropic removal of the water from the aqueous THPC solution starts at a boiling point of 102°C. After removal of this water (53.5 parts), the boiling point of the xylene is 130°C. A further 22 parts of water are now removed azeotropically by additional treatment at 130°C, after which the condensation product forms a highly viscous mass. This is cooled to 90°C, the product is dissolved by adding 200 parts of water and the xylene is largely siphoned off. The aqueous solution is completely evaporated in vacuo at 70°C. 168 parts of a highly viscous, colorless condensation product, which gives a copious precipitate with ammonia in aqueous solution, are obtained.

Example 2: 244 parts of a 78% strength aqueous solution of THPC (1 mol of THPC) and 10.5 parts of cyanamide (0.25 mol) are treated for 2 hours at 100° to 105°C internal temperature in a stirred vessel of 500 parts by volume capacity which is equipped with a thermometer and reflux condenser. At the beginning of the treatment, the temperature transiently rises to 110°C. After completion of the treatment, the water is removed in vacuo at 70°C. 199 parts of a colorless, clear precondensate of low viscosity are obtained.

Example 3: The procedure described in Example 2 is followed, but 21 parts of cyanamide (0.5 mol) are used. 185 parts of a yellowish, viscous precondensate are obtained. To facilitate handling on further use, the product is again diluted with water to 70% active substance content.

Example 4: The procedure described in Example 2 is followed, but additionally 21 parts of 37% strength aqueous formaldehyde solution (0.25 mol) are weighed in. After completion of the reaction, the mixture is cooled to room temperature. The reaction product is a clear, colorless solution which contains 74% of active substance.

Example 5: 244 parts of a 78% strength aqueous solution of THPC (1 mol) are neutralized to a pH value of 7.2 by means of 55.6 parts of 30% strength sodium hydroxide solution in a stirred vessel of 500 parts by volume capacity, which is equipped with a thermometer and reflux condenser, and 21 parts of cyanamide (0.5 mol) are subsequently added. Thereafter, the mixture is condensed for 2 hours at 100° to 110°C internal temperature. After cooling, a yellow, clear precondensate of low viscosity is obtained, which contains 63% of active substance.

Example 6: 244 parts of a 78% strength aqueous THPC solution (1 mol), 10.5 parts of cyanamide (0.25 mol) and 200 parts of benzene are heated to the boil, with rapid stirring, in a stirred vessel of 500 parts by volume capacity which is equipped with a water separator and thermometer. A total of 57.5 parts of water are removed azeotropically over the course of 8 hours at a boiling point of 75° to 78°C. Further separation of water is not observed. The reaction product is a very highly viscous mass. The benzene is then largely siphoned off and the resin is dissolved in 200 parts of water. The aqueous solution containing benzene is completely evaporated in vacuo at about 70°C. 194 parts of a viscous condensation product are obtained and are diluted with water to 80% active substance content in order to facilitate handling.

Example 7: Fabrics of polyester/cotton, 67:33 (PES/CO), polyester (PES), cotton serge (CO) and wool gabardine (W) are padded with the liquors as shown in Tables 1 and 3, dried at 80° to 100°C and subsequently cured for 5 minutes at 150°C (thermofixing process). In addition to the thermofixing process, the ammonia fixing process can also be used. The padded fabric is dried at 80°C (not completely), gassed with ammonia for 5 minutes, then padded in a liquor which contains 300 ml of a 24% strength aqueous ammonia solution per liter and, after a dwell time of 5 minutes in air, dried at about 80°C.

The wool fabric and the polyester fabric are washed for 5 minutes at 40°C in a liquor which per liter contains 4 g of sodium carbonate and 1 g of a 25% strength aqueous solution of a condensation product of 1 mol of p-tert-nonylphenol and 9 mols of ethylene oxide. Afterwards, the fabric is rinsed and dried. The polyester-cotton fabric and the cotton fabric are respectively washed for 5 minutes at 60°C or 5 minutes at 95°C with a liquor which per liter contains 5 ml of hydrogen peroxide (35% strength), 3 g of aqueous sodium hydroxide solution (30% strength) and 1 g of a 25% strength aqueous solution of a condensation product of 1 mol of p-tert-nonylphenol and 9 mols of ethylene oxide. Afterwards, the fabric is rinsed and dried. The fabrics are then washed up to 40 times for 45 minutes and at temperatures of 40°C, 60°C, or 95°C in a domestic washing machine, using a liquor which contains 4 g/l of a domestic detergent (SNV 198,861 wash). The individual fabrics are then tested for their flameproof character

(DIN 53,906 vertical test; ignition time, 6 seconds). Untreated fabrics burn away. For pure polyester fabric, the test of the flameproof character is carried out in accordance with the AATCC Test Method 34-1969. The degree of fixing indicates the amount of the flameproofing agent after rinsing, as a percentage of the original uptake. The results are summarized in Tables 2 and 4.

TABLE 1

	Treated with Liquor						
	A. Thermofixing Process				B. Ammonia Fixing Process		
Constituents	**PES/CO**	**PES**	**CO**	**W**	**PES/CO**	**CO**	**W**
Product according to Example 1, g/l	440	430	189	325	440	250	325
Dimethylolmelamine, g/l	103	120	96	70	–	–	–
Condensation product,* g/l	–	–	–	2	2	2	2
pH value of the liquor (adjusted with NaOH)	5.5	5.5	5.5	5.5	5.5	5.5	5.5
Liquor uptake, %	73	80	80	66	73	77	68
Degree of fixing, %	75	81	81	68	88	55	69

*Condensation product of p-tert-nonylphenol + 9 mols of ethylene oxide.

TABLE 2

	Smoldering Time (sec)/Tear Length (cm)						
	A. Thermofixing Process				B. Ammonia Fixing Process		
Flameproof Character	**PES/CO**	**PES**	**CO**	**W**	**PES**	**CO**	**W**
After rinsing	0/12	3/13.5	0/7	0/4	0/11.5	0/4.5	0/2
After 5 washes, 40°C	–	–	–	0/3	–	–	0/1.5
After 10 washes, 40°C	–	–	–	9/7	–	–	6/6
After 20 washes, 60°C	0/10.5	–	0/8	–	0/11.5	0/5	–
After 40 washes, 60°C	0/8.5	–	0/7	–	0/10	0/5	–

TABLE 3

	Treated with Liquor									
	C			D		E		F		G
Constituents	**PES/CO**	**PES**	**CO**	**PES/CO**	**PES**	**PES/CO**	**PES**	**PES/CO**	**PES**	**PES/CO**
Product according to										
Example 2, g/l	440	432	189	–	–	–	–	–	–	–
Example 3, g/l	–	–	–	650	685	–	–	–	–	–
Example 4, g/l	–	–	–	–	–	745	780	–	–	–
Example 5, g/l	–	–	–	–	–	–	–	655	850	–
Example 6, g/l	–	–	–	–	–	–	–	–	–	485
Dimethylolmelamine, g/l	103	120	96.5	96.5	120	96.5	120	72.5	120	96.5
Condensation product,* g/l	–	–	–	–	0.5	–	–	–	–	–
Silicone oil emulsion, 40% strength	30	–	–	–	–	–	–	–	–	–
pH value of liquor, adjusted with NaOH	5.5	5.5	5.5	5.5	5.5	5.5	5.5	5.5	5.5	5.5
Liquor uptake, %	70	90	80	75	80	75	75	70	70	75
Degree of fixing, %	69	75	84	75	78	66	73	68	60	73

* Condensation product of p-tert-nonylphenol + 9 mols of ethylene oxide.

TABLE 4

	Smoldering Time (sec)/Tear Length (cm)									
	C			D		E		F		G
Flameproof Character	PES/CO	PES	CO	PES/CO	PES	PES/CO	PES	PES/CO	PES	PES/CO
After rinsing	0/11	13/12.5	0.75	0/9.5	3/12.5	0/14	–	0/11	6/14.5	0/10.5
After 20 washes, 60°C	0/9.5	–	–	0/9.5	–	0/11	–	0/10	–	0/8.5
After 40 washes, 60°C	0/9.5	–	–	0/9.5	–	0/11	–	–	–	0/10

Condensation Products of Amidophosphates, Triazines and Formaldehyde

H. Nachbur and P. Rohringer; U.S. Patent 4,059,532; November 22, 1977; assigned to Ciba-Geigy Corporation describe phosphorus-containing reaction products which are formed from: (a) an amidophosphate of the formula

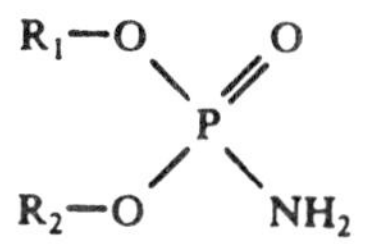

in which R_1 and R_2 each are lower alkyl, halogenoalkyl or alkenyl or R_1 and R_2 together are lower alkylene; (b) a 1,3,5-triazine substituted by at least 2 primary amino groups; (c) formaldehyde and, optionally, (d) a lower alkanol.

Suitable as constituent (b) or as 1,3,5-triazines are, e.g., acetoguanamine, formoguanamine, benzoguanamine or, in particular, melamine. In the case of constituent (c), this is preferably formaldehyde itself, or an agent releasing formaldehyde such as paraformaldehyde. The constituent (d) is preferably not used. It may, however, be desirable to use in some cases products of which the free methylol groups are etherified. Suitable as etherification constituent (d) are in this case, e.g., n-butanol, n-propanol, ethanol or methanol.

The phosphorus-containing reaction products can be produced by conventional methods known per se. An advantageous procedure is to react the constituents (a), (b) and (c) with each other at a temperature of 70° to 180°C. All three constituents are simultaneously reacted together in the so-called single vessel process. The phosphorus-containing reaction products are suitable, in particular, for use as effective and permanent fireproofing agents for cellulose-containing fiber materials. The following examples illustrate the process. Percentages and parts are by weight.

Example 1: 61.2 parts of hexamethylolmelamine (0.2 mol) and 153 parts of O,O'-diethylamidophosphate (1 mol) as well as 250 parts of benzene are heated to boiling. After boiling for 4 hours, the calculated amount of 18 parts of water distills off azeotropically. The boiling temperature is initially 76°C and, at the end of the reaction, 81°C. The reaction mixture is cooled to 20°C and the benzene-insoluble, higher-condensed parts are removed by filtration. There are obtained 21 parts of benzene-insoluble white powder.

The benzene solution is concentrated under reduced pressure. There are obtained 174 parts of a highly viscous product which is soluble in the cold state both in water and in trichloroethylene. The separation of the product by means of gel permeation chromatography (GPC) gives the following composition: approximately 35% of condensation product of molecular weight ~1,000, approximately 25% of condensation product of molecular weight ~1,400, approximately 30% of condensation product of molecular weight >2,000, approximately 5% of O,O'-diethylamidophosphate, approximately 5% of benzene. For the purpose of better handling, there is prepared from the highly viscous product an 80% aqueous solution (P content = 12.9%).

Example 2: A cotton fabric is padded with the aqueous liquors A and B given in the following table. The liquor absorption is 80%. The material is dried for 30 minutes at 80°C and then cured for 4½ minutes at 160°C. A portion of the fabric is subsequently washed at 95°C for 5 minutes in a solution containing per liter of water 4 g of anhydrous sodium carbonate and 1 g of an addition product of 1 mol of 4-nonylphenol and 9 mols of ethylene oxide; the material is afterwards rinsed and dried. A further portion of this fabric is now washed 20 or 30 times in a solution at 95°C, which solution contains 2 g of anhydrous sodium carbonate and 5 g of soap per liter of water. The individual pieces of fabric are subsequently tested with respect to their fireproofness (vertical test according to DIN 53906). The results of these tests are given in the table below.

	Untreated	. Treated with .	
		A	B
Constituents, g/l			
Product of Example 1	–	290	290
Ditrimethylolmelamine	–	120	–
Trimethylolmelaminedimethyl ether	–	–	178
2-Amino-2-methyl-1-propanol hydrochloride	–	40	40
pH of bath	–	5.8	5.3
Degree of fixing, %	–	62	75
Fireproofness*			
After subsequent washing			
BT	burns	0	0
SL	–	12.5	12
After 20 washings			
BT	burns	0	0
SL	–	12.5	12
Handle**after subsequent washing	0	½	2¼

*BT is burning time in seconds and SL is spreading length in centimeters.
**Handle scale: 0 = unchanged, 1 = trace stiffer than 0, 2 = somewhat stiffer than 0, 3 = stiff, and 4 = very stiff.

Quaternary Phosphonium Salts

R.G. Weyker and W.F. Baitinger, Jr.; U.S. Patent 4,026,711; May 31, 1977; assigned to American Cyanamid Co. describe stable, aqueous, flame retardant finishes for cellulosic textile materials employing tetrakis(hydroxymethyl) phosphonium phosphate, tetrakis(hydroxymethyl) phosphonium carboxylate and urea or a urea-formaldehyde condensate. Of the combined phosphonium salts used in the flame retardant finish of this process, about 30 to 70 mol percent is tetrakis(hydroxymethyl) phosphonium phosphate and about 70 to 30 mol percent is tetrakis(hydroxymethyl) phosphonium carboxylate. The finish also contains 1.0 to 1.5 mols of urea and/or urea-formaldehyde condensate per mol of combined phosphonium salts.

The aqueous solutions of tetrakis(hydroxymethyl) phosphonium phosphate and tetrakis(hydroxymethyl) phosphonium carboxylate can be prepared by reacting in aqueous medium 1 mol of tris(hydroxymethyl) phosphine with at least 1 mol of formaldehyde and essentially 1 mol of combined orthophosphoric and carboxylic acid. The phosphoric acid should represent between 10 and 70 mol percent of the combined acids, the remainder is carboxylic acid. The reaction temperature should be between 20° and 100°C, preferably between 25° and 50°C.

The aqueous flame retardant finishes are applied to the textile material by padding, dipping, spraying, etc. The materials are then dried at a temperature between about 75° and 210°C or higher. The drying time can range from several minutes at the lower temperature to as briefly as 15 seconds at the higher temperature. The finish is then cured at a temperature of between about 125° and 225°C, preferably between about 150° and 200°C. The time required for curing the finish is between 5 and 1 minute, depending on the temperature and the weight and structure of the textile material. The following examples illustrate the process.

Example 1: To a reaction kettle containing 1,240 g (10.0 mols) of tris(hydroxymethyl) phosphine (about 1,308 g of 95% material) and 390 g (13.0 mols) of formaldehyde (as 1,050 g of 37% aqueous formaldehyde) there was added, at a temperature of 45° to 50°C, 392 g (4.0 mols) of orthophosphoric acid (461 g of 85% H_3PO_4), followed by 360 g (6.0 mols) of acetic acid (361 g of 99.8% glacial acetic acid), while at all times maintaining a temperature of 25° to 50°C. The reaction mixture was stirred at 50°C for 1 hour and then cooled to below 30°C. The product was an aqueous solution of essentially tetrakis(hydroxymethyl) phosphonium acetate, tetrakis(hydroxymethyl) phosphonium phosphate and formaldehyde in the calculated molar ratio of 0.6/0.4/0.3. The solution contained about 9.7% active phosphorus; i.e., phosphine-derived phosphorus.

Example 2: An aqueous pad bath was prepared containing 31.1% of a product prepared by the process of Example 1, (9.65% active phosphorus) and 9.0% of urea. The pad bath, containing 3.0% active phosphorus, was applied to cotton sheeting by a standard padding procedure, obtaining a wet pickup of 95%. The padded fabric, containing 2.85% owf of active phosphorus, was dried at 107°C (225°F) for 4 minutes and then cured at 163°C (235°F) for 4 minutes. The durability of the flame retardant finish to laundering was determined by (1) repeatedly washing the fabric in an automatic washing machine using a commercial detergent and water at 140°F, and (2) measuring the flame resistance of the dried fabric after about every 10 launderings by a vertical flame test according to Standard Test Method AATCC 34-1966. The limit of practical durability on a cotton fabric is reached when the char length is 6 inches. After 100 launderings, the char length was about 4.3 inches, which indicates that the flame retardancy of a finish of this process on a 100% cotton sheeting is durable for at least 100 normal home launderings in an automatic washing machine.

Example 3: An aqueous pad bath was prepared containing 36.3% of a product prepared by the process of Example 1 (9.65% active phosphorus) and 9.0% urea. The pad bath, containing 3.5% active phosphorus, was applied to a 50/50 polyester/cotton fabric obtaining a wet pickup of 78%. The padded fabric, containing 2.73% owf of active phosphorus, was dried at 107°C for 4 minutes and then cured at 163°C for 4 minutes. The durability of flame retardant finish was measured by the procedure of Example 2. The practical limit of wash durability of polyester/cotton fabric is reached when the char length is about 7.0 inches. The flame retardant finish was durable for 70 to 80 home launderings.

Example 4: Two aqueous pad baths were prepared of the following compositions.

Pad Bath Composition	A	B
Product of Example 1 (9.65% active P)	25.9%	25.9%
% Active P in Bath	2.5%	2.5%
Resin A[1]	6.0%	—
Resin B[2]	—	6.0%
Urea	4.0%	9.0%

[1]Resin A is 75% highly methylated hexamethylol melamine and 25% highly methylolated urea.
[2]Resin B is dimethylol melamine.

The pad baths were applied to 100% cotton fabrics of various weights and structures by a standard padding procedure. The padded fabrics were dried at 107°C for 4 minutes and cured at 163°C for 4 minutes. The fabrics were then process washed with water at about 50°C and dried. The durability of the finishes to laundering was determined by the procedure of Example 2. The results, including percent wet pickup, percent active phosphorus on the fabric and durability of the finishes, are shown in the table below.

	Cotton Fabric	% Wet Pickup	% Active P OWF[1]	Durability - No. of Washes Pad Bath A	Pad Bath B
a.	Poplin 96 × 40, 2.62 oz.	79	1.97	—	85
b.	Print Cloth 64 × 40, 5.35 oz.	128	3.21	>100	>100
c.	Print Cloth 80 × 80	87	2.17	—	>100
d.	Tubular Rib Knit	118	2.95	>100	>100
e.	Flannel	114	2.85	85	>100
f.	Sheeting	99	2.47	85	>100

[1]On weight of fabric

This example demonstrates that finishes of this process provide flame retardancy on 100% cotton fabrics of various weights and structures, and that the finishes are very durable to repeated launderings.

Phosphoalkylated Polyamines

N.S. Marans and A. Gluecksmann; U.S. Patent 4,045,396; August 30, 1977; assigned to W.R. Grace & Co. describe a method of imparting simultaneously flame retardant and wrinkle resistance characteristics to cellulosic materials by impregnating the materials with the phosphoalkylated reaction product of a polyisocyanate and an alkylene amine. The polyisocyanate, which is the reaction product of polyethylene glycol and an aryl isocyanate, e.g., tolylene diisocyanate, is treated with diethylene triamine or triethylene tetramine in ketone solution, followed by removal of the ketone, then phosphoalkylation either by addition of phosphorous trichloride, water and aldehyde or with phosphorous acid and then aldehyde. Cellulosic fabric is then dipped into the water-soluble reaction product in the presence of formaldehyde or other conventional crosslinking agents, such as 1,4-dichloro-2-butene, 1,4-dibromo-2-butene and epichlorohydrin, allowed to soak and then heated. The resulting fabric has excellent wrinkle resistance and flame retardant properties. The following examples illustrate the process.

Example 1: (A) Diethylene triamine and anhydrous acetone are mixed in a mol ratio of 1.0 to 6.75 respectively at 25°C for 24.0 hours. The resulting product is designated as (A).

(B) Polyethylene glycol of molecular weight 600 and a mixture of 2,4- and 2,6-tolylenediisocyanate are allowed to react by addition over a period of 45 minutes of the glycol to the diisocyanate held at a temperature between 30° and 60°C. The mol ratio of the glycol to the diisocyanate is 1 to 2 respectively. The resulting product is designated as (B).

(C) (A) is charged to a flask and upon vigorous stirring and cooling to maintain temperature below 10°C (B) is added to it over a period of 35 minutes. The mol ratio of (A) to (B) is 2 to 1 respectively. The resulting product is designated as (C).

(D) Acetone is removed from reaction product (C) by distillation. To 98.5 g of the above prepared polyamine, 200 cc of distilled water is added followed by the gradual addition upon cooling to maintain temperature below 30°C over a period of 10 minutes, of PCl_3 which in turn is followed by the gradual addition over a period of 5.0 minutes of a 37% solution of CH_2O.

A sodium hydroxide solution containing 3 mols of NaOH per mol of PCl_3 is used to trap the HCl which evolves in the course of the PCl_3 addition. The reaction mixture is then heated up to 65°C and held at that temperature for 2½ hours. The molar ratio of PCl_3 to CH_2O to reaction product (C) is respectively 6.0:6.0:1.0.

Example 2: The same procedure is followed as in Example 1 with the exception that as alkylene amine, triethylenetetramine is used and thus the mol ratio of the amine to anhydrous acetone is 1.0 to 13.5 and the molar ratio of PCl_3 to CH_2O to the reaction product analogous to (C) in Example 1 is respectively 8.0:8.0:1.0.

Example 3: After bringing the pH of the final reaction product of Example 1 up to about 5.0 with a solution of NaOH, 15 drops of 37% formaldehyde are added to 23.2 g of the above product. A minor portion of the formed mixture was placed on a watch glass and heated at 150°C for 30 minutes to evaporate the water. A crosslinked self-supporting film of polymer resulted. A cotton cloth sample is allowed to soak in the remaining portion of the mixture for 3.0 minutes followed by drying of the sample at 100°C for 10 minutes in a vacuum oven.

Flame testing reveals that the sample thus treated will not support combustion when removed from the hottest part of a Bunsen burner flame. Also, within the flame, the sample will char but will not visibly burn away as is the case with an identical untreated sample; the untreated samples will ignite within 8 seconds when held 2½ inches above flame's tip; the treated sample will not ignite and only after 20 seconds will it start to gradually turn black. The approximate % add-on after cure is between 15 and 35%. The wrinkle resistance characteristic is as expected equivalent to that of clothes conventionally polyurethane treated for that purpose.

Example 4: In this example, samples of cotton cloth were allowed to soak in the reaction product prepared as described in Example 2, identical qualitative and quantitative evaluation techniques were used as outlined in Example 3 and as expected no significant differences were noted with regards to the flame retardance and wrinkle resistance characteristics as compared to those of Example 3.

Example 5: The reaction product from Example 2, 0.294 mol based on the amount of triethylenetetraamine, was treated with an aqueous solution of 6.0 N NaOH to give a pH of 5.0. To this solution, there was added 1.0 mol of 1,4-dichloro-2-butene. The resultant mixture was used to impregnate cotton and the impregnated cotton was heated for 30 minutes at 100°C. The cotton sample did not support combustion.

Stable Oil and Water Emulsions

R.J. Berni, M.M. Smith, and R.R. Benerito; U.S. Patent 4,084,027; April 11, 1978; assigned to U.S. Secretary of Agriculture describe a method of applying flame retardant water-soluble phosphorus- and nitrogen-containing compounds and water-insoluble halogenated flame retardant compounds to a cotton-polyester fabric. This method comprises:

(a) Incorporating an emulsifying agent of high Hydrophilic-Lipophilic Balance (HLB) Number of at least 14, in the aqueous phase of an aqueous phase-hydrocarbon phase two phase system, and an emulsifying agent of low HLB Number of 9 or less in the chlorinated hydrocarbon phase, thus providing a stable emulsion of HLB Number of about from 9.6 to 15.1. The ratio of aqueous phase to chlorinated hydrocarbon phase is about from 3:1 to 2:1. The stable emulsion has a suitable shelf life for finishing simultaneously the cotton and the polyester fiber when the final emulsion contains, respectively, the phosphorus and nitrogen flame retardants in the water phase and the halogenated flame retardant in the water-insoluble phase;

(b) Impregnating the fabric with the stable emulsion; and

(c) Curing the impregnated fabric.

The phosphorus derivatives applicable to the formulations are tetrakis(hydroxymethyl) phosphonium chloride (THPC) and its neutralization adducts such as tetrakis(hydroxymethyl) phosphonium hydroxide, -oxylate, -citrate, -phosphate, etc., that impart flame retardancy to cotton. Urea and N-methylolated ureas are used to react with THPC reagent so that some nitrogen becomes part of the polymer finish, while N-methylolated melamines are used to make the polymer and an unreacted tris(2,3-dibromopropyl) phosphate in the cotton-polyester blended textile more durable to laundering. Any of a number of methylolated melamines can be used; more particularly, trimethylolated melamine (TMM) is preferred. Sodium hydroxide is added to help neutralize the THPC, but any strong base could be used, including potassium or calcium hydroxide or tetramethyl-, ethyl-, propyl-, or butylammonium hydroxides. The water-insoluble halogenated reagent can be any of a number of hydrocarbon phosphate type compounds, including tris(2,3-dibromopropyl) phosphate and phosphates containing mixtures of brominated and chlorinated species known to impart flame retardancy to polyester fibers and/or fabrics.

Suitable emulsifying reagents in the present process include the hydrophilic polyoxyethylene sorbitan lauric-, palmitic-, stearic-, oleic-acid esters and the like used in conjunction with the lipophilic (hydrophobic) sorbitan lauric-, palmitic-, stearic-, and oleic-acid esters. The following examples illustrate the process.

Example 1: A cotton-polyester 50/50 blended flannelette sample weighing 19 g was padded twice to 95% wet pickup with 250 g of an emulsion of the following compositions:

Component The Water Phase:	Percent by Weight of Final Emulsion
THPC	19.0
Urea	5.2
TMM*	3.5
NaOH	3.2
H_2O	42.9
Tween 40**	1.0
The Oil Phase:	
Tris(2,3 dibromopropyl) phosphate	18.4
Perchloroethylene	6.1
Span 40**	0.7

*TMM refers to trimethylolated melamine.

**Tween and Span 40 are commercially available emulsifiers and are principally polyethylene sorbitol palmitic acid ester and sorbitol palmitic acid ester, respectively.

The two phases were prepared separately and then stirred in a Waring blender for 2 minutes at high speed setting. The impregnated fabric was cured at 180°C for 2 minutes, rinsed in hot tap water for 15 minutes, soaked in 3% hydrogen peroxide solution for 30 minutes, rinsed in cold tap water for 5 minutes, and then dried in a forced draft oven at 60°C for 7 minutes. Each sample was allowed to equilibrate overnight before being weighed, tested, and analyzed.

The treated fabric had a weight gain of 39.3%, and nitrogen, phosphorus, and bromine contents, respectively, of 3.4%, 3.0%, and 6.8%. The char length before laundering was 0.9 inch, as compared to 10 inches for an untreated control fabric; the oxygen index (OI) value before laundering was 32.4% compared to a control value of 21.3%. The breaking strength retained was 107% of the original value or 30.9 pounds, compared to an untreated fabric of 28.9 pounds. After 20 AATCC home launderings, the char length was 2.3 inches while the OI value was 28.0%. After 50 home launderings, the char length was 3.2 inches and the OI value was 26.7%. The char length evaluation was done by the Department of Commerce FF 3-71 procedure. The 3.2 inches figure is acceptable in evaluating a durable flame retardant treatment for sleepwear. The breaking strength after 50 home launderings was still 96.5% of the original control value. Durable flame retardancy is defined as the ability of a treated fabric to pass the Department of

Commerce DOC FF 3-71 test after 50 home launderings without bleach. A char length of less than 7 inches after 50 launderings will pass the DOC FF 3-71 test. Oxygen Index values of greater than 26% are acceptable for cotton-polyester blended fabrics. The emulsions prepared by the formulations of this process were allowed to stand for 2 weeks and found to be without separation of layers. This is a suitable shelf life for these emulsions. In comparison to the emulsions, the use of self-emulsifiable, commercially available tris(2,3-dibromopropyl) phosphate formulation resulted in separation after a few days.

Example 2: A series of five cotton-polyester blend fabric samples were treated as in Example 1 with the exception that the ratio of emulsifying reagents (Tween 40/Span 40) were varied, in order to vary the final HLB Number of the emulsions and cure time was 3 minutes in every instance. All ratios within this range gave stable emulsions and fabrics treated with these emulsions exhibited durable flame retardancy as shown by their char lengths of less than 7 inches after 50 launderings. Results are shown in the following table.

Sample No.	% Tween 40	% Span 40	HLB No.	% Gain	% N	% P	% Br	Char Len. (in.)	% OI
A	32	68	9.55	39.0	3.6	3.2	7.0	1.3	32.0
B	52	48	11.32	37.9	3.5	3.0	5.8	1.9	31.4
C	72	28	13.08	36.8	3.4	2.9	6.6	1.5	31.7
D	87	13	14.47	37.1	3.4	2.9	6.0	1.3	31.6
E	94	6	15.10	36.1	3.5	3.0	5.6	1.4	31.6

Nitrilo Methylol-Phosphorus Polymers

D.J. Donaldson, G.L. Drake, Jr., J.V. Beninate, W.A. Reeves, and D.J. Daigle; U.S. Patent 4,067,687; January 10, 1978; assigned to U.S. Secretary of Agriculture describe a procedure for textile treating compositions containing the elements phosphorus, nitrogen, and bromine and the processes for imparting flame resistance to organic fibrous materials. This is accomplished by impregnating the textile treating compositions consisting of: (1) tris(hydroxymethyl) phosphine mixed with nitrogenous compounds containing at least two members of the group hydrogen and methylol radicals ($-CH_2OH$) attached to trivalent nitrogen atoms, or (2) soluble methylol phosphine adducts which contain one or more methylol phosphine radicals ($P-CH_2OH$) prepared by reacting tris(hydroxymethyl) phosphine with an amide or amine which contains one or more hydrogens attached to a trivalent nitrogen with an organic bromine containing material with at least 10% bromine in the form of bromine atoms attached to carbon atoms. The proportions are selected so that generally the phosphorus contained in structures of the group $(-CH_2)_3P$ amount to a small portion of the total weight of the mixture. The mixture is heated to cause insolubilization in or on the fibrous structure. The following examples illustrate this process.

Example 1: Twill fabrics were padded with solutions containing both a THP-cyanamide (1:1) adduct solution prepared by allowing a mixture of 62 parts THP, 42 parts 50% cyanamide and 90 parts of water to stand at room temperature for 1 hour and various brominated compounds. The fabrics were dried at 85°C for 3 minutes and cured at 155°C for 3 minutes.

Solution A contains 10 parts adduct solution, 25 parts 50% polyvinyl bromide, 5 parts H_2O and 10 parts alcohol. The ratio of adduct to polyvinyl bromide is therefore about 1:2:5. Solution B contains 5 parts adduct solution, 20 parts brominated triallylphosphate (prepared by reacting 22 parts triallylphosphate, 51 parts bromoform and 1 part potassium persulfate). The ratio of adduct to brominated triallylphosphate is therefore about 1:8.3. Solution C contains 10 parts adduct solution, 20 parts tris(2,3-dibromopropyl) phosphate, 10 parts H_2O. The ratio of adduct to tris(2,3-dibromopropyl) phosphate is therefore about 1:4.

Solution Used	Type Fabric	% Add-on	Match Test Angle	Color
A	Cotton	15.9	110	Sl. Brown
	Polyester/cotton	18.4	180	"
B	Cotton	4.0	0	White
	Polyester/cotton	2.0	0	"
C	Cotton	18.0	110	"
	Polyester/cotton	15.0	135	"

All of the treated fabrics exhibited good hand and strength and formed a black carbonaceous char when exposed to flame, indicating the presence of phosphorus.

Example 2: Twill fabrics were padded with solutions containing both a THP-Cyanamide (2:1) adduct solution prepared by heating on a steam cone for 10 minutes a mixture consisting of 62 parts THP, 21 parts 50% cyanamide and 90 parts H_2O and various brominated compounds. The fabrics were dried at 85°C for 3 minutes and cured at 155°C for 3 minutes.

Solution A contains 10 parts adduct solution, 20 parts 50% polyvinylbromide, 10 parts alcohol. The ratio of adduct to polyvinylbromide is therefore about 1:2. Solution B contains 5 parts adduct solution, 20 parts brominated triallylphosphate (prepared by reacting 22 parts triallylphosphate, 51 parts bromoform and 1 part potassium persulfate). The ratio of adduct to brominated triallylphosphate is therefore about 1:8. Solution C contains 10 parts adduct solution, 20 parts tris(2,3-dibromopropyl) phosphate, 10 parts H_2O. The ratio of adduct to tris(2,3-dibromopropyl) phosphate is therefore about 1:4.

Solution Used	Type Fabric	% Add-on	Match Test Angle	Color
A	Cotton	14.5	110	Sl. Brown
	Polyester/cotton	14.0	90	"
B	Cotton	7.0	45	White
	Polyester/cotton	5.0	30	"
C	Cotton	15.6	145	"
	Polyester/cotton	18.1	135	"

All of the treated fabrics exhibited good hand and strength and formed a black carbonaceous char when exposed to flame, indicating the presence of phosphorus.

Example 3: Twill fabrics were padded with solutions containing both a THP-Urea (2:1) adduct solution prepared by allowing a mixture of 62 parts THP, 30 parts urea and 90 parts H_2O to stand at room temperature for 4 hours and various brominated compounds. The fabrics were dried at 85°C for 3 minutes and cured at 155°C for 3 minutes.

Solution A contains 10 parts adduct solution, 20 parts 50% polyvinylbromide, 10 parts alcohol. The ratio of adduct to polyvinylbromide is therefore about 1:2. Solution B contains 5 parts adduct solution, 20 parts brominated triallylphosphate (prepared by reacting 22 parts triallylphosphate, 51 parts bromoform and 1 part potassium persulfate). The ratio of adduct to brominated triallylphosphate is therefore about 1:8. Solution C contains 10 parts adduct solution, 20 parts tris-2,3(dibromopropyl) phosphate, 10 parts H_2O. The ratio of adduct to tris-2,3(dibromopropyl) phosphate is therefore about 1:4. Solution D contains 30 parts adduct solution, 10 parts brominated melamine (prepared by dissolving in 20 parts 37% formaldehyde the product obtained by slowly adding 16 parts bromine to 10.3 parts diallylmelamine dissolved in acetic acid. The ratio of adduct to brominated melamine is therefore about 1:0.6. Solution E contains 30 parts adduct solution, 10 parts brominated acrylamide (prepared by heating for 1 hour at 80°C 7.1 parts acrylamide, 15.2 parts bromoform and 0.6 part potassium persulfate). The ratio of adduct to brominated acrylamide is therefore about 1:0.6.

Solution Used	Type Fabric	% Add-on	Match Test Angle	Color
A	Cotton	18.1	120	Sl. Brown
	Polyester/cotton	17.6	110	"
B	Cotton	8.0	0	Sl. Yellow
	Polyester/cotton	8.0	0	"
C	Cotton	16.0	110	White
	Polyester/cotton	14.0	135	"
D	Cotton	19.0	180	"
	Polyester/cotton	24.0	135	"
E	Cotton	28.0	180	"
	Polyester/cotton	28.0	180	"

All of the treated fabrics exhibited good hand and strength and formed black carbonaceous char when exposed to flame, indicating the presence of phosphorus.

Example 4: Cotton or polyester-cotton twill fabric was treated with a mixture containing 5 parts THP, 5 parts trimethylolmelamine, 61 parts brominated triallylphosphate (prepared as in Example 3) and 15 parts H_2O. The fabrics were dried at 85°C for 3 minutes and cured at 160°C for 3 minutes. The cotton sample had 19.7% add-on, poor strength and a metal test angle of 90° while the polyester-cotton sample had 6% add-on, good strength and a match test angle of 45°.

Example 5: 10 parts of a THP-cyanamide adduct, prepared by allowing to stand 1 hour a solution containing 62 parts THP, 42 parts 50% cyanamide and 90 parts water, and 25 parts polyvinylbromide, 5 parts H_2O and 10 parts alcohol were heated in an oven at 150°C for 1 hour. The resultant product was a black hard polymer which resisted burning when exposed to flame. The polymer was insoluble in alcohol and H_2O.

Example 6: A THP-cyanamide adduct was produced by heating on a steam cone for 10 minutes a solution containing 62 parts THP, 21 parts 50% cyanamide and 90 parts H_2O. 10 parts of this adduct was mixed with 20 parts of tris(2,3-dibromopropyl) phosphate and 10 parts H_2O. The mixture was heated 1 hour at 155°C to form a yellow rubber-like polymer which resisted burning when exposed to flame and formed a black char. The polymer is insoluble in alcohol, and at most, very slightly soluble in H_2O.

Derivatives of Diaminotrihalopropyltriazines

A process described by *L.H. Chance and J.D. Timpa; U.S. Patent 4,055,720; October 25, 1977; assigned to U.S. Secretary of Agriculture* relates to diaminotrihalopropyltriazines, to their methylol derivatives, to the preparation and to processes for imparting durable flame retardancy and wrinkle recovery to cellulosic materials, protein fibers, blends of cotton and polyester fibers by treating these materials with the methylol derivatives.

As examples of the process, 2,4-diamino-6-(3,3,3-tribromo-1-propyl)-1,3,5-triazine and 2,4-diamino-6-(3,3,3-trichloro-1-propyl)-1,3,5-triazine were prepared by the reaction of the corresponding methyl or ethyl γ-tribromobutyrate or methyl or ethyl γ-trichlorobutyrate with biguanide. The N-methylol derivatives of the diaminotrihalopropyltriazines were prepared by the reaction of the triazines with aqueous formaldehyde. Two pure tetramethylol derivatives were prepared, namely: 2,4-bis[bis(hydroxymethyl)amino]-6-(3,3,3-tribromo-1-propyl)-1,3,5-triazine and 2,4-bis[bis(hydroxymethyl)amino]-6-(3,3,3-trichloro-1-propyl)-1,3,5-triazine.

Example 1: Preparation of Ethyl γ-Tribromobutyrate – γ-Tribromobutyronitrile (576.5 g, 1.89 mols), water (34.52 g, 1.92 mols), and 1,000 ml of absolute ethanol were placed in a 2-liter flask equipped with a mechanical stirrer, a gas inlet tube, and a reflux condenser topped with a calcium sulfate drying tube. While stirring, dry hydrogen chloride gas was bubbled slowly into the mixture for about 50 minutes.

The mixture became warm spontaneously. At this point, HCl addition rate was increased to a rapid stream. After about 5 minutes, all of the nitrile dissolved, the solution refluxed vigorously and NH_4Cl began to precipitate. Rapid addition was continued for 10 minutes more and the HCl addition rate was reduced to a slow stream and continued for a total addition time of 2 hours. Stirring was continued for 2 hours more. Then, the mixture was heated slowly on a water bath until reflux began. It was allowed to cool to room temperature.

The NH_4Cl was removed by filtration. The ethanol was removed by rotary evaporation. The crude liquid ester remained as a residue containing a small amount of solid. The ester was dissolved in 300 ml of benzene and the resulting solution filtered to remove the solid. The benzene solution of the ester was shaken in a separatory funnel with 300 ml of water containing 10 g of Na_2CO_3. The benzene layer was separated and dried over anhydrous Na_2SO_4. The benzene was removed by vacuum distillation leaving a 100% yield of crude ethyl γ-tribromobutyrate. The crude ester was vacuum distilled, a yield of 98.6% of a clear colorless liquid was obtained, BP 73°C/0.13 mm. It had a pleasant odor characteristic of esters. Analysis–Calculated for $C_6H_9Br_3O_2$: C, 20.42%; H, 2.57%; Br, 67.94%. Found: C, 20.52%; H, 2.57%; Br, 68.03%.

Example 2: Preparation of 2,4-Diamino-6-(3,3,3-Tribromo-1-Propyl)-1,3,5-Triazine (DABT) – Absolute methanol (1,300 ml) was placed in a 3-neck flask equipped with a mechanical stirrer and a reflux condenser topped with a soda lime trap to exclude atmospheric CO_2. Metallic sodium (28.3 g, 1.23 g atoms) was added portion-wise in small pieces. This resulted in a methanolic solution of sodium methylate. (An alternative procedure would be to dissolve 1.23 mols of crystalline sodium methylate in the methanol.) Anhydrous biguanide sulfate (184.7 g,

0.615 mol) was added to the solution. (The anhydrous biguanide sulfate was obtained by drying the dihydrate of biguanide sulfate to constant weight at 110°C.) The mixture was refluxed for 3 hours and then cooled to room temperature. The resulting solution contained the free base, biguanide, and sodium sulfate as a white precipitate. It was not necessary to filter off the Na_2SO_4 because it does not interfere with the next step. Crude ethyl γ-tribromobutyrate (434.7 g, 1.23 mols) was added in a slow stream, with stirring, over a period of about 20 minutes. Stirring was continued at room temperature for about 18 hours. The white precipitate, which contained both sodium sulfate and the DABT, was filtered. The filter cake was washed with 250 ml of fresh methanol and dried in an oven at 70° to 85°C.

In order to separate the DABT from the sodium sulfate, the precipitate was stirred with 725 ml of water, filtered, and the filter cake washed with 500 ml of water. Finally, the filter cake was washed with 125 ml of cold methanol and dried at 65°C. The DABT was a white powder, weighing 304.6 g, a yield of 63.5%. The melting point was 170° to 171°C, with decomposition. The yield of DABT was no better when pure distilled ethyl γ-tribromobutyrate was used. Analysis–Calculated for $C_6H_8Br_3N_5$: C, 18.48%; H, 2.07%; Br, 61.49%; N, 17.96%. Found: C, 18.26%; H, 2.05%; Br, 61.56%; N, 18.02%. DABT was prepared in the same manner as described above except that methyl γ-tribromobutyrate was used. A yield of 50.5% was obtained, which was significantly lower than when the ethyl ester was used.

Example 3: Preparation of 2,4-Bis[bis(hydroxymethyl)amino]-6-(3,3,3-Tribromo-1-Propyl)-1,3,5-Triazine, or Tetramethylol-2,4-Diamino-6-(3,3,3-Tribromo-1-Propyl)-1,3,5-Triazine (TM-DABT) – DABT (80.0 g, 0.205 mol) and 37.3% aqueous formaldehyde (240.0 g, 2.98 mols) were placed in a flask and adjusted to pH 10.0 by adding, with good stirring, 1.05 g of 20% NaOH. The mixture was stirred at room temperature until a clear solution was obtained. For convenience, the mixture was stirred overnight for a total of about 20 hours. The solution was then added in a slow stream with stirring to 600 ml of water.

The TM-DABT precipitated as a fine white powder. After standing for a while to allow complete precipitation, the white solid was collected on a filter. The TM-DABT was washed while on the filter with 175 ml of water. It was then air dried at room temperature. It weighed 97.2 g, a 93% yield, MP 117° to 118°C dec. Analysis–Calculated for $C_{10}H_{16}Br_3N_5O_4$: C, 23.55%; H, 3.16%; Br, 47.01%; N, 13.73%; formaldehyde, 23.55%. Found: C, 23.37%; H, 3.01%; Br, 46.92%; N, 13.74%; formaldehyde, 22.18%.

Example 4: Pure crystalline TM-DABT (4.0 g) was dissolved in dimethylformamide (DMF). Cotton flannelette (4 oz) impregnated with this solution (110% wet pickup) was dried for 5 minutes at 85°C and cured for 5 minutes at 140°C, rinsed, and dried. The resin add-on to the fabric was 22.1%. The match test angle was 150°. After boiling for 4 hours in water containing 0.5% $Na_3PO_4 \cdot 12 H_2O$ and 0.1% wetting agent, the match test angle was 135°, indicating good durability to laundering. Similar results were obtained when the solvent was dimethylacetamide.

Example 5: The preferred solvent for treating textiles with TM-DABT is a DMF/H_2O mixture. A typical pad bath is prepared as follows: crystalline TM-DABT (75 g) was dissolved in 171 g of DMF (Reagent ACS grade). An alkylaryl polyether alcohol wetting agent (0.3 g) was dissolved in 54 g of distilled water and added with stirring to the DMF solution. The resulting solution contained 25% TM-DABT, 57% DMF, and 18% water by weight and had pH 5.0. The pH of the pad-bath varied about from 5.0 to 7.8 depending on the particular batch being prepared and the age of the solution. The solutions were quite stable and suitable for use for up to at least 3 weeks after preparation.

A typical procedure for treating fabric with the TM-DABT/DMF/H_2O solution described above consisted of padding the sample on a laboratory padder, drying the fabric and curing at specified temperatures, rinsing, and drying. The results on 4 oz cotton flannelette cured at various temperatures are shown in the table below. All samples were dried 5 minutes at 85°C.

Add-on, %	Cured min./° C	Flame retardancy, match test angle, degrees
20.0	10/120	180
19.5	3/130	180
20.9	3/140	180
20.8	5/140	180
19.2	3/150	180
20.1	2/160	180

All of the samples of cotton flannelette had very good flame retardancy as shown by 180° match test angles.

A sample of 8.8 oz O.D. cotton sateen was padded to a wet pickup of 74% with a solution containing 22% TM-DABT, 59% DMF, and 19% water. The sample was dried 5 minutes at 85°C and cured 5 minutes at 140°C, rinsed, and dried. It had an add-on of 11.3% and a match test angle of 180°.

Reaction of Soluble Adduct Polymers with Nitrogenous Compounds

D.J. Daigle, A.B. Pepperman, Jr., and S.L. Vail; U.S. Patent 3,975,560; August 17, 1976; assigned to U.S. Secretary of Agriculture describe a process for flameproofing organic textiles. This comprises reacting a hydroxymethyl phosphorus compound with a nitrogenous compound to form an insoluble polymeric precipitate. The precipitate is dissolved in the presence of formaldehyde and an inorganic or organic acid with a mol ratio of 1:4 to 4:1 of formaldehyde to phosphorus contained in the precipitate and a 1:10 to 2:1 ratio of acid to phosphorus contained in the precipitate. The soluble polymer adduct is then reacted with a nitrogenous compound to produce a flame resistant organic textile; the textile structure is wetted with the above solution and dried from 90° to 170°C for a period of time necessary to promote reaction of the adduct polymers and the nitrogenous compound, within or on the textile structure. The dried fabric is then exposed to sufficient vapors of ammonia or an amine to react and insolubilize the adduct polymer. The following examples illustrate the process.

Example 1: The original or phosphine polymers were prepared by reacting the appropriate phosphorus compounds with the amine or diamine as described in Table 1. All of the polymers were filtered from the mother liquor, washed with water, ethanol, and methanol and left to air dry for 8 days. The polymers were then bottled and used to prepare the adduct polymers. In Table 2, the preparation of the adduct polymer from formaldehyde, acid, and polymer is described.

Organic fibrous materials were impregnated with a solution of the adduct polymer and then dried in an oven. The dried fibrous structures were then exposed to either gaseous alkyl amine or ammonia to react and insolubilize a polymer in or on the fibers. All of the products contained phosphorus and nitrogen after washing in water and drying and exhibited flame resistance as measured by the match test [see *Textile Research J.* 27, 294-299 (1957)]. In Table 3 are shown the fibrous materials treated, the polymer adduct solution of Table 2 used to treat the fibrous product, the amine used to react and insolubilize a polymer, the polymers formed, and a brief description of each product.

Example 2: The nitrogenous agents were added to solutions of the adduct polymer as described in Table 2 to produce solutions as described in Table 4. These solutions were used to (a) make insoluble polymers and (b) produce flame resistant organic fibrous materials. Aliquots of the solutions were heated from about 120° to 150°C to produce flame resistant polymers. These are described in Table 5. To produce flame resistant fibrous materials, the textile structures were impregnated with the solutions of Table 4, then heated to produce insoluble polymers in or on the fibrous structure. In Table 6 are shown the fibrous materials treated, the solution of Table 4 used to treat the fibrous material and a brief description of each product.

TABLE 1: PHOSPHINE POLYMERS

Polymer No.	............Reactants, g............		% P in Polymer
1	THPOH* (40%)	1,500	31.8
	NH_4OH (29% NH_3)	205	
2	Thpc** (80%)	238	23.4
	H_2O	238	
	NH_4OH (29% NH_3)	175	
3	THP*** (80%)	160	29.9
	Water	200	
	Hexamethylenetetramine	105	
4	THPOH (40%)	430	35.4
	Hexamethylenetetramine	105	
5	THPOH (40%)	573	29.5
	Ethylenediamine (98%)	61.2	
6	THPOH (40%)	573	21.5
	1,6-Hexamethylenediamine (70%)	162	

*THPOH refers to a solution prepared by neutralizing any tetrakis(hydroxymethyl)phosphonium salt with any suitable base, inorganic or organic, e.g., sodium hydroxide or triethanolamine.

**Thpc refers to the compound tetrakis(hydroxymethyl)phosphonium chloride.

***THP refers to the compound tris(hydroxymethyl)phosphine $(HOCH_2)_3P$.

TABLE 2: SOLUBLE ADDUCT POLYMERS

Designation Adduct Polymer Solution	..........Reactants, g..........		Time Required for Complete Solution (hr)
A	Polymer No. 1	120	4
	Acetic acid	30	
	Formalin (37.5%)	74.4	
	Water	75.6	

(continued)

TABLE 2: (continued)

Designation Adduct Polymer Solution	Reactants, g.		Time Required for Complete Solution (hr)
B	Polymer No. 2	40	4
	Formalin (37.5%)	24.2	
	H_3PO_4 (85%)	10.3	
	Water	25.5	
C	Polymer No. 3	40	4
	Formalin (37.5%)	30.8	
	Hydrochloric acid (37%)	19.2	
	Water	10	
D	Polymer No. 4	43.8	4
	Oxalic acid	11.2	
	Formalin (37.5%)	40	
	Water	14.5	
E	Polymer No. 5	40	24
	Acetic acid	10	
	Formalin (37.5%)	25	
	Water	25	
F	Polymer No. 6	40	100
	H_3PO_4 (85%)	10	
	Formalin (37.5%)	25	
	Water	65	

TABLE 3: FLAME RESISTANT FIBROUS ORGANIC MATERIALS

Fibrous Material Treated	Adduct Polymer Solution (Table 2)	Amine Used to Cure Adduct	Approximate Add-On of Insoluble Polymer (%)	Properties of Treated Fibrous Products: Hand	Strength	Color	Match Test Angle
Cotton flannelette	A	NH_3*	27	crisp	good	white	180
PE/cotton (50/50)	A	NH_3*	27	crisp	good	white	180
Cotton sateen	A	NH_3**	19	crisp	good	white	180
Cotton flannelette	A***	NH_3**	29	crisp	good	white	135
PE/cotton (50/50)	A***	NH_3**	27	crisp	good	white	135
Cotton sateen	A***	NH_3**	21	crisp	good	white	180
PE/cotton (50/50)	B	NH_3**	12	good	good	white	180
Cotton printcloth	D	NH_3**	4	good	good	white	90
Cotton sateen	B***	NH_3**	12	crisp	good	white	180
PE/cotton (50/50)	D***	NH_3**	5	good	good	white	45

*Atmosphere.
**Forced through.
***Solution has 38% solids and was neutralized to pH of 6.5.

TABLE 4: SOLUTIONS OF ADDUCT POLYMER AND NITROGENOUS COMPOUNDS

Designation of Copolymer Solution	Composition of Solution
G	50 g Solution A 10 g urea 40 g water

(continued)

TABLE 4: (continued)

Designation of Copolymer Solution	Composition of Solution
H	20 g Solution A 4 g thiourea 16 g water
I	20 g Solution A 4 g acrylamide 16 g water
J	20 g Solution A 4 g trimethylolmelamine 16 g water
K	20 g Solution A 8 g cyanamide (50%) 12 g water
L	25 g Solution B 5 g urea 20 g water
M	25 g Solution C 5 g thiourea 20 g water
N	25 g Solution E 5 g urea 20 g water
O	30 g Solution F 5 g urea 15 g water

TABLE 5: POLYMER DESCRIPTIONS

Solution Used	Description of Polymer*
G	Clear, white hard polymer insoluble in water and ethanol; flame and glow resistant; contains nitrogen and phosphorus.
H	Clear, white hard polymer insoluble in water and ethanol; flame and glow resistant; contains nitrogen and phosphorus.
I	Tan, foamy polymer insoluble in water and ethanol; flame and glow resistant; contains nitrogen and phosphorus.
J	Clear, white hard polymer insoluble in water and ethanol; flame and glow resistant; contains nitrogen and phosphorus.
K	Clear, white hard polymer insoluble in water and ethanol; flame and glow resistant; contains nitrogen and phosphorus.

(continued)

TABLE 5: (continued)

Solution Used	Description of Polymer*
L	Clear, white hard polymer insoluble in water and ethanol; flame and glow resistant; contains nitrogen and phosphorus.
M	Clear, white hard polymer insoluble in water and ethanol; flame and glow resistant; contains nitrogen and phosphorus.
N	Clear, rubbery, red polymer insoluble in water and ethanol; flame and glow resistant; contains nitrogen and phosphorus.
O	Clear, hard, gold polymer insoluble in water and ethanol; flame and glow resistant; contains nitrogens and phosphorus.
A**	White hard polymer insoluble in water and ethanols; flame and glow resistant; contains nitrogen and phosphorus.
A***	White hard polymer insoluble in water and ethanol; flame and glow resistant; contains nitrogen and phosphorus.

*Resulting from heating the designated solution from about 120° to 150°C for about 1 to 10 minutes.
**Polymer was formed by addition of ammonia or ammonium hydroxide to Solution A.
***Same as ** except Solution A was neutralized to pH 6.5 before addition of ammonia or ammonium hydroxide.

TABLE 6: PREPARATION OF FLAME RESISTANT TEXTILES

Fabric Treated	Solution Applied to Fabric	Approximate Add-On of Retardant*	. . . . Properties of Treated Fabric. . . . Hand	Strength	Color	Match Test Angle
PE/cotton (50/50)	G	7.0	good	good	white	75
Rayon	G	15	good	good	white	120
Cardboard	G	30	good	good	no change	180
Paper	G	12	good	good	white	100
Wool	G	13	good	good	white	180
Jute	G	87	good	good	no change	100
Cotton printcloth	H	18	good	good	white	90
PE/cotton (50/50)	H	18	good	good	white	90
Cotton sateen	I	18	crisp	good	white	120

(continued)

TABLE 6: (continued)

Fabric Treated	Solution Applied to Fabric	Approximate Add-On of Retardant*	. Properties of Treated Fabric. . Hand	Strength	Color	Match Test Angle
PE/cotton (50/50)	I	22	crisp	good	white	120
Cotton printcloth	J	26	crisp	fair	white	135
PE/cotton (50/50)	J	31	crisp	good	white	180
Cotton printcloth	K	18	crisp	good	white	110
PE/cotton (50/50)	K	20	crisp	good	white	180
Cotton printcloth	L	9	crisp	good	white	120
Cotton sateen	M	10	good	good	white	135
Cotton printcloth	N	11	crisp	good	yellow	100
Cotton printcloth	O	16	crisp	good	white	110

Note: The conditions for depositing insoluble polymer in fabric was 150°C for 2 minutes.

*After washing the treated textile.

In related work, *D.J. Daigle, A.B. Pepperman, Jr., and S.L. Vail; U.S. Patent 4,017,462; April 12, 1977; assigned to U.S. Secretary of Agriculture* describe a process whereby phosphorus- and nitrogen-containing polymers, solubilized by hydroxymethylphosphonium salts and formaldehyde, were allowed to react with phenol, alkyl-substituted phenols, or halogen-substituted phenols which contain one or more hydrogens ortho or para to the phenolic hydroxyl group to (a) prepare flame-resistant polymers and (b) prepare flame-, glow-, wrinkle-, and shrink-resistant woven and knit textiles which contain 25% or more cellulosic fibers.

Dicarbamic Acid Diesters

A process described by *H.-D. Block, U.-W. Hendricks, and K. Walz; U.S. Patent 4,022,757; May 10, 1977; assigned to Bayer AG, Germany* relates to new dicarbamic acid diesters containing phosphorus of the formula:

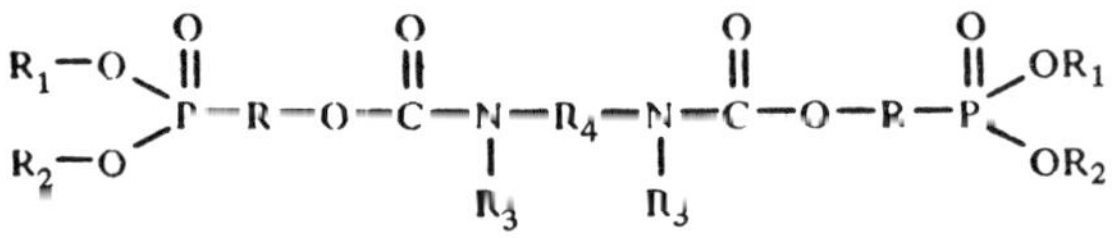

R represents a straight-chain or branched C_{1-4}-alkylene radical which is optionally substituted by one or more halogen atoms, preferably chlorine atoms. R_1 and R_2, independently of one another, represent a C_{1-4}-alkyl radical which is optionally substituted by halogen atoms, preferably chlorine or bromine atoms, or together with the oxygen atoms and the phosphorus atom form a 5-membered to 7-membered heterocyclic structure.

R_3 is hydrogen, a hydroxymethyl or a C_{1-4}-alkoxymethyl group. R_4 represents a divalent radical with at least 2 carbon atoms, which is free from hydroxyl groups, and of which the two bonds must start from different carbon atoms. The phosphorus content of the diester is more than 5% by weight.

The majority of the compounds of the above formula are highly viscous, colorless or slightly colored liquids or crystalline products. They are outstandingly suitable for the flameproof finishing of fiber materials consisting of natural or synthetic fibers, and of plastics. Using these compounds, an excellent wash-resistant flameproofing finish is achieved on textiles of fiber materials containing cellulose, such as cotton, linen, viscose or rayon and polyester-cotton mixtures. The dicarbamic acid diesters are employed in an amount of 10 to 40% by weight based on the fiber material when used as flameproofing agents for textile materials. The following examples illustrate the process.

Example 1: 924 parts of 2-hydroxyethanephosphonic acid dimethyl ester are mixed with 2 parts of triethylamine, the mixture is heated to 70° to 80°C and 504 parts of hexamethylenediisocyanate are added slowly while stirring. After completion of the addition, the reaction mixture is stirred for a further 5 hours at 80°C. 1,420 parts of a colorless viscous product which corresponds to the formula:

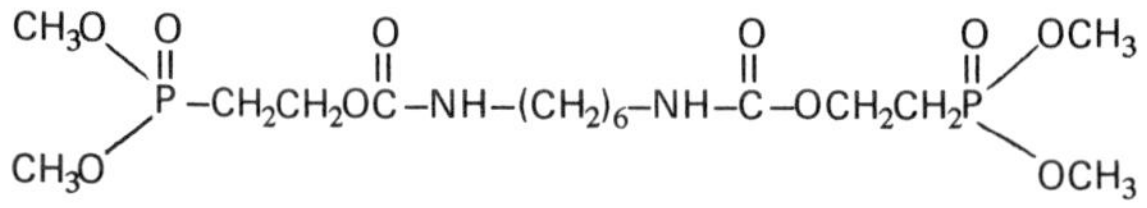

are obtained. On being left to stand, the product solidifies to colorless crystals of MP 56° to 58°C.

Example 2: 141 parts of the product described in Example 1 are mixed with 50 parts of water and brought to pH 8 with 3 parts of concentrated sodium hydroxide solution. The mixture is heated to 50° to 60°C and 48 parts of 37% strength formaldehyde solution are then added. The mixture is stirred for a further ½ hour at 50° to 60°C and is then freed from water in vacuo at down to 2 mm Hg. 156 parts of the compound of the formula:

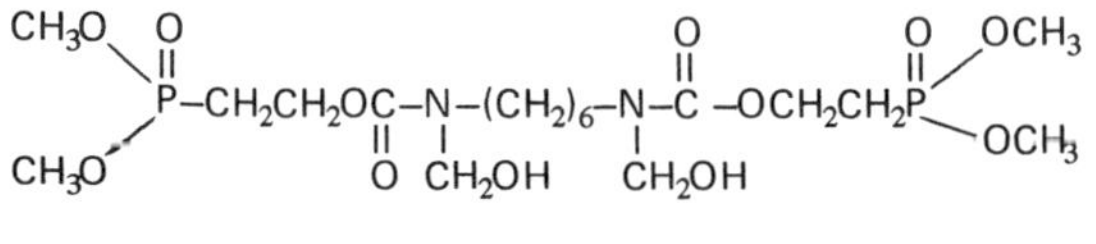

are obtained as a colorless, highly viscous liquid. $n_D^{20} = 1.474$.

Example 3: A cotton decorative fabric is impregnated with an aqueous liquor which contains, per liter, 300 g of the product described in Example 2, 150 g of hexamethylolmelamine, 20 g of phosphoric acid, and 1 g of a reaction product of 1 mol of nonylphenol with 10 mols of ethylene oxide. The impregnated fabric is squeezed off to a weight increase of 90 to 100%, dried for 10 minutes at 100°C and condensed for 5 minutes at 165°C. The fabric is then washed with a dilute sodium carbonate solution at 40°C, and dried. To test its flame-retarding properties, the fabric finished in this way was subjected to the vertical test according to DIN 53,906. The results of the test are summarized in the table which follows.

		....Burning Length....	
		Warp, cm	**Weft, cm**
A	Untreated	..... burnt away	.....
B	Finished according to Ex. 3	7.4	6.9
C	B after 10 machine washes at the boil	7.2	10.9

Condensation of Tetrakis(Hydroxymethyl)Phosphonium Chloride with Tris-(N′,N″,N‴-Methyl)Phosphoramide

D.A. LeBlanc and R.B. LeBlanc; U.S. Patent 4,020,262; April 26, 1977; assigned to LeBlanc Research Corporation describe a water-soluble product suitable for rendering textile materials flame retardant produced by condensing (a) at least one hydroxymethyl phosphorus compound selected from the group consisting of $(CH_2OH)_4P{-}Y$ and $(CH_2OH)_3P$ with (b) about 0.33 to 3 times the molar amount of at least one substituted phosphoramide of the formula $PO(NR^1CH_3)(NR^2CH_3)(NR^3CH_3)$ where R^1 and R^2 each independently is H or CH_2OH, R^3 is H, CH_2OH or CH_3, and Y is an equivalent amount of at least one anion of an acid, such as chloride, bromide, carbonate, nitrate, sulfate, phosphate or carboxylate. The hydroxymethyl phosphorus compound is tetrakis(hydroxymethyl)phosphonium chloride (THPC) or tris(hydroxymethyl)phosphine, present in about 1 to 3 times the phosphoramide, tris(N', N'', N'''-methyl)phosphoramide $[OP(NHCH_3)_3]$ (TMPA).

The materials or substrates to which this process is applicable include textiles or webs formed of cotton, linen, regenerated cellulose, rayon, partially etherified and esterified cellulosic materials; textile blends of these fiber types with other fibers, such as polyester, nylon, acrylic, modacrylic, vinyon, wool, silk, etc.; other forms of cellulose, such as wood and paper products; and proteinaceous textiles, such as wool and silk. The textile materials may be in the form of fibers, yarns, fabrics (nonwoven, woven, and knitted), webbings, and the like. The following examples illustrate the process.

Example 1: A 2.5:1 molar THPC:TMPA condensation product was prepared by adding 152 parts of 80% aqueous THPC and 50 parts of 70% aqueous TMPA to a three-necked flask fitted with a thermometer, mechanical stirrer and water-cooled reflux condenser. Upon mixing, the temperature rose from 18° to about 40°C. The temperature of the reaction mixture was increased to 80°C using external heating and was maintained at 80°C for 1 hour. The reaction mixture was cooled to room temperature. Infrared spectra and thin layer chromatography confirmed formation of a condensation product. The resulting water-soluble product contained 13.7% total phosphorus.

Example 2: A 2.5:1 molar THPC:TMPA condensation product was prepared by adding 152 parts of 80% aqueous THPC to a flask fitted with a water-cooled reflux condenser, a thermometer, an addition funnel and a mechanical stirrer. The THPC was heated to 50°C using a heating mantle. The heating mantle was removed and 50 parts of 70% aqueous TMPA were placed in the addition funnel and added to the THPC over a period of 20 minutes. During addition, the temperature rose to 68°C. The heating mantle was reapplied to maintain the temperature of the reaction at 65° to 70°C for 1 hour. The reaction mixture was cooled. The aqueous solution was analyzed by infrared spectrophotometry and by thin-layer chromatography to verify formation of a condensation product.

Example 3: A 0.44:1 molar THPC:TMPA condensation product was prepared by adding 143 parts of 80% aqueous THPC, 265 parts of 70% aqueous TMPA, and 75 parts water to a flask fitted with a water-cooled reflux condenser, a thermometer and a mechanical stirrer. A heating mantle was used to heat the mixture to 60° to 65°C and to maintain it at that temperature range for 1 hour. The water solution of the condensate contained 12.6% total phosphorus.

Example 4: Condensates were prepared in the mol ratios shown in the table below by adding the indicated parts of THPC and TMPA to a flask fitted with a water-cooled condenser, a thermometer and a mechanical stirrer. The temperature was raised to 60° to 65°C and maintained there for 1 hour using a heating mantle. The products were then cooled to room temperature and compared by thin layer chromatography using silica gel plates and a 65/35 chloroform/methanol mixture as eluant. The chromatogram was developed in iodine vapor. The TLC work indicated that, with decreasing mol ratios, material of higher molecular weight was formed.

Run	THPC:TMPA Mole Ratio	Parts 80% THPC	Parts 70% TMPA
A	2.5:1	228	75
B	2.3:1	224	80
C	2.1:1	218	85
D	2.0:1	219	90
E	1.9:1	208	90
F	1.7:1	197	95
G	1.5:1	183	100
H	1.3:1	166	105

Example 5: A treating bath was prepared by mixing 60 parts of the product from Example 1, 7 parts of Resloom HP (a product which is substantially trimethylolmelamine), 3.5 parts of urea, 0.1 part of Triton X-100 (a wetting agent which is an alkyl, probably octyl, phenoxy polyethoxyethanol), and 29.4 parts of water. The treating bath was applied to a 3.4 oz/yd^2 50/50 polyester/cotton shirting fabric by padding to a 67% wet pickup (WPU). The fabric was dried for 5 minutes at 104°C, cured for 5 minutes at 149°C, and then oxidized in a washing machine with water at 60°C containing 0.1% of H_2O_2, 0.01% NaOH and 0.01% of Na_2SiO_3. The fabric had a dry add-on of 24.9%. It had a soft hand. It was laundered and tested according to FF 3-71 (the Children's Sleepwear Standard) with flame test results at intervals as follows:

Launderings	Average Char Length (in)
0	4.0
10	3.8
25	4.0
50	3.7

Example 6: Five treating baths were made by mixing the indicated amounts of condensation products with 6 parts of Resloom HP, 3 parts of urea, 0.1 part of Triton X-100 and water to a total of 100 parts: Run 1, 28.8 parts of the product from Run A of Example 4 and 27.7 parts of the product from Run B of Example 4; Run 2, 26.6 parts of the product from Run C of Example 4, 17.1 parts of the product from Run D of Example 4 and 13.3 parts of the product from Run E of Example 4; Run 3, 57.7 parts of the product from Run F of Example 4; Run 4, 55.1 parts of the product from Run G of Example 4; and Run 5, 16.3 parts of the product from Run G of Example 4 and 40.7 parts of the product from Run H of Example 4.

The baths were each padded onto a 5 oz/yd^2 50/50 polyester/cotton fabric at a 68 to 73% wet pickup. The fabric was dried at 99°C for 5 minutes, cured at

149°C for 3.5 minutes and oxidized as in Example 5. The add-ons and FF 3-71 flame test results after various laundering intervals are given below.

Run	% add-on	Char Lengths (in.) at Laundering Intervals	
		0	25
1	23.3	2.7	2.3
2	22.5	2.5	2.5
3	21.2	2.9	1.8
4	23.2	2.8	2.3
5	23.8	2.7	2.4

Example 7: A treating bath was prepared by mixing 35 parts of the product from Example 3, 20 parts of Resloom HP, and 3 parts of Catalyst AC (an aqueous solution of 2-amino-2-methyl-1-propanol hydrochloride). The treating bath was applied to a 3.5 oz/yd² 50/50 polyester/cotton muslin fabric at an 86% wet pickup. The fabric was dried at 104°C for 5 minutes, cured at 163°C for 5 minutes, and oxidized as in Example 5. The fabric was tested by AATCC Test Method 34-1969 and showed a char length of 5.3 inches initially, 6.0 inches after 10 launderings, and 6.0 inches after 25 launderings.

Methyl Phosphonic Diamide

T.O. Sistrunk; U.S. Patent 3,997,699; December 14, 1976; assigned to Ethyl Corporation describes a flame resistant substrate comprising cotton and polyester fibers. The polyester fibers have incorporated in them tetrabromobisphenol A ethoxylate flame retardant, and the substrate has affixed on it a flame retardant amount of a phosphorus-containing flame retardant compound of the formula:

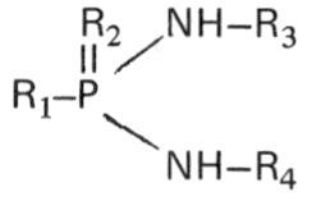

where R_1 is methyl, ethyl, propyl, phenyl, $ClCH_2$ or $BrCH_2$; R_2 is oxygen or sulfur; R_3 is hydrogen, methyl, ethyl or propyl; and R_4 is hydrogen, methyl, ethyl or propyl. The substrate may contain from 15 to 85% by weight polyester, the remainder being cotton.

Several fabrics composed of cotton/polyester blends were treated by dipping the fabric into a solution of the phosphorus-containing flame retardant followed by drying, curing, washing and redrying. The fabrics were then tested according to the procedure of Department of Commerce Flammability Test FF 3-71. The results are shown in the following table. In the table, the following explanations are used:

Fabric

A. 50% cotton/50% woven polyester (by weight) blend fabric, 2.6 oz/yd², in which the polyester fiber was made flame retardant prior to making the fabric by incorporating a bromine-containing compound.

B. 35% cotton/65% polyester blend of materials of Fabric A.

Treating Solution

I. One part crude MPDA (47% MPDA, 53% NH_4Cl) and two parts water. Parts and percentages by weight. (MPDA is Methyl Phosphonic Diamide.)

II. Two parts crude MPDA (47% MPDA, 53% NH_4Cl) and three parts water. Parts and percentages by weight.

Drying and Curing Conditions

Where drying is indicated, the fabric was dried at 90°C to constant weight.

Where curing is indicated, the fabric was heated at 340° to 350°F for 1 minute.

MPDA Efficiency–this term refers to the weight increase of the fabric because of fire retardant treatment divided by the weight of MPDA in the treating solution absorbed by the fabric multiplied by 100.

Vertical Flame Test–the test employed was the procedure of Department of Commerce Flammability Test FF 3-71. W refers to samples tested in the warp direction. F refers to samples tested in the fill direction. BEL is an acronym meaning burned entire length of the 10-inch specimen. Data are given in the table in the following order: (1) direction of test, W or F meaning warp or fill; (2) char length, in inches; and (3) time of burning, in seconds, e.g., W 4.8 25. The samples were tested as produced and the results indicate initial values.

Example No.	1		2	3	4
Fabric	A		A	A	B
Dry Weight, g	38.6		10.5	27.0	41.4
Wet Weight, g	73.9			54.3	75.1
Wet Pick-up, wt %	92			101	81
Dry Weight, g	50.3		13.9	37.4	54.5
Redry Weight, g	50.4		14.0	37.4	54.5
Post Cure Weight, g	49.7		13.6	36.5	53.7
Weight After Wash and Dry, g	41.1		11.2	28.8	43.6
Redry Weight, G	—		11.2		
% DAO,* MPDA	6.5		6.2	6.6	5.3
Analysis, % Phosphorus	2.03		**	2.66	1.75
% Bromine	3.30		**	3.26	4.22
MPDA Efficiency, %	45		—	40	39
Treating Solution	I		I	I	I
		(2nd Series)			
Vertical Flame Test	W 4.8 25	W 5.0 25	W 4.5 22	3.8 18	BEL 34
	W 6.9 39	W 3.9 21	W 4.1 17	4.3 18	BEL 31
	W 5.6 29	W 5.7 28	W 4.6 25	BEL 23	BEL 31
	W 3.7 20	W 4.9 28		BEL 21	BEL 32
	F BEL 18	F 5.1 27		4.4 18	BEL 33

*DAO is dry add on.

**Analysis originally performed was erroneous; additional fabric sample unavailable for reanalysis; original fabric Br wt % range is 3.3 to 3.6%.

Although acceptance criteria for Flammability Test DOC FF 3-71 require that the average char length for 5 specimens be less than 7 inches and no specimen burns the entire length (BEL), when a BEL result is obtained the burning time is an important indication of relative difficulty of burning compared to other specimens also having BEL results. Therefore, when two specimens are reported as BEL, the one with a longer burning time is considered relatively more resistant to flame. Specimens of this process which burn the entire length can be considered more flame resistant, based on burning time, than the comparative specimens. For example, the specimens of Example 4, having BEL results with burning times of 23 and 21 seconds would be considered more resistant to flame than the comparative specimens, all of which gave BEL results and burned more than twice as fast, i.e., from 9 to 13 seconds.

For comparative purposes, samples not in accord with this process were tested in the Vertical Flame Test for initial results. One sample used a blend of 50% cotton, 50% polyester, by weight, blend without treatment. All specimens burned the entire length within 12 to 13 seconds. Another sample was the same Fabric A described as a 50% cotton/50% polyester in which the polyester has been flame retarded with a bromine-containing compound, but without treatment according to this process. All specimens burned the entire length in 9 to 10 seconds.

REGENERATED CELLULOSE AND RAYON

Synergistic Effects of Phosphorus Nitrides with Phosphoric Acid Alkylesters

J. Hüpfl, M. Czermak, H. Teichmann, and J. Paul; U.S. Patent 4,063,883; Dec. 20, 1977; assigned to Hoechst AG and Chemiefaser Lenzing AG, Austria describe a process for the manufacture of flame retardant, regenerated, cellulose fibers by adding one or more flame retardant phosphorus compounds to viscose, extruding the viscose-containing mixture obtained into a spinning bath, stretching and aftertreating the resulting filaments or staple fibers. The flame retardant fibers are made by using as the flame retardant agent a combination of compounds of the general formula, PN_xO_y, in which x stands for a number between 0.9 and 1.7 and y stands for a number between 1.2 and 0, preferably approaching zero as the values of x increase towards 1.7 (component A), with the following compounds.

These compounds are selected from phosphoric acid alkylesters, preferably halogen-containing phosphoric acid alkylesters; cyclophosphazenes, polyphosphazenes; tetrakis(oxymethyl)phosphonium chloride or hydroxide or its condensation products with ammonia, oxymethylamine, urea or other amines or amides, vinyl compounds containing halogen, phosphonitrogen or phosphorus; and tris(1-aziridinyl)-phosphine oxide or N-(oxymethyl)-3-phosphonopropionamide, if desired in further combination with a condensation resin (component B).

The preferred component A representatives are phosphorus nitride, triphosphorus pentanitride, tetraphosphorus hexanitride, and/or phosphorus nitride oxide. Preferred component B representatives include tris(2,3-dibromopropyl) phosphate, tris(2,2'-dichloroisopropyl) thiophosphate, and poly(isopropoxyphosphazene).

The combination of the above component A compounds with the component B compounds enables satisfactory flame retardant effects to be produced with the use of the combined agent in low proportions within the range of about 5 and 22%, based on the quantity of cellulose, i.e., in proportions considerably smaller than those needed for the separate use of the A and B components, respectively. Experiments have shown this synergism to be based on the fact that the component

B compounds commence pyrolyzing activity in the cellulose at temperatures within the range about 220° and 250°C, i.e., at lower temperatures than the component A compounds (about 300°C). Such early pyrolysis causes the activity of the component A compounds, which contain considerable nitrogen and phosphorus and, accordingly, critically determine the flame retardant effect, to be initiated and promoted. The following examples illustrate the process.

Example 1: 15 wt % (based on cellulose) of triphosphorus pentanitride, P_3N_5, in pigment form (component A) and 5 wt % (based on cellulose) of liquid poly(isopropoxy phosphazene), $\langle -N{=}P[(CH_3)_2CHO]_2 - \rangle_n$, where n stands for a number between 3 and 9 (component B) were metered continuously into a viscose containing 6.5 wt % of cellulose, 6 wt % of alkali, 34 wt % of CS_2, based on cellulose, and 2 wt % of a modifying agent, based on cellulose. The viscose had a viscosity of 80 fbsec and a spinning ripeness of 54 gamma units. By means of a homogenizer, the A and B components were very finely distributed in the viscose, which was spun at 40°C into a spinning bath containing 80 g/l of H_2SO_4, 120 g/l of Na_2SO_4, and 50 g/l of $ZnSO_4$.

The spinning tow, which had a total titre approaching 200,000 dtex, was drawn at a rate equal to 100% of its initial length in a second and more dilute spinning bath at 90°C and thereafter cut up in a cutting machine to form staple fibers, which were subjected to the after-treatment customary in the spinning of viscose fibers. The conditioned fibers had the following textile properties: Titre, 1.3 dtex; strength of fiber, 24.8 p/tex; elongation, 17%; loop strength, 8.1 p/tex. To determine the flame retardant effect, a fabric which had a superficial density of 150 g/m² was subjected to a vertical burning test. The following results were obtained:

	Ignition time 3 sec.	Ignition time 15 sec.
Burning time in sec.	0	0
Glow time in sec.	0	0
Tear length in cm	2.9	8.1

The fabric was washed 50 times at 60°C in a domestic washing machine. The flame retardant properties of the fabric so washed were tested once again, but could not be found to have been impaired.

Example 2: 14 wt % (based on cellulose) of phosphorus nitride oxide (PNO) in pigment form (component A) and 4 wt % (based on cellulose) of a condensation product (which was also used in pigment form) of tetrakis(oxymethyl)phosphonium chloride (component B) with ammonia were added to the viscose of Example 1. The other conditions were the same as those used in Example 1. The conditioned fibers had the following textile characteristics: Titre, 1.7 dtex; strength of fiber, 25.2 p/tex; elongation, 16.5%; loop strength, 8.5 p/tex. To determine the flame retardant effect, a fabric which had a superficial density of 170 g/m² was subjected to a vertical burning test. The following results were obtained:

	Ignition time 3 sec.	Ignition time 15 sec.
Burning time in sec.	0	0
Glow time in sec.	0	0
Tear length in cm	2.6	7.8

The wash test described in Example 1 was repeated. The flame retardant properties of the fabric could not be found to have been impaired.

Example 3: 12 wt % (based on cellulose) of triphosphorus pentanitride (P_3N_5) (component A) was added to a viscose the same as that used in Example 1, which was spun in the manner described therein. The second bath, however, was maintained at a lower temperature and contained less acid so that the spinning tow coming from the second bath presented a residual gamma value of more than 1. The tow was cut to staple fibers which were suspended so as to form a fiber fleece. The fleece was impregnated with an iron(II) sulfate catalyst, squeezed off and then impregnated with a hydrogen-peroxide-containing aqueous emulsion of diethylphosphatoethylmethacrylate (component B), of which 4 wt % (based on cellulose) was applied to the fleece. Following this, the fleece was conveyed through a zone maintained at high temperatures and subjected to the after-treatment customary in the spinning of viscose fibers.

The conditioned graft-polymerized fibers had the following textile characteristics: Titre, 1.8 dtex; strength of fiber, 25.8 p/tex; elongation, 16.5%; loop strength, 7.9 p/tex. To determine the flame retardant effect, a fabric having a superficial density of 190 g/m^2 was subjected to a vertical burning test. The following results were obtained:

	Ignition time 3 sec.	Ignition time 15 sec.
Burning time in sec.	0	0
Glow time in sec.	0	0
Tear length in cm	3.5	8.4

The fabric was washed in the manner described in Example 1. Its flame retardant properties could not be found to have been impaired.

Example 4: 15 wt % (based on cellulose) of triphosphorus pentanitride, P_3N_5, (component A) was added to a viscose the same as that used in Example 1, and the whole was treated under the conditions described therein. The resulting fibers were made into a fabric having a superficial density of 170 g/m^2. The fabric so made was impregnated on a two roll calender with a solution containing N-(oxymethyl)-3-(diethylphosphono)propionamide (component B), a fixing agent and a catalyst. 4 wt % (based on cellulose) of component B was found to have been absorbed by the fabric. Following this, the fabric was dried for 60 seconds at 125°C, set for 60 seconds at 160°C, washed out and dried once again. To determine the flame retardant effect, the fabric was subjected to a vertical burning test. The following results were obtained:

	Ignition time 3 sec.	Ignition time 15 sec.
Burning time in sec.	0	0
Glow time in sec.	0	0
Tear length in cm	2.5	6.8

Phosphorus Nitrides

In related work, *J. Cremer, H. Harnisch, J.A. Krässig, J. Hüpfl, and W. Gschaider; U.S. Patent 3,974,251; August 10, 1976; assigned to Hoechst AG, Germany and*

Chemiefaser Lenzing AG, Austria describe the use of one or more phosphorus nitrides as agents for flame retarding fibers of regenerated cellulose. The following example illustrates this.

Example: 10 wt %, based on the cellulose, of triphosphorus pentanitride (P_3N_5) in aqueous suspension was continuously metered into a viscose containing 6 wt % of cellulose, 6 wt % of alkali, 35 wt % of CS_2, based on the cellulose, and 3 wt % of a modifying agent, based on the cellulose. The viscose had a viscosity of 75 kgfsec and a spinning ripeness of 57 gamma units. By means of a homogenizer, the suspension was very finely distributed in the viscose, which was spun at 42°C into a spinning bath containing 70 g/l of H_2SO_4, 100 g/l of Na_2SO_4, and 60 g/l of $ZnSO_4$. The spinning tow, which had a total titre approaching 200,000 dtex, was drawn at a rate equal to 110% of its initial length in a second and more dilute spinning bath at 92°C and thereafter cut up in a cutting machine to form staple fibers.

The staple fibers were completely regenerated in a series of successive baths with hot dilute sulfuric acid, desulfurized by means of hot dilute sodium sulfide solution, bleached by means of dilute sodium hypochlorite solution and revived or brightened. The fibers were dried, and were found to have the following textile characteristics: Titre, 3.3 dtex; strength of fiber, conditioned and with 11% moisture, 25.0 p/tex; elongation of fiber, conditioned, 15.8%; strength of fiber, moist, 16.2 p/tex; elongation of fiber, moist, 17.0%; loop strength, 7.8 p/tex. To determine the flame retardant effect, a fabric having a superficial density of 300 g/m^2 was subjected to a vertical burning test. The following results were obtained:

	Ignition time 3 sec.	Ignition time 15 sec.
Burning time in sec.	0	0
Glow time in sec.	0	0
Tear length in cm	2.3	4.8

A similar fabric made from non-flame-retardant fiber was subjected to a control test, in which the test surface was found to have been completely burned.

Derivatives of Phosphoric Acid as Flameproofing Agents

C. Mauric and R. Wolf; U.S. Patent 4,062,687; December 13, 1977; assigned to Sandoz Ltd., Switzerland describe a process whereby flameproofed cellulose includes as a flameproofing agent a compound selected from esters, ester-amides and amides of pyrophosphoric, of mono-, di- and trithiopyrophosphoric, of triphosphoric, and of mono-, di-, tri- and pentathiotriphosphoric acids. The cellulose employed is regenerated cellulose.

Prior to the regeneration, cellulose is brought into solution, e.g., by such known processes as converting it into a soluble derivative by the xanthate method or through formation of the tetramine cupric hydroxide complex. The flameproofing agent is then added to the cellulose solution, e.g., by itself or as a fine dispersion in water. When added alone, the agent may be introduced into the cellulose solution either continuously or discontinuously, i.e., in batches, and thereafter vigorous stirring of the cellulose solution containing the compound may be applied to distribute it uniformly in the solution. The same technique may also be adopted for the addition of an aqueous dispersion of the agent. The weight of the flameproofing compound present in the cellulose solution from which flameproofed regenerated cellulose is to be produced is in the range 8 to 25% of the weight of the cellulose starting material, e.g., α-cellulose. In all cases, it can be advantageous to add conventional dispersion stabilizers and/or dispersion agents to the cellulose medium to promote uniformity of distribution of the compound in the cellulose solution. Additionally, natural fibrous materials, including natural cellulose, e.g., cotton, are treated by coating the flameproofing agent suitably present in a coating liquor, e.g., a solution or aqueous dispersion, onto the substrate. The following examples illustrate the process.

Example 1: A suspension of 44.4 parts of diphosphorus pentasulfide and 104.5 parts of 2,2-dibromomethyl-1,3-propanediol in 200 parts of toluene is warmed to 90°C over a period of 2½ hours. The reaction is continued at this temperature for 7 hours, after which the mixture is heated for a further 2 hours under reflux. A clear solution results, from which the solvent is removed by evaporation. The residue is crystallized from carbon tetrachloride, yielding a solid with a MP of 117° to 120°C. 64 parts of the solid product obtained as above are suspended in 250 parts of water and 12.2 parts of a 25% ammonia solution. To the suspension is added, over a period of 6 hours, a solution of 22.65 parts of iodine and 44.95 parts of potassium iodide in 178.8 parts of water. The mixture is stirred at room temperature for a further hour.

The precipitate obtained is collected by filtration, dried and suspended in 150 parts of benzene. To this suspension at 30°C is added dropwise a solution of 14 parts of triphenylphosphine in 60 parts of benzene during a period of 20 minutes, and the mixture is maintained at 30°C for a further 8 hours. Thereafter, the benzene solvent is removed by evaporation until the volume is reduced by half, and the precipitate is collected by filtration, dried, washed with warm water at 70°C and finally dried. Produced is the compound having the following structure:

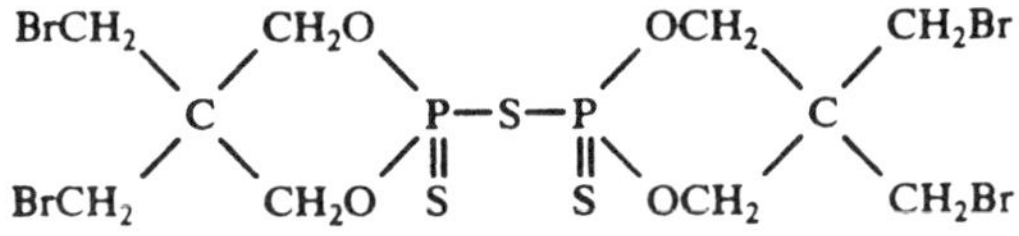

Example 2: To 200 parts of a cellulose solution based on xanthate, which solution contains 18 parts of α-cellulose, are added with stirring 15.7 parts of a 23% aqueous dispersion of the compound with the following structure:

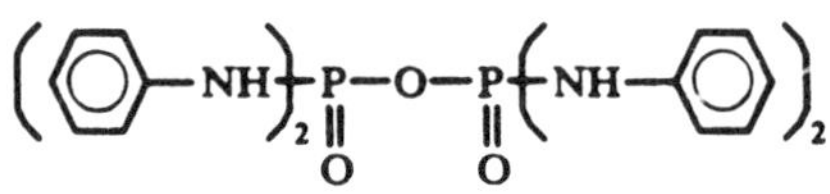

The dispersion is produced in the following way. 40 parts of the above compound, 10 parts of a dispersing agent based on sodium naphthalene sulfonate and 110 parts of water are ground with sand in the presence of 160 parts of quartzite beads for 4 hours. The grinding is effected with ice-cooling at 1,500 revolutions per minute of the grinder. After removal of the quartzite beads by filtration, 140 parts of a dispersion containing 23% of the active ingredient are obtained. A cellulose solution containing the active ingredient is extruded through spinerettes according to the conventional procedure into a precipitation bath containing per liter 125 g of sulfuric acid, 240 g of anhydrous sodium sulfate and 12 g anhydrous zinc sulfate. The fibers produced are thoroughly washed and formed into knitted products. According to the method of Fenimore and Martin, the knitted products are tested for flameproofing properties by measurement of the Limiting Oxygen Index (LOI).

Phosphonitrilate Polymers

C.W. Lanier and J.T.F. Kao; U.S. Patent 4,081,593; March 28, 1978; assigned to Ethyl Corporation describe a process for producing fire retardant phosphonitrilate polymers by reacting phosphonitrilic halide with a hydroxyl compound, such as an aliphatic or aromatic alcohol, at not more than about 40°C in the presence of an acid acceptor, such as pyridine. The conditions are such that a relatively low concentration of the hydroxyl compound is present during the initial ½ to 3 hours of the reaction and the reaction conditions are maintained at from 20° to 50°C for a period of ½ to 120 hours. Optionally the reaction mixture is further heated to a temperature of 40° to 80°C for ½ to 8½ hours. The phosphonitrilate polymers are suitable for fire retarding cellulosic materials, such as regenerated cellulose, rayon and the like.

Example: Into a glass-lined, 30-gallon reactor was charged 148 pounds of monochlorobenzene and 65 pounds of phosphorus trichloride. The reactor stirrer was started and the reactor contents heated to about 27°C and the pressure was atmospheric. To the reactor was charged 36 pounds of gaseous chlorine over a period of 1 hour and 10 minutes. The reactor temperature increased to about 55°C and was maintained during chlorine feed at from 55° to about 60°C. The reactor pressure was controlled at from about 2 to 3.8 psig.

The reactor was pressurized with HCl up to about 13 psig. Gaseous ammonia, 8.4 pounds, was added over a period of 4 hours and 10 minutes. The temperature was maintained at about 140°C and the pressure at 20 psig by venting HCl to a water scrubber. The ammonia feed was stopped and the reactor pressure and temperature were maintained for 30 minutes. Then, the reactor was vented for 20 minutes to remove HCl and the pressure was decreased to 20 psig during the venting period. The reactor contents were then cooled. The unfiltered product phosphonitrilic chloride weighed 201.5 pounds. The product was then filtered to remove NH_4Cl and the monochlorobenzene solvent was stripped from the phosphonitrilic chloride and 128 pounds of solvent were recovered. Stripping was carried out for 2 hours and 25 minutes at temperatures from about 24° up to 77°C and pressure of from 50 mm Hg to 1 mm Hg.

Then, 160 pounds of pyridine was added to the reactor. The reactor contents were heated to 22°C and 42.5 pounds of propanol was fed into the reactor with stirring over a period of 1 hour and 47 minutes. The temperature varied from about 17° to about 22°C during the feed period. The propanol feed was completed and the reaction mixture held at 22° to about 30°C for 15 hours. At this point, another 8 pounds of propanol was fed into the reaction mixture during 10 minutes at 30°C. This amounted to about 10 wt % over theoretical amount of propanol. The reaction mixture was further heated at about 57° to 63°C for 4½ hours to finish off the reaction. The reactor contents were cooled to about 22°C and left with stirring for 64 hours. Temperature ranged from 22° to about 27°C during this period. Then, the reactor contents were cooled to ambient temperature and 140 pounds of monochlorobenzene was added. After stirring, there was added 168 pounds of 16 wt % sodium hydroxide solution to neutralize the reaction mixture. The final mixture had a pH of 7.6 after neutralization. The bottom brine solution was drained and weighed 204 pounds containing water, salts and rag layer. A second wash using 25 pounds of water was conducted and 48 pounds of pyridine and water were recovered from the top layer. The product solution was heated to 101°C under vacuum of 10 to 1 mm Hg for 3 hours. Recovery of pyridine and monochlorobenzene from the condenser was 272.5 pounds. The phosphonitrilate polymer produce weighing 46.8 pounds was recovered from the reactor.

The phosphonitrilate polymer prepared above was added to viscose at several concentrations, spun into filament and processed conventionally. The viscose mixture was spun into aqueous coagulating and regenerating bath containing aqueous sulfuric acid and other conventional components, e.g., metal sulfates. Further processing includes washing, desulfurizing, bleaching, etc. The flame retardant effectiveness of the product made above was evaluated by the Children's Sleepwear Test, DOC FF 3-71 in both filament-knit fabric and staple yarnwoven fabric. The char lengths recorded below are the average of 5 samples tested.

Viscose Loading (wt %)	Fab. Wt. (oz/yd²)	Type Fabric	Avg. Char Length (in.) After 50 Launderings
13.1	6–7	Filament-Knit	3.8
16.7	6–7	Filament-Knit	0.6
20.0	6–7	Filament-Knit	1.0
20.0	4.0	Filament-Knit	2.9

In related work, *J.T.F. Kao; U.S. Patent 4,011,089; March 8, 1977; assigned to Ethyl Corporation* describes a phosphazene composition prepared by reacting a fully hydrocarboxylated phosphazene with a partially hydrocarboxylated phosphazene under conditions whereby an organic halide is evolved. At least one of the fully and partially hydrocarboxylated phosphazenes are prepared by adding an alkali metal hydrocarboxide to a phosphonitrilic halide whereby the final phosphazene composition contains some –P–ONa bonds, as well as –P–O–P–bond configurations. The phosphazene compositions are useful as fire retardants for cellulosics.

Substituted Polyphosphonates

K. Mimura, A. Kawai, Y. Kametani, and T. Nakahama; U.S. Patent 4,066,730; January 3, 1978; assigned to Mitsubishi Rayon Co., Ltd., Japan describes a process whereby regenerated cellulose fibers, having very good flameproofness, are obtained by adding a polyphosphonate having the following formula:

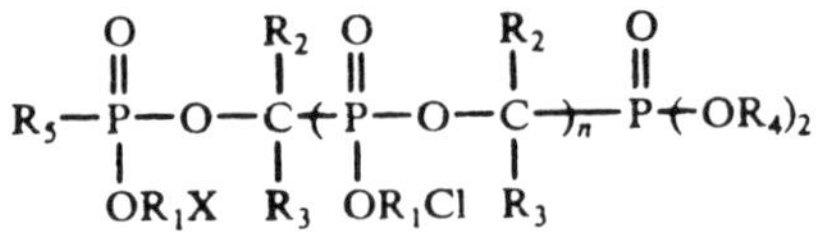

where R_1 is ethylene or propylene; R_2 and R_3 are methyl or ethyl which may be the same or different; R_4 is an alkyl group having 1 to 4 carbon atoms, or aralkyl in which the alkyl substituent has 1 to 4 carbon atoms and the hydrogen atoms in the alkyl and aralkyl groups may be substituted with chlorine or bromine;

R_5 is an aromatic group having 1 to 20 carbon atoms or an unsaturated or saturated aliphatic or alicyclic group having 1 to 20 carbon atoms or a combination of these and R_5 may contain carbonyl group, sulfonyl group or nitrogen; X is halogen, and n is an integer from 1 to 1,000, to a viscose in an amount of 5 to 40% by weight of the cellulose contained in the viscose; and then spinning the viscose in a coagulation bath.

The polyphosphonates of the above formula have excellent hydrolysis resistance and are obtained by fixing one terminal of the polyphosphonate while simultaneously regulating the polymerization degree of the polyphosphonate with the use of a triorganophosphite as a polymerization regulator. In addition, the other terminal of the polymer having a cyclic phosphite structure which remains unreacted, can then be reacted with an active halide to convert the trivalent phosphorus to pentavalent phosphorus and thereby fix this terminal of the polymer.

Furthermore, in connection with the production of a polymer having a narrow molecular weight distribution, the active halide which effectively fixes one terminal of the polymer also functions as a polymerization catalyst. Consequently, it is not necessary to conduct the polymerization reaction in an acidic medium which constitutes an extremely severe reaction medium and, therefore, it is possible to conduct the polymerization under mild conditions. Therefore, the molecular weight distribution of the obtained polymer has excellent quality. The following examples illustrate the preparation and use of these flame retarding agents.

Example 1: A 1-liter separatory flask provided with a stirrer, a thermometer, a reflux condenser, a dropping funnel, and a nitrogen flow inlet device was charged with 253 parts of purified ethylene chlorophosphite, 386 parts of purified methylchloroform, and 15.5 parts of purified tris-2-chloropropyl phosphite. Then, a dropping funnel was charged with 116 parts of purified acetone and 1.9 parts of purified tert-butyl chloride.

While under a sufficient flow of dried nitrogen gas, the outside of the flask was cooled in an ice-water bath and the mixed solution of acetone and tert-butyl chloride was dropped with stirring into the flask during which addition the temperature within the flask was kept at 5° to 10°C. After addition of the reagent was completed, the reactants were allowed to stand at room temperature for about 48 hours. The polymer obtained existed as 384.9 parts of a solid portion, had a mean molecular weight of 6,000 and had the following structure:

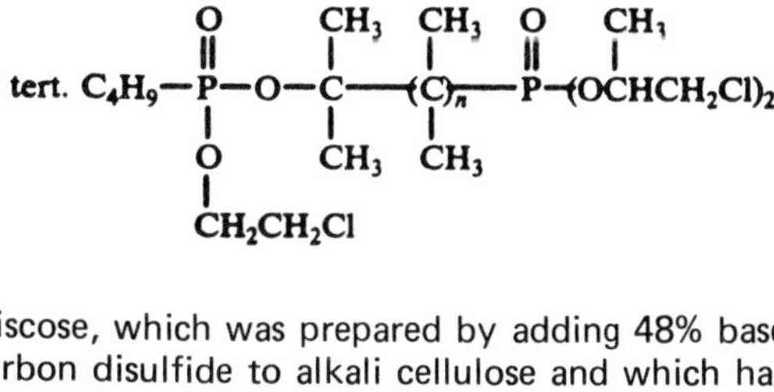

Example 2: A viscose, which was prepared by adding 48% based on the weight of cellulose of carbon disulfide to alkali cellulose and which had a cellulose concentration of 7.5% and an alkali concentration of 4.2%, was ripened until the viscosity reached 200 poises and the salt point reached 16. To 100 parts of the prepared viscose was added 3.5 parts of a solution of the polyphosphonate pre-

pared in Example 1 in methylchloroform. The solution was mixed to homogeneity. Thereafter, the viscose was spun into a coagulation bath containing 18 g/l of sulfuric acid, 70 g/l of sodium sulfate, and 0.4 g/l of zinc sulfate at 30°C. The filaments withdrawn from the coagulation bath were stretched by 100% in a second bath containing 2 g/l of sulfuric acid at 80°C and were successively treated in a third bath containing 5 g/l of sulfuric acid at 60°C to complete regeneration. Thereafter, the filaments were subjected to the conventional aftertreatments such as desulfurization, bleaching, acid treatment, lubrication, and drying.

Essentially, no sticky materials were recognized as forming during the spinning step which is a problem when conventional polyphosphonates are used. Moreover, the retention of the polyphosphonate (flameproofing agent) in the fibers was 95.5% which is extremely high. Furthermore, there were no troubles in the yarn spinning operation of the obtained fibers and the limit of the oxygen index (LOI) of the fibers was 27.5 which indicates that the fibers had excellent flameproofness.

Polymeric Phosphazene Esters

B.R. Franko-Filipasic, E.F. Orwoll, and J.F. Start; U.S. Patent 3,990,900; Nov. 9, 1976; assigned to FMC Corporation describe a process consisting of reacting a phosphonitrilic chloride with an alcohol or alkoxide to 12 carbon atoms to produce a partially esterified phosphonitrilic chloride, removing impurities from the chloroester and heating the ester to form a condensed polymeric phosphazene with P–O–P oxygen linkages. This is performed by adding the phosphonitrilic chloride to a stirred slurry of the alkoxide. A stoichiometric insufficiency of the alkoxide is used to produce a chloroester containing 1 to 15% residual chlorine. The esterification is done in a solvent or diluent and the esterification reaction is generally conducted at the reflux temperature of the solvent and/or diluent selected as the reaction medium.

Ordinarily, phosphonitrilic chloride is added to a slurry of insufficient alkoxide, such as propoxide, and the esterification is completed in the usual way. The reaction proceeds in a manner so that first phosphonitrilic chloride molecules which react are completely esterified, thus producing a mixture of completely esterified or essentially completely esterified material mixed in with unreacted polymeric phosphonitrilic chloride. However, a reverse addition process, in which the alkoxide is added to a suitable liquid phosphonitrilic chloride, produces a particularly efficacious product. The reverse addition process produces more chloroester and little or no unreacted phosphonitrilic chloride is left unreacted. This class of polymer phosphazenes, with P–O–P oxygen linkages, is unusually efficacious in rendering rayon flame retardant.

Example: A suspension of sodium proproxide was formed in 5.2 liters of mixed xylenes by charging 1,140 ml of n-propanol to 308 g (13.4 mols) of agitated molten sodium at 110° to 115°C. The reaction was completed by stirring at reflux for an additional ½ hour. A chloroester was prepared by charging a 19.3% stoichiometric excess (928 g, 8 mols) of cyclic phosphonitrilic chloride in chlorobenzene solvent to the propoxide suspension at reflux temperatures. The chloroester formed in 3 hours. The chloroester was isolated by water-washing the cooled reaction mixture and concentrating the organic layer in vacuo (85°C, 16 torr). The chloroester was then heated in a rotating flask at 220°C, 15 torr, for 8 minutes.

Analysis and properties of the product follows. Values in parentheses are for untreated chloroester.

Analysis		
Carbon	38.46	(40.07)
Hydrogen	7.39	(7.32)
Nitrogen	9.86	(9.06)
Phosphorus	22.85	(20.69)
Chlorine	0.55	(3.54)
Molecular weight	923	(612)
Acid No.	13.4	(14.7)
Viscosity at 25°, cs	6993	(44.2)
Refractive index at 21°	1.4748	(1.4645)

The phosphorus-containing product was evaluated for its flame retardant effect in rayon yarn produced from a filament-forming viscose comprising 8.6 wt % cellulose, 6.2 wt % sodium hydroxide and 33.0% carbon disulfide, based on the weight of the cellulose, and having a viscosity at spinning of 6,000 cp at 18°C. The phosphorus-containing flame retardant compound was injected into the viscose stream at the desired rate based on the weight of the cellulose in the viscose and the viscose mixture passed through a high shear blender. This provided viscose having the flame retardant dispersed as fine liquid particles of from 1 to 10 microns in size.

The viscose was spun into a conventional aqueous acid spin bath comprising 9.8 wt % sulfuric acid, 3.0 wt % zinc sulfate and 17.5 wt % sodium sulfate at a bath temperature of 50°C. The yarn was wet stretched about 35% of its original length. Yarns, having a denier of 240 and 40 filaments, were processed by passing them through a series of baths including water wash, desulfurization, bleach, bleach acid, antichlor, and soft finish bath. The yarns were dried, transferred to packages, and finally knit into circular knit fabrics. The regenerated cellulose yarns prepared in this manner were made up of individual filaments having fine liquid flame retardant particles locked in the cellulose matrix. The fabric retained 15% by weight of the phosphazene and the fabric was found to be flame resistant.

In related work, *B.R. Franko-Filipasic and J.F. Start; U.S. Patent 3,986,882; October 19, 1976; assigned to FMC Corp.* describe regenerated cellulose filaments having dispersed in them a flame retardant amount of phosphonitrilic esters containing a substantial amount of cyclic structures, dimerized, oligomerized, or polymerized, that are prepared by esterification of a phosphonitrilic chloride polymer with ethylene glycol.

WOOD

Ammonium Bromide, Urea Phosphate and Hexamine Bonding Compositions

E. Wurman; U.S. Patent 4,012,558; March 15, 1977; assigned to Dead Sea Bromine Company Ltd. describes a flame retardant mixture adapted for incorporation into the bonding agent used in the production of chipboard, plywood and the like. This consists essentially of ammonium bromide, urea phosphate and hexamethylenetetraamine. The hexamethylenetetraamine is present in a sufficient amount to inhibit the rate of gelling of the bonding agent and the ammonium bromide and urea phosphate are present in flame retarding amounts.

The bonding agents are either sprayed on wood particles prior to the preparation of the loaf from which the chipboard itself is prepared by pressing it, or the individual wooden plates are passed through a glue bath before being combined in the press to form the plywood. In the bonding, flame retardant compositions of the process, urea formaldehyde is preferred as the bonding agent. Since glue material requires a hardening agent (hardener) in any case, the ammonium bromide and/or urea phosphate act as hardeners, and the amount of hexamine is adjusted to impart proper work-life (pot-life) and hardening speed (gel time) to the glue-flame retardant combination.

Deactivated Solution of Chlorosulfonic Acid Containing Digested Cellulose

S. Hartman; U.S. Patent 4,063,003; December 13, 1977; assigned to Champion International Corporation describes a process whereby a cellulosic substrate is rendered fire resistant through application of an alcohol deactivated solution of chlorosulfonic acid and digested plant material. First, a plant material, commonly wood, is digested in chlorosulfonic acid. Incident to digestion, the cellulose, lignin, and other constituents of the plant material are solubilized in the acid. Thereafter, the solution containing plant material is treated with alcohol to decompose the remaining chlorosulfonic acid, thus yielding a treating solution which is deactivated with regard to its ability to digest cellulose.

The plant material used is present in an amount of 4 to 12% by weight of chlorosulfonic acid. Cellulosic substrates usefully treated in accordance with this process include any of the normally combustible cellulosic materials. This includes hemicellulose and other essentially carbohydrate polymers containing glucose or glucose derivative units. Particularly, however, the process is for woody materials commonly employed as structural members in the construction art–lumber, particleboard, cork, and other building materials.

Example: 60 g of Douglas-fir wood flakes are dried for 1 hour at 100°C and then added to 85 g of chlorosulfonic acid. This slurry is allowed to digest for 1 hour at 23°C during which essentially all of the wood is dissolved. The solution is then filtered and 170 g of isopropyl alcohol is added slowly to decompose the remaining chlorosulfonic acid, and to form a fire retardancy imparting treating solution. Dry fir wood chips are then submerged in the treating solution for 2 minutes before being removed and dried at 100°C for 15 minutes. Upon being inserted in the flame of a Bunsen burner, these treated chips exhibit only a slow smouldering of the wood in the flame area; no flame develops. After removal from the flame, only minor charring of the chip surface is observed. In contrast to the foregoing, however, untreated fir chips which were heated in the same manner burst into flame and continued to burn until they were converted to ash.

Monomethylol Dicyandiamide, Melamine and Phosphonic Acid Based Compositions

W.J. Oberley; U.S. Patent 3,986,881; October 19, 1976; assigned to Koppers Company, Inc. describes a composition useful for imparting nonhygroscopic, leach resistant fire retardant properties to wood comprising a 6 to 80% by weight aqueous solution of partially reacted monomethylol dicyandiamide, melamine, and phosphoric acid. The molar ratio of monomethylol dicyandiamide to melamine is from about 11.5:1.0 to 3.0:1.0. The molar ratio of the phosphoric acid to monomethylol dicyandiamide plus melamine is from about 1.0:1.0 to 1.0:1.5. The wood is impregnated with a solution of the partially reacted ingredients and the impregnated wood is, thereafter, dried and cured by heating to a temperature of from about 60° to 110°C. The wood treated in accordance with this process has leach resistant fire retardant properties, has improved dimensional stability and has reduced hygroscopicity. In addition, the treated wood does not bloom.

Example: An aqueous wood treating solution of partially reacted monomethylol dicyandiamide, melamine, and phosphoric acid was prepared as follows. A solid composition containing monomethylol dicyandiamide (MMD) and melamine was prepared by mixing 152 lb of monomethylol dicyandiamide (1.35 lb-mols) and 32 lb of melamines (0.25 lb-mols). The resulting solid composition had a mol ratio of MMD to melamine of 5.4:1.0 and was stable for at least 1 year at room temperature and may be easily shipped to a wood treatment plant. 184 lb of the solid composition was dissolved in 184 lb of water at temperatures of from about 50° to 80°C. Then, the solution was diluted with cold water until the resulting solution was about 8.0% by weight active solids. The dilute solution was cooled to about 30°C and then 261 lb of 75% phosphoric acid (1.25 lb-mols) was slowly added. The resulting solution contained 14.8% active ingredients and was very useful in treating wood to impart leach resistant, nonhygroscopic fire retardant properties to the wood.

Potassium Hydroxide and Potassium Carbonate as Intumescent Agents

R.C.F. de Lissa and W.G. Schwarze; U.S. Patent 3,973,074; August 3, 1976; assigned to MacMillan Bloedel Limited, Canada describe a flameproof product comprising a cellulosic or wood base substrate having at its surface at least one fire retardant chemical compound selected from the group consisting of potassium hydroxide and potassium carbonate. The compound is provided in an amount of at least 75% by weight based on the total weight of fire retardant chemicals applied to the surface by drying a solution of the compound. The amount of compound at the surface is sufficient such that when the surface is subjected to high temperatures, the surface foams with decomposing cellulosic material to form a layer of relatively noncombustible intumescent carboniferous char.

An important advantage is the inherent ability of the treatment solution to penetrate the surface of the cellulosic or wood based material. It is found that absorption of the fire retardant into cellulosic based materials is promoted by the alkalinity of the fire retardant with a preferred solution pH of 9 or 10 or greater giving maximum penetration. Such an ability allows intimate contact of the fire retardant chemicals with the cellulosic portion of the substrate and, subsequently, more efficient formation of the protective intumescent foam.

Example: Hand-felted particle boards bonded with urea-formaldehyde resin were prepared in the laboratory. The surface material was treated with KOH and some with K_2CO_3 before the board preparation. The following flame-spread ratings (FSR) were obtained on the Monsanto version of the 2-foot tunnel test (H.L. Vandersall, "The Use of a Small Flame Tunnel for Evaluating Fire Hazards," *Journal of Paint Technology*, Vol. 39, No. 51, 494-500, 1967) which has been correlated with the ASTM-E 84 25-foot tunnel test. A lower number indicates a better rating than a higher number. The chemical loading was based on the weight of chemical added to each surface divided by the total board weight.

		% Treatment	FSR
	Control board	0	>140
(a)	KOH treatment		
	1.	6.6	68
	2.	8.75	36
(b)	K_2CO_3 treatment		
	1.	4.8	32
	2.	10	5
	3.	10	8

Boric Acid and Boron Oxide

The process of *B.M. Short and J.W. Rayfield; U.S. Patent 4,076,871; February 28, 1978; assigned to Masonite Corporation* relates to a method of impregnating wood with boric acid or boron oxide. In particular, it relates to a method of introducing boric acid into wood products by application thereto of boric acid esters of alcohols and thereafter hydrolyzing the esters to deposit boric acid within the wood product. The wood products to which this process can be applied include lumber, plywood, hardboard, particle board, luan board, fiberboard and similar materials.

Often it is advantageous to introduce boric acid or boron oxide into untreated wood products, but usually it is desirable to combine the boric acid treatment with other fireproofing systems in order to provide wood products with maximum fire resistance. It has been found desirable to treat wood products which contain alumina, preferably in the form of alumina trihydrate, with the boric acid esters to produce wood products which contain both alumina and boric acid. Such wood products are highly flame and smoke resistant and qualify for Class 1A ratings in Underwriters' Laboratory tests. The borate ester used can be produced from boric acid and any of the higher organic alcohols such as butanol, pentanol, hexanol, nonanol, etc.

Ordinarily, these esters are produced by heating a mixture of the organic alcohol and boric acid to temperature in excess of 100°C for a period of time varying from ½ to 5 hours. In general practice, it is desirable to add an azeotropic agent such as toluene, xylene or kerosene which aids in the removal of water produced as a by-product of the esterification reaction and thus tends to force the reaction toward esterification. It is desirable that the boric acid ester contains as high a percentage of boron as possible so that the amount of boron introduced into the wood product can be kept to a maximum. Thus, mono esters of boric acid are desirable materials for impregnation of wood products.

In treating wood products to increase the fire retardant and flameproof properties, the boric acid ester in a fluid form is applied to the surface of the wood products with any of the usual coating equipment such as a direct roll coater, a wier pan, etc. The amount of boron in wood, calculated as boric acid, varies from 3 to about 20%, depending upon the special qualifications and properties desired in the wood product.

After the ester is applied to the surface of the wood, the wood product is passed through an oven or kiln at a temperature of 100° to 200°C to remove any solvent or any excess alcohol and to cause the borate ester to penetrate into the fibers of the wood product. During this operation, there may occur esterification between the boric acid of the borate ester and the hydroxyl groups of the cellulose in the wood fibers. Finally, the wood product is treated with moist air or steam, usually at elevated temperature in the range of 50° to 100°C for a time sufficient to cause hydrolysis of a major part of the boric acid ester and to cause the deposition of boric acid or boron oxide in the fibers of the wood product.

Alumina Hydrate and Ulexite Synergistic Composition

E. Panusch, I. Sobolev, and S.C. Carniglia; U.S. Patent 4,076,580; February 28, 1978; assigned to Kaiser Aluminum & Chemical Corporation describe a procedure whereby cellulosic boards, such as hardboards, fiberboards and particle boards, are flame retarded by incorporation of a synergistically acting alumina trihydrate-ulexite composition in the boards. The alumina trihydrate-ulexite flame retarding agent provides a board of satisfactory strength properties with a flame spread at least equivalent to that of a commercial Class 1-rated hardboard when the loading of the flame retarding agent in the board is kept at 50% or below of the total weight of the flame retarded board.

Alumina trihydrate refers to an alumina compound generally defined by the formulas of $Al_2O_3 \cdot 3H_2O$ or $Al(OH)_3$. The hydrated, boron-containing mineral ulexite, employed as one of the essential components of the synergistically performing flame retarding agent, is naturally occurring and has the generally accepted formula of $NaCaB_5O_9 \cdot 8H_2O$. This mineral is only slightly soluble in water and is found in large deposits, e.g., in Nevada and California.

The flame retarding composition is an essentially uniform mixture of ulexite and alumina trihydrate particles. The mixture, to be efficient as a synergistically performing flame retarding agent, generally contains from about 30 to 50% by weight ulexite, with the balance being alumina trihydrate. Concerning the particle size of the ulexite-alumina trihydrate mixture, it has been found that satisfactory results can be obtained when the average particle size of the mixture is kept within the range of about 7 to 200 microns (0.007 to 0.200 mm).

Example 1: A series of flame retarded hardboards was made from dry unbleached wood fibers by incorporating varying quantities of ulexite-alumina hydrate compositions in each individual board. All boards of the series were made according to the process described below with the exception of the ulexite-alumina hydrate quantity which was changed from board to board. For each board, 260 g of fluffed fiber were charged into a container of 19 liter (5 gallon) capacity. The container was closed, then tumbled at an angle, while through an aperture, an aqueous binder composition, containing 27 g of phenol-formaldehyde resin dispersed in 130 ml water, was introduced.

Subsequently, the ulexite-alumina hydrate composition, containing a 50:50 mixture (by weight) of these components, was incorporated in the fibers under vigorous mixing. The mixture was then transferred to a sheet mold of 25.4 x 25.4 cm (10" x 10") size and compacted with a metal plate. The fiber mat was further compacted in a platen press without employing heat, then the compacted mat was transferred to a perforated aluminum plate on which it was pressed into a board. For board making, a temperature of 215°C (419°F) was used with an initial force of about 19 tons (380 psi) for 1.5 minutes. This was followed by an approximately 4-minute pressure treatment at 7.5 tons (140 psi) at the same temperature. The finished board was stored for about 12 to 14 hours, then specimens of about 8 x 25 cm (3¼" x 10") were cut for flame retardancy testing. To establish the effect of loading of the alumina hydrate-ulexite composition on flame-retardancy, the loading in this series of board was varied between 0 and 60%, based on the total weight of the cellulose-ulexite-alumina hydrate composition in the finished flame retarded board. The flame retardancy of these boards, expressed as oxygen index, is graphically depicted in Figure 9.1.

FIGURE 9.1: EFFECT OF LOADING LEVELS ON FLAMMABILITY OF HARDBOARD

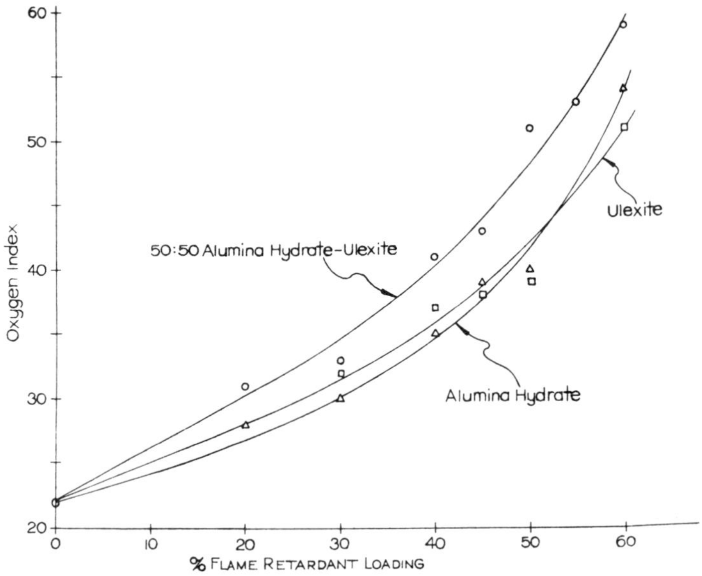

Source: U.S. Patent 4,076,580

Example 2: Another series of hardboards was prepared according to the process described in Example 1. In this series, the flame retarding effect of the individual components of the synergistic mixture was tested. Thus, boards were made containing only alumina hydrate at loadings between 0 and 60% by weight based on the total weight of the board and the tests were repeated using only ulexite at the same 0 to 60% by weight loading. All the board test specimens were subjected to oxygen index measurements and the results are reported graphically in Figure 9.1.

The synergistic flame retarding effect attained by the use of a ulexite-alumina hydrate composition can be clearly observed. Moreover, nearly all of the ulexite containing boards exhibited afterglow.

Cyclic Sulfonium Zwitterions as Leach Inhibitors of Flame Retardant Salts

F.L. Brown; U.S. Patent 4,049,849; September 20, 1977; assigned to The Dow Chemical Company describes a process for inhibiting the leaching of a fire retardant salt from a wood substrate. This comprises impregnating the wood substrate with an aqueous solution comprising between about 5 and 40 wt % solids. The solids consist of the fire retardant salt and a cyclic sulfonium zwitterion at a fire retardant salt:cyclic sulfonium zwitterion weight ratio between about 1:1 and 20:1. Impregnation can also be achieved subsequent to impregnating the wood substrate with the fire retardant salt, with an aqueous solution comprising between about 1 and 40 wt % of the cyclic sulfonium zwitterion. The resulting cyclic sulfonium zwitterion-impregnated wood substrate is then cured. This process is applicable to wood, such as southern and ponderosa pine (softwood species), and maple and oak (hardwood species); wood composites, such as particle and fiberboard, plywood, and reconstituted products; and wood-derived products, such as veneer and paper.

Example: Cyclic sulfonium zwitterions (CSZ) of the formulae:

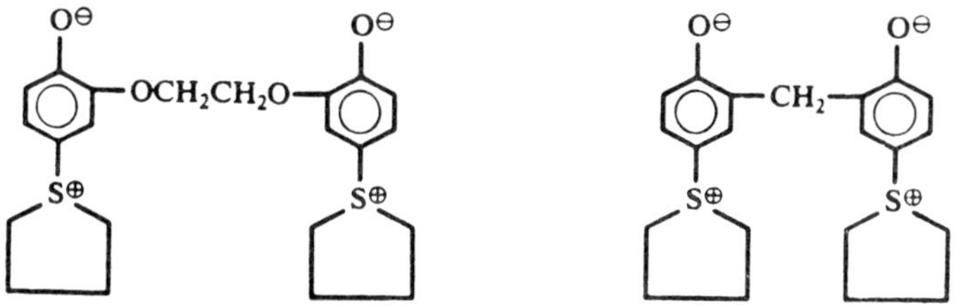

hereinafter respectively referred to as resorcinol S^+ and novalac S^+, were aqueously formulated with Fire Retardant A (FR-A, described as a 40/60 mixture of mono-/di-ammonium phosphate). The aqueous formulations were prepared at 1:1, 2:1, 4:1 and 6:1 weight ratios of FR-A solids:CSZ monomer solids. The aqueous formulations were tested at 15 and 30% solids.

Ponderosa pine sapwood specimens (½" x ½" x 3") were impregnated with the various aqueous formulations by placing them in an impregnating chamber, evacuating the chamber to approximately 29.5" of mercury for 30 minutes, admitting the various aqueous formulations to cover the specimens (which were restrained from floating while maintaining vacuum), and then exposing the submerged specimens to atmospheric pressure (by returning the chamber's atmosphere). The submerged specimens were then held overnight to allow a period of diffusion. Specimens were subsequently cured for about 16 hours in an oven at 105°C, cooled over desiccant and weighed to determine oven-dry, cured weight.

Leach resistance was then determined by reimpregnating (per the above procedure) the cured specimens with water followed by overnight soaking. This procedure will remove 50 to 75% of FR-A from an FR-A impregnated wood substrate if it (FR-A) is used alone. The leached specimens were oven-dried a second time and then conditioned at 50% relative humidity to achieve a 7±3% moisture content as required in the American Society for Testing and Materials' (ASTM) E160-50 crib test. The conditioned specimens were then burned according to the crib test. The data is set forth in the table on the following page.

Cyclic Sulfonium Zwitterion Monomer	Fire Retardant Salt	Weight Ratio Solids Salt/Monomer	Solids in Treating Solution %	Solids Retained at Treatment %	Solids Lost During Leaching %	Phosphorus Retained at Time of Burn % (1)	Phosphorus Retained at Time of Burn % (2)	Weight Loss During Burn %
Resorcinol S⊕	FR-A	1/1	30	74.6(1)	6.65	9.33	5.35	16.9
			15	34.0	7.16	4.25	3.11	19.4
		2/1	30	72.9	11.0	12.2	6.55	14.4
			15	33.4	15.4	5.60	3.49	26.6
		4/1	30	71.6	14.6	14.3	7.47	13.8
			15	34.1	23.2	6.82	3.85	24.8
Novalac S⊕	FR-A	1/1	30	71.3	14.8	7.66	5.84	17.0
			15	33.2	6.8	3.52	3.54	36.6
		6/1	30	70.3	16.6	12.7	8.41	12.8
			15	33.8	26.0	6.24	4.48	25.7

[1] Based upon oven-dry, untreated weight.
[2] Based upon oven-dry, treated, leached weight.

MISCELLANEOUS

Reaction Products of Bis(Hydroxyethyl) Terephthalate with Phenyl Phosphonyl Dichloride

A.B. Conciatori and R.W. Stackman; U.S. Patent 4,030,933; June 21, 1977; assigned to Celanese Corporation describe a process for preparing phosphorus and halogen containing polymers useful as flame retardant additives in cellulose acetate or cellulose triacetate fibers or films. The phosphorus and halogen containing polymers may be prepared by reacting a halogenated derivative of bis(hydroxyethyl) terephthalate with a halogen containing phosphorus monomer having the formula:

```
   O
   ||
X–P–X
   |
   R
```

R represents a radical selected from the group consisting of (a) alkyl, (b) bromoalkyl, (c) chloroalkyl, (d) alkylamido and (e) dialkylamido radicals, where the alkyl substituents may contain from 1 to 10 carbon atoms, (f) cyclohexyl, (g) phenyl, (h) benzyl, (i) tolyl, and (j) xylyl, and X represents a halogen atom (preferably chlorine or bromine). Cellulose acetate and cellulose triacetate fibers and films containing as little as 2% by weight of the polymeric flame retardant additives are rendered both nonburning and resistant to loss of the polymeric flame retardant additive during use. The halogenated derivatives of bis(hydroxyethyl) terephthalate may be prepared in accordance with the procedures set forth in U.S. Patent 3,652,647.

These polymeric products may be intimately mixed with cellulose acetate or cellulose triacetate so as to produce cellulose acetate or cellulose triacetate fibers or films, and in each case, the polymeric additive acts as a flame retardant when used in amounts of generally from about 5 to 15% by weight of the total composition. The cellulose acetate or cellulose triacetate is correspondingly present in an amount of generally from about 85 to 95% by weight of the total composition. Thus, the cellulose acetate and cellulose triacetate fibers and films are rendered nonburning when treated with these polymeric additives.

Example 1: This example illustrates the preparation of the flame retardant polymeric additives by reacting tetrachloro bis(hydroxyethyl) terephthalate with phenyl phosphonyl dichloride. The tetrachloro bis(hydroxyethyl) terephthalate used in forming the polymeric flame retardant additive may be prepared as follows: 20 g of finely divided (passes through 325 mesh U.S.) animal bone carbon catalyst are added to 275 g of tetrachloro terephthalic acid, 100 g of ethylene oxide and 2 liters of chlorobenzene solvent in a 2.5 liter autoclave fitted with a steam jacket and mechanical stirrer. The autoclave is then sealed and heated with stirring to 160°C for about 2 hours after which the contents of the autoclave are flushed to remove ethylene oxide. The remaining slurry is filtered hot to remove the unreacted tetrachloro terephthalic acid and the carbon catalyst, and finally the tetrachloro bis(hydroxyethyl) terephthalate is recovered by cooling the mother liquor to about 25°C so as to effect crystallization of the diester from solution.

60% of the tetrachloro terephthalic acid charged to the reactor is converted to tetrachloro bis(hydroxyethyl) terephthalate, having a melting point of 197° to 200°C. Gel permeation chromatography of the product shows that only monomeric material is present. 19.6 g (0.05 mol) of the tetrachloro bis(hydroxyethyl) terephthalate prepared above, 100 ml of benzene, 8 g of pyridine, and 12 g of triethylamine are then added to a 250 ml, three-neck, round bottom flask provided with a stirrer, condenser, nitrogen inlet, and dropping funnel. To this mixture there is added dropwise from the dropping funnel a solution comprising 9.75 g (0.05 mol) of phenyl phosphonyl dichloride dissolved in 25 ml of benzene.

This mixture is allowed to reflux for 3 hours after the addition of all of the phenyl phosphonyl dichloride. Then, the mixture is cooled to room temperature and filtered through a 25 centimeter Buchner funnel using Whatman No. 1 filter paper. The filtrate is then evaporated to dryness and about 24 g (93% yield, based on phenyl phosphonyl dichloride) of product are obtained. A minor amount of pyridine hydrochloride is also present. The product is a pale cream colored solid having an intrinsic viscosity of 0.30, a weight average molecular weight of about 10,000, and a melting point of about 50°C. The polymeric product contains 6% by weight phosphorus and 27% by weight chlorine based upon the weight of the polymer.

Example 2: This example illustrates the preparation of a flame retardant cellulose acetate film containing 5% by weight of the polymeric flame retardant additive prepared in Example 1. 5 g of the polymeric flame retardant additive prepared in Example 1 are mixed with 650 ml of acetone. The solid readily dissolves. Then, 95 g of cellulose acetate are added to this solution and the mixture is stirred for 24 hours until the solution is clear and uniform. The viscous solution is then filtered through a fine porosity sintered glass funnel under pressure to give a clear polymer dope.

A portion of this solution is then poured onto a glass plate and spread with a doctor knife to a thickness of 15 mils. Upon drying, the film measures approximately 3 mils and is clear and transparent. This film containing the polymeric additive prepared in Example 1 is subjected to the following flammability test. A strip of the film 2" wide and 10" long is rolled into a cylinder 10" long, having a 1/16" diameter. The flame of a match is applied to the bottom of a vertically held cylinder for 3 seconds and then removed. The sample extinguishes within 2 seconds after the removal of the flame. The polymeric flame retardant additive also does not wash out of the film after 50 washings in water at 50°C for 15 minutes each.

Anionic Complex Containing Nitrogen, Phosphorus, Oxygen and Aluminum

E. Stossel; U.S. Patent 4,076,540; February 28, 1978 describes a fire extinguishing composition having enhanced efficiency and effectiveness as a result of the cooperating and mutual reinforcing interaction of the ingredients. This comprises (A) an ionic complex containing in an anionic portion of the molecule complexed nitrogen, phosphorus, oxygen, and aluminum, and having a P:N molar ratio of 10:1 to 2:6, a P:Al molar ratio of 1:1 to 100:1, an acid number less than 400, and a pH in a 2% aqueous solution above 2.5; (B) a boric acid compound selected from the group consisting of orthoboric acid, metaboric acid, and boric anhydride, and (C) urea. The proportions of these ingredients in the composition are 30 to 80 parts by weight of the ionic complex, 3 to 4 parts by weight of boric acid compound, and 10 to 300 parts by weight of urea.

Flammable objects, particularly objects made of cellulose, are reduced in flammability by being pretreated with the composition. For example, cotton batting can be passed through a treating bath containing an aqueous solution of boric acid, urea, and ionic complex prepared according to this process, and then squeezed between rollers and oven dried to give a fire resistant batting that can be stored for long periods of time in either dry or humid atmospheres without dusting or loss of the ability to pass the cigarette burning test for fire resistant mattresses.

Furthermore, the fire extinguishing composition can be used to douse fires as a dry chemical extinguisher or dissolved in water, and in either case can be mixed with additional fire extinguishing components such as sodium bicarbonate. The fire extinguishing composition can also be applied to objects such as structural members, building materials, paper, textiles, or other combustible polymers, which it is desired to protect from the destructive effect of fire.

Example 1: 1,400 ml of a rinse solution from the bright-dipping of aluminum containing per liter 350 g of phosphoric acid and 20 g of dissolved aluminum are stirred and heated to boiling. With continued heating, 41 g of crude North African aluminum phosphate (analysis: Al_2O_3 31, P_2O_5 22.5, Ca 0.9, Fe_2O_3 7, SiO_2 2.5, TiO_2 1.5, moisture 10%) are sprinkled into the boiling solution. Stirring and evaporation of water are continued until the volume of the solution is reduced to 800 ml, at which point the mixture is filtered to remove insoluble impurities and then concentrated to the boiling temperature of 125°C.

After cooling to 95°C, 250 g of urea are stirred in along with 50 mg FC-98 acid-resistant fluorochemical surfactant and the mixture transferred to a shallow pan in an oven heated at 200°C. The mixture foams on reaching 150°C and by 30 minutes of foaming has turned into a friable sponge and can be removed. The yield is 625 g of anionic complex analyzing 54.5% P_2O_5, 14.1% N, 7.7% Al_2O_3 for a P_2O_5/Al_2O_3 weight ratio of 7.1.

Example 2: A fire-preventing impregnating solution for 1.8 cm thick ceiling tiles made of sugar cane bagasse compressed with a gypsum binder is prepared by dissolving in 10 liters of water 2 kg boric acid, 5 kg urea and 4.4 kg of anionic complex prepared as in Example 1. A vertical extinguishing rating in the burning test of ASTM D1692 is achieved with a quantity of this solution that provides a 12% dry add-on. Compositions lacking one or more of the essential ingredients required the following dry add-on quantities by weight: boric acid alone, 90%; urea alone, 77%; anionic complex alone, 18%; anionic complex with urea in equal weight proportions, 29%; anionic complex with boric acid (2 to 1 by weight), 25%; boric acid and urea (1 to 2 by weight), 75%. The results of this experiment show the mutual reinforcement action of the essential ingredients of the composition of this process upon one another in providing improved fire prevention.

Synergistic Effects Between Boric Acid and Sulfate Constituents

S.M. Draganov; U.S. Patent 3,983,040; September 28, 1976 describes a fire retardant composition of matter produced as the dried and pulverized reaction product of treating a calcium, sodium, magnesium, or potassium borate ore with sulfuric acid. The composition is especially useful for application as a dry powder to cellulosic materials such as cotton batting and ground waste paper insulation. It may also be made into a liquid suspension for spray application or into a paste for some fire retardant applications.

Example: 2,000 pounds of a calcined and pulverized (minus 200 mesh) calcium borate ore, analyzing 50% B_2O_3 as colemanite mineral with the major impurity being limestone, was charged into a stainless steel ribbon blender. Over a period of 30 minutes of operation of the blender, 1,273 pounds of sulfuric acid (66° Baume) were sprayed onto the agitated ore material from multiple nozzles. Immediately following this period of acid addition and blending, 876 pounds of water was added through the spray nozzles over a time period of 20 minutes while operation of the blender continued for mixing purposes. Mixing was continued for another 20 minutes while observing the reaction in the mixer. At such times as excessive dryness of the mix indicated the need for additional water, more water was supplied.

The heat of the reaction promoted rapid drying, and at the end of the mixing period a hot and dry reaction product was discharged from the reaction vessel for further drying during atmospheric cooling. The cool and dry reaction product was then pulverized to provide a dry, pulverous composition of matter containing 48% boric acid and 40% calcium sulfate as well as minor impurities originally present in the ore as gangue. This composition was mixed with ground waste paper insulation material on the basis of 25% of the composition to 75% of the insulation material, yielding excellent fire retardant properties. Added to cotton batting on the basis of 12.5% of the composition to 87.5% of the batting, it was found to be as effective as the use of 10% pure boric acid. All percentages were based on weight.

Cellulose Thiocyanate

S. Gladstone and C.R. Neumoyer; U.S. Patent 4,076,918; February 28, 1978; assigned to Argus Chemical Corporation describe a process for preparing cellulose thiocyanate by direct reaction of cellulose with thiocyanic acid in the presence of dimethyl formamide or dimethyl sulfoxide at elevated temperatures, the thiocyanic acid being formed in situ from a thiocyanate salt and sulfuric acid. Any cellulosic material can be used as the source of natural cellulose. Cellulose fibers such as cotton linters can be used, as well as cellulose pulp. The product cellulose thiocyanate has the following characteristics: carbon content, 38.00 to 44.4%; nitrogen content, 0.10 to 6.30%; sulfur content, 0.23 to 14.50%; flammability, self-extinguishing; and solubility, insoluble in cold water, sparingly soluble in hot water, insoluble in aromatic solvents such as toluene and xylene, aliphatic solvents such as hexane and alcohols, and heterocyclic solvents such as dioxane, pyridine and piperidine. The cellulose thiocyanate's lack of flammability and its ability to impart self-extinguishing properties to other solids make it useful as a flame retardant for plastics, fibers and films, textiles, and other flammable materials.

Example: To 15 g (0.0925 mol) of cellulose (medical grade absorbent cotton) were added 200 ml of dimethyl formamide, 30 g (0.394 mol) of ammonium thiocyanate and 25 ml (0.44 mol) of concentrated sulfuric acid. The mixture was stirred and held at 110°C for 6 hours. The cellulose thiocyanate-containing product, a brownish white powder, was separated in a Buchner funnel, and thoroughly washed with cold water until all salts and liquids had been washed out. The residue was then extracted with hot boiling water for 72 hours in a Soxhlet extractor. The cellulose thiocyanate was dissolved in water and recovered by precipitation by cooling the extract solution. The residue in the Soxhlet was sulfur. The cellulose thiocyanate recovered by precipitation from the extract was a translucent white solid falling within the following analysis: carbon content, 38.00 to 44.40%; nitrogen content, 0.10 to 6.30%; sulfur content, 0.23 to 14.50% (corresponding to 0.015 to 1 SCN group per anhydroglucose unit of the cellulose, or a degree of substitution of 0.015 to 1); it was self-extinguishing.

Urea/THPC Precondensate

R. Cole; U.S. Patent 4,078,101; March 7, 1978; assigned to Albright & Wilson Ltd., England describes a process for flameproofing cellulosic textile fabric which comprises impregnating the fabric with an aqueous solution containing a precondensate of urea and a tetrakis(hydroxymethyl)phosphonium salt together with any excess of the tetrakis(hydroxymethyl)phosphonium salt necessary to make up a molar ratio of urea to tetrakis(hydroxymethyl)phosphonium lying between 0.05:1 and 0.25:1. The solution is neutralized by the addition of an alkali or base to a pH in the range 5 to 8. The impregnated fibers are then treated with ammonia to form a crosslinked polymer.

The impregnating solution is made up by mixing the requisite quantities of urea and THPC in water and refluxing to cause them to react to form the precondensate. Alternatively, a solution with a molar ratio of urea to THP salt of 0.5:1 may be refluxed and the necessary quantity of THPC to obtain the correct ratio of urea to THP added afterwards. However it is made, the solution is neutralized to preferably a pH of 5.8 to 7. This is conveniently effected by adding caustic soda, but other alkalis such as sodium carbonate can be used.

The THP salt is normally the chloride, but other halides such as the bromide, or other salts such as the acetate, the sulfate or phosphate may be used. The concentration of the solution of precondensate is preferably 20 to 40% by weight. Preferred values are 25 to 30%. It is convenient to make initially a solution of about 50% concentration, which is then diluted shortly before use. Preferably, the neutralization is effected at this stage rather than at the manufacturing stage. The treatment with ammonia may be carried out by processes described in U.S. Patent 2,983,623 or British Patents 1,439,608 and 1,439,609.

GENERAL UTILITY ADDITIVES

HALOGEN COMPOUNDS

Halogenated Polythiols

C.R. Morgan and D.E. Kramm; U.S. Patent 4,081,598; March 28, 1978; assigned to W.R. Grace & Co. describe halogen-containing polythiols which impart flame retardancy to radiation cured thiol-ene compositions. The solid cured polythioether polymer products resulting from this process may have many and varied uses. Examples of some uses include adhesives; caulks; elastomeric sealants; coatings such as wire coatings; electrical circuits over resists, photoresists and the like; encapsulating or potting compounds; liquid castable elastomers; thermoset resins; impregnants for fabric, cloth, fibrous webs and other porous substrates; laminating adhesives and coatings; mastics; glazing compounds; fiberglass reinforced composites; sizing or surface finishing agents; filleting compounds; cured in place gasketing compounds; rocket fuel binders; foamable thermosetting resins or elastomers; molded articles such as gaskets, diaphragms, balloons, automobile tires, etc. The following examples illustrate the process.

Example 1: 180.5 g of 1,4,5,6,7,7-hexachloro-5-norbornene-2,3-dimethanol and 111.3 g of β-mercaptopropionic acid along with 500 ml of dichloroethane as solvent and 9.64 g of p-toluenesulfonic acid as catalyst were charged into a 1-liter flask equipped with a stirrer, thermometer and reverse Dean Stark trap. The mixture was heated to reflux. Water of esterification was removed by azeotropic distillation. The amount of water collected was about 15 ml in 6.5 hours. The reaction solution was cooled to room temperature and washed once with 500 ml of water; two times with 500-ml portions of 5% sodium bicarbonate solution; and then one time with 500 ml of water.

The resulting dichloroethane solution containing the product was dried over 40 g of magnesium sulfate, treated with 1 g of Norit A decolorizing carbon, and then filtered. After filtration, the solvent was removed by vacuum stripping to give 161 g of a polythiol of the formula shown on the following page, with a thiol content of 3.66 meq/g (theory = 3.72 meq/g). This polythiol will hereafter be referred to as polythiol 1.

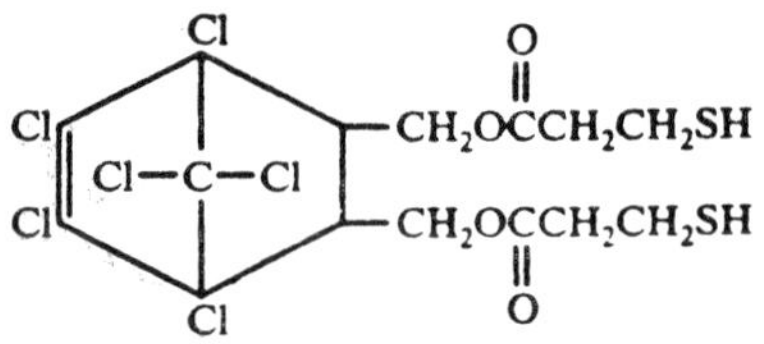

Example 2: 948 g of the bis(2-hydroxyethyl ether) of tetrabromobisphenol A, 349.5 g of β-mercaptopropionic acid, 1,200 ml of dichloroethane as solvent, and 26 g of p-toluenesulfonic acid as catalyst were charged into a reaction flask equipped with a stirrer, thermometer, and a reverse Dean Stark trap. The mixture was heated to reflux and the water of esterification removed by azeotropic distillation. The amount of water collected was about 61 ml in 3.5 hours. The reaction solution was cooled to room temperature and washed once with 1,800 ml of water.

The resulting dichloroethane solution containing the product was dried over 120 g of magnesium sulfate, treated with 1 g of Norit A decolorizing carbon, and then filtered. After filtration, the solvent was removed by vacuum stripping to give 932 g of a polythiol of the formula:

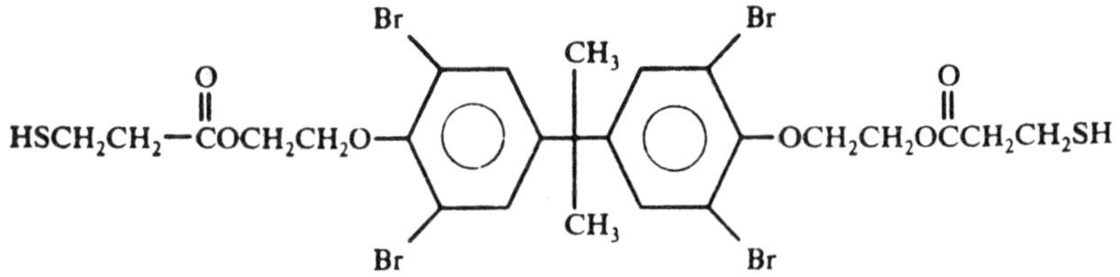

with a thiol content of 2.46 meq/g (theory = 2.48 meq/g). This polythiol will hereafter be referred to as polythiol 2.

The following example shows the flame retardancy obtained with halogen-containing polythiols as compared with commercially available polythiol.

Example 3: The following compositions were compounded by mixing the ingredients with stirring until a homogeneous mixture was obtained:

	Grams
Composition A	
Triallyl cyanurate	1,610
Pentaerythritol tetrakis(mercaptopropionate)	2,390
Benzophenone	0.8
Phosphorous acid	80
Hydroquinone	0.2
Pyrogallol	0.2
Composition B	
Triallyl cyanurate	23
Polythiol 1	76.7
Benzophenone	2
Phosphorous acid	0.02
Hydroquinone	0.005
Pyrogallol	0.005

(continued)

	Grams
Composition C	
Triallyl cyanurate	24.8
Polythiol 2	115.2
Benzophenone	2.8
Phosphorous acid	0.07
2,6-di-tert-butyl-4-methylphenol (Ionol)	0.14
Octadecyl-β-(4-hydroxy-3,5-di-tert-butyl phenyl)-propionate (Irganol 1076)	0.28

Thereafter the viscous mixtures were poured onto a Mylar backing sheet and drawn down to a uniform thickness of 20 mils. The coatings were then exposed to a mercury vapor lamp for 1 minute at a surface intensity on the coatings of 10,000 $\mu W/cm^2$. The cured coatings were removed from the Mylar sheet, cut to the proper dimensions and measured for oxygen indexes. The results of the oxygen indexes were as follows:

Composition	Oxygen Index (%)
A	21.0
B	28.9
C	30.2

From this Oxygen Index data it can be seen that the halogen-containing polythiols of Compositions B and C significantly improve flame retardancy as measured by the increased Oxygen Index of the cured polyene/polythiol compositions.

Bis(Tetrahalophthalimidocarbonyl)Benzene

S.B. Richter and G.D. Kyker; U.S. Patent 4,001,179; January 4, 1977; assigned to Velsicol Chemical Corporation describe chemical compounds having the following formula

(1)

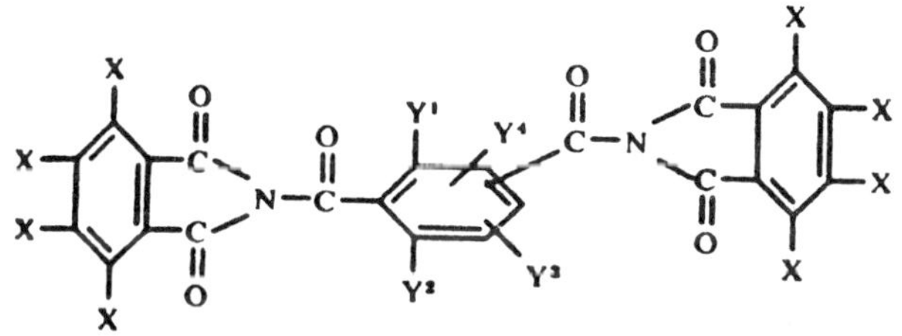

where X is chlorine or bromine and each of Y^1, Y^2, Y^3 and Y^4 are hydrogen, chlorine or bromine.

The compounds can be prepared by reacting an isophthaloyl dichloride or terephthaloyl dichloride of the formula below, where Y^1, Y^2, Y^3 and Y^4 are as described above, with 2 molar amounts of a tetrahalophthalimide of the formula shown on the following page.

(2)

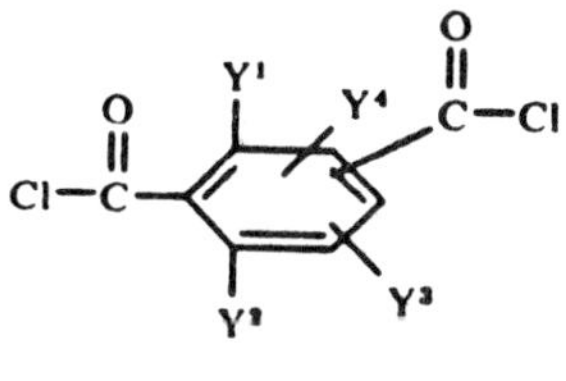

In the following formula X is a halogen as described in formula (1).

(3)

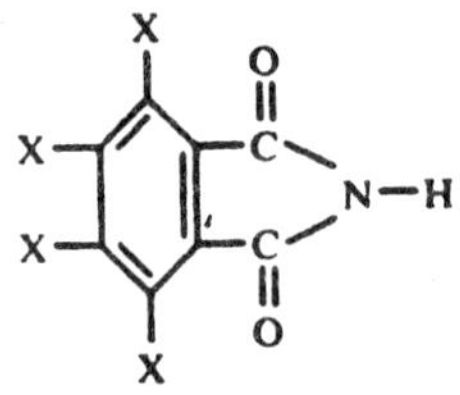

This reaction can be effected by combining the compounds of formulas (2) and (3) in an inert organic reaction medium in the presence of an acid acceptor such as a tertiary amine. The reaction mixture can then be heated at reflux for a period of from about 1 to 24 hours to ensure completion of the reaction. After this time, the mixture can be filtered and the filtrate can be stripped of solvent to yield the desired product. This product can be used as such or can be further purified by conventional techniques such as washing, recrystallizing and the like. The manner in which the compounds can be prepared is illustrated in the following examples.

Example 1: 1,4-Bis(3,4,5,6-Tetrachlorophthalimidocarbonyl)Benzene – 0.04 mol of 3,4,5,6-tetrachlorophthalimide, 500 ml benzene and 0.045 mol triethylamine are charged into a glass reaction vessel equipped with a mechanical stirrer, thermometer and reflux condenser. 0.02 mol terephthaloyl dichloride is added dropwise with stirring to the reaction mixture at room temperature.

After the addition is completed, the reaction mixture is heated at reflux for a period of about 2 hours. After this time, the mixture is filtered and the filtrate is washed with water. The washed solution is then dried over anhydrous magnesium sulfate, filtered and stripped of solvent to yield the desired product, 1,4-bis(3,4,5,6-tetrachlorophthalimidocarbonyl)benzene.

Example 2: 1,3-Bis(3,4,5,6-Tetrabromophthalimidocarbonyl)Benzene – 0.04 mol of 3,4,5,6-tetrabromophthalimide, 500 ml benzene and 0.045 mol pyridine are charged into a glass reaction vessel equipped with a mechanical stirrer, thermometer and reflux condenser. 0.02 mol isophthaloyl dichloride is added dropwise with stirring to the reaction mixture at room temperature.

After the addition is completed, the reaction mixture is heated at reflux for a period of about 2 hours. After this time, the reaction mixture is filtered and the filtrate is washed with water. The washed solution is then dried over anhydrous magnesium sulfate, is filtered and stripped of solvent under reduced pressure to yield the desired product, 1,3-bis(3,4,5,6-tetrabromophthalimidocarbonyl)-benzene.

Example 3: 1,4-Bis(3,4,5,6-Tetrachlorophthalimidocarbonyl)-2,6-Dichlorobenzene – 0.04 mol of 3,4,5,6-tetrachlorophthalimide, 500 ml benzene and 0.045 mol pyridine are charged into a glass reaction vessel equipped with a mechanical stirrer, thermometer and reflux condenser. 0.02 mol 2,6-dichloroterephthaloyl dichloride is added dropwise with stirring to the reaction mixture at room temperature. After the addition is completed, the reaction mixture is heated at reflux for a period of about 2 hours. After this time, the reaction mixture is filtered and the filtrate is washed with water. The washed solution is then dried over anhydrous magnesium sulfate, is filtered and stripped of solvent under reduced pressure to yield the desired product 1,4-bis(3,4,5,6-tetrachlorophthalimidocarbonyl)-2,6-dichlorobenzene.

The compounds of this process can impart fire retardant properties to a variety of combustible polymers. Exemplary of such polymers which can be used in a mixture with the compounds to form the fire retardant polymeric compositions are the homopolymers and copolymers of unsaturated aliphatic, cycloaliphatic, and aromatic hydrocarbons such as polyethylene, polypropylene, polybutene, ethylene-propylene copolymers, copolymers of ethylene or propylene with other olefins, polybutadiene, polymers of butadiene, polyisoprene, polystyrene, polyvinylidene, and polymers of pentene, hexene, heptene, etc.

Bis(Guanidine)Tetrabromodian

H. Jenkner; U.S. Patent 4,022,698; May 10, 1977; assigned to Chemische Fabrik Kalk GmbH, Germany describes an organic bromine compound possessing flame retardant properties. The compound has the formula

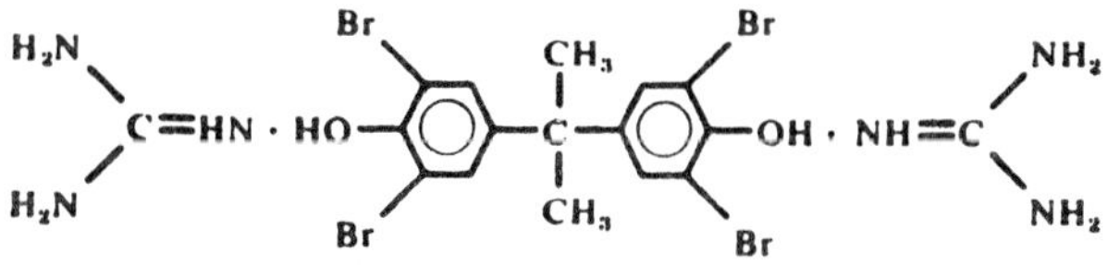

The bis(guanidine)tetrabromodian is an excellent additive for flame retardation in plastics, in particular in polyolefins such as polyethylene, polypropylene, polystyrene, and so on. A very good flame retardant effect is caused with an amount of bis(guanidine)tetrabromodian of 1 to 20% by weight (relative to the weight of the final product) incorporated in the polyolefin. The following example illustrates the process.

Example: To a reaction vessel, which is provided with a stirrer, thermometer, heating and cooling apparatus, 1,000 parts by weight of water at ambient temperature are added, and 816 parts by weight of tetrabromobisphenol A are suspended in the water while stirring. Subsequently and with further stirring, 120 parts by weight of solid sodium hydroxide are added.

After a brief time a clear solution develops which is then filtered in order to separate slight residues of insoluble substances. The slightly yellowish filtered solution is again put into the reaction vessel, heated to a temperature of 56°C and is adjusted to a pH value of 9 by careful addition of aqueous hydrochloric acid with 10% by weight of HCl content.

A solution of 270 parts by weight of guanidine carbonate in 640 parts by weight of water, which had previously been heated to a temperature of 56°C, is then added drop by drop while stirring vigorously; and a colorless crystalline product is precipitated. In order to complete the precipitation, the reaction mixture is cooled slowly to a temperature of 18°C while stirring and is filtered at this temperature.

After washing it three times with 150 ml of water and drying at 70°C there is obtained 857 parts by weight of bis(guanidine)tetrabromodian. This quantity corresponds to a yield of 86.6% of theory. Chemical analysis shows a bromine content of 46.0% by weight and a nitrogen content of 12.45% by weight. The corresponding theoretical values are 48.35% Br and 12.68% N.

Tris(Halophenoxy)-1,3,5-Triazine

H. Klinkenberg, E. Petersen and W. Schmidt; U.S. Patent 4,039,538; August 2, 1977; assigned to Dynamit Nobel AG, Germany describe a process for the preparation of tris(halophenoxy)-1,3,5-triazines of the formula

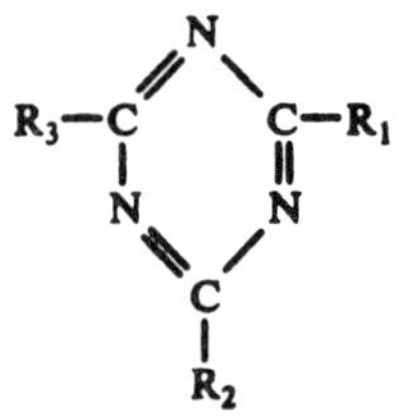

in which R_1, R_2 and R_3 are identical or different phenoxy substituents, halogenated in the nucleus with 1 to 5 halogen atoms.

These triazines are used as the flameproofing agents, in amounts of 5 to 25% by weight, with respect to the polymers in the plastic. Furthermore, such compounds are thermally very stable, easily miscible with the most important thermoplastics, and in addition they have a strong flame-inhibiting effect. Flame-inhibiting effects can be further improved by adding certain suitable substances; these substances include Sb_2O_3. For example, if only 10% of a tris-pentabromophenoxy-s-triazine together with 4% of an antimony compound such as Sb_2O_3 are added to a polyester, such as polybutyleneterephthalate, the mixture is already outstandingly flame resistant and attains the best score, SE 0, in the UL 94 test.

The compounds of the process have a considerable range of application. For example, the flameproofing agents added have a strong flame inhibiting action in polyesters, i.e., condensation products of diols and dicarboxylic acids such as polyethylene or polybutylene terephthalates, polyamides, low-pressure and high-pressure polyethylenes, their mixtures with chlorinated polyethylenes or ethylene-vinyl ester copolymers, polypropylene, polystyrene, ternary copolymers or graft polymers such as those of acrylonitrile, butadiene and styrene (ABS) or methacrylic acid, butadiene and styrene (MBS) and other thermoplastics, and also in thermosetting plastics. Thus, plastics can be made which are slow-burning or even self-extinguishing, depending on the proportions. The following examples serve to illustrate the properties of the triazines.

Example 1: Preparation of Tris(2,3,4,5,6-pentachlorophenoxy)-s-Triazine — In a three-necked round flask, equipped with stirrer and reflux condenser, 90.7 g (0.3 mol plus 5% excess) of technical pentachlorophenol sodium is dissolved at about 50°C in 900 ml of ethylene glycol monomethyl ether and this solution is heated to about 60°C.

At this temperature, 18.4 g (0.1 mol) of technical cyanuric acid chloride is poured into the phenolate solution with good stirring, over a period of 2 to 3 minutes, the temperature of the reaction mixture rising to 70° to 73°C due to the reaction heat. When all the cyanuric chloride has been added, the mixture is refluxed for an additional 30 minutes with stirring, a temperature of approximately 110°C establishing itself in the reaction mixture.

After cooling, the triazine derivative which has already precipitated at the boiling temperature is suction filtered, washed first with some methanol and then with water to remove the sodium chloride that has formed, until there is no chloride in the filtrate, and the tris(pentachlorophenoxy)triazine is dried.

85 g is obtained, corresponding to 97% of the theoretical amount, melting point 265° to 268°C. When twice recrystallized from trichloroethylene, an analytically pure preparation is obtained having a melting point of 277° to 278°C. Elemental analysis: $C_{21}Cl_{15}N_3O_3$ (MW = 874). Calculated: 28.85% C; 60.85% Cl; 4.81% N; 5.49% O. Found: 29.18% C; 60.45% Cl; 4.69% N; 5.63% O.

From a signal series of 832 to 846 in the mass spectrum it becomes apparent that the molecular weight of the substance obtained must be greater than 832, thus confirming the presence of the tris compound.

Example 2: Preparation of Tris(2,3,4,6-Tetrachlorophenoxy)-s-Triazine — In the same apparatus as in Example 1, 37.0 g of potassium hydroxide (0.66 mol) is dissolved in 900 ml of ethylene glycol monomethyl ether at 75°C; and then at this temperature 152.5 g (0.66 mol) of 2,3,4,6-tetrachlorophenol is poured in; immediately thereafter 40.5 g (0.22 mol) of cyanuric acid chloride is stirred into the phenolate solution over a period of 1 to 2 minutes, whereupon the temperature of the reaction mixture rises to 84°C.

The reaction is allowed to run to completion for 30 minutes at the refluxing temperature, and the product is separated as described in Example 1. Yield: 134 g (79%), melting point 165° to 168°C. When recrystallized from benzine (BP 80° to 100°C), the melting point was 167° to 169°C. Elemental Analysis: $C_{21}H_3Cl_{12}N_3O_3$ (MW = 770.69). Calculated: 32.70% C; 0.39% H; 55.22% Cl; 5.45% N; 6.23% O. Found: 32.98% C; 0.42% H; 55.47% Cl; 5.33% N; 6.09% O.

The following examples contain formulas for the preparation of flame resistant plastics.

Example 3: The following components were mixed in a double-screw extruder with mixing section and kneading block at a temperature of 240° to 265°C, and extruded in strand form and granulated: 86 parts of polybutylene terephthalate; 10 parts of tris-pentabromophenoxy-s-triazine; 4 parts of Sb_2O_3. The granules were molded on a small injection molding machine to form test specimens measuring 1.6 x 20 x 240 mm.

The specimen thus prepared was given a score of 1 in the test, which is the score for the best flame resistance. A confirmatory UL 94 test gave similar results: the specimen is self-extinguishing, drips do not ignite cotton. It is accordingly classified SE 0.

Example 4: Using the following components test specimens are produced as in Example 1: 81 parts low-pressure polyethylene, 15 parts tris-tribromophenoxy-s-triazine, 4 parts Sb_2O_3. In the fire test the mixture is given a score of 2, for good fire protection. On the basis of UL 94, the specimen is classified SE II. Similar results are obtained with high-pressure polyethylene.

Example 5: The following components are mixed together at 135°C in a correspondingly heated roller mixer, then granulated and injection molded to form test specimens: 85 parts polystyrene, 12 parts tris-pentabromophenoxy-s-triazine, 3 parts Sb_2O_3. In the fire test the plastic proves to be incombustible and receives a score of 1.

Example 6: The following components are processed as in Example 1: 80 parts polyethylene terephthalate, 16 parts tris-pentachlorophenoxy-s-triazine, 4 parts Sb_2O_3. The fire test gives a score of 2, and according to the UL 94 test the specimen is self-extinguishing, with a score of SE II.

Bicyclo and Tricyclo Dicarboxylic Acids

S.A. Cerefice and E.K. Fields; U.S. Patent 4,005,122; January 25, 1977; assigned to Standard Oil Company (Indiana) describe 7,7-dihalobicyclo[4.1.0] hept-4-ene-2,3-dicarboxylic acid compounds (esters and acids) and 5,5,8,8-tetrahalotricyclo-[5.1.0.0^{4,6}] octane-2,3-dicarboxylic acid as compounds suitable for use as fire retardants. The compounds of the process can be represented by the structural formulas:

(1)

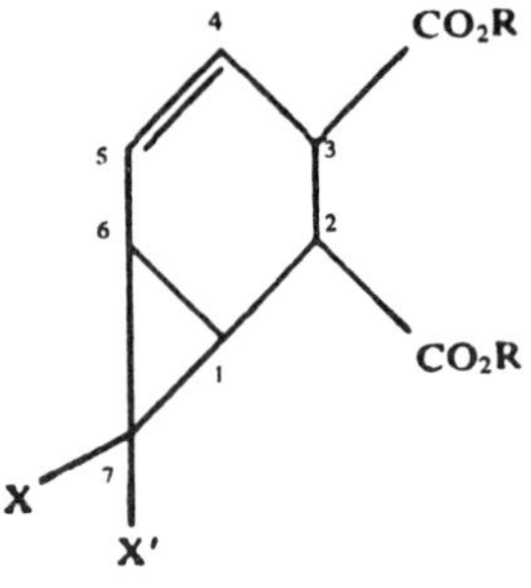

(2)

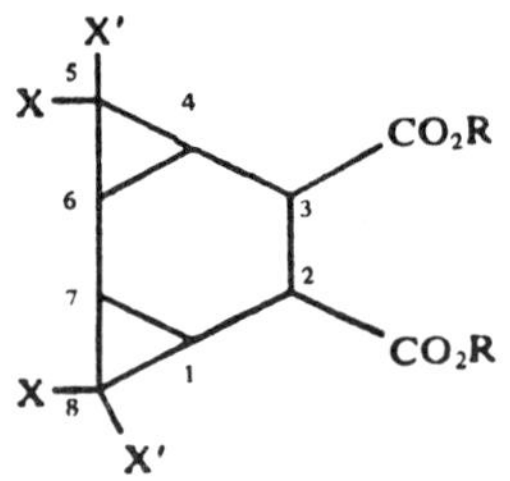

In the above structures, each X and X' is halogen, such as bromine or chlorine; and R represents the residue of a monohydroxy compound (monovalent aromatic or aliphatic group whose free valence is bonded to –OH) or hydrogen. For convenience, the dihalocarbene compound containing one dihalocarbene group (structure 1) is referred to as a bicyclo compound, while those compounds containing two dihalocarbene groups (structure 2) are referred to as tricyclo compounds. These bicyclo and tricyclo compounds can be added as fire retardants to resinous polymers, such as polyethylene, polypropylene, polystyrene, polyvinyl chloride, etc.

Example 1: 9.48 g phenyl(trichloromethyl)mercury (0.049 mol) were added to a solution of 4.9 g dimethyl 1,2-dihydrophthalate (0.025 mol) in 125 ml benzene at 80°C and contained in a 200-ml flask equipped with stirrer and reflux condenser. After the reactants were refluxed at about 80°C for 5 days, the insoluble phenyl mercurous chloride was filtered off and the benzene removed from the filtrate on a rotary evaporator. The insoluble phenyl mercurous chloride was dispersed in 100 ml hexane and filtered, yielding 12.43 g phenyl mercurous chloride (83% of the starting mercury compound).

Gas chromatographic analysis showed that dimethyl 1,2-dihydrophthalate had been converted quantitatively into a mixture of 40% by weight dimethyl 7,7-dichlorobicyclo[4.1.0] hept-4-ene-2,3-dicarboxylate (two stereoisomers in a 7:4 ratio) and 60% by weight dimethyl 5,5,8,8-tetrachlorotricyclo[5.1.0.$0^{4,6}$] octane-2,3-dicarboxylate (two isomers in a 3:1 ratio).

The bicyclo and tricyclo compounds were separated chromatographically on silica gel (12 g on 160 g Davidson 60 silica gel) using hexane. The unreacted phenyl(trichloromethyl) mercury eluted first followed by the tricyclo isomers and and then the bicyclo isomers. After the major bicyclo isomer was purified by recrystallization from hexane, it melted at 100° to 101°C. Chemical analysis for the bicyclo compound was as follows: Calculated for $C_{11}H_{12}O_4Cl_2$: C, 47.34%; H, 4.34%; Cl, 25.40%. Found: C, 47.20%; H 4.12%; Cl, 25.4%.

After the major tricyclo isomer was purified by recrystallization from hexane and sublimation at 80°C (0.4 mm Hg) it melted at 136° to 136.5°C. Chemical analysis for the tricyclo compound was as follows: Calculated for $C_{12}H_{12}O_4Cl_4$: C, 39.81%; H, 3.34%; Cl, 39.17%. Found: C, 39.61%; H, 3.27%; Cl, 39.9%.

Example 2: This example illustrates the preparation of dimethyl 7,7-dibromobicyclo[4.1.0] hept-4-ene-2,3-dicarboxylate and dimethyl 5,5,8,8-tetrabromotricyclo[5.1.0.$0^{4,6}$] octane-2,3-dicarboxylate. Example 1 was repeated using 5.3 g dimethyl 1,2-dihydrophthalate (0.27 mol) and 32.0 g phenyl(tribromomethyl) mercury in 125 ml benzene at 80°C for five days.

17.1 g phenyl mercurous bromide (80% of the starting mercury compound) was removed by filtration. Column chromatography on silica gel, using hexane and benzene eluents, yielded 63% weight dimethyl 7,7-dibromobicyclo[4.1.0] hept-4-ene-2,3-dicarboxylate (two stereoisomers in a 7:5 ratio) and 37% by weight dimethyl 5,5,8,8-tetrabromotricyclo[5.1.0.$0^{4,6}$] octane-2,3-dicarboxylate (two isomers in a 7:3 ratio).

After the major tricyclo isomer was purified by recrystallization from diethyl ether/hexane, it melted at 153° to 154°C. Chemical analysis for the tricyclo compound was as follows: Calculated for $C_{12}H_{12}Br_4O_4$: C, 26.90%; H, 2.26%; Br, 59.6%. Found: C, 26.95%; H, 2.39%; Br, 59.0%.

Halogen-Containing Amides

H. Braus and J.R. Woltermann; U.S. Patent 3,979,545; September 7, 1976; assigned to National Distillers and Chemical Corporation describe the flame resistance of synthetic fibers and solid polymers by treatment with a halogen-containing amide having the formula

$$(C_nH_{2n+1-y}X_y)_{2-m}\overset{H_m}{\underset{|}{N}}-\overset{O}{\overset{\|}{C}}-R$$

where R is lower alkoxy; lower alkoxy carbonyl;

$$-N(H_m)(C_nH_{2n+1-y}X_y)_{2-m};$$

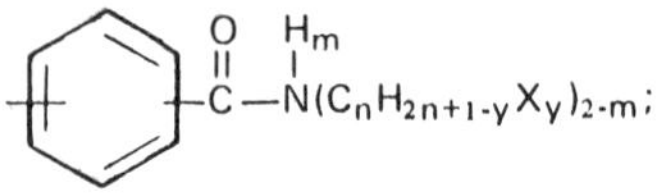

$$\text{or } -C(=O)-N(H_m)(C_nH_{2n+1-y}X_y)_{2-m};$$

X is chlorine or bromine; m is 0 or 1; n is 1 to 6; and y is 1 to 13.

Illustrative of lower alkoxy groups which can be R in the formula are methoxy, ethoxy, propoxy, etc. Bis(β,γ-dibromopropyl) ethyl carbamate,

$$C_2H_5O-\overset{O}{\overset{\|}{C}}-N(CH_2CHBrCH_2Br)_2$$

is exemplary of such compounds.

The halogen-containing amides are employed in solution form or in the form of an aqueous emulsion. In either case, the amides can vary in concentration over a wide range of from 5 to 25 wt %, and preferably 10 to 20 wt %. Suitable solvents for the halogen-containing amides include the halogenated hydrocarbons such as trichloroethylene, perchloroethylene, methylene chloride, and the like. Suitable emulsifying agents for maintaining the amides in aqueous emulsions include anionic, cationic and nonionic dispersing agents or surfactants. Suitable nonionic surfactants include the alkyl phenoxy poly(ethyleneoxy)ethanols, and the dialkyl phenoxy poly(ethyleneoxy)ethanols.

The synthetic fibers which can be treated in accordance with this process include the polyamides such as 6-, and 6,6-nylons, polyesters, polyolefins such as polypropylene or polyethylene, acrylics, triacetates such as Arnel, and blends of natural and synthetic fibers such as cotton/polyester and wool/polyester. Mixed fabrics and fibers may be similarly treated.

The fire retardant compounds can be applied to the synthetic textile fibers by a variety of procedures and using different kinds of equipment. While the individual steps and equipment can vary with the different procedures, in each case the fiber is immersed in the solution or emulsion and the compound is deposited by removing the solvent by evaporation or vaporization.

In one suitable procedure, a multitude of synthetic textile yarn ends from a creel are passed through a container in which is found the solution of the halogen-containing amide. Thereafter, the yarn ends are passed between squeeze rolls to remove excess solution and then passed over a multitude of drying cans which are usually maintained at a temperature of about 150° to 230°F. Similarly, the fibers can be treated by dipping them into the solution or emulsion, followed by drying.

The temperature at which the fibers are immersed in the solution or emulsion can vary over a wide range of from 30° to 100°C. The immersion time can similarly vary over a wide range from ¼ to 2 hours. The fibers are maintained immersed in the solution or emulsion until an effective flame resistant amount of the halogen-containing amide, generally 5 to 20 wt % of the dried impregnated fibers, has been picked up by the fibers. The halogen-containing compounds can also be used to impart flame resistance to solid thermoplastic polymers such as polyethylene, polypropylene, polyesters and copolymers and blends, thereof, nylons, polyurethanes, etc.

The following examples illustrate the preparation of the compounds employed and their use.

Example 1: Bis(β,γ-dibromopropyl) ethyl carbamate was prepared by the bromination of N,N-(diallyl) ethyl carbamate as follows: Charged into a 1-liter, three-necked flask equipped with the necessary adjuncts were 48.6 g (0.5 mol) diallylamine; 50.6 g (0.5 mol) triethylamine and 200 ml of ether. This was cooled to 0°C and 54.3 g (0.5 mol) ethyl chloroformate was added slowly dropwise with constant stirring. The addition took 3½ hours, maintaining a temperature of 0°C. An additional 350 ml of ether was added to maintain fluidity.

After warming to room temperature, the solid amine salt was filtered and the ether distilled from the filtrate. The resulting product was purified by distillation under reduced pressure: BP 40.5° to 41°C at 0.1 mm Hg; yield 71%.

Charged into a 1-liter, three-necked flask equipped with a stirrer, N_2 inlet, thermometer, dropping funnel and condenser were N,N-diallyl ethyl carbamate, 57.8 g (0.342 mol), and 300 ml of chloroform. This was cooled to 0°C and with vigorous stirring 109.3 g (0.684 mol) bromine was added dropwise. When the addition was complete, the bath was warmed to room temperature and allowed to stand overnight. The chloroform was then distilled and finished at 0.5 mm pressure; yield 81.2%.

A sample was submitted for mass spectra which showed a compound of mass 485 containing 4 bromines. This confirms the structure of the compound.

Bis(β,γ-dibromopropyl) ethyl carbamate, 40 g, is dispersed in 80 g ethylene glycol using as emulsifier 2 g Duponal OS, an amine long-chain alcohol sulfate. The mixture is heated with stirring until a homogeneous dispersion is obtained. The solution is then cooled to room temperature. Before use, this dispersion is emulsified with water, 150 g, preferably using a colloid mill.

Example 2: N,N-bis(β,γ-dibromopropyl) ethyl oxamate was prepared by the bromination of N,N-diallyl ethyl oxamate as follows: 146.1 g (1.0 mol) ethyl oxalate was charged into a three-necked flask equipped with a mechanical stirrer, dropping funnel, condenser and drying tube. Diallylamine, 97.2 g (1.0 mol), was slowly added with constant stirring. There was a moderate exotherm during the addition until approximately one half the amine was added, at which time the temperature declined. When the addition was complete, the bath was heated to 80° to 90°C for 4 hours and then cooled by standing overnight.

The product was then fractionated and vacuum distilled and the portion boiling at 105° to 106°C at 1 mm Hg was collected. The product was submitted for mass spectra determination, which showed a mass of 197 and the expected fragments of the desired product, N,N-diallyl ethyl oxamate.

77.3 g (0.39 mol) of N,N-diallyl ethyl oxamate dissolved in 100 ml of chloroform was charged to a 1-liter, three-necked flask equipped with a stirrer, thermometer, dropping funnel, condenser and drying tube. The solution was cooled to 4°C and 124.7 g (0.78 mol) bromine was slowly added. When all the bromine was added, the chloroform was distilled at atmospheric pressure. The remaining chloroform was removed under reduced pressure by water pump. The residue was taken up in benzene, filtered and evaporated under a stream of N_2;

MP 160°C (with decomposition). Mass spectra analysis confirmed the desired compound. N,N-bis(β,γ-dibromopropyl) ethyl oxamate, 30 g, is dissolved in 180 ml perchloroethylene using 5 g Triton X-100, which is an octylphenoxy-polyethoxy ethanol.

Example 3: The compound bis(β,γ-dibromopropyl) ethyl carbamate is most readily applied to a fabric material by exhaustion from an aqueous system. The solution is prepared as shown in Example 1. A cloth of polyester is placed in the bath which is then heated to 90° to 100°C. A polyester softener such as sorbitol, sulfonated castor oil, low molecular weight polyethylene or silicone textile lubricants, may be added.

The absorption of the carbamate by the fibers is low at low heat and high at higher temperatures. The fabric is kept submerged for 1 to 2 hours, after which the fabric is removed and dried in an air-circulating oven at 100° to 110°C. The cloth is weighed to determine the add-on, usually 5 to 10% increase is sufficient to render the fabric flame retardant. The pick-up can be regulated to yield the desired degree of flame retardance.

Example 4: A weighed nylon cloth (6,6) is submerged in the solution prepared in Example 2 and soaked for about 15 to 20 minutes. The cloth is then dried in an air-circulating type oven and weighed for total pick-up. 7% was the pick-up on weight of goods. The cloth, when held vertically, would not support combustion when the source of flame was removed.

Example 5: Example 4 was repeated except that the flame retardant compound is tetrakis(β,γ-dibromopropyl)urea and the fabric used is cellulose triacetate. The treated fabric is rendered flame retardant as measured by the vertical flame test.

Brominated Biphenols

According to *C.M. Orlando and F.A. Lavallee; U.S. Patent 3,989,531; Nov. 2, 1976; assigned to General Electric Company* bromobiphenols comprising 2,2',-6,6'-tetraalkyl-3,3',5,5'-tetrabromo-1,1'-biphenols and its derivatives are employed as fire retardant additives for normally flammable resins yielding fire retardant polymeric compositions. These fire retardant compositions are thermally stable and can be molded or formed into fire retardant articles such as films, sheeting, fibers, laminates, or reinforced plastics by conventional techniques.

The additive is selected from halogenated monomeric and polymeric organic compounds of the following formula where each R independently is selected from primary alkyl groups; each R^1 is independently selected from hydrogen and organic groups; and n is a positive integer of at least 1:

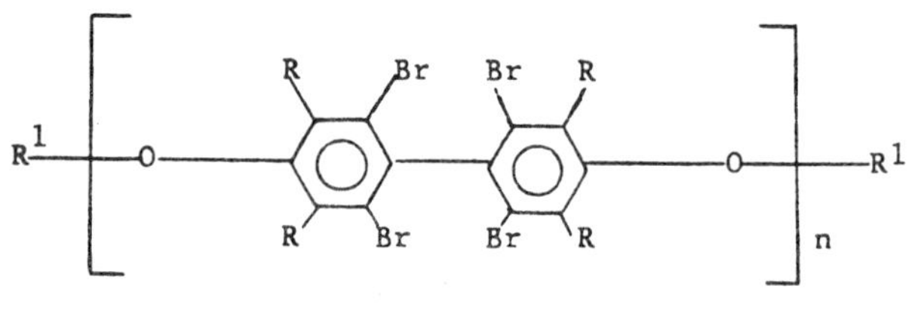

The normally flammable resins which are employed are such materials as acrylonitrile-butadiene-styrene resins, acetal resins, acrylic resins, alkyd resins, allyl resins, amino resins, cellulosic resins, epoxy resins, chlorinated polyether resins, fluoroplastics, polyimides, polybenzimidazoles, etc.

In general, the amount of fire retardant additive used is not critical so long as it is present in an effective, preferably minor, proportion based on the amount of the total polymeric composition. However, the amount of fire retardant additive employed should be at least sufficient to render the normally flammable resin nonburning, self-extinguishing, or to increase the resin oxygen index above the resin oxygen index in the absence of the additive. The amount of additive will be generally from 0.5 to 50 parts by weight per 100 parts of resin. If synergists, e.g., antimony oxide, are used, they will be employed at about 2 to 5 parts by weight per 100 parts of resin.

The following examples illustrate the preparation of fire retardant polymeric compositions. In the examples, the following glossary of abbreviations is frequently employed for purposes of brevity.

A	=	polyphenylene oxide (PPO)
B	=	polystyrene (PS)
C	=	polyethylene terephthalate (PET)
D	=	polybutylene terephthalate (PBT)
F	=	copolymer (BPA/TTB)
G	=	copolymer (BPA/TBBPA)
		BPA = bisphenol A
Q	=	antimony oxide (Sb_2O_3)
T	=	tetrabromotetramethylbiphenol (TTB)
V	=	tetrabromobisphenol A (TBBPA)
W	=	triphenylphosphate

Example 1: Polymer blends of polyphenylene oxide and polystyrene were evaluated both in the presence and in the absence of 2,2',6,6'-tetramethyl-3,3',5,5'-tetrabromo-1,1'-biphenol (TTB) as well as in the presence of other known effective fire retardant additives for these polymer blends. The polymer compositions contained other ingredients which are nonessential to the determination or the efficacy of the fire retardant properties of the compositions, such as minor amounts of processing additives, antioxidants, etc.

The effect of the addition of a fire retardant supplement, such as antimony oxide, to fire retardant polyphenylene oxide polystyrene-TTB fire retardant compositions was also evaluated. The characteristics of the fire retardant compositions were evaluated in accordance with Underwriters' Laboratory Bulletin 94 and in accordance with the Oxygen Index test standards and procedures set out in ASTM D 2863.

A tabulated summary of the resin, fire retardant additive, fire retardant supplement, bromine content, burning characteristics, polymer properties and polymer processing conditions of the various polymer compositions tested are set out in Table 1 on the following page. The data illustrate that polyphenylene oxide-polystyrene polymeric materials have improved fire retardancy when they contain tetrabromotetramethylbiphenol (TTB) as an additive without deleteriously affecting the basic polymer properties associated with polyphenylene oxide-polystyrene blends.

TABLE 1:

	RUN NUMBERS	1	2	3	4
(a)	Resin Type				
	1. Ingredients-pbw	A-50	same	same	same
	2. Ingredients-pbw	B-50	same	same	same
(b)	Fire Retardant Additive Type				
	1. Ingredients-pbw	none	W-3.0	T-8.6	T-8.6
(c)	Fire Retardant Supplement				
	1. Ingredients-pbw				Q-2.5
(d)	Bromine Content Composition-%	none	none	4.5	4.4
(e)	Burning Characteristics				
	1. U.L. 94		V-1		V-1
(f)	Polymer Properties				
	1. Tensile-yield, psi		8700	9450	8900
	2. Tensile Strength-ultimate, psi		8300	8600	
	3. Elongation, yield-%		66	62	45
	4. Izod Impact-ft.lbs/in. notch		4.8	4.0	3.1
	5. Heat Distortion Temperature-° F.		248	254	
	6. Gardner Impact-in.lb.		190	180	
(g)	Polymer Processing Conditions				
	1. Extrusion temperature range-° F.		520–600° F	same	same
	2. Molding temperature range-° F.		200° F	same	same

Example 2: A series of runs were carried out evaluating the fire retardant properties of polyethylene terephthalate resins and polybutylene terephthalate resins in the presence of TTB/bisphenol A polycarbonate and TBBPA/bisphenol A polycarbonate fire retardant additives as well as in the absence of any flame retardant additive for control purposes.

The polycarbonate TTB/bisphenol A was prepared by the addition of molar equivalents of bisphenol A bischloroformate in a solution of methylene chloride to a solution of TTB, aqueous sodium hydroxide, benzoyl triphenylphosphonium chloride and sodium hydrosulfite. The TBBPA/bisphenol A polycarbonate was prepared by a similar procedure, where TTBPA was substituted for TTB. The resulting TTB/bisphenol A and TBBPA/bisphenol A polycarbonates, respectively, had powder I.V.s ($CHCl_3$) of 0.417 dl/g and 0.367 dl/g; 22.9 and 27.0% bromine; 0.024 and 0.029% hydroxyl; a melt viscosity of 45,870 and 15,610; and a Tg_{TOA} of 209° and 191°C.

The burning characteristics were determined in accordance with the procedures referred to in Example 1. The results of the evaluation of the fire retardant properties of the polyester compositions are set out in Table 2.

TABLE 2.

	RUN NUMBERS	5	6	7	8	9	10
(a)	Resin Type						
	1. Ingredients-pbw	C-100	C-100	C-100	D-100	same	same
	2. Ingredients-pbw		F-20.5	G-20.5		F-20.5	G-20.5
(b)	Fire Retardant Additive Type						
	1. Ingredients-pbw						
(c)	Fire Retardant Supplement						
	1. Ingredients-pbw		Q-7.5	Q-7.5	none	Q-7.5	Q-7.5
(d)	Bromine Content Composition-%	6.0					
(e)	Burning Characteristics						
	1. U.L. 94	burns	V-0	V-0	burns	V-0	V-0
	2. O.I.	26.9	35.2	37.1	22.4	28.8	27.9
(f)	Polymer Properties						
	1. Tensile-yield, psi	8940	9450	9680	7990	7400	7830
	2. Tensile Strength-ultimate, psi	5890	7020	5260	7260	7820	7980
	3. Elongation, yield-%				15	12	12
	4. Elongation, ultimate-%	490	490	250	690	30	~55
	5. Flex Modulus-psi × 10^5	3.61	3.46	3.61	3.44	3.95	3.81
(g)	Polymer Processing Conditions						
	1. Extrusion temperature range-° F	270–290	same	same	same	same	same
	2. Molding temperature range-° F	160–270	same	same	same	same	same

The data illustrate that polyester compositions have improved fire retardancy when they contain TTB polycarbonates as a fire retardant additive without deleteriously affecting the basic properties associated with the polyester compositions.

Stable Emulsions Containing 1,1,2,3,4,4-Hexabromobutene-2

According to *E.P. Di Bella, D.A. Keyworth and M. Rosen; U.S. Patent 3,981,806; September 21, 1976; assigned to Tenneco Chemicals, Inc.* emulsions containing 1,1,2,3,4,4-hexabromobutene-2 can be used to impart flame retardant characteristics to a wide variety of textile materials without adversely affecting their physical properties. These emulsions, whose discontinuous phase is a solution containing about 5 to 25% by weight of 1,1,2,3,4,4-hexabromobutene-2 in a solvent mixture that contains 1 to 5 parts by weight of an aromatic hydrocarbon per part by weight of an N-alkylpyrrolidinone whose alkyl group has 1 to 5 carbon atoms and whose continuous phase is water, are particularly effective as the flame retardant for wool, cellulose acetate, polyester and polyamide fibers and fabrics. 1,1,2,3,4,4-Hexabromobutene-2, which has the structural formula

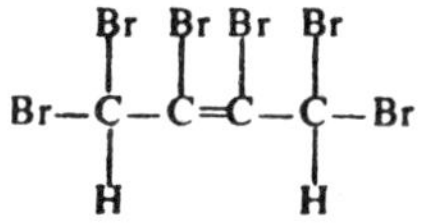

has a combination of properties that makes it valuable as a flame retardant for many textile materials. It is a crystalline compound that melts at about 180°C and contains 90% by weight of bromine. Because of its solid form, it is very well suited for use at relatively high levels in those applications in which plasticizer activity is undesirable because it has a deleterious effect on the hand and other properties of the fabrics. Because it is nonvolatile, stable, and insoluble in water and in most common organic solvents, hexabromobutene-2 cannot be separated from textile materials by leaching, washing or evaporation. In addition, hexabromobutene-2 is nontoxic, nonirritating, and biodegradable.

1,1,2,3,4,4-Hexabromobutene-2 can be prepared easily and in good yield by the bromination of diacetylene. This reaction can be conveniently and safely carried out by contacting a gas stream that contains about 20 to 50 mol percent of diacetylene in an inert gas such as nitrogen with a dilute solution of bromine in water, carbon tetrachloride or another inert solvent.

The product of this reaction is a mixture of polybrominated compounds that contains about 75 to 80% by weight of 1,1,2,3,4,4-hexabromobutene-2, 15 to 23% by weight of 1,1,2,4-tetrabromobutene-2, and small amounts of dibromobutadienes. The crystalline hexabromobutene-2 may be separated from the product mixture by filtration. Hexabromobutene-2 prepared in this way contains 85 to 95% of the trans isomer and 5 to 15% of the cis isomer, and usually about 90% of the trans isomer and 10% of the cis isomer. This isomer mixture can be used without further treatment in emulsions that are used to impart flame retardance to textile fibers and fabrics.

The emulsions can be used to impart flame retardance to a wide variety of normally flammable textile materials. They include, for example, fabrics and fibers of cellulose acetate and other cellulose esters; polyethylene terephthalate and other cellulose esters; nylon and other polyamides; polyolefins such as polyethylene, polypropylene, and polyisobutylene; polyacrylonitrile; polyacrylates; etc. The amount of the emulsion that is applied to a textile material is that which will provide the desired degree of flame retardance. It is generally preferred to treat the textile material with an amount of the emulsion that will provide about 15 to 100 parts by weight of 1,1,2,3,4,4-hexabromobutene-2 per 100 parts by weight of the textile material.

The preferred emulsifying agents are ester-type nonionic emulsifying agents, but other nonionic emulsifying agents can be used. Examples of the ester-type nonionic emulsifying agents include polyoxyethylene monocarboxylates such as polyoxyethylene monolaurate, polyoxyethylene monopalmitate, and polyoxyethylene monostearate; and polyoxyethylene sorbitan monocarboxylates such as polyoxyethylene sorbitan monolaurate and polyoxyethylene sorbitan monostearate.

The amount of water that is added to the hexabromobutene-2 solution or to the emulsifiable concentrate is that which will bring the hexabromobutene-2 content to the desired level. It is usually from 1 to 5 parts, preferably from 1 to 2 parts, per part by weight of the solution or concentrate.

The resulting stable emulsions contain from 2 to 10% by weight of 1,1,2,3,4,4-hexabromobutene-2, with the maximum hexabromobutene-2 content dependent upon the temperature at which the emulsion is prepared and maintained. It is generally preferred that the emulsion be prepared and stored at approximately the temperature at which it will be applied to textile materials.

Example 1: To a solution of 82.5 g of 1,1,2,3,4,4-hexabromobutene-2 in 100 g of N-methylpyrrolidinone at 75° to 80°C was added 300 g of toluene. The resulting solution was stirred and maintained at 75° to 80°C while a solution of 19 g of polyoxyethylene sorbitan monolaurate (Tween 20) in 500 g of water was added to it. There was obtained an emulsion that was stable at 75° to 80°C and that contained 8.2% by weight of hexabromobutene-2.

Example 2: To a solution of 31 g of 1,1,2,3,4,4-hexabromobutene-2 in 100 g of N-methylpyrrolidinone at 50°C was added 300 g of toluene. The resulting solution was stirred while a solution of 4 g of polyoxyethylene sorbitan monolaurate in 500 g of water was added to it. There was obtained an emulsion that contained 3.8% by weight of hexabromobutene-2 and that was stable at 25° to 30°C for about 1 month.

Example 3: To a solution of 31 g of 1,1,2,3,4,4-hexabromobutene-2 in 100 g of N-methylpyrrolidinone at 50°C was added 300 g of toluene. The resulting solution was stirred while 4 g of polyoxyethylene sorbitan monolaurate was added to it. The resulting emulsifiable concentrate was mixed with 500 g of water to form an emulsion that contained 3.8% by weight of hexabromobutene-2 and that was stable at 25° to 30°C for about 1 month.

Example 4: Weighed 1 x 12-inch samples of fabrics were immersed for 5 minutes in the 1,1,2,3,4,4-hexabromobutene-2 emulsion whose preparation is described in Example 1, while the emulsion was gently agitated and maintained at 70° to 80°C. The treated samples were squeezed free of excess liquid and air dried at room temperature. Each sample was reweighed and then subjected to the match test to determine its flammability qualitatively.

In the match test, the sample is suspended vertically lengthwise from a clamp. A mark is made in the middle of the sample. A wooden match is lit and used to ignite the bottom edge of the sample. This edge remains in the flame as long as the match burns, which is usually about 15 seconds. If the flame is extinguished within 5 seconds after the match is consumed and the sample does not burn beyond the center mark, the fabric is considered to be self-extinguishing.

For comparative purposes, fabric samples were treated with an emulsion that was prepared by the procedure of Example 1 except that it did not contain hexabromobutene-2 and then evaluated by the match test. The fabrics tested and the results obtained are summarized below.

Fabric	Hexabromobutene-2 % by wt	Flammability Rating*
Polyester	42.9	SE
Polyester	37.6	SE
Polyester	0**	B
Polyester, flat type	60.8	SE
Polyester, flat type	51.3	SE
Polyester, flat type	0**	B
Polyester-double knit, ribbed type	92.8	SE
Polyester-double knit, ribbed type	88.2	SE
Polyester-double knit, ribbed type	0**	B
Polyamide-taffeta	22.6	SE
Polyamide-taffeta	0**	B
Polyamide	70.9	SE
Polyamide	65.5	SE
Polyamide	0**	B
Cellulose acetate	51.6	SE
Cellulose acetate	50.3	SE
Cellulose acetate	0**	B
Cellulose acetate-taffeta	25.7	SE
Cellulose acetate-taffeta	22.5	SE
Cellulose acetate-taffeta	0**	B
Wool	72.0	SE
Wool	67.4	SE
Wool	58.3	SE
Wool	0**	B

*B = Burns
SE = Self-extinguishing
**Control

From the foregoing data, it will be seen that 1,1,2,3,4,4-hexabromobutene-2 is an effective flame retardant for cellulose acetate, polyamide, polyester, and wool fabrics. Each of the other textile materials shown can also be rendered flame retardant by applying to it an emulsion containing a flame-retarding amount of 1,1,2,3,4,4-hexabromobutene-2.

Aromatic and Multicyclic Halogenated Compounds

F. Arai, M. Kira, S. Kokuryo and T. Ueshima; U.S. Patent 4,059,561; Nov. 22, 1977; assigned to Showa Denko KK, Japan describe a flame resistant composition that has good thermal stability. The composition comprises (a) a ring-opening polymerization product of a norbornene derivative containing at least one polar group; and (b) a flame-retarding compound in sufficient amounts to elevate the flame resistance of the composition without reducing the mechanical strength. The flame-retarding compound is selected from the group consisting of the halogen-containing multicyclic compounds of the formulas (1) through (3), or the bromine-containing aromatic compounds of the formulas (4) through (6) as follows:

(1) 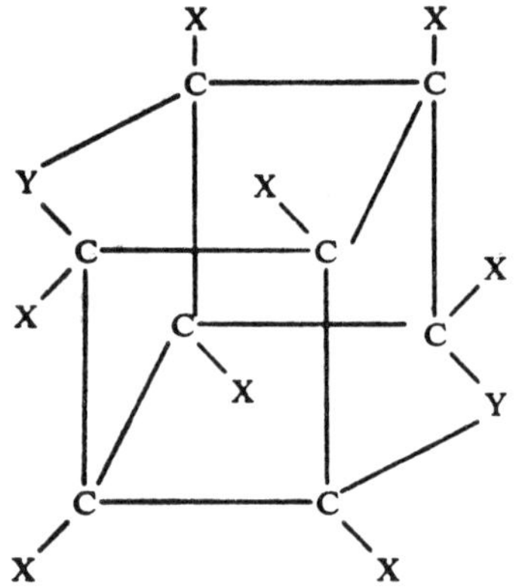

(2) 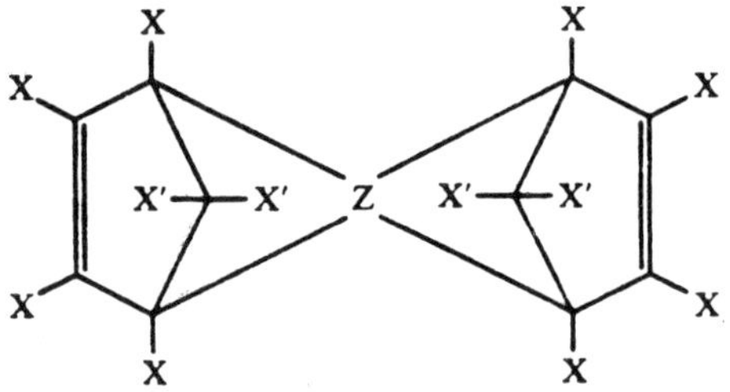

(3) 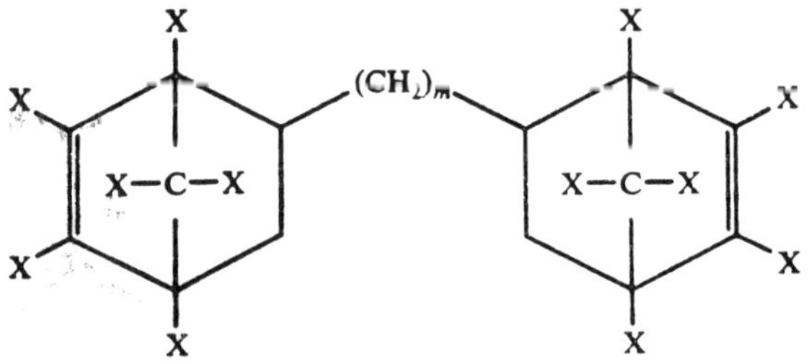

In the above, X is chlorine or bromine, X' is fluorine, chlorine, bromine, alkyl, or alkoxy having 1 to 20 carbon atoms; Y is

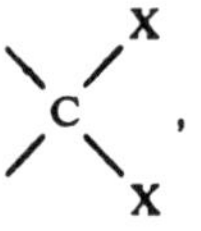 , 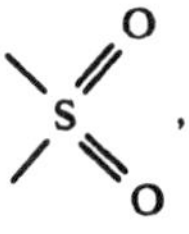 , or 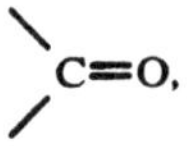

Z is a tetravalent cyclic hydrocarbon group having at least 5 carbon atoms; and m is an integer from 4 to 16.

(4)

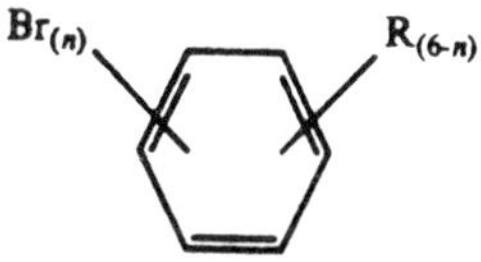

(5)

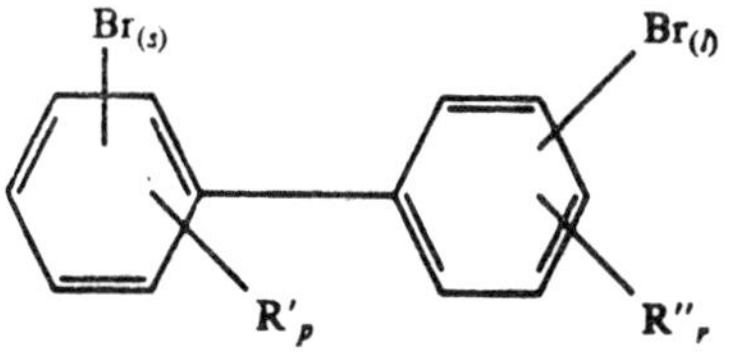

(6)

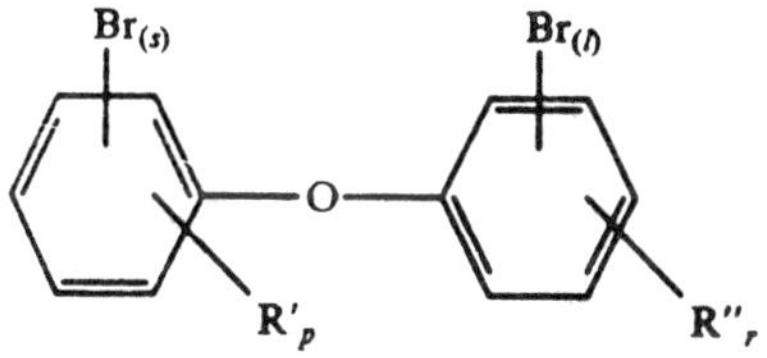

In the above, R is hydrogen, alkyl having 1 to 6 carbon atoms, bromoalkyl, or hydroxy where one R, but not more than one R, is hydroxy; R' and R'' are hydrogen or alkyl having 1 to 6 carbon atoms; n is an integer from 3 to 6; l, s, p and r are each integers; the sum of l plus s is an integer from 2 to 10; and the sum of l, s, p and r is 10.

The compositions obtained by this process can be formed into films, sheets, tapes, boards, rods, balls, containers, pipes, and many other shaped articles employing conventional molding methods, such as extruding, injection molding, blow molding, compression molding, and casting. Typical shaped articles include machine parts, vehicle parts, building materials, and the like. The compositions are particularly useful when formed into parts for electrical appliances (for example, cabinets for televisions and radios). Such parts require thermal stability and flame and heat resistance.

Bisphenoxy Compounds in Polymer Compositions

A.L. Anderson; U.S. Patents 4,058,501; November 15, 1977; and 4,082,725; April 4, 1978; both assigned to Velsicol Chemical Corporation describes polymer compositions containing a three-component system consisting of an elastomer; certain bisphenoxy compounds having the formula

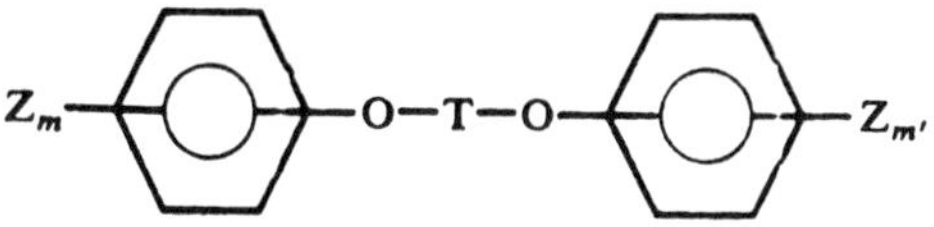

where Z is bromine, m and m' are each integers having a value of 1 to 5 with the proviso that the total bromine atom content is from 6 to 10, and T is a straight or branched chain carbon group having 1 to 4 carbon atoms; and an enhancing agent comprising the oxides and halides of Groups IV-A and V-A of the Periodic Table.

The amount of bisphenoxy compound employed in the compositions is from about 1 to 25% by weight based on the total weight of the composition. The second component of the three-component system is certain compounds which, when used with the bisphenoxy compounds, promote a cooperative effect and thus enhance the flame retardancy of the resultant polymer compositions. The amount of enhancing agent employed is 0.5 to 15% based on the total weight of plastic composition.

The third component of the system is an elastomer polymeric composition. This includes materials such as natural rubber, butadiene polymers, 2-chlorobutadiene polymers, isoprene polymers, silicone rubber, chlorosulfonated polyethylene and polyethers.

The experimentation is described in the polyester chapter (see page 94).

Poly(Aryloxyphosphazenes)

R.L. Dieck and E.J. Quinn; U.S. Patents 4,026,838; May 31, 1977; and 4,055,545; October 25, 1977; and R.L. Dieck, A.B. Magnusson and E.J. Quinn; U.S. Patent 4,055,523; October 25, 1977; all assigned to Armstrong Cork Company describe poly(aryloxyphosphazene) polymers and foams having di(C_{1-10})alkylaminophenoxy groups. The polymers consist of the units which occur in a nonregular fashion and are represented by the following general formula

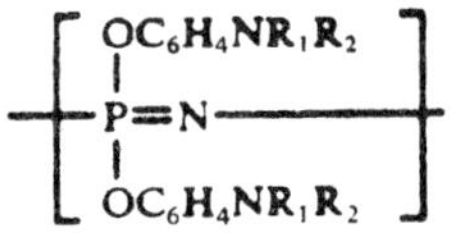

where R_1 and R_2 are the same or different and are C_{1-10} linear or branched alkyl radical. Copolymeric dialkylaminoaryloxy-aryloxyphosphazenes are disclosed, represented by units of the general formulas

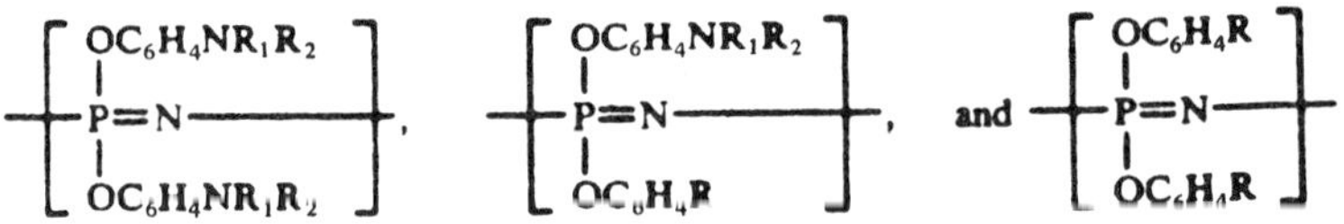

where R_1 and R_2 are defined above and R represents hydrogen, halogen, C_{1-10} linear or branched alkyl radical, or C_{1-4} linear or branched alkoxy radical substituted on any sterically permissible position on the other phenoxy group. The radical NR_1R_2 is substituted in the meta or para position on the phenoxy radical.

The polymers are elastomers, have desirable tensile strengths and may be used to form flexible or semirigid films, sheets, and foams. The polymers are extremely fire retardant and produce low smoke loads when heated in an open flame. They readily form salts with mineral acids that are water soluble. The following examples illustrate the process.

Example 1: Preparation of $-\!\!\left(NPCl_2\right)\!\!-_n$ – 250 parts of phosphonitrilic chloride trimer, previously recrystallized from h-heptane, were degassed and sealed in a suitable thick-walled reaction vessel at 10^{-2} torr and heated to 250°C for 6 hours.

Polymerization was terminated at this time since a glass ball, ½ inch in diameter ceased to flow due to the increased viscosity of the molten mass, when the vessel was inverted. Termination was effected by cooling the vessel to room temperature. The resulting polymeric mixture was then dissolved in toluene to form an anhydrous solution.

Example 2: Preparation of $\langle NP[OC_6H_4\text{-}m\text{-}N(C_2H_5)_2]_2\rangle_n$ – The anhydrous toluene solution of poly(dichlorophosphazene) formed in Example 1 containing 0.47 equivalent of poly(dichlorophosphazene) was added to an anhydrous diglyme-benzene solution of 1.10 equivalents of $NaOC_6H_4\text{-m-}N(C_2H_5)_2$ at a temperature of 95°C with constant stirring. After the addition, benzene was distilled from the reaction mixture until a temperature of 115° to 116°C was attained.

The reaction was then heated at reflux for 50 to 65 hours. At the end of this time the homopolymer was precipitated by pouring the reaction mixture into excess methyl alcohol. The polymer was stirred in methyl alcohol for 24 hours. Next, it was added to a large excess of water and stirred for an additional 24 hours.

The resulting product (43% yield) was an off-white plastic material having a glass transition temperature of -7.1°C. The product was soluble in benzene, toluene, tetrahydrofuran and dimethylformamide. The polymer was then pressed into a tough film which did not burn and was water repellent.

The homopolymer had an Oxygen Index of 26.1 as determined according to the procedure described in ASTM D-2863-74. By this method, material samples are held in a U-shaped frame and the burning of the samples under a specific set of conditions is measured. This technique measures the lowest oxygen concentration in an atmosphere which will just prevent sustained burning of a top-ignited sample. The Oxygen Index values also have been related to the temperature at which a mixture of fuel and a controlled flow of oxygen will just burn when the fuel is composed of volatile pyrolytic products or fragments. Analysis – Calculated for homopolymer of $\langle NP[OC_6H_4\text{-m-}N(C_2H_5)_2\rangle_n$: C, 64.31%; H, 7.57%; N, 11.25%; P, 8.29%. Found: C, 62.08%; H, 7.70%; N, 10.85%; P, 8.12%.

Example 3: Preparation of $\langle NP[OC_6H_4\text{-}m\text{-}N(CH_3)_2](OC_6H_5)\rangle_n$ – The procedure of Example 2 was followed, except that 0.47 equivalent of poly(dichlorophosphazene) was added to 0.56 equivalent each of $NaOC_6H_5$ of $NaOC_6H_4\text{-m-}N(CH_3)_2$.

The resulting product (31% yield) was an off-white, fibrous solid having a glass transition temperature of 0.0°C. The copolymer was soluble in benzene, toluene, tetrahydrofuran and dimethylformamide. Films of this material do not burn and are water resistant.

Analysis – Calculated for copolymer of $\langle NP[OC_6H_4\text{-m-}N(CH_3)_2](OC_6H_5)\rangle_n$: C, 61.30%; H, 5.52%; N, 10.22%; P, 11.29%. Found: C, 61.16%; H, 5.49%; N, 10.09%; P, 11.40%.

Example 4: Preparation of $\langle NP[OC_6H_4\text{-}m\text{-}N(C_2H_5)_2](OC_6H_4\text{-}p\text{-}Cl)\rangle_n$ – The procedure of Example 2 was followed, except that 0.47 equivalent of poly(dichlorophosphazene) was added to 0.56 equivalent of $NaOC_6H_4\text{-p-Cl}$ and 0.56 equivalent of $NaOC_6H_4\text{-m-}N(C_2H_5)_2$.

The resulting product (47% yield) was a brown, tough elastomeric solid having a glass transition temperature of -4.9°C. The copolymer was soluble in benzene, toluene, tetrahydrofuran, and dimethylformamide. Gel permeation chromatography indicated that the copolymer had a MW of 811,000. Films of this material did not burn and were water resistant.

PHOSPHORUS COMPOUNDS

Monomeric and Polymeric Phosphonates

Alcohols preferably having a pKa below that of propanol-1 are heated with certain heterocyclic phosphites to give phosphonates and with certain heterocyclic phosphonites to to give phosphinates in a process described by *I. Hechenbleikner; U.S. Patent 3,978,166; August 31, 1976; assigned to Borg-Warner Corporation.* If small amounts of alcohols are employed, the products are generally polymeric in nature whereas if larger amounts of the alcohol are employed, there is generally ring opening to the simpler monomer, or isomerization to the phosphonate or phosphinate. These products are useful as fire retardants.

Phosphoramidomethyl Phosphonates

P. Golborn and J.J. Duffy; U.S. Patent 4,018,560; April 19, 1977; assigned to Hooker Chemicals & Plastics Corp. describe several compounds of the formula

$$(R)_m\overset{O}{\overset{\|}{P}}NHCH_2\overset{O}{\overset{\|}{P}}(OR')_2$$

where R' is phenyl, lower alkenyl and halogen substituted and unsubstituted lower alkyl of 1 to 6 carbon atoms; m is an integer from 1 to 2 where when m is 2, R is alkoxy of 1 to 8 carbon atoms and when m is 1, R is alkylene dioxy of 2 to 8 carbon atoms. The compounds are useful for imparting flame retardancy to textiles and thermoplastic, thermosetting and elastomeric resin materials. The flame retardant agents may be applied to various textiles such as cellulosic or proteinaceous materials. Cellulosic materials include cotton, rayon, paper, regenerated cellulose and cellulose derivatives which retain a cellulose backbone of at least one hydroxy substituent per repeating glucose unit. Proteinaceous materials include those textile materials comprising the functional groups of proteins such as the various animal wools, hairs and furs.

The flame retardant additives may also be incorporated into thermoplastic resin compositions by any known method. They may be added to the resin by milling the resin and the additive on, for example, a two-roll mill, or in a Banbury mixer, etc., or it may be added by molding or extruding the additive and resin simultaneously, or by merely blending it with the resin in powder form and thereafter forming the desired article. Additionally, the flame retardant may be added during the resin manufacture, i.e., during the polymerization procedure by which the resin is made, provided the catalysts, etc. and other ingredients of the polymerization system are also inert thereto. Generally, the compounds may be incorporated into the thermoplastic resin in flame-retarding amounts, i.e., generally amounts ranging from 5 to 50% by weight, preferably from 20 to 40% by weight, based on the weight of the polymer, have been found sufficient.

The following examples illustrate the process. All parts and percentages are by weight unless otherwise indicated.

Example 1: Preparation of

A 250-ml flask, fitted with a reflux condenser, mechanical stirrer and thermometer, was charged with 83 g (0.5 mol) triethyl phosphite and was heated to 100°C. 73.2 g (0.4 mol) of diethylphosphoramidate was added to the flask over a period of 15 minutes and the reaction mixture was thereafter heated at 115°C for 2 hours. After cooling, the reaction mixture was stripped at 70°C under a reduced pressure of 0.5 mm Hg, to give 109 g of a pale-yellow liquid. Infrared and nuclear magnetic resonance spectroscopy confirmed the structure to be essentially pure.

Example 2: Preparation of

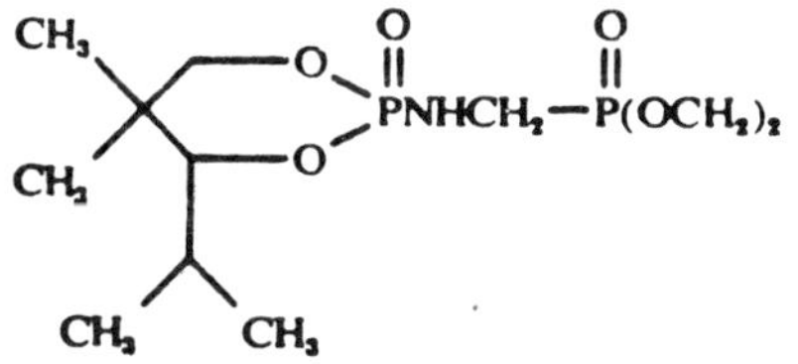

N-hydroxymethyl-(1-isopropyl-2,2-dimethyl)-1,3-propylidenyl phosphoramidate, 59.3 g (0.25 mol), was mixed with 37 g (0.3 mol) of trimethyl phosphite and 80 ml diglyme in a 250-ml flask. The reaction mixture was slowly heated to 115°C and held at this temperature for 15 hours. After cooling, volatile material was removed from the reaction mixture by stripping under a vacuum of about 1 mm Hg. 70 g of product, a viscous oil, was obtained which, under infrared and nuclear resonance spectroscopy, was shown to be essentially pure.

Example 3: To 70 parts of polypropylene was added 30 parts of N-(diethylphosphonomethyl) diethylphosphoramidate and dry blended for about 5 minutes. The resulting mixture was then brought to a melt and thoroughly mixed for about 15 minutes. After cooling, the resulting solid was cut into small pieces and slowly added to a 9-mm glass tube immersed in a hot salt bath. The temperature of the salt bath is maintained above the melt temperature of the polymer mixture added. After all the pieces had been melted, a steel rod with a weight attached is placed in the glass rod and the mixture allowed to cool. The resulting rod of polymer and additive (150 to 200 mm length) is then removed and tested by modified ASTM tests D-2863-70 and D-635-68. The test results are shown in Table 1.

ASTM D-2863-70 generally provides for the comparison of relative flammability of self-supporting plastics by measuring the minimum concentration of oxygen in a slowly rising mixture of oxygen and nitrogen that will support combustion. The procedure encompasses supporting cylindrical test specimens (70 to 150 mm by 8 mm) vertically in a glass tube fitted with controlled upward oxygen/nitrogen gas flow. The top of the specimen is ignited and oxygen flow is adjusted until it reaches that minimum rate at which the specimen is extinguished before burning 3 minutes or 50 mm whichever happens first. The Oxygen Index (n) in percent is then calculated as $n = (100 \times O_2)/(O_2 + N_2)$, where O_2 is the volumetric flow of oxygen, at the minimal rate, and N_2 is the corresponding volumetric flow rate of nitrogen.

A modification of ASTM D-635-68 generally provides for the comparison of burning rates, self-extinguishment and nonburning characteristics of plastics in the form of sheets, bars, plates or panels. The procedure encompasses preparing 150 x 8 mm cylindrical samples with and without the subject flame retardant additive. Each sample is marked at points 1 inch and 4 inches from its end and held, marked end in the flame, at a 45-degree angle in a controlled burner flame (1 inch flame length) for two 30-second attempts. The movement of the flame up the length of the sample through the two points is measured for rate of burning, nonburning, or self-extinguishing characteristics. A sample is rated SE (self-extinguishing) if the flame burns through the first point but extinguishes before reaching the second point. A sample is rated NB (nonburning) if, upon ignition, it does not burn to the first point.

Examples 4 through 11: Various phosphonates are mixed with polymers of the preparations and according to the process of Example 3. Testing under modified ASTM tests D-2863-70 and D-635-68 are recorded in Table 1.

Examples 12 through 18: Various polymers are treated by the process of Example 3, with the exception that no phosphate additives are mixed. Test results are recorded in Table 1.

TABLE 1

Example	Additive	Polymer	Flammability Tests: Oxygen Index	D-635	Percent Additive
3	$(CH_3CH_2O)_2P(=O)-NH-CH_2-P(=O)(OCH_2CH_3)_2$	Polypropylene	23.2	NB	30
4	$(CH_3CH_2O)_2P(=O)-NH-CH_2-P(=O)(OCH_2CH_2Cl)_2$	Polystyrene	23.0	NB	30
5	$(CH_3CH_2O)_2P(=O)-NH-CH_2-P(=O)(OCH_2CH=CH_2)_2$	ABS	24.3	NB	30
6	$(CH_3CH_2O)_2P(=O)-NH-CH_2-P(=O)(O\phi)_2$	Epoxy	27.2	NB	30
7	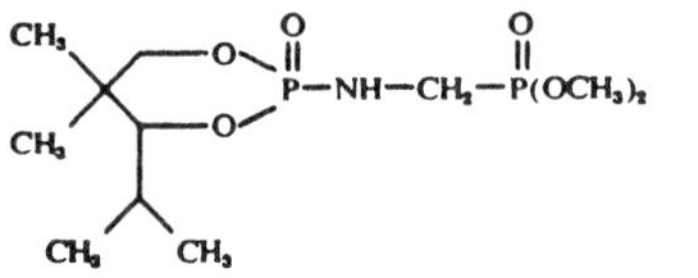	Polypropylene	19.8	NB	30
8	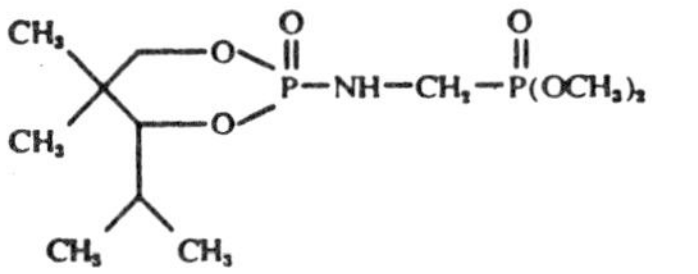	ABS	22.0	NB	30

(continued)

TABLE 1: (continued)

Example	Additive	Polymer	Flammability Tests: Oxygen Index	Flammability Tests: D-635	Percent Additive
9	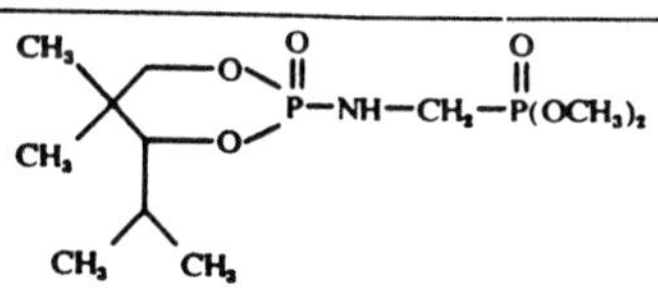	Nylon	29.0	NB	30
10	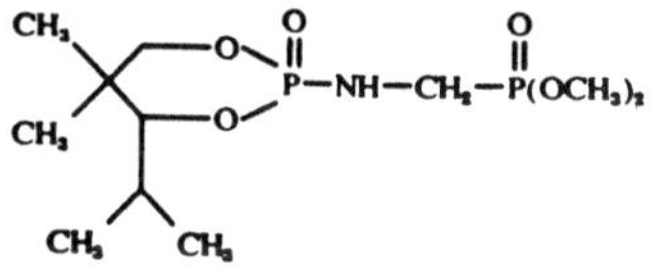	Polyethylene terephthalate	25.0	NB	30
11	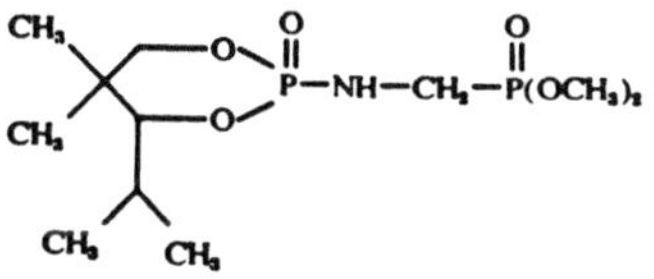	SBR	24.8	NB	30
12		Polypropylene	17.4	B	
13		Polystyrene	18.0	B	
14		ABS	19.0	B	
15		Nylon	22.1	SE	
16		Polyethylene terephthalate	22.7	B	
17		Epoxy	22.5	B	
18		SBR	19.6	B	

Polyphosphazenes

A.H. DeEdwardo, F. Zitomer, R.W. Stackman and C.E. Kramer; U.S. Patent 4,042,561; August 16, 1977; assigned to Celanese Corporation describe a flame retardant composition comprising a normally flammable organic material and a flame retardant amount of a polyphosphazene having the general formula:

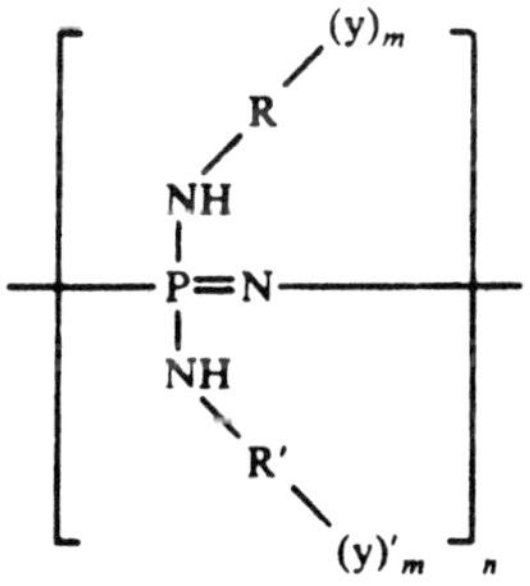

where n is at least 10; R and R' may be the same or different and represent aryl groups; y and y' may be the same or different and represent halogen and halogen substituted and unsubstituted lower alkyl (e.g., 1 to 10 carbon atoms), aryl, alkoxy, and aryloxy groups; and m is 0 to 10.

Blends of the above-described polyphosphazenes, preferably at levels of about 10%, in various organic polymers show a minimal effect on thermal stability. This allows the flame retardant to function without polymer degradation.

The polyphosphazene compounds described are sufficiently stable to allow use even at elevated temperatures encountered during critical processing operations such as blending and spinning.

Specific examples of the polyphosphazene additives include poly[bis(p-fluoroaniline)phosphazene]; poly[bis(anilino)phosphazene]; poly[bis(p-chloroaniline)phosphazene]; poly[bis(p-bromoaniline)phosphazene]. The actual blending of the polyphosphazene additives with selected polymeric substrate may be accomplished by means of any convenient procedure which will result in an intimate mixture of the additive within the mass of the substrate polymer. Thus, for example, an aqueous suspension containing the particles of the phosphazene additive may simply be blended or otherwise mixed with the substrate polymer.

The polyblends may be utilized in any of the coating, impregnating and molding applications where it is desired to provide fire retardancy to the resulting end product. For example, these compositions may be used for preparing such diverse items as calendered films, blow-molded bottles, extruded and blown films, extruded and shaped articles such as panels, tubes, sheets, rods, fibers, particularly with polyester fibers, and textiles based on these fibers; and in carrying out such processes as injection molding, fluidized bed coatings, electrostatic powder spraying and rotational coating, etc.

Blending of the polymers with polyphosphazenes typically results in a polymer with reduced flammability, as shown by the increased critical oxygen concentration index values. COC is defined as the lowest oxygen concentration in nitrogen at which steady combustion, without dripping, takes place on a vertically positioned polymer specimen after ignition at the top surface. The ignition flame is provided from a 22-gauge syringe through which passes H_2 gas which is ignited, yielding a flame having a temperature of about 3000°C. Dripless combustion is attained by using a specimen of small diameter containing a wick.

The COC values are determined in an all-glass testing chamber containing O_2 and N_2 under quiescent flow conditions. The composition of the O_2/N_2 mixtures can be regulated and monitored continuously. Test specimens are obtained from solution cast or compression molded films cut into strips approximately 0.125 inch wide, 2 inches long and 0.04 inch in thickness. These strips are wound loosely with a stainless steel wire of 0.007 inch diameter. Actual COC values are measured on samples suspended vertically in the testing chamber and ignited at the top surfaces.

Examples 1 through 22: A polyphosphazene derivative corresponding to the following formula (where n is at least 10) is prepared by the melt ring-opening polymerization of hexachlorocyclotriphosphazene to form polydichlorophosphazene at a temperature of about 250°C and reduced pressure:

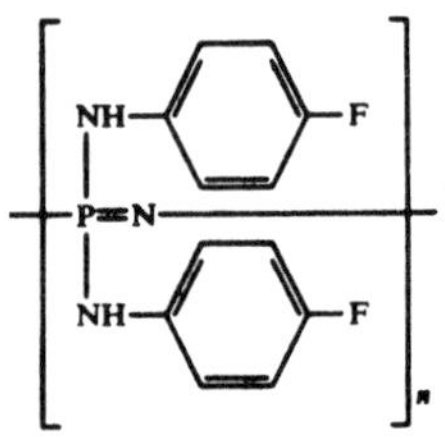

To 100 parts by weight of the polydichlorophosphazene are added 191.8 parts by weight of p-fluoroaniline and 174.6 parts by weight of triethylamine and the resultant slurry is refluxed in tetrahydrofuran for about 24 hours to form poly-[bis(p-fluoroaniline)phosphazene]. The resulting phosphazene polymer is then precipitated into ethanol (nonsolvent), washed and dried. In a similar manner, poly[bis(anilino)phosphazene] $[(\phi\text{-NH})_2P{=}N]_n$ is prepared.

As representative of the normally flammable synthetic organic polymer polyethylene terephthalate, cellulose acetate, cellulose triacetate and polystyrene are prepared and tested with the polyphosphazene compounds described above. These polymers are prepared by known methods.

Blends comprising 10% by weight of each of the two above-described anilino substituted polyphosphazene derivatives and 90% by weight of the polyethylene terephthalate are then prepared. Thus, the polyethylene terephthalate and each of the phosphazene additives are individually ground to an extent sufficient to enable them to pass through a 60 mesh screen. A 10:90 batch of each phosphazene:PET blend is manually mixed and passed several times through a grinding mill fitted with a 60 mesh screen to form a powdered blend. The blend is then compression molded into a film of about 0.04 inch in thickness.

The cellulose acetate, cellulose triacetate, and polystyrene compounds are blended in the same amounts with the same two anilino substituted phosphazenes and molded into a film in the same manner as above. The blends are then subjected to the above-described COC index test. The resulting values are shown in the table below.

Example	Polymer	% Additive	Polyphosphazene	COC Value*	Burn Type
1	PET	10	$[(F{-}\phi{-}NH)_2P{=}N]_n$	22.9	Char
2	CTA	10	$[(\phi{-}NH)_2P{=}N]_n$	21.3	Char
3	CTA	10	$[(F{-}\phi{-}NH{-})_2P{=}N]_n$	23.4	Char
4	CA	10	$[(\phi{-}NH{-})_2P{=}N]_n$	21.1	Char
5	CA	10	$[(F{-}\phi{-}NH)_2P{=}N]_n$	22.7	Char
6	PS	10	$[(\phi{-}NH{-})_2P{=}N]_n$	17.4	Melt
7	PS	10	$[(F{-}\phi{-}NH)_2P{=}N]_n$	20.1	Char/Melt

*Note all values represent percent by volume.
CA = cellulose acetate
CTA = cellulose triacetate
PET = polyethylene terephthalate
PS = polystyrene

To provide a comparison with various other phosphazene compounds, several compounds conforming to the recurring structural formula

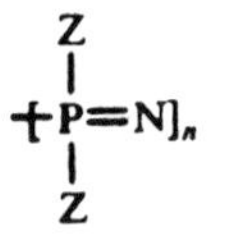

where Z represents various alkoxy, phenoxy, and a piperidino group (as described in the table on the following page) are also provided, blended, and tested in accordance with the procedures outlined above. The resulting COC values obtained are set out in chart form on the following page.

Comparative Example	Polymer	% Additive	Z Group	COC Value*	Burn Type
8	PET	0	—	18.8	Melt
9	PET	10	ϕ—O—ϕ—O	20.5	Melt
10	PET	10	CH_3—O—ϕ—O	22.0	Melt
11	PET	10	(Cl)CH_3—ϕ—O—	22.6	Melt
12	PET	10	Br—ϕ—O—	22.8	Melt
13	CTA	0	—	19.3	Melt
14	CA	0	—	19.0	Melt
15	CTA	10	F—ϕ—O—	21.2	Char
16	CA	10	F—ϕ—O	20.3	Char
17	CA	10	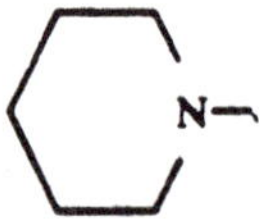	21.8	Char
18	CA	10	Cl—ϕ—O—	20.9	Melt
19	PS	0	—	16.9	Melt
20	PS	10	CF_3—CH_2—O—	19.3	Melt
21	PS	10	F—ϕ—O—	18.2	Melt
22	PS	10	Cl—ϕ—O—	18.3	Melt

*Note all values represent percent by volume.
CA = cellulose acetate
CTA = cellulose triacetate
PET = polyethylene terephthalate
PS = polystyrene

The data presented show that the various synthetic polymers, by themselves (as expected), show the lowest COC values while the presence of the halogenated anilino phosphazenes yields the highest increase in the COC values, up to about 23.4.

Solutions of Silicates and Hydrogen Phosphates

A process described by *H.-J. Scholl, P. Markush and D. Dieterich; U.S. Patent 4,067,815; January 10, 1978; assigned to Bayer AG, Germany* relates to aqueous solutions of alkali silicates and alkali hydrogen phosphates comprising, by weight, 4 to 18% K_2O, 2 to 11% Na_2O, 3 to 10% P_2O_5, and 12 to 21% SiO_2, based on the sum total of the constituents including water, where the ratio of the sum totals of the parts by weight of $(K_2O + Na_2O):(P_2O_5 + SiO_2)$ is from 0.9 to 1:1.7, and has a solids content of from 30 to 50%, by weight. The solutions are homogeneous and stable and are particularly useful in improving the burning properties of organic/inorganic plastics.

The stable solutions may optionally contain various soluble additives such as those which influence the stability, viscosity, hardening and burning properties of the plastics materials produced using the solutions. Examples of such additives include soluble inorganic salts such as alkali borates, alkali zincates, alkali chromates, alkali fluoborates; soluble salts of ammonia, hydrazine, guanidine, guanamine such as guanidine carbonate; soluble polyols such as glycerol, pentaerythritol, cane sugar, glucose, fructose and formose; and alkali-soluble polymers such as maleic acid-styrene copolymers. The quantity of such additives should generally not exceed 30%, by weight, of the total solids content of the solutions.

Inert fillers may also be added to the stable solutions or to the individual silicate and phosphate solutions before they are mixed. The fillers may be either inorganic or organic provided they are inert with respect to alkali silicate solutions. Examples of such fillers include zeolites, mixed silicates, calcium silicates, aluminum silicate, glass powder, hollow glass beads, sawdust, ground shale, hollow polymer beads and melamine resin powder.

The solutions of this process may be used as fire-retarding coating compositions. They are particularly suitable for use as the inorganic component in the manufacture of organic/inorganic plastics. The following examples illustrate the process.

Examples 1 through 8: 500 parts by weight of a 40% aqueous dipotassium hydrogen phosphate solution are introduced at room temperature into 500 parts by weight of a 44% aqueous sodium silicate solution having an SiO_2:Na_2O ratio of 2.0. The homogeneous solution formed is stable in storage and has the following composition: 10.8% K_2O, 7.5% Na_2O, 8.2% P_2O_5, 14.5% SiO_2.

Several mixtures having different alkali silicate/dipotassium hydrogen phosphate ratios were prepared in accordance with the above. The results of these tests are shown in the table below. Corresponding mixtures outside the ranges give nonhomogeneous solutions, as shown in Examples 6 through 8.

Example	K_2O (%, by weight)	Na_2O (%, by weight)	P_2O_5 (%, by weight)	SiO_2 (%, by weight)	(K_2O+Na_2O):	(P_2O_5+SiO_2)	Solids content (%, by weight)	Property of the solution
1	10.8	7.5	8.2	14.5	1	1.2	41	homogeneous
2	10.3	6.8	7.8	18.5	1	1.5	45	homogeneous
3	10.8	8.2	8.2	16.1	1	1.2	44	homogeneous
4	9.7	6.4	7.3	12.4	1	1.2	35.6	homogeneous
5	7.4	9.2	5.6	17.8	1	1.4	40	homogeneous
6	5.4	11.3	4.1	21.8	1	1.6	43	non-homogeneous
7	2.1	13.6	1.5	26.4	1	1.8	44	non-homogeneous
8	18.2	3.4	13.7	6.6	1	0.9	42	non-homogeneous

Dehydrocondensation of 10-Hydroxymethyl-9,10-Dihydro-9-Oxa-10-Phosphophenanthrene-10-Oxide

T. Saito, M. Kitani, K. Mori and S. Izawa; U.S. Patent 4,086,206; April 25, 1978; assigned to Sanko Kaihatsu Kagaku Kenkyosho and Asahi-Dow Limited, Japan describe phosphorus-containing condensation products exhibiting a flame-retarding effect on a normally inflammable synthetic resin without lowering its thermal deformation or degradation temperature.

With amounts of about 2 to 20 parts by weight per 100 parts by weight of resin, flame retardancy is obtained. Examples of the plastics are AS resins, ABS resins, polystyrenes, polycarbonates, polyesters, polyamides, phenoxy resins, and others. When mixed with thermosetting phenolic resins or aminoplasts, the compounds of formula (1), below, participate in a curing reaction to form a part of the cured resin.

(1)

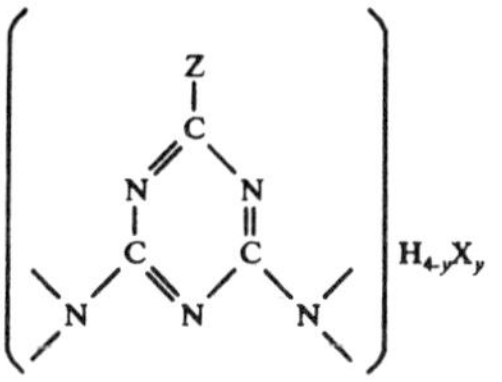

In the formula (1) structure on the preceding page, Z represents phenyl or $-NH_{2-x}X_x$, and X represents radical

(2)

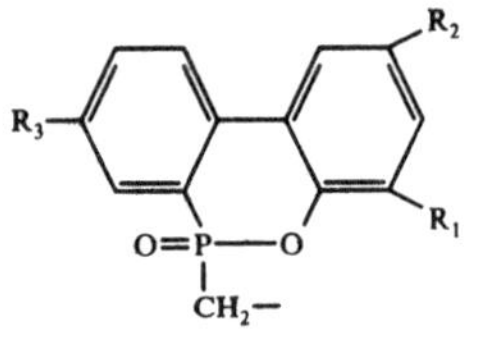

where R_1, R_2 and R_3 may be hydrogen, a halogen atom, an alkyl group of 1 to 8 carbon atoms, an aralkyl group, cyclohexyl or phenyl; x is 0, 1 or 2; and y is from 1 to 4.

The process for the production of phosphorus-containing condensation products of formula (1) comprises conducting dehydrocondensation of organophosphorus compounds represented by formula (3)

(3)

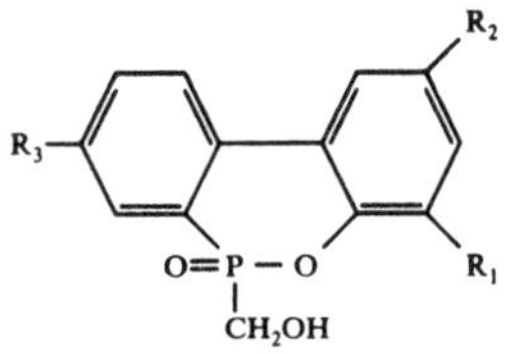

where R_1, R_2 and R_3 are as defined above with melamine or benzoguanamine. Alternatively, between organophosphorus compounds represented by formula (4)

(4)

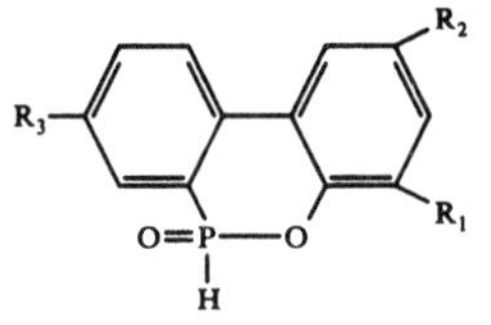

where R_1, R_2 and R_3 are as defined above and alkoxymethyl melamines or alkoxymethyl benzoguanamines, is a heating condensation conducted to obtain phosphorus-containing condensation products of formula (1).

Compounds of formula (3) may be prepared by effecting an addition reaction of formaldehyde to compounds of formula (4). For example, formalin or paraform containing 1.0 to 1.2 mols of formaldehyde is added to 1.0 mol of the compound of formula (4) and reaction is effected at temperatures of 120° to 160°C for about 2 hours. Afterwards the remaining water and formaldehyde are removed from the reaction mixture under a vacuum. Thus, compounds of formula (3) are obtained with high purity.

Examples of the compound of formula (3) include: 10-hydroxymethyl-9,10-dihydro-9-oxa-10-phosphophenanthrene-10-oxide; 6,8-dichloro-10-hydroxymethyl-9,10-dihydro-9-oxa-10-phosphophenanthrene-10-oxide; 2,6,8-tri-tert-butyl-10-hydroxymethyl-9,10-dihydro-9-oxa-10-phosphophenanthrene-10-oxide; 2-methyl-6-tert-butyl-8-methyl-10-hydroxymethyl-9,10-dihydro-9-oxa-10-phosphophenanthrene-10-oxide; 8-phenyl-10-hydroxymethyl-9,10-dihydro-9-oxa-10-phosphophenanthrene-10-oxide; 6,8-dicyclohexyl-10-hydroxymethyl-9,10-dihydro-9-oxa-10-phosphophenanthrene-10-oxide.

Examples of compounds of formula (4) are 9,10-dihydro-9-oxa-10-phosphophenanthrene-10-oxide; 6,8-dichloro-9,10-dihydro-9-oxa-10-phosphophenanthrene-10-oxide; 6-methyl-9,10-dihydro-9-oxa-10-phosphophenanthrene-10-oxide; 2,6,8-tri-tert-butyl-9,10-dihydro-9-oxa-10-phosphophenanthrene-10-oxide; 6-phenyl-9,10-dihydro-9-oxa-10-phosphophenanthrene-10-oxide; 6,8-dibenzyl-9,10-dihydro-9-oxa-10-phosphophenanthrene-10-oxide.

The compound of formula (3) alone or in mixture is subject to dehydrocondensation with melamine or benzoguanamine. The dehydrocondensation is almost completed at temperatures of 150° to 250°C during the period of about 2 to 10 hours. Although the reaction velocity is accelerated by the presence of a catalyst such as an acid or alkali, the condensation usually advances without catalysts. As the reaction advances, the infrared absorption at wave number 3170 identifying a hydroxyl group of the compound of formula (3) disappears or decreases and the infrared absorption at wave number 3390 identifying N–H bonds of melamine or benzoguanamine decreases. This shows that the reaction products have the structure represented by formula (1) or (2).

The compounds of this process may be used in synthetic fibers such as polyacrylonitrile, polyester and polyamide as a stretch-spinning modifier, a dyeing assistance, a light-fastness modifier of fibers after dyeing, a heat stabilizer or thermal-coloration preventive, a decolorizing agent, a flame retardant material and a flame retardant assistance.

Example 1: Preparation of 9,10-Dihydro-9-Oxa-10-Phosphophenanthrene-10-Oxide (HCA) – 6,800 g of o-phenylphenol and 14 g of zinc chloride are charged into a four-neck flask of 10,000 ml in capacity, provided with a stirrer, a thermometer, a dropping funnel and a reflux condenser, and elevated to a temperature of 80°C. 6,500 g of phosphorus trichloride are dropped from the dropping funnel at the same temperature while stirring the contents of the flask.

After completion of the dropping, the temperature is elevated to 180°C and phosphorus trichloride is continuously added to such a degree that reflux of phosphorus trichloride is slowly effected. When the inside temperature of the flask is lowered by the addition of phosphorus trichloride and no longer elevated, addition of phosphorus trichloride is discontinued and then heating is effected for 10 hours. The contents are distilled under reduced pressure of 17 mm Hg and 10-chloro-9,10-dihydro-9-oxa-10-phosphophenanthrene is obtained. When 1,500 g of water are added to this compound at 130°C and after 30 minutes the water is removed under reduced pressure of 10 mm Hg, HCA is obtained.

Preparation of 10-Hydroxymethyl-9,10 Dihydro 9 Oxa 10 Phosphophenanthrene-10-oxide (10-Hydroxymethyl-HCA) – 1,517 g (7 mols) of HCA obtained above are charged into a 2,000-ml, four-neck flask provided with a thermometer, a dropping funnel, a water exit and a stirrer, and after elevating to 130°C, stirring is effected. 624 g (7.7 mols) of a 37% formalin are dropped from the dropping funnel at 150°C. Reaction occurs immediately and water is distilled out.

After completion of the dropping, the reaction system is maintained at 160°C for 0.5 hour. Next, the remaining water and formaldehyde are removed at 160°C under reduced pressure of 18 mm Hg. Thus, 1,727 g of 10-hydroxymethyl-HCA having the formula shown on the following page are obtained.

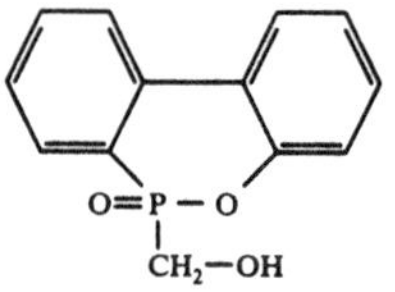

800 g of 10-hydroxymethyl-HCA obtained above and 30 g of melamine are charged into a three-neck, 1,000 ml flask provided with a stirrer, a thermometer and a water exit, and elevated to 170°C and then stirring is effected. After elevating the temperature to 210°C, a dehydration reaction is effected under reduced pressure for 4 hours. The reaction is completed in such a manner that the final pressure reached about 30 mm Hg.

When the reaction mixture is dissolved in chloroform and subjected to thin layer chromatography on silica gel (developer chloroform), it is separated into two parts, one being a component containing no nitrogen, and the other being a component containing nitrogen and phosphorus. According to elementary analysis, the latter component is a single compound of formula (1) where an atomic ratio of nitrogen:phosphorus is 1:1, x is 2 and y is 4.

Example 2: In the same procedure as Example 1, a single compound of formula (1) where Z is phenyl and y is 4 is obtained using 800 g of 10-hydroxymethyl-HCA and 50 g of benzoguanamine.

Example 3: 50 g of 10-hydroxymethyl-HCA, 150 g of melamine and dimethylformamide are charged into the same flask as in Example 1 and heated to boiling temperature of the contents. At this time, the resulting water is distilled out together with a small amount of dimethylformamide. After 4 hours reaction, the reaction mixture is separated by thin layer chromatography in the same manner as in Example 1. Thus, a single compound of formula (1) where Z is NH_2, x is 0 and y is 1 is obtained. According to elementary analysis, the atomic ratio of N:P is 6:1.

Polymeric Phosphine Oxide

K. Moedritzer; U.S. Patent 4,072,653; February 7, 1978; assigned to Monsanto Company describes a process for preparing polymeric phosphine oxides having a repeating P(O)—R' linkage in the backbone of the polymer where R' is phenylene or naphthylene.

These polymers have been found to be a useful combination with organic polymers, e.g., ester or amide products, so that the modified organic polyester or polyamide can be melted and spun from an orifice to yield smooth fibers which are readily stretched and washed for utilization in weaving operations. The modified organic polymers are also characterized by improved flame retardance properties. The polymeric phosphine oxides existing as polymeric resins are characterized by the general molecular structure as polymer products

$$[-R'-\underset{\displaystyle R}{\overset{\displaystyle O}{\overset{\|}{\underset{|}{P}}}}-]_n$$

where R is an alkyl group of 1 to 20 carbon atoms or an aryl group of 6 to 20 carbon atoms; R' is phenylene or naphthylene; and n is from 2 to 50. The successive repeating units are provided by a P(O)R' linkage which is desirable for fire retardant properties and thermal stability. The R and the aromatic R' (e.g., naphthylene or phenylene) backbone groups may also have chlorine or bromine substituents. The R' group linked to the phosphorus and the continuation of the polymeric chain may have such linkages in o, m, or p positions for the phenylene, and also in any positions (e.g., 1,4 or 2,6, etc.) of the naphthylene group.

The general process for preparing the polyphosphine oxides employs an Arbuzov rearrangement, catalyzed by a salt at a temperature of 50° to 300°C involving a monomeric compound having the general formula

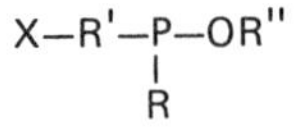

where X is chlorine or bromine, and R is an alkyl group of 1 to 20 carbon atoms or an aromatic group of 6 to 20 carbon atoms, and R" is an alkyl group of 1 to 20 carbon atoms. Examples of the monomer are

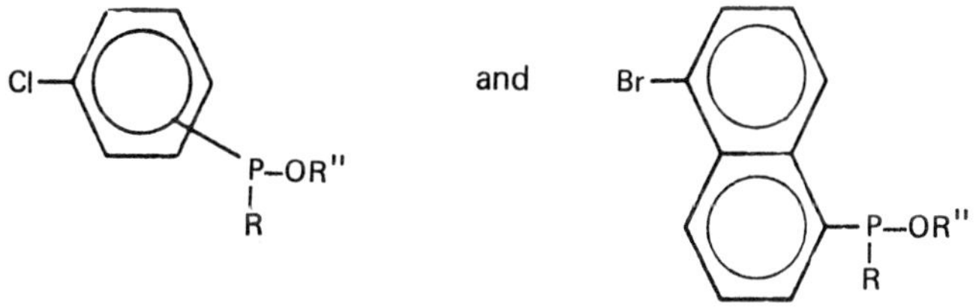

These condensation products are useful in combination with organic polymers to reduce combustibility. The normally flammable organic polymers which are rendered flame retardant may be natural or synthetic but are preferably a solid synthetic polymer. Examples of the polymer are cotton, wool, silk, paper, natural rubber, and paint; and also the high molecular weight homopolymers and copolymers of amides, e.g., nylon 66 and nylon 6.

Example 1: In carrying out the preparation of the starting material for the polymeric phosphine oxides, a reactor is charged with 1 mol (63.9 g) of n-butyllithium in 2.1 mols of hexane as a solvent, and the charge cooled to -60°C. Slow addition of 1 mol (235.9 g) of p-dibromobenzene is conducted over a period of 4 hours, with the mixture cooled to about -60°C, thus forming

as a white slurry. Dropwise addition of 1 mol (215 g) $(C_6H_5)P(NEt_2)Cl$ diluted with an equal weight of tetrahydrofuran is then commenced to the above lithium compound. The symbol Et represents the ethyl group. The addition is carried out in about 4 hours of stirring. The reaction is as follows:

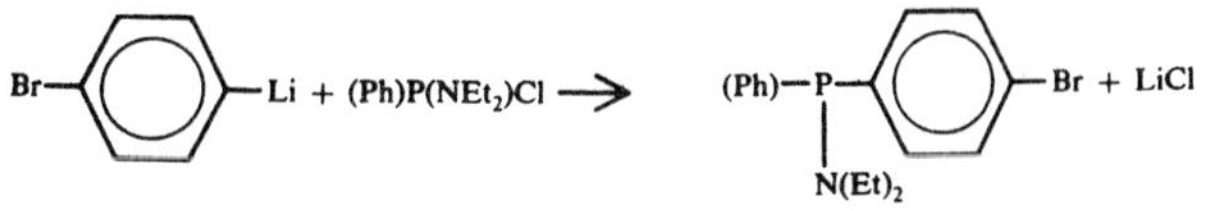

Continued addition of the phosphorus compound changes the cooled mixture from white to a beige color, with further addition giving a precipitate. The mixture is then warmed to room temperature and heated to 65°C with refluxing and stirring. During the course of the addition of the phosphorus compound some light colored solids, lithium chloride, settle out and are subsequently filtered from the liquid layer. Vacuum distillation of the liquid phase gives the product p-bromophenyl(phenyl)phosphinous diethylamide,

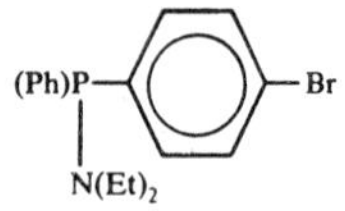

It is soluble in methanol, acetone, dichloromethane, trichloromethane, and dimethylformamide. The product distills at 170° at 0.5 mm Hg.

Example 2: The transformation of the intermediate of Example 1 to a phosphinite product proceeds according to the following equations, where Ph is phenyl:

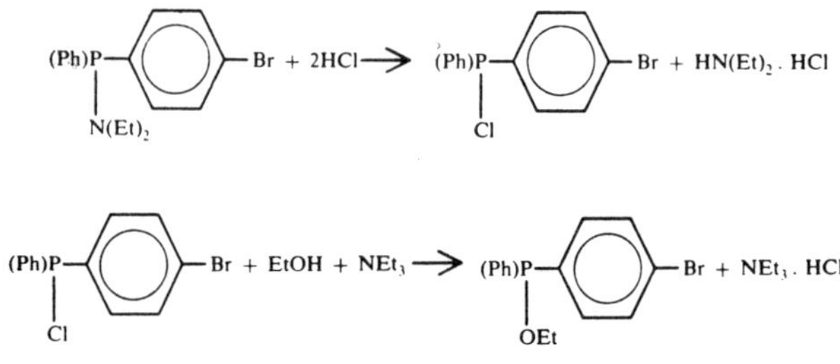

The reaction is carried out in a glass reactor by diluting the phosphorus derivative (0.406 mol, corresponding to 136.5 g) with 1.5 liter of dry hexane. The original phosphorus compound solution is a murky brown color, although the addition of HCl (1 mol, approximately 36 g) results in a white precipitate being formed. Absorption of the theoretical amount of HCl is monitored by watching for the appearance of HCl in the gases leaving the condenser of the reaction vessel.

The reaction mixture heats up from room temperature to about 39°C during a 2.5-hour period, during which the solution becomes a viscous white slurry. The slurry is further stirred for 1.5 hours while purging with a slow stream of nitrogen gas. The reaction mixture is then cooled to +5°C. Dropwise addition continues of 35 cc ethyl alcohol (0.6 mol) containing 83.6 cc triethyamine (0.6 mol) with further dilution with dry hexane to obtain 2,500 cc total volume of the solution to be added. Addition of this solution is continued over a 0.5-hour period with a further thickening of the white slurry and an exothermic reaction.

Settling of the reaction mixture overnight results in a white precipitate settling out. The precipitate is filtered and washed with ethyl ether, after which the filtrate is dried and distilled in vacuo and the above product identified, boiling at 130° to 135°C at 0.1 mm Hg.

Example 3: Polymerization of ethyl phenylbromophenylphosphinite is carried out utilizing a nickel bromide catalyst. A solution of 0.4 g of nickel bromide and 25 cc of mesitylene is added in a reaction vessel together with 7.2 g of the monomeric product of Example 2. The solution is maintained at refluxing conditions, with polymerization taking place at 165° to 180°C. After 12 hours of heating, the reaction mixture is cooled and the green solid filtered off and purified. The polymer is dried and the molecular weight determined as 743, corresponding to 4 molecular units as shown below (more severe temperature, time and concentration conditions in general increase the molecular weight).

The polymer is soluble in ethanol (but reprecipitated by the addition of water) and in dimethylformamide; and insoluble in ethyl ether, toluene, acetone and water. The polymer has the formula shown below.

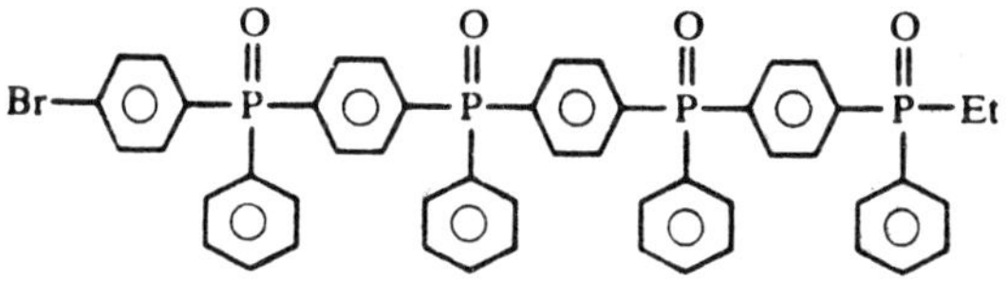

Tetrametaphosphimic Acid

D.F. Lawson; U.S. Patent 4,072,651; February 7, 1978; assigned to The Firestone Tire & Rubber Company describes the use of tetrametaphosphimic acid and its salts as flame retardants. This relates to the use of such compounds with both synthetic resins and natural materials to impart reduced flammability and increased fire resistance.

Tetrametaphosphimic acid is considered to have the formula $[HNPO(OH)]_4 \cdot XH_2O$ where X is any small integer. As indicated by the formula tetrametaphosphimic acid and salts contain high percentages of phosphorus and nitrogen and contain water which may be liberated on heating, adding to their efficiency as flame retardants and are relatively stable.

Furthermore, tetrametaphosphimic acid and its derivatives can form molten glasses within a treated article upon strong heating, as in burning, thus promoting and strengthening char, the combined effect of which is insulation and surface protection of the treated article.

The formation of mineral acids is beneficial to dehydration reactions; hence, tetrametaphosphimic acid and its derivatives are useful promoters for intumescent coating formations when combined with conventional hydrated spumifiers and char-forming ingredients. Tetrametaphosphimic acid and its diammonium salt exhibit low water solubility.

Tetrametaphosphimic acid is prepared from octachlorocyclotetraphosphazene by stirring an ether solution of the phosphazene in the presence of an equal volume of water or aqueous potassium acetate. To prepare tetraammonium tetrametaphosphimate $(NH_4^+ O^- PONH)_4 \cdot 4H_2O$, the solid acid is suspended in water and treated with an excess of ammonium hydroxide. The solution is heated for about ½ hour, then cooled, and the solid salt is filtered out. The diammonium salt of tetrametaphosphimic acid can be prepared from the tetraammonium salt by acidification of a neutral solution or suspension with acetic acid. The variously prepared derivatives were tested for flame retardancy in silica-filled polybutadiene thermoset formulations.

Compositions were prepared and tested as follows. 100 parts by dry weight of low to medium molecular weight high vinyl (>90%) butadiene rubber resin was mixed in a bowl-type mixer with a suitable solvent such as hexane, and with 380 parts by weight of silica (-325 mesh), 3 parts by weight of a mold lubricant and release agent such as calcium stearate, 3 parts by weight of dicumyl peroxide or equivalent curing agent, and 3 parts by weight of a commercial coupling agent. After thorough mixing, the tetrametaphosphimic acid (or its salt) was added and mixed into the batch. The mixture was then poured into a dish or tray, heated mildly to evaporate the hexane or other solvent, and was then molded to the desired shape, and cured at 350°F in the mold, under pressure. Except for the tetrametaphosphimic acid, the formulation and molding are conventional.

The product was cut into test strips 6" x ½" x ⅒" and tested for self-extinguishment in accordance with ASTM D-635-68, a horizontal burn test, with the following results:

	Burn Rate (in/min)
Control (No. Fr.R.)	0.12
Tetrametaphosphimic acid*, 15 phr	0.03
Tetraammonium salt of TMPA, 15 phr	0.02
Tetraammonium salt of TMPA, 10 phr	0.00

Note: *TMPA
phr = parts per 100 parts of rubber

Similar reduced flammability is found with the synthetics including polyurethanes, polyesters, polyvinyl and polyvinylidene resins, and with natural materials such as cotton, wool, wood and other cellulosics, when treated with between 12 and 25 parts of tetrametaphosphimic acid or its salts per 100 parts of resin or natural product.

HALOGENATED PHOSPHORUS COMPOUNDS

Copolymers of Bis(Hydrocarbyl) Vinyl Phosphonate

R.H. Brunner and P. Kraft; U.S. Patent 4,035,571; July 12, 1977; assigned to Stauffer Chemical Company describe fire retardant copolymers of (1) a bis(hydrocarbyl) vinyl phosphonate, (2) a monomer which contains one ethylenically unsaturated bond, and (3) acrylic or methacrylic acid. These copolymers are useful in a variety of applications, including as coatings and as additives for enhancing the fire retardancy of thermoplastics. The type of bis(hydrocarbyl) vinyl phosphonate to be used in forming the copolymer has the structure:

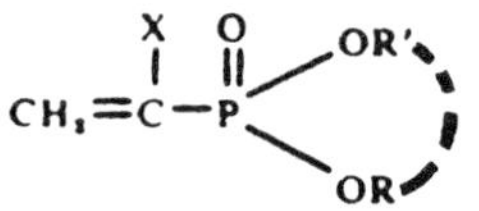

where X is selected from the group consisting of hydrogen, halogen, cyano, aryl; and R and R' are hydrocarbyl and substituted hydrocarbyl groups consisting essentially of hydrogen and carbon and containing up to about 18 carbon atoms inclusive with the proviso that R and R' may be the same, different or conjoint, i.e., R and R' may form one single radical.

Representative bis(hydrocarbyl) vinyl phosphonates which may be used in preparing the fire retardant vinyl chloride copolymer latexes are: bis(β-chloroethyl) vinyl phosphonate, bis(β-chloroethyl)-1-methyl vinyl phosphonate, bis(β-chloroethyl)-1-cyano vinyl phosphonate, etc.

The types of monomers that contain one ethylenically unsaturated bond include the vinyl halides, vinylidene halides, α-olefins, vinyl esters of carboxylic acids and esters of acrylic and methacrylic acid, etc.

The copolymer latexes may be prepared by means of free radical initiated emulsion polymerization processes. In these procedures, the various monomers and catalysts are emulsified in water by means of one or more surface-active emulsifiers, whereupon the polymerization reaction is then initiated.

Suitable water-soluble, free radical initiating catalysts for use in preparing the copolymer latexes include sodium, potassium and ammonium persulfate. These catalysts should be present in a concentration of from 0.1 to 5% by weight of the total monomer charge. With respect to the emulsifier or surfactant, it is possible to utilize one or more anionic, cationic or nonionic emulsifiers, for example, the alkyl carboxylic acid salts; the alkyl sulfate, sulfonate, phosphate, or sulfosuccinate salts.

The actual polymerization reaction will ordinarily be conducted at a temperature of from 30° to 80°C for a period of from 6 to 24 hours, depending upon such factors as the particular monomer, catalysts, surfactants and polymerization apparatus that is being utilized. The latexes resulting from this polymerization process will ordinarily have a resin solids content of from 20 to 50% by weight, where the particles range in size from 0.02 to 2 microns.

Polymerization may ordinarily be initiated by heating the system to a temperature in the range of from 20° to 150°C for a period of from 5 to 24 hours, with agitation being applied throughout the course of the reaction. The resulting product will comprise an aqueous suspension of the copolymers which will be in the form of particulate solids having a resin solids content in the range of 5 to 60% by weight. These copolymer particles will have a particle size of from 20 to 250 microns when they are to be used as additives for preparing flame retardant thermoplastic polymers.

Example 1: A latex according to the process was prepared by mixing 558 g of water, 1.5 g of sodium lauryl sulfate and 2.0 g of Triton X-200 (a sodium sulfonate salt of an ethoxylated alkyl phenol anionic emulsifier) in a three-neck flask under nitrogen and heating the mixture to 80°C. 40 g of a catalyst solution of ammonium persulfate (2% solution) was added. To the resulting solution was then added a monomer mixture comprising 60 g of methyl methacrylate, 60 g of bis(β-chloroethyl) vinyl phosphonate and 30 g of methacrylic acid in a dropwise manner over a period of 1 hour while maintaining the temperature at 80°C. The reaction medium was then heated to 90°C for 5 minutes with no refluxing acrylic monomer noted. The medium was cooled and an essentially gum-free latex was filtered therefrom.

The latex was neutralized to a pH of 7 to 9 with dilute ammonium hydroxide and was cast into a film. The dried film exhibited an LOI of 26 using the test procedure described by Fenimore and Martin in the November 1966 issue of *Modern Plastics.*

Example 2: A water-soluble copolymer was formed by adding the following components to an appropriate reaction vessel:

	Grams
Water	200
Bis(β-chloroethyl) vinyl phosphonate	40
Vinyl chloride	40
Acrylic acid	20
Surfactant (10% Sipex UB)	15
Triton X-200	2.5
Ammonium bisulfite (2% solution)	70

This reaction mixture was heated at 50°C for 5 hours with agitation (at 20 rpm) and had a 26% solids content. After neutralization with ammonium hydroxide to a pH of 7.0, a clear water-soluble system was produced. When added to water and applied as a surface coating, a hard flame resistant coating was produced.

Copolycondensed (2-Haloalkyl) Vinyl Phosphonates

E.D. Weil; U.S. Patent 4,067,927; January 10, 1978; assigned to Stauffer Chemical Company describes copolycondensed vinyl phosphonates which are prepared by the cocondensation of: (1) a (2-haloalkyl) vinyl phosphonate or an α-methyl vinyl phosphonate, preferably (2-chloroethyl) vinyl phosphonate, and/or its halohalide adduct precursor which may be, for example, a 2-haloethyl or 2-haloisopropyl phosphonate; with (2) at least one pentavalent phosphorus ester of the structure ROP(O)XY where R is selected from the class consisting of C_{1-20} alkyl, C_{1-20} alkenyl and C_{1-20} haloalkyl; and X and Y are groups selected from the class consisting of RO–, C_{1-20} alkyl, C_{1-20} alkenyl, aryl, aryloxy, amino, C_{1-20} alkyl- or aryl-substituted amino groups and C_{2-20} alkylene, C_{2-20} alkyleneoxy.

The resulting copolycondensed vinyl phosphonates possess excellent flame-retarding properties. They may be used as monomers for the preparation of flame retardant resins or, if so desired, they can be used to prepare flame retardant finishes for many different types of flammable substrates such as unsaturated polyesters and particularly for textiles.

These copolycondensates yield fire retardant, resinous textile finishes which are characterized by their outstanding degree of resistance to laundering and dry cleaning. This resistance is achieved without any adverse effects upon the softness, hand, or other desirable physical properties of the treated textiles.

All types of textiles may be treated so as to provide them with durable fire retardant finishes. Thus, one may treat textiles derived from natural fibers such as cotton, wool, silk, sisal, jute, hemp and linen; and from synthetic fibers including nylon and other polyamides; polyolefins such as polypropylene; polyesters such as polyethylene terephthalate; cellulosics such as rayon, cellulose, acetate and triacetate; fiber glass; acrylics and modacrylics.

Example 1: Part A — In a reactor fitted with a stirrer, vertical reflux condenser and effluent line from the latter to a dry ice cooled receiver, there are placed 2,796 g (6 mols) of bis(2-chloroethyl) vinyl phosphonate, 1,736 g (7 mols) of dimethyl methyl phosphonate and 10 g of sodium carbonate. The reactor is stirred and heated. At 135° to 140°C the evolution of methyl chloride begins and the temperature is then raised over 5 hours to 190°C and held at this point until the rate of collection of distillate, i.e., of methyl chloride and a minor percentage of ethylene dichloride, in the dry ice trap becomes less than 10 cc per hour. The reaction mixture is then sparged briefly with nitrogen. By titration in water, the 3,317 g of residual product is found to contain 1.4 meq/g of total acidity.

Part B – The above-described reaction product is treated at 100° to 110°C with 182 g of methanol over 1 hour, thus reducing the acidity to 0.4 meq/g. The acidity is then reduced to a negligible level, i.e., to less than 0.1 meq/g, by passing in ethylene oxide over 3 hours at 100° to 110°C. The resultant product, after brief sparging with nitrogen to remove dissolved ethylene oxide, comprises 3,498 g of a pale yellowish clear liquid, n_D^{25} 1.4735, acid number 0.42 mg KOH/g, which is soluble in water and has a chlorine content of only 0.42%. The nuclear magnetic resonance spectrum shows 1.16 $CH_3P(O)$ groups per $CH_2{=}CH{-}P(O)$ group.

Example 2: Part A – A mixture of 2 mols (280 g) of trimethyl phosphate, 1 mol (233 g) of bis(2-chloroethyl) vinyl phosphonate, 1 g of sodium carbonate as a catalyst and 0.1 g of di-tert-butylhydroquinone as a polymerization inhibitor is heated at 178° to 180°C for 7 hours until substantially the theoretical amount of methyl chloride by-product (2 mols, 101 g) is collected in a dry ice cooled trap.

The residual product is a colorless syrup, soluble in water, and its infrared spectrum is consistent with the structure $CH_2{=}CHP(O)[CH_2CH_2OP(O)(OCH_3)_2]_2$ except for the presence of OH bonds. Titration of the product in aqueous medium shows 1.1 meq of acid or acid-yielding compounds per gram. Heating the product with 25 g of trimethyl orthoformate at 100°C for 1 hour is partially successful in reducing the acidity to 0.3 meq/g. The product has a refractive index of 1.4460 (n_D^{25}). Nuclear magnetic resonance shows 2 OCH_2CH_2O groups and 5.4 $CH_3{-}O{-}P$ groups per $CH_2{=}CH{-}P$ group, approximating the structure $CH_2{=}CHP(O)[OCH_2CH_2OP(O)(OCH_3)_2]_2$. The following cocondensation reactions are prepared by means of a reaction procedure comparable to that set forth in Part A.

Part B – The reaction is carried out between 2 mols of bis(2-chloroethyl) vinyl phosphonate and 4 mols of triethyl phosphate evolving principally ethyl chloride and a small amount of ethylene dichloride. A chlorine-free, water-soluble product remains.

Part C – The reaction is carried out between 3 mols of bis(2-chloroethyl) vinyl phosphonate and 4 mols of dimethyl methyl phosphonate, evolving principally methyl chloride and a small amount of ethylene dichloride.

Part D – The reaction is carried out between 3 mols of bis(2-chloroethyl) vinyl phosphonate and 4 mols of tris(2-chloroethyl) phosphate which, in this case, evolves ethylene dichloride as the volatile product.

Example 3: This example illustrates the preparation of durable, fire retardant textile finishes with a number of copolycondensed vinyl phosphonates. It also demonstrates the superiority of these finishes when compared with the finishes derived from a homopolycondensed vinyl phosphonate.

A series of four aqueous padding baths are prepared each of which contains 33% by weight of a 60% aqueous solution of N-methylolacrylamide, 0.4% by weight of octylphenoxypolyethyleneoxyethanol as a wetting agent, 0.5% by weight of potassium persulfate as a catalyst and 30% by weight, respectively, of each of the following vinyl phosphonate condensates:

Bath No. 1	The copolycondensed vinyl phosphonate prepared in Part A of Example 2.
Bath No. 2	The copolycondensed vinyl phosphonate prepared in Part B of Example 2.
Bath No. 3	The copolycondensed vinyl phosphonate prepared in Part C of Example 2.
Bath No. 4	A homopolycondensed vinyl phosphonate described in U.S. Patent 3,822,327.

In the preparation of a homopolycondensed vinyl phosphonate (Bath No. 4) a total of 8 mols (1,864 g) of bis(2-chloroethyl) vinyl phosphonate, 8 g of sodium carbonate and 0.2 g of tert-butylhydroquinone are heated at a temperature of 170° to 185°C for 3 hours and then sparged with nitrogen gas until a total of 475 g of ethylene dichloride has been distilled off. This corresponds to the formation of a condensation product having the idealized average formula:

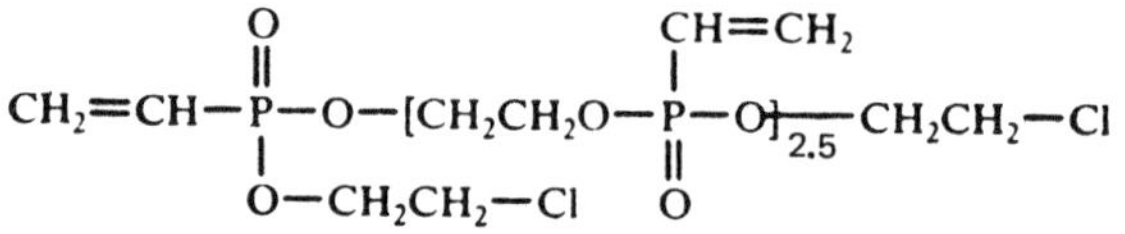

Samples of cotton flannel cloth having a weight of 3.5 oz/yd^2 are padded through each of the above-described baths, cured at 280° to 300°F, whereupon they are subjected to a hot water (60°C) wash followed by a series of 10 hot water (60°C) detergent washes in a home washing machine containing 50 g of a strong detergent and 200 ppm of water hardness [calculated as $CaCO_3$ using $Mg(NO_3)_2 \cdot 6H_2O$ and $Ca(NO_3)_2 \cdot 4H_2O$] and 8 bath towels as ballast. The flammability of the treated cloths is evaluated by means of the LOI test as well as by means of the vertical char length test. The LOI test is described in the procedure in ASTM D-2863. The complete details of the vertical char length test are described in the Federal Flammability Standard of July 27, 1971, (35 *Federal Register* 146).

As a control for this experiment, an untreated sample of the cotton flannel cloth is utilized. The data obtained from each of the test procedures are described below:

	Fabric Treated in Bath No.				
	1	2	3	4	Control
LOI	29.09	31.94	31.53	26.32	18–19
Vertical Char Length	—	3"	3"	6.25"	BEL*

*Burned over its entire length.

In another series of evaluations, Bath No. 3 and Bath No. 4 are each modified by substituting 20% by weight of acrylamide for the 33% of N-methylolacrylamide. Cotton flannel cloth treated in this version of Bath No. 3 is found to have a char length of 3.5 inches, whereas the cloth treated in the modification of Bath No. 4 is found to be far more flammable, with a char length of 6.75 inches.

The results of the above-described evaluations are, therefore, indicative of the superior flame retardant properties displayed by the textile finishes derived from the copolycondensed vinyl phosphonates of the process as compared with a textile finish derived from a homopolycondensed vinyl phosphonate.

Example 4: This example illustrates the excellent resistance to chlorine bleaching which is displayed by the textile finishes derived from the copolycondensates. A sample of 3.8 oz/yd^2 cotton flannel fabric is padded through a bath consisting of 47 parts water, 1 part of a 10% aqueous solution of octylphenoxypolyethyleneoxyethanol, 20 parts of an ethylene oxide neutralized copolycondensate of 6 mols of bis(2-chloroethyl) vinyl phosphonate and 7 mols of dimethyl methyl phosphonate, 22 parts of a 60% aqueous solution of N-methylolacrylamide and 10 parts of a 5% aqueous solution of $K_2S_2O_8$ and subsequently cured for 5 minutes at 300°F.

The treated fabric is then evaluated for its resistance to chlorine bleaching before and after 10 detergent washes. Thus, in each detergent washing cycle there is introduced 1 cup of bleach containing 5.25% available chlorine. The results of this evaluation are: a vertical char length before washing was 3.0 inches; after 10 detergent washes containing chlorine bleach the vertical char length was 4.5 inches.

The above data reveal that the textile finish of this process retains substantially most of its flame retardancy, even after having been exposed to ten wash cycles containing a chlorine bleach.

Emulsions Containing Haloalkyl Phosphates

R.E. Smith and J.A. Albright; U.S. Patent 4,015,037; March 29, 1977; assigned to Michigan Chemical Corporation describe a flame retardant finish containing a fire retardant compound of the generic formula:

(1) 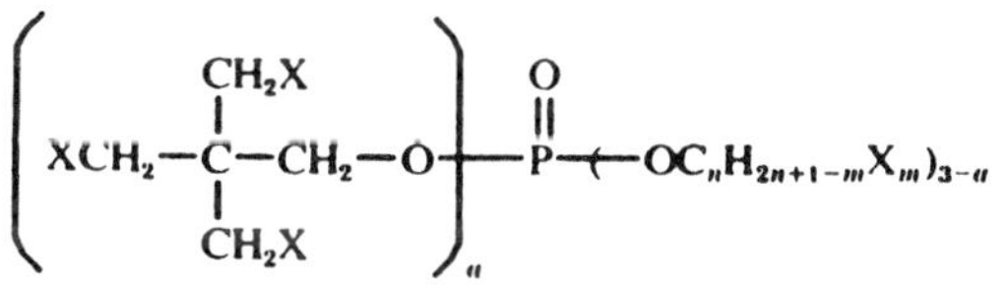

The flame retardant finish is composed of several constituent parts. One of the constituent parts is a flame retardant compound which comprises from 16.0 to 30.0% of the flame retardant finish. The flame retardant compound is of the formula (1), above, where each X is independently selected from chlorine, bromine and hydrogen; where a is an integer from 1 to 2; n is an integer from 1 to 9; and m is an integer from 0 to n, provided that when n is 1, m is 0. Exemplary compounds include bis(2,3-dibromopropyl)-3-bromo-2,2-bis(bromomethyl)propyl phosphate; bis(2,3-dibromopropyl)-3-chloro-2,2-bis(methyl)propyl phosphate; and bis(2,3-dichloropropyl)-3-chloro-2,2-bis(chloromethyl)propyl phosphate.

A second constituent is an organic solvent which comprises 3.2 to 4.8% of the flame retardant finish. The organic solvent is substantially water-immiscible and is further characterized by having a flash point of at least 80°F and a boiling point within the range of 300° to 350°F.

A third constituent part is an emulsifying agent which comprises from 3.2 to 4.8% of the flame retardant finish. The emulsifying agent possesses a hydrophile-lipophile balance value of from 10 to 14. The emulsifying agent is selected from the group comprising (a) a nonionic/anionic blend of an isopropyl amine salt of dodecylbenzene sulfonic acid and an ethoxylated alcohol containing from 10 to 18 carbon atoms; and (b) an oil-soluble metal sulfonate and a polyoxyethyl ether blended in proportions such that the emulsifying agent possesses a hydrophile-lipophile balance of 10 to 14.

A fourth constituent part is a water-soluble quaternary phosphonium salt which comprises from 27.0 to 41.0% of the flame retardant finish. The water-soluble quaternary phosphonium salt is selected from the group comprising tetrakis(hydroxymethyl)phosphonium and tetrakis(methylhydroxymethyl)phosphonium salts, where the anion is derived from organic or inorganic, mono- or polybasic acids and blends.

A fifth constituent part is a water-soluble organic nitrogen-containing compound which comprises from 10.0 to 15.0% of the flame retardant finish. The nitrogen-containing compound is selected from the group consisting of

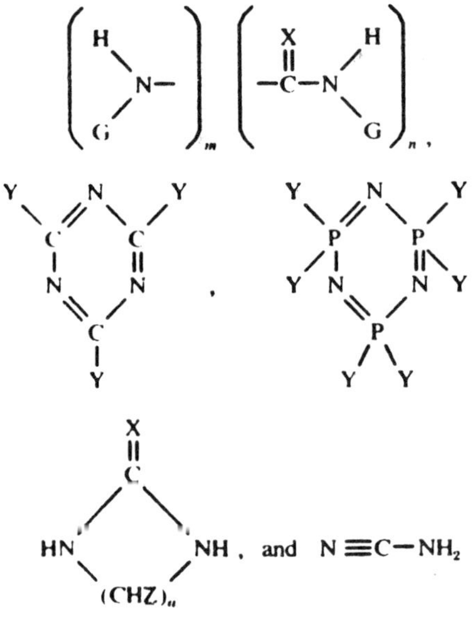

where each G is independently selected from the group comprising hydrogen, hydroxymethyl, alkyl containing 1 to 6 carbon atoms, amino, and cyano; X is selected from the group comprising oxygen, sulfur, =NH, and =NC≡N; m is an integer from 0 to 1; n is an integer from 1 to 2 with the provision that m plus n equals 2; a is an integer from 2 to 3; each Y independently is –NHG, where G is defined above; and Z is selected from the group comprising hydrogen and hydroxyl. Exemplary compounds include urea, thiourea, guanidine, dicyandiamide, melamine, trimethylol melamine, aminocyclophosphazene, N-methylolcyclophosphazene, ethylene urea, propylene urea, cyanamide and oxamide.

A sixth constituent is water which comprises from 20.0 to 31.0% of the flame retardant finish. The retardant finish may optionally have incorporated a wetting agent. If the wetting agent is a constituent part of the finish, the wetting agent would comprise from 0.1 to 1.0% of the flame retardant finish. The wetting agents employed may be selected from the group comprising anionic, nonionic, and nonionic-anionic blend wetting agents. Exemplary wetting agents include an anionic phosphate surfactant in free acid form, a nonionic nonylphenyl polyethylene glycol ether, a nonionic octylphenoxy polyethoxy ethanol, a nonionic trimethyl nonyl polyethylene glycol ether, and a nonionic polyethylene glycol ether of linear alcohol.

A method of making the flame retardant finish involves adding the desired amounts of the various constituents in the following sequence: (1) Dissolve the water-soluble organic nitrogen-containing compound in water; (2) add the wetting agent, if used, to (1) while keeping the temperature of the water-soluble organic nitrogen-containing compound-water-wetting agent solution below 40°C; (3) add an aqueous solution of the desired water-soluble quaternary phosphonium salt to (2); and (4) finally add to (3) an emulsion concentrate comprising the flame retardant compound, the solvent, and the emulsifying agent.

The flame retardant finish may be applied to textile fabrics by a pad, dry, cure and oxidation afterwash procedure. Many textile fabrics may be treated with this finish. Examples of such textile fabrics include cellulosics, rayon, acrylics, polyesters, acetates, nylon, and textile fabrics derived from animal fibers such as wool and mohair, and blends of these. The following examples illustrate the process.

Example 1: Samples of a 50/50 spun blended polyester and cotton poplin were treated with finishing formula A by a pad, dry, cure and oxidative afterwash procedure:

Formula A

	Percent
Tetrakis(hydroxymethyl)phosphonium oxalate, 65.0% aqueous solution	52.0
Water	7.4
Wetting agent	0.2
Urea	12.4
Emulsion concentrate	28.0
Emulsifying agent	14.3
Solvent	14.3

The wetting agent employed was a nonionic trimethyl nonyl polyethylene glycol ether having an HLB value of 11.7. The emulsion concentrate consisted of 71.4% bis(2,3-dibromopropyl-3-bromo-2,2-bis(bromomethyl)propyl phosphate. The emulsifying agent was an anionic emulsifier comprising a blend of oil-soluble metal sulfonates with polyoxyethylene ethers having an HLB value of 12.5. The solvent had a flash point of about 110°F and a boiling point of about 315°F.

The wet pick-up of finish was 74%. The fabrics were dried for 5 minutes at 105°C and then cured for 2 minutes at 205°C. The cured fabrics were oxidized using hydrogen peroxide to insure conversion of all the phosphorus to the +5 valence state. The oxidation was completed in a washing machine using the following conditions: (1) Soft water at 60°C, set at a pH of 10 to 11 with sodium carbonate; (2) 5% hydrogen peroxide (100%) based on the weight of fabrics (wof); and (3) add treated fabrics and run regular wash cycle, high water level, rinsing at 40° to 44°C. After oxidation, the fabrics were tumble dried in a an electric dryer.

Example 2: The same fabric used in Example 1 was treated according to the processing procedure as described in Example 1, using the following formula B:

Formula B

	Percent
Tetrakis(hydroxymethyl)phosphonium oxalate, 65% aqueous solution	52.0
Water	35.4
Wetting agent*	0.2
Urea	12.4

*same as Formula A, Example 1

Example 3: The flame retardancy of the treated fabrics of Examples 1 and 2 was evaluated using the procedures established by the "Standard for the Flammability of Children's Sleepwear," U.S. Department of Commerce FF 3-71. The durability of the flame retardant treatment was determined by measuring the char lengths of the treated fabrics after multiple laundering and drying cycles as set forth in DOC FF 3-71. Char-length data, in inches, for the untreated fabric and the treated fabrics of Examples 1 and 2 are shown in the table below.

	Laundering and Drying Cycles					
	0	10	20	30	40	50
Untreated	BEL					
Example No. 1	2.6	2.5	2.6	3.7	3.4	5.5
Example No. 2	BEL					

BEL - Specimen burned the entire length.

Example 4: Samples of the same fabric used in Example 1 were treated with the finishing formula C, below, by a pad, dry, cure and oxidative afterwash procedure. The wet pick-up of the finish was 77%.

Formula C

	Percent
Tetrakis(hydroxymethyl)phosphonium acetate, 65.0% aqueous solution	52.0
Water	7.4
Wetting agent*	0.2
Urea	12.4
Emulsion concentrate*	28.0

*same as Formula A, Example 1

The treated fabric samples were dried 5 minutes at 105°C and cured 5 minutes at 160°C. The samples were oxidized as follows: (1) Pad, 3 times, with a solution of 5% hydrogen peroxide (100%) at 80°C; (2) rinse in water at 60°C to remove excess peroxide using a washing machine set on regular wash cycle, high water level, and rinse water at 40°C; and (3) repeat above cycle using a 0.05% solution of sodium carbonate at 60°C in the wash cycle to neutralize any residual acidity; complete rinses at 40°C and tumble dry.

In addition to bis(2,3-dibromopropyl)-3-bromo-2,2-bis(bromomethyl)propyl phosphate, other flame retardant compounds of formula (1), shown previously, e.g., bis(2,3-dibromopropyl)-3-chloro-2,2-bis(bromomethyl)propyl phosphate, bis(2,3-dibromopropyl)-3-chloro-2,2-bis(methyl)propyl phosphate, and bis(2,3-dibromopropyl)-tris(2,2,2-chloromethyl)ethyl phosphate can be used in formula C with comparable results being obtained.

Bis(2,3-Dibromopropyl) Phosphates

H. Nachbur and A. Maeder; U.S. Patent 4,055,689; October 25, 1977; assigned to Ciba-Geigy Corporation describe a process for flameproofing organic material which comprises applying to the material a preparation containing a phosphate of the formula

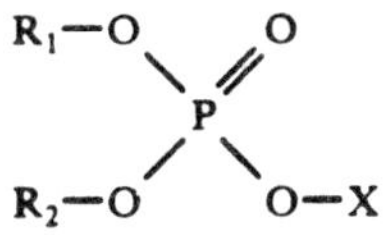

where X is phenyl or substituted phenyl and R_1 and R_2 each is bromoalkylene, and then drying the material and subjecting it to heat treatment.

The phosphates are manufactured by esterifying a phosphoric acid dihalide with at least one brominated alkanol or alkenol. These products are then used to flameproof organic fibers. The preparations or agents which are used for flameproofing the material can be in the form of aqueous emulsions or suspensions or organic solutions.

Depending on the desired coating, the preparations for finishing the textiles contain 5 to 25% by weight of the compound. The preparations which contain the compound can be applied to the material by conventional methods of spraying, padding, exhaustion or printing. Preferably, the preparations are applied by the padding method. After the phosphates have been applied, the material is dried at temperatures below 100°C and then subjected to a thermosol process at 175° to 220°C. This thermosol process generally lasts about 10 to 200 seconds.

The phosphates can be used as flame retardants in many kinds of applications, i.e., as additives to plastics, for example, polystyrene molding compounds, polyester systems, polyurethane foams, epoxide resin compositions, polyvinyl chloride systems, paper (together with a suitable binder), varnish preparations and polyamide molding compositions.

Example 1: (Bis-2,3-Dibromopropyl) Phenyl Phosphate – A solution of 84.4 parts (0.4 mol) of phenyl phosphoric acid dichloride in 100 parts by volume of dry benzene is added dropwise over the course of 30 minutes to an ice-cold (0° to 1°C) solution of 174.4 parts (0.8 mol) of 2,3-dibromopropyl alcohol and 63.2 parts of pyridine in 400 parts by volume of dry benzene.

The mixture is subsequently stirred for 5 hours, in the course of which the temperature gradually rises to about 25°C. The pyridine hydrochloride which has formed is then isolated and the filtrate benzene phase is washed in succession with 1,300 parts of water, then with 1,300 parts of a 5% sodium carbonate solution and finally again with 1,300 parts of water.

The benzene phase is dried with anhydrous sodium sulfate, the benzene then removed and the residue purified from unreacted starting materials by treatment in vacuo at 120°C/0.03 to 0.05 torr. Yield: 192 parts (83½% of theory) of a clear, colorless oil which corresponds to formula (1) below:

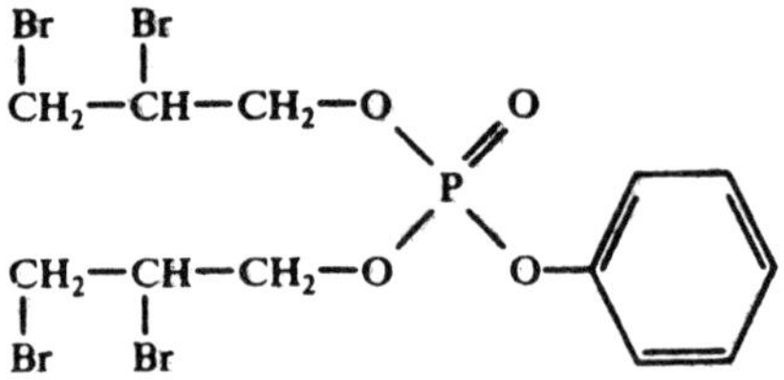

The infrared spectrum shows the characteristic aromatic substance bands at 1,580 and 1,480 cm^{-1}, the P=O band at 1,270 cm^{-1}, the P–O–aryl band at 1,190 cm^{-1}, the P–O–alkyl band at 1,030 cm^{-1} and the 5 adjacent aromatic hydrogen atoms which are in positions 2,3,4,5 and 6 of the phenyl group at 760 cm^{-1}. In the mass spectrum, the parent peak can be determined for the molecular weight 574. The indicated structural formula is clearly evident from the degradation products. The gel permeation chromatogram shows that the substance is homogeneous.

Example 2: 1,4-Phenylene-[Bis(2,3-Dibromopropyl) Phosphate] – A suspension of 68.6 parts (0.2 mol) of 1,4-phenylenephosphoryl dichloride in 300 parts by volume of acetonitrile is cooled to -10°C and then treated over the course of 5 minutes with a solution of 63.2 parts (0.8 mol) of anhydrous pyridine in 100 parts by volume of acetonitrile. Then a solution of 174.4 parts (0.8 mol) of 2,3-dibromopropanol in 100 parts by volume of acetonitrile is added dropwise at 0° to 10°C over the course of 30 minutes. The reaction mixture is subsequently kept for 1 hour at 0°C.

The resulting clear solution is freed from acetonitrile in vacuo at 50°C to yield 315 parts of a white, greasy product which is stirred in 200 parts by volume of dried toluene using a homogenizing device, and is then cooled to -5°C. The pyridine hydrochloride is filtered off and washed with 50 parts by volume of dried toluene. The toluene phase is concentrated in vacuo at 100°C and the residue is freed from unreacted 2,3-dibromopropanol at 160°C and 14 torr. Yield: 161 parts (75% of theory) of a viscous, brownish oil which corresponds to formula (2), below:

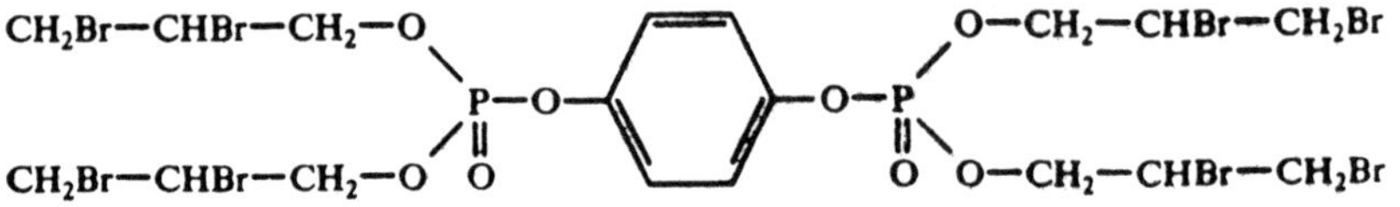

The infrared spectrum shows the characteristic aromatic substance bands at 1,500 cm^{-1}, the P=O band at 1,280 cm^{-1}, the P–O–aryl band at 1,190 cm^{-1}, the P–O–alkyl band at 1,050 cm^{-1} and the two adjacent aromatic hydrogen atoms in positions 2,3 and 5,6 of the phenyl group at 840 cm^{-1}. The gel permeation chromatogram shows that the substance is homogeneous.

Example 3: A fabric of 100% polyester (polyethylene glycol terephthalate) weighing 150 g/m^2 is padded, dried, subjected to a thermosol treatment, washed, rinsed and dried in accordance with the particulars given in the following table, and tested for its flame resistance according to DOC FF 3-71. The washing is carried out over the course of 5 minutes at 40°C in a bath which contains 2 g/l of anhydrous sodium carbonate and 1 g/l of a condensation product of 1 mol of p-tert-nonylphenol and 9 mols of ethylene oxide.

The flameproofing test is performed before and after the thermosol treatment and after 25 and 50 machine washes. The machine washes are carried out in a domestic washing machine at 40°C over the course of 45 minutes and in a liquid which contains 4 g/l of a household detergent. The results are summarized in the following table:

Constituents Conditions	un- treated	treated with liquor 1	2	3	4
g of product of formula 1 per kg of liquor	100	200	100	100	
solvent		tri-chloro-ethyl-ene	tri-chloro-ethyl-ene	tri-chloro-ethyl-ene	tri-chloro-ethyl-ene
liquor pick-up (%)		100	100	100	100
thermosol treat-ment temperature in °C		200	200	200	200
thermosol treat-ment time (sec.)		20	20	60	60
deposit after final wash (%)		11.1	21	13	21
Flame resistance (time: sec./tear length: cm)					
before thermosol treatment	16/11	0/7.5	0/0.5	0/7.5	0/7
after thermosol treatment	burns	0/7	0/6.5	0/6	0/7
after 25 washes	burns	5/7.5	1/7.5	3/8	0/8
after 50 washes	burns	1/7	1/7	4/7.5	1/7

The flameproofing test is carried out as follows: 5 pieces of fabric, each measuring 8.9 x 25.4 cm, are clamped into a testing frame and dried with circulating air for 30 minutes at 105°C in a drying cabinet. The pieces of fabric are then conditioned in a sealed container over silica gel for 30 minutes and subsequently subjected to the actual flameproofing test in a combustion chamber. The pieces of fabric are each ignited with a methane gas flame for 3 seconds in the vertical position. The test is considered as having been passed if the average charred zone is not more than 17.5 cm in length and no single sample exhibits a charred zone of over 25.4 cm in length and the individual smouldering times are not longer than 10 seconds.

Example 4: A polyurethane soft foam is prepared by mixing the following materials: 100 g of a polyhydroxyl compound of a polyether suitable for polyurethane soft foam formation and having a molecular weight of about 3,000 (manufactured from trimethylol propane and propylene oxide); 1 g of siloxane-oxyalkylene copolymer; 0.32 g of tin(II) octoate (catalyst); 3.5 g of water; 48.2 g of toluylene diisocyanate (80:20 mixture of 2,4:2,6 isomers); 6 g of the phosphate according to formula (3):

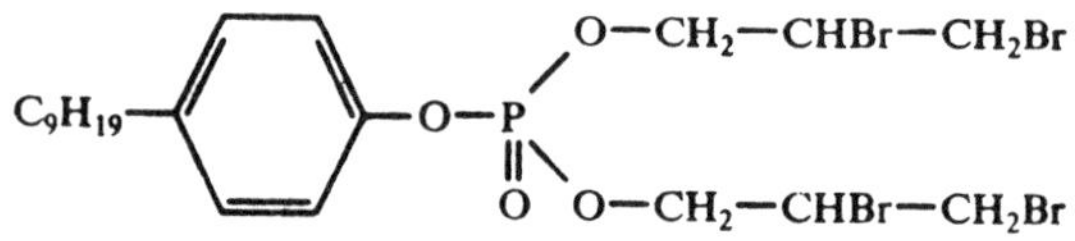

The resulting polyurethane soft foam is tested for its flame resistance according to ASTM D-1692 test method, the test being repeated after the test samples have been stored for 7 days at 140°C in a dry atmosphere and at 90°C in a humid atmosphere as well as after 5 hours storage at 120°C in an atmosphere saturated with steam. Simultaneously, a polyurethane soft foam specimen which was manufactured without addition of a phosphate is tested. The results are summarized in the following table:

	without storage	after 7 days storage at 140° C (dry)	after 7 days storage at 90° C (humid)	after storage at 120° C (saturated steam) in hours
test samples with phosphate added BZ/RC	7.5/9.3	5.0/12.8	9.5/10.7	8.5/10.3
test samples without phosphates	burned through/ 27	burned through/ 19	burned through/ 17	burned through/ 18

BZ - burned zone in cm; RC = rate of combustion in cm/min.

The ASTM D-1692 test is carried out as follows: One each of the test samples, measuring 150 x 50 x 13 mm is fixed with the 50 x 13 mm edge in the horizontal position. Markings are made at 24 and 100 mm. The sample is ignited at its bottom end with a dovetail burner. The ignition time is 60 seconds. If the burned zone is not longer than 25 mm, the foam is termed noninflammable. If the sample burned over the 25 mm mark and the burned zone is smaller than 125 mm, the foam is termed self-extinguishing. The length of the burned zone is indicated in centimeters. If the sample burns over the 125 mm mark, the foam is termed inflammable and the rate of combustion is indicated.

Aqueous Emulsions of Tris(2,3-Dibromopropyl) Phosphate

A process described by *G.P. Guzzardo; U.S. Patent 4,047,959; September 13, 1977; assigned to Pennwalt Corporation* is directed to stable flame retardant emulsions of tris(2,3-dibromopropyl) phosphate. This involves a nonionic aqueous emulsion system of tris(2,3-dibromopropyl) phosphate which is both stable and compatible with either anionic, cationic or nonionic fluorochemical oil and water repellents. These nonionic emulsions consist essentially of about 40 to 50% by weight of tris(2,3-dibromopropyl) phosphate flame retardant, about 6 to 12% by weight of one or more C_{5-7} aliphatic or alicyclic ketones, about 3 to 6% by weight of one or more particular nonionic surfactants. These emulsions impart excellent flame retardancy and when combined with a fluorochemical oil and water repellent give good water repellency to fabrics with high hydrostatic water pressure values.

The tris(2,3-dibromopropyl) phosphate flame retardant can be prepared by reacting 2,3-dibromopropanol with phosphoric acid, phosphorus oxychloride, phosphoric acid esters or phosphorus pentoxide in appropriate solvents.

The emulsifying agents are selected from three classes of nonionic emulsifiers. One or more of the same or different nonionics can be used to emulsify the flame retardant. The emulsifiers which are satisfactory must have a hydrophile-lipophile balance (HLB) within the range of about 10 to 14. If only a single emulsifier is used, it must be within the range of 10 to 14. If more than one emulsifier is used the HLB values can range above or below the range of 10 to 14, provided the blend of emulsifiers has an HLB value within the range of 10 to 14.

The first class of emulsifiers which are useful are the sorbitan fatty acid esters and the ethoxylated sorbitan fatty acid esters. Examples of nonionics in this class are sorbitan monolaurate, sorbitan trioleate, sorbitan monopalmitate, ethoxylated sorbitan monolaurate having 20 ethylene oxide units, ethoxylated sorbitan monolaurate having 4 ethylene oxide units, ethoxylated sorbitan trioleate having 20 ethylene oxide units and ethoxylated sorbitan monostearate having 20 ethylene oxide units.

The second class of nonionic emulsifiers are the ethoxylated glycol tallates in which the ethylene oxide content can range from about 4 to 20 mols or units. An example of this class is polyoxyethylene glycol tallate having about 15 mols of ethylene oxide.

The third class of nonionic emulsifiers are the series of polyoxyethylene polyoxypropylene glycols (Pluronic L and Pluronic F). These emulsifiers are formed by ethoxylating a condensate of propylene oxide and ethylene glycol. The ethylene oxide content added to the condensate will range between 4 to 20 mols or units and the molecular weight of the glycols will range from 1,100 to 15,000. Suitable Pluronics used in the emulsions are F68, L43, L63, L64 and L121. The L designation are liquids, while the F68 is a solid.

The flame retardant phosphate esters used are very difficult to emulsify in water and form stable compositions. In order to assist the emulsifiers in forming stable emulsions, it is necessary to employ about 6 to 12% by weight of a ketone solvent. The ketones useful in making the stable emulsions are the C_{5-7} aliphatic and alicyclic ketones.

The final ingredient of the flame-retarding emulsions is water, which is added to balance the composition at 100% by weight. Generally the water will be present at about 51 to 32% by weight.

Emulsion preparation is carried out at room temperature in standard homogenizing apparatus. The ingredients are added to the homogenizer and homogenization is continued until a stable emulsion is obtained. The stability of an emulsion is determined by allowing it to remain quiescent in a vertical glass container and observing any phase separation after 24 hours or any longer period. The emulsions are readily pourable and dispersable in all proportions in typical pad bath formulations.

For imparting flame retardancy to textiles, the emulsions are dispersed with water in a pad bath for wetting the fabrics. The pad bath compositions is formulated to give about 40 to 70% by weight of the tris(2,3-dibromopropyl) phosphate emulsion in water. The fabric moves through the pad bath, usually at ambient temperature, and exits through squeeze rollers. The speed of the fabric and the pressure on the squeeze rollers determine the amount of retention of the pad bath composition on the fabric. Generally, the pad bath is regulated to obtain about a 70 to 80% by weight liquid pick-up based on the dry weight of the fabric.

After leaving the pad bath, the fabric goes to a dryer where the water and ketone are evaporated, leaving the tris(2,3-dibromopropyl) phosphate distributed throughout the fabric. The pad bath is operated to leave at least about a 10% weight increase on the fabric on a dry basis as an effective amount of flame retardant. Air temperatures for hot air dryers are generally within the range of 275° to 300°F.

Example 1: The flame retardant emulsion is prepared using the following ingredients, all percents by weight: 50% tris(2,3-dibromopropyl) phosphate; 3% Pluronics F68/L121 (31.25/68.75), MW of F68 of 8,350 and MW of L121 of 4,400; 6% methylisobutyl ketone; and 41% water.

After homogenization and storage, this emulsion shows only 3⁄16 inch phase separation after 6 days of storage. The individual surfactant HLB values are 29 for F68 and 5 for L121, while the blend in the proportions used has an HLB of 12.5.

Example 2: The emulsions are compatible with water repellents of all types. Formulation ranges are shown below:

	% by Wt
Emulsion of Example 1	40-70
Fluorochemical oil/water repellent (Pentel GH-28)	10-20
Ammonium dihydrogen phosphate	2.5-4.0
Water - to make 100%	

The ammonium dihydrogen phosphate is useful in preventing after-glow. The above formulations show less than 5% phase separation after standing 24 hours in an unagitated pad bath, and agitation readily restores complete homogeneity.

Application of these formulations to tent fabrics made of cotton, polyester, or blends, imparts flame retardancy as well as oil/water repellency properties and resistance to water penetration. For example, tent fabrics treated with these formulations give superior hydrostatic pressure ratings typically 24 cm or more. The test (AATCC 127-1968) involves forcing water under pressure through a fabric just until water starts to pass through it. Good results are also obtained in the Bag test (Army code LP/P DES-48-68) which requires that a 12 x 12 inch sample, when formed into a bag, does not leak over a 24-hour period when it contains 250 ml of water.

Examples 3 through 6: The following ingredients are mixed and homogenized in standard equipment to give stable emulsions.

Ingredient	Example No. 3	4	5	6
	Percent by Weight			
Tris(2,3-dibromopropyl)phosphate	50	50	50	50
Pluronics F68/L121 (31.25/68.75)	3	6	3	6
Methylisobutyl ketone	6	6	12	12
Water	41	38	35	32

Example 7: Pad baths were prepared by adding 70 parts each of the emulsions of Example 3 through 6 to 12.5 parts water, 2.5 parts $NH_4H_2PO_4$ and 15.0 parts Pentel. These pad baths were used to treat swatches of 10.38 oz/yd^2 undyed cotton (white greige) under the following conditions: 70 to 80% wet pick-up; a drying time of 9 minutes at 230°F; and a curing time of 2 minutes at 302°F.

The treated fabrics were tested for flame retardancy using the CPAI-84 flame test. For this weight fabric, the test allows a maximum char length of 4 inches. The fabrics are also tested by the previously described bag test and hydrostatic pressure tests, as well as oil repellency using the AATCC Method 118-1972 and spray rating test using AATCC Method 22-1974. Stability of the pad bath was observed by measuring the amount of separation of 200 g in 24 hours at room temperature. The results are summarized in the table below.

Example	Shelf Stability*	Oil Repellency Rating	Spray Rating	Hydrostatic Pressure	Pad Bath Stability**	Flame Retardancy Char Length, Inches CPAI-84	Bag Test
3	3/16	6	70+	25.7	1/8	2-15/16	No leak
4	3/16	5	70+	24.2	1/8	2-7/8	No leak
5	3/16	5+	70+	25.1	1/8	2-5/16	Leak after 24 hours
6	1/8	5+	70+	24.0	1/8	2-15/16	Leak after 24 hours

* inches separation after 6 days - refers to emulsion alone.
** inches separation after 24 hours - refers to pad bath.

Tribromoneopentyl Esters of Phosphorus Acids

A process described by *E.D. Weil; U.S. Patent 4,049,754; September 20, 1977; assigned to Stauffer Chemical Company* relates to tribromoneopentyl esters of phosphorus acids which exhibit reactive flame retardant properties. These compounds can be used as flame retardant additives, but have the particular feature of being flame retardant reactive components for textile finishing and, in some instances, as monomers which have flame retardant properties.

These compounds have the structure $[(BrCH_2)_3CCH_2O]_2P(O)R$; where R can be $-CH{=}CH_2$, $-CH_2X$, $-OR'OC(O)CH{=}CH_2$, $-OROC(O)C(CH_3){=}CH_2$, $-O(R'O)_nH$, $-OH$, $-Cl$, and $-NR''R'''$, where R' is a lower alkylene or haloalkylene group, R'' and R''' are the same or different and are selected from the group consisting of hydrogen, lower alkyl, phenyl, hydroxy-lower-alkyl and allyl groups, where X is a halogen and where n is an integer between 1 and 20.

These compounds have reactive functional groups for R which are well adapted to serve as means for attaching the compounds to textiles. When R is vinyl or terminates in an acrylate or methacrylate group, the compounds are polymerizable by free radical mechanisms and can be polymerized or copolymerized in situ on textiles by means of graft polymerization, as described in U. S. Patent 3,695,925. When R is CH_2X, where X is a halogen, the compound is sufficiently reactive as an alkylating reagent to allow for attachment via an ether linkage to cellulose that has been treated with alkali in a manner similar to that described by Drake et al in *Textile Research Journal,* Vol. 29:270,884 (1959). When R is –NR''R''', the compounds can phosphorylate cellulose in a manner described for phosphoramides by Morris et al in *Textile Research Journal,* Vol. 43:336 (1973), or when either R'' or R''', or both, are hydrogen or hydroxyalkyl, they can be bonded to textiles in the manner described in U.S. Patents 2,838,228 and 3,746,572.

Example 1: A mixture of 129.5 g of tribromoneopentyl alcohol, 35 g of vinyl phosphonyl chloride, 78 ml of triethylamine, and 300 ml of methylene chloride was refluxed for 1 day, then evaporated to dryness. The residue was redissolved in methylene chloride, was washed with water and sodium bicarbonate solution, and then dried over magnesium sulfate; was filtered and then evaporated at 100°C and 1 mm Hg. The product was a viscous syrup which crystallized upon standing. The nuclear magnetic resonance (nmr) spectrum showed a ratio of about 5.3 aliphatic protons/vinyl proton (theory: 16:3 or 5.33:1 for the indicated product).

After recrystallization from slightly aqueous ethanol, the product, bis(tribromoneopentyl) vinyl phosphonate, was a crystalline solid having a melting point of 71° to 72°C. Analysis – Calculated for $C_{12}H_{19}O_3BrP$: Br, 66.5%; P, 4.3%. Found: Br, 67.0%; P, 4.3%.

Example 2: A mixture of 33.5 g vinyl chloride, 16.5 g of the bis(tribromoneopentyl) vinyl phosphonate, 15 ml of water, 15 ml of 1% polyvinyl alcohol solution, and 1.5 ml of 5% isopropyl peroxydicarbonate in heptane was charged to a polymerization vessel, and the mixture was agitated for 16 hours at 45°C. The suspension copolymer of vinyl chloride and bis(tribromoneopentyl) vinyl phosphonate thus obtained was dried. A quantitative yield of polymer was obtained and an infrared spectrum on the material did not show any residual unsaturation from the phosphonate.

Comparison was made with representatives of other flame retarded acrylonitrile-butadiene-styrene (ABS) resins, those containing Sb_2O_3 and various chlorinated organic compounds and the ABS-polyvinyl chloride-Sb_2O_3 polyblends. The former tend to have reduced deformation temperatures under load. Several formulations containing the copolymer formed earlier in this example were milled to obtain comparison data with flame retardant ABS. The results are summarized in the table on the following page.

The bromine-containing copolymer produced flame retardancy in ABS which compared favorably with the types of flame retardant ABS resins already known. The copolymer additive did not detract from the impact strength of the ABS, whereas the Dechlorane produced a brittle sheet.

Sample No.	Blend Composition (*)	LOI(***)	Physical Appearance
1	ABS** (Control)	17.49	Tough Sheet
2	ABS** + 22 phr. (parts per hundred) Dechlorane†	23.49	Brittle
3	ABS** + 21 phr. copolymer from Example 2 and 2.3 phr. Sb_2O_3	21.39	Tough
4	ABS** + 21 phr. copolymer from Example 2 and 8 phr Sb_2O_3	24.14	Tough
5	60% ABS** + 40% copolymer from Example 2	24.64	Tough

*All the materials milled well and had sufficient heat stability. All parts per hundred (phr) are based on the ABS resin as 100.

**Lustran 420, Monsanto.

***Limiting Oxygen Index: Test procedure (ASTM D-2863) is described in Fenimore et al, *Modern Plastics,* Nov. 1966. A higher LOI indicates better flame retardancy.

†The Diels-Alder diadduct of hexachlorocyclopentadiene and cyclooctadiene-1,5 (Hooker Chemical).

Example 3: To a stirred solution of 64.8 g tribromoneopentyl alcohol and 18.5 g chloromethyl phosphonic dichloride in 200 cc of benzene was added 34 ml of triethylamine. After the exotherm had subsided, the mixture was refluxed for 2 hours on the steam bath, was cooled, and 1 liter of water added. The organic layer was separated and dried over magnesium sulfate, was filtered, and was diluted with heptane. The diluted filtrate was then partly evaporated until crystals began to form, was then refrigerated overnight, and the crystalline product then filtered off.

The product, bis(tribromoneopentyl) chloromethyl phosphonate, was colorless and had a melting point of 118°C. The infrared spectrum showed no OH band, but a phosphonate P=O multiplet centered at 1,252 cm^{-1}. When applied at 15% add-on to alkali-treated cellulose and heated at 150°C for 1 hour, a wash-durable flame retardant finish is produced.

One or more of the compounds formed by this process can be used to flame retard a wide variety of normally flammable substrates by being incorporated either in or on the substrates as either an additive or as a reactant with the substrate in an amount which is effective to flame retard the substrate. Examples are fibrous materials, which for purposes of this process, include cotton, wool, silk, linen, rayons, cellulose acetate, acrylic, polyester, nylon, polyolefin, including polypropylene, blends such as cotton-polyester, and other organic fibers, yarns, and fabrics.

ANTIMONY COMPOUNDS

Antimony Complexes

E. Radlmann, H. Schaffner, G. Lorenz and G. Nischk; U.S. Patent 4,018,809; April 19, 1977; assigned to Bayer AG, Germany describe a process for the production of a solution of a substantially water-insoluble complex antimony compound in a polar organic solvent. This comprises esterifying an α-hydroxy polycarboxylic acid with a polyol in a molar ratio of the carboxyl groups to the hydroxyl groups in the polyol of at least 2:1, and in the presence of an acid transesterification catalyst. An excess of the esterification product is then reacted with an antimony oxide in a polar organic solvent at a temperature in the range of 100° to 200°C to form a clear solution still containing free carboxyl groups. A monoisocyanate is added to the solution at a temperature below 100°C in such a quantity that the free carboxyl groups are converted into the corresponding amide groups, the monoisocyanate corresponding to the general formula: $R_1-N=C=O$, in which R_1 is a straight-chain or branched alkyl group having from 1 to 20 carbon atoms, or

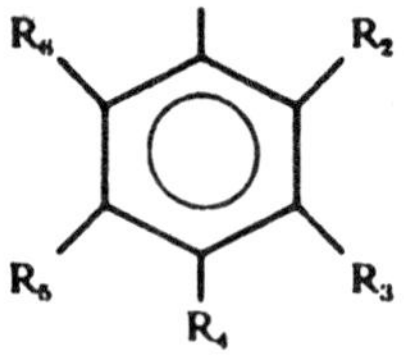

in which R_2, R_3, R_4, R_5 and R_6 may be the same or different from one another and may represent a member selected from the group consisting of hydrogen, an alkyl having from 1 to 5 carbon atoms, fluorine, chlorine or bromine.

Example 1: Solution A – 480 parts by weight of tartaric acid and 99.2 parts by weight of ethylene glycol in the presence of 3.2 parts by weight of an acid ion exchanger are condensed, with stirring for 5 hours at 140°C, in a condensation apparatus while nitrogen is passed over and until the stoichiometric quantity of 57.6 parts by weight of water has distilled off.

The esterification product is then dissolved in 768 parts by weight of dimethyl formamide. Following the addition of about 20 parts by weight of active carbon, the solution is filtered off from the ion exchanger and the active carbon. 116.6 parts by weight of antimony trioxide are added to the solution obtained, followed by heating with stirring at 150°C, while nitrogen is passed over, until after about 40 minutes a clear solution is obtained and the stoichiometric quantity of water has distilled off.

The solution is then cooled at 20°C and phenyl isocyanate is added dropwise at a temperature in the range of 20° to 40°C, until no more carbon dioxide is given off. The consumption of phenyl isocyanate amounts to 182 parts by weight. The solution is then stirred for about 2 hours at 40°C.

The solution obtained is readily miscible and compatible with solutions of copolymers of acrylonitrile and vinyl or vinylidene chloride. The complex antimony compound precipitates on the addition of water to the dimethyl formamide solution.

Example 2: Solution B – In a condensation apparatus, 300 parts by weight of tartaric acid, together with 118 parts by weight of 1,6-hexane diol and 3 parts by weight of an acid ion exchanger, are heated with stirring for about 10 hours at 140°C, while nitrogen is passed over and until the stoichiometric quantity of water (36 parts by weight) has distilled off. The reaction product is dissolved in 516 parts by weight of dimethyl acetamide and the solution is freed from the acid catalyst by filtration. 102.9 parts by weight of antimony trioxide are then added and the dispersion is heated with stirring for approximately 50 minutes at 150°C, while nitrogen is passed over, until a clear solution is obtained. After cooling to approximately 25°C, a solution of 34.2 parts by weight of methyl isocyanate in 100 parts by weight of dimethyl acetamide is added dropwise in such a way that the temperature does not exceed 40°C. The solution is then stirred for 2 hours at 40°C. A slight deposit may form from the almost colorless solution in the event of prolonged standing, although it is completely dissolved on mixing with a solution of an acrylonitrile-vinyl chloride copolymer. The complex antimony compound precipitates on the addition of water to its solution.

Example 3: Solution C – Following the procedure of Example 2, 210 parts by weight of citric acid and 62 parts by weight of ethylene glycol in the presence of 2 parts by weight of an acid ion exchanger are condensed under nitrogen for 6 hours at 140°C. The condensation product is then dissolved in 285 parts by weight of dimethyl formamide and filtered while still warm. Following the addition of 43.7 parts by weight of antimony trioxide, the mixture is heated for approximately 40 minutes at 150°C until a clear solution is obtained. The solution is then cooled to room temperature, followed by the addition of portions of 75.2 parts by weight of 3,4-dichlorophenyl isocyanate. The temperature should not exceed 40°C. The solution is then stirred for 2 hours at 40°C. The complex antimony compound is insoluble in water, while its solution is compatible with solutions of copolymers of acrylonitrile and vinyl or vinylidene halides.

Example 4: Solution D – In an apparatus of the kind described in Example 1, 1,012 parts by weight of glycerol and 4,950 parts by weight of tartaric acid in the presence of 33 parts by weight of an acid ion exchanger are condensed with stirring under a nitrogen atmosphere for about 6 hours at a bath temperature of 170°C until no more water distills off.

The product of condensation is then dissolved in 18,565 parts by weight of dimethyl formamide and filtered, following the addition of 500 parts by weight of active carbon. 1,202 parts by weight of antimony trioxide are added to the filtered solution, followed by heating at 140° to 150°C, while nitrogen is passed over, until a clear, pale yellowish solution has formed from the dispersion. After cooling to room temperature, 4,653 parts by weight of 3,4-dichlorophenyl isocyanate are added in portions with evolution of carbon dioxide in such a way that the temperature does not exceed 50° to 60°C.

The antimony compound precipitates on the addition of water. A solution of the complex antimony compound in dimethyl formamide is compatible in any mixing ratios with dimethyl formamide solutions of copolymers of acrylonitrile and vinyl or vinylidene chloride.

Example 5: The complex antimony solutions, A, B, C and D, described in Examples 1 through 4, are added in various quantities to a 35% dimethyl formamide solution of an acrylonitrile-vinylidene chloride copolymer of 58.6% acrylonitrile, 38.5% vinylidene chloride and 2.9% sodium methallyl sulfonate with a K-value according to Fikentscher (*Cellulosechemie 13*, 1932) of 75.3. The resulting polymer solutions with the antimony additives are spun into filaments by conventional dry-spinning processes, stretched in hot water, and washed in the usual way. The results of the vertical burning test according to DIN 53 906 carried out on pieces of knitting produced from the resulting fiber yarns are set out in the following table:

Solution added	Antimony content [%] Original	Found in the piece of the knitting	Vertical burning test according to DIN 53 906 Ignition time [secs.]	Burning time* [secs.]	Distance burnt [cm] max: 35 cm
			3	48	32.5
Comparison test without any addition			3	29	32.5
			3	27	20.0
			15	25	32.5
			15	27	32.5
			15	26	32.5
			3	16	15.0
			3	16	13.0
A	0.5	0.5	3	19	14.5
			15	12	9.5
			15	15	11.0
			15	17	11.5
			3	—	1.0
			3	—	1.0
A	2.0	1.9	3	1	1.5
			15	—	1.5
			15	1	2.0
			15	—	1.0
			3	—	1.0
			3	—	1.5
B	2.5	2.5	3	—	1.0
			15	—	1.5
			15	—	1.5
			15	—	1.0
			3	—	1.5
			3	1	3.5
B	1.5	1.4	3	—	2.0
			15	1	3.0
			15	—	1.5
			15	1	2.0
			3	1	2.5
			3	—	1.5
C	1.8	1.6	3	—	1.0
			15	—	1.0
			15	1	1.5
			15	1	1.5
			3	—	1.0
			3	—	1.0
D	1.7	1.7	3	—	1.0
			15	1	2.0
			15	—	1.0
			15	2	1.5

*following removal of the ignition flame

Antimony Pentoxide Dispersions in Polar Organic Solvents

C.E. Crompton and A.M.Z. Kazi; U.S. Patents 4,017,418; April 12, 1977 and 4,059,540; November 22, 1977; both assigned to Chemetron Corporation describe a process for making a colloidal dispersion of hydrous antimony pentoxide particles which comprises introducing particles of an oxidizable antimony oxide into a polar organic solvent, contacting the particles with aqueous hydrogen peroxide in the presence of an aliphatic α-hydroxy carboxylic acid and a halogen acid at a temperature of 0°C to the decomposition temperature of the reaction mixture for a length of time sufficient to convert at least a portion of the particles to colloidal particles of hydrous antimony pentoxide.

The colloidal sols are formed in a polar organic solvent dispersion medium which contains a quantity of, usually from 5 to 20% by weight water. The polar organic solvents that are useful are formamide, methyl formamide, dimethylformamide, acetamide, methylacetamide, dimethylacetamide, dimethylsulfoxide, formic acid, tetramethyl urea, methanol, ethanol, and the like. The halogen acids that are used are hydrogen chloride and hydrogen bromide. The halogen acid is preferably added to the reaction mixture as an aqueous solution for convenience although gaseous hydrogen halide may be introduced into the reaction mixture if desired.

The concentration of the aqueous solution of hydrogen peroxide which may be employed is not critical. However, a commercially available aqueous solution containing 35% hydrogen peroxide by weight is used. If a colloidal dispersion containing a concentration of water of less than about 10% by weight is desired it is preferable to introduce the hydrogen peroxide as an aqueous solution containing 70% hydrogen peroxide by weight. The stoichiometric ratio of hydrogen peroxide to oxidizable antimony is generally employed.

The concentration of antimony, as antimony metal, in the sols produced in this process is usually from about 0.01 to 10% of the total weight of the dispersion. The colloidal sol which results from the contacting of antimony oxide particles with the hydrogen peroxide is a stable colloidal dispersion of antimony pentoxide particles having an average size of from about 50 to 200 A.

An aliphatic α-hydroxy carboxylic acid is employed as a stabilizer. The carboxylic acid is usually present at a molar ratio of stabilizer to antimony of from 0.01:1 to 4.0:1. Increasing amounts of stabilizer cause corresponding decreasing particle sizes in the products sols until the amount of stabilizer reaches a molar ratio of stabilizer to antimony of 4:1.

The order of addition of the six components of the colloidal dispersion has no effect upon the dispersion produced if the introduction of all of the components into the reaction medium is accomplished within a reasonable length of time. Thus, the antimony component, polar organic solvent, hydrogen peroxide solution, stabilizer, halogen acid and water may be combined in any sequence within a short period of time, such as one minute, and still produce equivalent dispersions of hydrous antimony pentoxide particles.

For use as a flame retardant it is preferable to use submicron sized particles of antimony oxide so that fibers which are treated with the antimony oxide are not delustered or plastics opacified. The colloidal dispersions are well suited for use in synthetic fibers and plastics because the average particle size of the sol is from 50 to 200 A and the sols are compatible with many organic solvents which are used for the spinning and film casting of many polymers. Examples of such solvents are methanol, ethylene glycol, acetic acid, dimethylformamide, dimethylacetamide, formamide, trimethylolpropane, and the like. Examples of polymers which may be treated with the products of this process include polyacrylonitrile, copolymers of acrylonitrile with vinyl and vinylidene halides, polyamides, polyesters, polyurethanes, polyacrylates, polyvinyl chloride, acrylonitrile-butadiene-styrene polymers, and the like. The following examples illustrate this process. In the examples all percentages are by weight, unless otherwise indicated.

Example 1: 30.3 g (0.23 mol) of malic acid is added to 110 g (1.5 mols) of dimethylformamide with stirring. 7.5 g (0.08 mol) of 37% aqueous hydrogen chloride is added to the stirring mixture, followed by the addition of 22.1 g (0.08 mol) of antimony trioxide.

The stirring mixture is heated to 70° to 80°C and 14.7 g (0.15 mol) of 35% aqueous hydrogen peroxide solution is added. The mixture is maintained at 70° to 80°C for 15 to 20 minutes and then cooled.

The product is a clear colloidal dispersion of hydrous antimony pentoxide particles having an average particle size of about 70 A. The sol has a molar ratio of hydrogen chloride to antimony of 0.5:1.

Example 2: 3.7 g (0.04 mol) of an aqueous solution containing 37% hydrogen chloride is added to 111 g (1.5 mols) of dimethylformamide, while stirring. 30.5 g (0.23 mol) of malic acid is added to the stirring solution followed by 22.1 g (0.08 mol) of antimony trioxide. The reaction mixture is heated to 60° to 70°C and 17.2 g (0.15 mol) of 30% aqueous hydrogen peroxide solution is added to the stirring mixture.

The product is a colloidal dispersion of hydrous antimony pentoxide particles. The sol has a molar ratio of carboxylic acid to antimony of 1.5:1 and a molar ratio of halogen acid to antimony of 0.25:1.

Example 3: 3.7 g (0.04 mol) of an aqueous solution containing 37% hydrogen chloride is added to 129 g (1.8 mols) of dimethylformamide, while stirring. 10.2 g (0.08 mol) of malic acid is added to the stirring solution, followed by 22.1 g (0.08 mol) of antimony trioxide. The reaction mixture is heated to 60° to 70°C and 17.2 g (0.15 mol) of 30% aqueous hydrogen peroxide solution is added with stirring. 10.2 g (0.08 mol) of malic acid is added to the stirring solution, followed by 22.1 g (0.08 mol) of antimony trioxide. The reaction mixture is heated to 60° to 70°C and 17.2 g (0.15 mol) of 30% aqueous hydrogen peroxide solution is added to the stirring mixture.

The product is a colloidal dispersion of hydrous antimony pentoxide particles. The sol has a molar ratio of carboxylic acid to antimony of 0.5:1 and a molar ratio of halogen acid to antimony of 0.4:1.

Example 4: 17.2 g (0.15 mol) of 30% aqueous hydrogen peroxide solution is added to 107.5 g (1.24 mols) of dimethylacetamide with stirring. 7.5 g (0.08 mol) of 37% aqueous hydrogen chloride is added to the stirring mixture, followed by the addition of 30.3 g (0.23 mol) of malic acid. The stirring mixture is heated to 79° to 87°C and 22.1 g (0.08 mol) of antimony trioxide is added.

The product is a colloidal dispersion of hydrous antimony pentoxide particles having an average size of less than 50 A. The sol has a molar ratio of carboxylic acid to antimony of 1.5:1 and a molar ratio of halogen acid to antimony of 0.5:1.

Antimony Pentoxide Dispersions in Polyhydroxy Alcohols

In related work on colloidal dispersions of antimony pentoxide, *C.E. Crompton and A.M.Z. Kazi; U.S. Patent 3,994,825; November 30, 1976; assigned to Chemetron Corporation* describe a process which comprises introducing particles of antimony trioxide into an aliphatic polyhydroxy alcohol having vicinal hydroxyl groups and contacting the particles with aqueous hydrogen peroxide at a temperature of from 0°C to the decomposition temperature of the reaction mixture. The contacting of the particles of antimony trioxide with aqueous hydrogen peroxide may also be done in the presence of a halogen acid selected from the group consisting of hydrogen chloride, hydrogen bromide and hydrogen iodide.

The colloidal sols are formed in a polyhydroxy alcohol containing a quantity of water. The polyhydroxy alcohols that are useful are glycerol, ethylene glycol, 1,2-propanediol, 1,2-butanediol, 2,3-butanediol, trimethylolpropane, and the like.

Example 1: 11.5 g (0.12 mol) of hydrogen peroxide solution is added over a 20-minute period to a stirring mixture of 100 g (1.67 mols) of ethylene glycol and 29 g (0.10 mol) of antimony trioxide while maintaining the temperature of the mixture at 90° to 95°C. The stirring is continued for an additional 1 hour and 40 minutes and a colloidal dispersion of hydrous antimony pentoxide is obtained. The size of the particles in the dispersion is about 150 to 200 A.

Example 2: 7.5 g (0.076 mol) of 37% hydrochloric acid is added to 140.3 g (2.34 mols) of ethylene glycol, with stirring. 22.1 g of antimony trioxide (0.0758 mol) is added to the stirring mixture, at room temperature, followed by the addition of 14.7 g (0.151 mol) of hydrogen peroxide. The temperature rises to about 90°C and is maintained at such temperature for about 30 minutes. A colloidal dispersion of hydrous antimony pentoxide particles having an average size of about 50 A is formed.

Example 3: 11.5 g (0.12 mol) of hydrogen peroxide solution is mixed with 50 g of ethylene glycol and added to a stirring mixture of 29 g of antimony oxide in 100 g of ethylene oxide at a temperature of approximately 65°C over a 30-minute period. The reaction mixture is allowed to stir for an additional 1 hour. A colloidal dispersion of hydrous antimony pentoxide particles having an average size of 70 to 120 A is formed.

Antimony Pentoxide Sol Composition

R.F. Langere, O.E. Loeffler and T.F. Santini; U.S. Patent 4,026,819; May 31, 1977; assigned to NL Industries, Inc. describe the preparation of an antimony pentoxide sol composition which is stable upon standing and possesses a concentration of antimony pentoxide from 10 to 50%. This process oxidizes antimony trioxide to antimony pentoxide by direct reaction of the trioxide with H_2O_2 without the need for solubilizing the trioxide values.

This sol also may contain up to 5.0% of an amine if desired. When the amine is present, the sol may be dried and redispersed in water or an organic solvent depending upon the amine additive to obtain again the sol composition. The dried powder contains from 85 to 95% antimony pentoxide and from 2.5 to 10% of an amine, the remainder being absorbed or combined water. This type of sol is particularly useful as a commercial product in plastics and textiles for improving the flame retardant properties.

Example 1: In this example an antimony pentoxide sol was prepared without adding any stabilizing additives. 200 g of water were charged into a reactor equipped with a reflux condenser. This water was heated to reflux temperature. Into another tank were added 146 g antimony trioxide, 97 g of 35% hydrogen peroxide and 96 g water. This mixture was stirred at room temperature to produce a well-dispersed antimony oxide slurry.

The antimony oxide slurry was then added slowly to the refluxing water with stirring over a period of 2¼ hours. After all of the antimony oxide slurry was added, the stirring and refluxing were continued for 2¾ hours. After refluxing, the mixture was allowed to cool to 50°C.

The sol obtained was stable upon standing, and it possessed the following properties: 30.0% Sb_2O_5; 11.0 lb/gal bulk; 1.8 pH; 10 cp viscosity; average particle size of 250 Å. The pH of this sol may be adjusted from pH of 2 to 10 without detrimentally affecting the stability of the sol.

The sol was used in a typical pad-on for a textile containing 50% cotton and 50% polyester blend. This application to the fabric imparted superior flame retardant properties to the textile material.

Example 2: This example utilizes the sol prepared in Example 1 in which an organic amine is employed to further increase the stability of the sol. 5.7 g of morpholine were added to the sol over a period of 5 minutes. The mixture was then stirred for 1 hour. The sol produced contained 1% of the amine and the pH was raised to 4.5.

These aqueous sols are commercial products which may be used in these forms to treat various textiles, such as canvas goods, polyester-cotton blends, nylon, and the like. This particular sol was used to treat the textile described in Example 1 and again superior flame retardant properties were obtained.

Using the stabilized sol containing the amine, another product was prepared as a a dry powder by spray drying the sol at 60°C. The spray dried product had the following properties: 91.5% Sb_2O_5; 3.5% amine additive; 4.14 g/cc density; 12 cp (dispersed) viscosity; an average particle size of 250 A (dispersed).

This spray dried powder possesses the ability to reform the aqueous sol by mixing the powder thoroughly with water. The reformed sol possessed the same properties as the sol before drying. This reformed sol was also as effective as a flame retardant agent as the sol before drying. The sol also remains stable over a pH range from 2 to 10. The reformed sol may contain from 10 to 75% antimony pentoxide, from 0.25 to 7.5% of an amine, and the remainder water.

MISCELLANEOUS

Molding Materials of Inorganic Compositions

H. Ueda, N. Suzuki and M. Nagao; U.S. Patents 3,988,402; October 26, 1976; and 4,013,476; March 22, 1977; both assigned to Sekisui Kagaku Kogyo KK, Japan describe a noncombustible molding composition which is capable of hardening after being dehydrated to a moisture content of 5 to 50%. This comprises:

(a) 100 parts by weight of a hydraulic inorganic mixture consisting of an inorganic substance composed predominantly of silica and an inorganic substance composed predominantly of calcium oxide;

(b) 10 to 200 parts by weight of mineral fibers; and

(c) a substance selected from the group consisting of 0.5 to 5 parts by weight of bitumen, 1 to 15 parts by weight of crystalline aluminum oxide, 1 to 10 parts by weight of sulfur, 1 to 10 parts by weight of metal sulfide, and 1 to 10 parts of vanadium oxide.

Various conventional methods can be utilized to perform the mixing of the above ingredients. One method comprises feeding a mixture of components (a) and (b) into a cylinder, while kneading the mixture, feeding an aqueous dispersion of the component (c) into the cylinder, and continuing the kneading to distribute the dispersion uniformly in the mixture. Specifically, this method is performed by using an extruder whose partial cross-sectional view is shown in Figure 10.1.

FIGURE 10.1: CROSS-SECTIONAL VIEW OF AN EXTRUDER

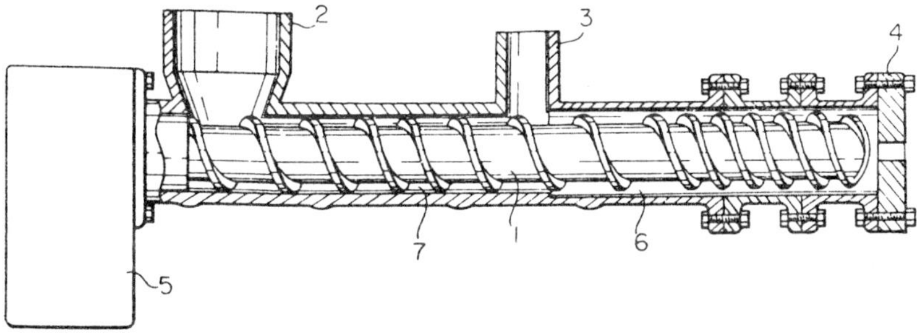

Source: U.S. Patent 3,988,402

Referring to the figure, mineral fibers are added to a powdery hydraulic inorganic mixture, and they are thoroughly mixed in a mixer. A desired amount of the resultant mixture is poured into a material feed inlet **2** of the extrusion-molding machine.

The mixture is then fed into a cylinder **7** and uniformly kneaded by the rotation of a screw **1** by a drive mechanism **5**. During this time, the mixture is being transported towards an extrusion outlet opening **4** formed at the forward end of the cylinder. About midway between the extrusion outlet opening and the inlet opening is provided a fluid feed inlet **3** adapted to feed an aqueous dispersion of the bitumen, the crystalline aluminum oxide, sulfur, metal sulfide and/or vanadium oxide into the cylinder. The dispersion is fed into the cylinder from a reservoir tank (not shown) through a conduit (not shown) by the action of a pump. It is preferred that the amount of the dispersion should be 50 to 160 parts by weight of the hydraulic inorganic mixture. Thus, water or the dispersion fed from the fluid opening is uniformly distributed in the composition consisting of the hydraulic inorganic mixture and mineral fibers, and the composition is transported in the cylinder while being continuously kneaded, and extruded from the extrusion outlet opening before the completion of hardening of the hydraulic inorganic mixture.

According to this method, the mineral fibers are oriented in the extrusion direction, and, therefore, a noncombustible molding material having very high flexural strength in the extrusion direction can be obtained. The surface of the resulting material is smooth.

The molded article obtained has very high flexural strength, low water imbibition and insignificant fuming, flaming and after-flaming properties. It can be formed into any desired shape, such as plate-like or curved, and is useful as wall, flooring and ceiling materials, or other articles which require noncombustibility.

The process is described by the following examples in which all parts are by weight. The various properties are determined by the following methods.

Noncombustibility: A 22 x 22 cm sample is burned for 3 minutes with a city gas burner in which air is fed at a rate of 1.5 l/min. Thereafter, it is additionally burned with an electric heater with nichrome wire having a capacity of 1.5 kWh for 17 minutes. The fuming, flaming and after-flaming are observed and the noncombustibility is determined from an overall consideration of the results of the observation.

Flexural Strength: A 50 x 120 x 30-80 mm sample is used. The measurement is made using an autoclave in a constant humidity and temperature room at 20±1°C and a humidity of 65±5%.

Impact Strength: Measured in accordance with JIS-A-5410.

Water Imbibition (absorption rate): Measured in accordance with JIS-A-5905.

Heat Shrinkage: On a sample, 30 x 200 x 60±1 mm are drawn indicators 100 mm long (measured with a microscope). The specimen is allowed to stand for 30 minutes in a gear oven at 100°C, removed from the oven and measured with a microscope.

Dimensional Variation: A sample, 30 x 200 x 60±1 mm, is allowed to stand at a temperature of 20±1°C and at a humidity of 65±1% for 1 hour, and its dimensions are measured. It is used as a blank. It is dipped in water at 20°C for 24 hours and its dimensions are remeasured. Percentaqe of dimension variation is calculated by the following formula:

$$\%\ \text{dimension variation} = \frac{\begin{array}{c}\text{dimension of sample}\\ \text{after 24 hr dipping}\end{array} - \begin{array}{c}\text{dimension}\\ \text{of blank}\end{array}}{\text{dimension of blank}} \times 100$$

Example 1: 800 parts of water were added to 100 parts of an incombustible molding material of the following composition: 57 parts diatomaceous earth, 43 parts slaked lime, 50 parts asbestos, 1, 2 and 5 parts coal tar. The diatomaceous earth was made up of: 78.56% SiO_2, 13.21% Al_2O_3, 4.11% Fe_2O_3, 1.51% CaO, 1.25%, MgO, 2 to 3% H_2O. It had 1.11% heat ignition loss, and a particle size of less than 300 mesh (Tyler).

The mixture was thoroughly kneaded in a kneader and put into a mold, 910 x 1,820 x 6 mm. The kneaded mixture was then molded under pressure, 200 kg/cm^2, and then aged under atmospheric pressure at 80°C. The mixture was then aged in an autoclave for 10 hours at 180°C under 7 atmospheres and hardened. The resulting molded plates were thoroughly dried and then flexural strength, impact strength and moisture absorption rate measured. Flexural strength and impact strength were excellent with low moisture absorption rate and excellent noncombustibility.

Examples 2 through 6: 800 parts water were added to 100 parts of a noncombustible molding material of the following formulation: 50 parts quartz rock, 50 parts slaked lime, 50 parts asbestos and 1, 2, 5, 10 and 15 parts crystalline aluminum oxide. The quartz rock was composed of 99.4% SiO_2, 0.44% Al_2O_3, 0.04% Fe_2O_3, and had 0.22% heat ignition loss.

The mixture was fully kneaded in a kneader and put into a 910 x 1,820 x 6 mm mold. The mixture was molded at a pressure of 200 kg/cm^2 to reduce its moisture content to 40%. The molded article was aged for 10 hours by heating at 80°C at atmospheric pressure and then 10 hours in an autoclave at 180°C and 7 atmospheres.

The resultant mixture was thoroughly dried and then observed by x-ray diffraction analysis and microscopic examination. It was confirmed that tobermolite and aluminum tobermolite were formed in the molded article. The flexural strength of the molded product was measured and the results are given in the table below. The molded articles were inferior in fuming, flaming and afterflaming properties and excellent in noncombustibility.

Examples	Amount of crystalline aluminum oxide (parts)	Flexural strength of molded articles (kg/cm^2)	Non-combustibility
2	1	370	Excellent
3	2	500	"
4	5	450	"
5	10	390	"
6	15	300	"

Fire Extinguishing Liquid Emulsion

R.H. Pierrou; U.S. Patent 3,975,277; August 17, 1976; assigned to Nitro-Nobel AB, Sweden describes a liquid composition which can be used as a fire-retarding and fire-extinguishing agent. The composition comprises in parts by volume:

(a) 1 part of a mixture of 1 part by volume of a plastic in 0 to 10 parts by volume of water, where the plastic is polyvinyl acetate, polyvinyl acrylate, polyvinyl chloride, polystyrene or copolymers of two or more of the plastics;

(b) 1 part of a solution of 0.5 to 3 parts by weight of an alkali metal sulfate in 1 to 10 parts by weight of water;

(c) 0.5 to 1 part of a solution of 1 part by weight of an alkali metal acetate in 4 to 20 parts by volume of water;

(d) 0.5 to 1 part of a solution of 0.25 parts by weight of an aluminum compound selected from aluminum fluoride, aluminum fluorosilicate, and alum in 10 to 40 parts by volume of water;

(e) 0.5 to 1 part of a solution of 0.5 to 3 parts by weight of an alkali metal thiosulfate in 1 to 10 parts by volume of water;

(f) 3 to 5 parts of waterglass of 38 to 42°Be;

(g) 0 to 10 parts of finely divided kaolin; and

(h) 0 to 10 parts of crushed pumice slag.

If desired, 1 part by volume of 6 to 60% perchloric acid may be added to 50 parts by volume of parts (a), (c), (d) and (f) in place of parts (b) and (e).

When this fire-retarding liquid is applied to, for example, wood, particle boards, fiberboard (porous as well as hard), they become highly incombustible. Also, paper and fabric are given a certain protection against ignition. Tests show that, for instance, particle board and board of different qualities, which have been immersed in or painted with the fire-retarding liquid of this process cannot burn. They can be burnt but not destroyed by fire. As soon as the flame is removed from the board, the fire will go out immediately. The effect of a heat shield is obtained. The fire-retarding liquid can be sprayed or coated on board already placed on walls or ceilings; it can also be applied to wallpaper at paper-hanging, being used as wallpaper paste in the usual manner.

This composition is especially suitable as a fire-retarding liquid for porous board. It is applied to sheets of porous board in the process line immediately after the last pair of rolls before the cutter and the dryer in an amount of 1 liter per square meter. This provides a fire protection approved according to British Class 1 and Swedish Class 2. Double-faced coating gives a better result.

Magnesium-Zinc Complex Salt

In a process described by *E.L. White, W.E. Robertson and J. Schwarcz; U.S. Patent 3,996,142; December 7, 1976; assigned to NL Industries, Inc.* magnesium oxide and a zinc salt are reacted with one another to form a magnesium-zinc complex salt which is added to a halogen-containing plastic composition to serve as a smoke and/or fire retardant during combustion of the plastic. This magnesium-zinc complex salt is normally added along with an antimony compound but may be used alone if desired.

These complex salts are useful for treating substantially any plastics. These plastics which are particularly applicable include step-growth polymers such as polyesters, polyamides, polyurethanes, polyanhydrides, polyacetals, phenol-formaldehyde resin, urea-formaldehyde resin, melamine-formaldehyde resin, epoxy resins, polyethers, polyethylene terephthalate, poly(arylene ethers), poly(arylene alkylidenes), polycarbonates, poly(carbodiimides); and chain-growth polymers such as polyethylene, polypropylene, polybutadiene, polystyrene, acrylonitrile-butadiene-styrene terpolymer or polyblend, polyvinylidene chloride, polyvinyl chloride, polyisoprene, polyvinyl acetate; acrylate polymers such as polymethyl methacrylate, polymethacrylic acid, and the like. With those polymers which do not contain halogen in their structure, an organic halogen-containing additive must be added.

Example 1: Preparation of a Basic Magnesium-Zinc Sulfate Complex — A normal zinc sulfate solution was reacted with a slurry of magnesium oxide to produce the magnesium-zinc complex salt. The mol ratio of magnesium oxide to zinc sulfate was 3.3 to 1. The normal zinc sulfate solution was prepared as follows:

62 g of 97.7% H_2SO_4 were added to 125 ml H_2O in a beaker immersed in an ice bath. The solution was cooled to 18°C. 50.4 g of zinc oxide were added to the bath over a 5-minute period. The temperature rose from 18° to 60°C. The solution was then allowed to cool to room temperature.

A slurry of magnesium oxide was prepared by adding 94 g of 88.4% magnesium oxide to 400 ml H_2O. The zinc sulfate solution was then added dropwise to the magnesium oxide slurry with agitation over a period of 50 minutes, during which time a thick slurry was formed. The slurry formed was then agitated for 5 hours. The pH was 9.7. The complex salt was then recovered by filtration, drying 18 hours at 105°C and hammer milling through a 0.02-inch screen. This complex salt obtained weighed 218 g and had the following composition: 18.4% zinc, 21.2% magnesium, 1.2% loss (2 hours at 105°C), and 3.10 mols magnesium/zinc.

This magnesium-zinc complex salt was used to treat a polyvinyl chloride composition as follows: 100 parts of polyvinyl chloride were mixed thoroughly with 50 parts of dioctyl phthalate, 1.5 parts of modified dibutyltin maleate, 3 parts antimony trioxide, and 3.5 parts of basic magnesium-zinc sulfate. After thorough mixing, the mill was set at 350°F and the plastic mixture was added to the rolls and milled to obtain a plastic sheet 40 mils in thickness. The plastic sheet was then tested for the following properties: maximum smoke density (Dm), Oxygen Index (O.I.), verticle bar test (UL-94) and oven heat stability. The operational details and results obtained are recorded in the table on the following page.

A control run (Control A) was carried out in which a mechanical mixture of magnesium oxide and zinc sulfate was used instead of the complex salt. The amounts employed were equal to the amounts used above in Example 1. The results were also recorded in the table. It is evident that the use of the complex employed in Example 1 is superior to the control, which used a mixture of magnesium and zinc compounds instead of the complex of Example 1.

Another control run (Control B) was carried out using a mechanical mixture of magnesium sulfate and zinc sulfate instead of using the complex salt of Example 1. The results are also recorded in the table. This Control B run, along with Control A run produced inferior results to those obtained when the complex salt was employed.

Example 2: In this example, the procedure of Example 1 was repeated except that basic magnesium-zinc fluoborate was prepared by reacting magnesium oxide with zinc fluoborate. This complex was added to polyvinyl chloride as described in Example 1 along with a control run (Control C) which was a mechanical mixture of magnesium oxide and monobasic zinc fluoborate. The results are also recorded in the table, and superiority is again shown using the complex salt over the mixture of the two compounds.

Example 3: Repeating the procedure of Example 1, the complex basic magnesium-zinc fluosilicate was prepared. This complex was used in polyvinyl chloride along with two controls: Control D, a mixture of magnesium oxide and normal zinc fluosilicate; and Control E, which contained no magnesium or zinc compounds. Control run E produced higher smoke density than any of the preparations in Examples 1 through 3. Again, results show that the plastic containing the complex is superior to the controls.

Examples 4 through 9: In these examples various magnesium-zinc complex salts were prepared according to Example 1, and these complexes were used as additives to another polyvinyl chloride composition. This plastic composition contained the following ingredients: 100 parts polyvinyl chloride, 50 parts dioctylphthalate, 2 parts barium/cadmium stabilizer, 3 parts epoxidized oil, 0.5 part stearic acid, and 3 parts antimony trioxide.

The results of these examples are also recorded in the table, along with two controls: Control F, a mechanical mixture of magnesium oxide and zinc sulfate; and Control G, which contains no magnesium or zinc compounds.

Example No.	Plastic Used	Agent Used	Amount of Complex PHR	Moles Mg-Zn	Calculated ZnO PHR	Smoke Density (Dm)	Heat Stability Onset of Black Spots	Heat Stability Complete Blackness	Oxygen Index
Example 1	Polyvinyl chloride	Complex Basic Mg-Zn sulfate	3.51	3.1	.75	292	80	> 100	
Control A	Polyvinyl chloride	Mixture ZnSO₄ MgO	1.64 1.40	3.4	.75	393	80	> 100	
Control B	Polyvinyl chloride	Mixture ZnSO₄ MgSO₄	1.64 3.40	3.1	.75	374	65	90	
Example 2	Polyvinyl chloride	Complex Basic Mg-Zn fluoborate	2.97	3.2	.75	282	40	60	
Control C	Polyvinyl chloride	Mixture Zn fluoborate MgO	2.01 1.40	3.3	.75	324	20	20	
Example 3	Polyvinyl chloride	Complex Basic Mg-Zn fluosilicate	3.34	4.4	.75	362	30	70	
Control D	Polyvinyl chloride	Mixture Zn fluosilicate MgO	3.05 1.40	3.2	.75	372	30	30	
Control E	Polyvinyl chloride	None	—	—	—	443	70	90	
Example 4	Polyvinyl chloride	Complex Basic Mg-Zn sulfate	3.51	3.1	.75	334	80	> 90	29.2
Control F	Polyvinyl chloride	Mixture ZnSO₄ MgO	1.64	3.4	.75	390	40	60	
Example 5	Polyvinyl chloride	Complex Basic Mg-Zn sulfate	3.08	2.5	.75	310	50	60	
Example 6	Polyvinyl chloride	Complex Basic Mg-Zn borate	3.77	3.3	.75	303	70	100	
Example 7	Polyvinyl chloride	Complex Basic Mg-Zn borate	3.08	2.6	.75	302	60	80	28.5
Control G	Polyvinyl chloride	None	—	—	—	402	70	> 100	29.0
Example 8	Polyvinyl chloride	Complex Basic Mg-Zn phosphite	3.53	3.3	.82	327		> 100	
Example 9	Polyvinyl chloride	Complex Basic Mg-Zinc phosphite	3.46	3.3	0.82	324		> 100	

3,3′-Sulfonylbis(6-Hydroxybenzene Sulfonic Acid)

J.H. Kozlowski; U.S. Patent 4,046,806; Sept. 6, 1977; assigned to Crown Zellerbach Corp. describes a method of preparing an ammonium or alkali metal disalt of 3,3'-sulfonylbis(6-hydroxybenzene sulfonic acid). 4,4'-dihydroxyldiphenyl sulfone is reacted with a sulfonating agent selected from chlorosulfonic acid and sulfur trioxide in the presence of sulfolane as a solvent to form 3,3'-sulfonylbis(6-hydroxybenzene sulfonic acid). Methanol or ethanol and a base soluble in the alcohol are added to the reaction mixture, the base being selected from hydroxides and salts of weak acids where the cation of the base is ammonium or an alkali metal, the base being present in an amount sufficient to form the disalt of 3,3'-sulfonylbis(6-hydroxybenzene sulfonic acid), and the alcohol being present in an amount sufficient to selectively precipitate the disalt. Disulfonated sulfonyl diphenol (SDP) and its salts are particularly useful as flame retardants. When exposed to a flame, the salts intumesce with the formation of voluminous black foam. They can be incorporated into polymers. The compounds can be included in coatings and incorporated into a water-absorbing material such as wood. The amount of the compound incorporated into a material is preferably from about 5 to 20% by weight.

Example 1: 1 mol of chlorosulfonic acid was added slowly to 1.5 mol of SDP with stirring. The mixture, which solidified after a short period of time, was heated overnight at 130°C and then dissolved in 1 liter of methanol. The resultant solution was poured into a solution of 100 g of potassium acetate in 1 liter of methanol. The dipotassium salt of disulfonated SDP precipitated immediately and was recovered at a yield of 88%.

Example 2: A reactor was charged with 1,326.6 g (5.45 mols) of SDP and 890 ml (1,123 g) of sulfolane. The contents were heated, with stirring, to 100°C as most, but not all, of the SDP dissolved. With the heat turned off, 1,331 g (11.42 mols) of chlorosulfonic acid were gradually added over a period of 2 hours. Due to the heat of the reaction, the temperature stayed between 100° and 108°C. The undissolved SDP dissolved within the first half-hour of the addition. Heat was then applied to maintain the temperature at 100°C for 1.5 hours while nitrogen was bubbled through the stirred mixture to remove HCl. The mixture was stirred for another half-hour without heat as the temperature dropped to 90°C and the solution become viscous.

At this point 1.5 liters of methanol at ambient temperature were added to the solution over a period of a half-hour. The contents of the reactor were then poured into 9.8 liters of methanol. A solution of 750 g (86.5% assay) of potassium hydroxide (11.59 mols) in 2.2 liters of methanol was added over a period of a half-hour, followed by the addition of 30 g of acetic acid (buffer). A salt precipitated and was recovered by filtration to produce 95+% of the dipotassium salt of disulfonated SDP.

Example 3: The end of a match stick was momentarily dipped into an aqueous solution of 20% by weight of the dipotassium salt of disulfonated SDP. The match was dried and dipped twice again before final drying. The match was then lit and proceeded to burn until the flame reached the salt, which intumesced and quickly extinguished the flame.

Hydrated Stannic Oxide

I. Obara and K. Yagami; U.S. Patent 4,041,144; August 9, 1977; assigned to Kohjin Co., Ltd., Japan describe a flame-resisting agent to be incorporated into an aqueous spinning dope containing chlorine-containing polymer as a main polymer ingredient for improving the flame resistant property of the fiber spun from the spinning dope. This includes a stannic oxide hydrate being substantially amorphous with respect to x-ray diffraction and having bound water in the range of 0.5 to 0.7 mol of H_2O per 1 mol of SnO_2.

The process comprises first forming a slurry containing β-stannic acid having a free water content of not more than 70% by weight. The slurry is adjusted to a pH in the range of 2 to 4.5 and then heated to a temperature in the range of 80° to 130°C in an atmosphere having a humidity of not less than 0.5 kg of steam to 1 kg of dry air. The resulting stannic oxide hydrate product is recovered, the product having a free water content of from 5 to 25% by weight and bound water in the range from 0.5 to 0.7 mol of H_2O per 1 mol of SnO_2.

The examples of the chlorine-containing polymer include homopolymers of chlorine-containing polymerizable monomers such as vinyl chloride and vinylidene chloride; copolymers of more than two of the monomers; copolymers of each of the above monomers and other polymerizable monomers such as vinyl acetate, vinyl propionate, acrylonitrile, vinyl acrylate, diallyl phthalate and divinylbenzene; and chlorinated polymers such as chlorinated polyethylene, chlorinated polypropylene and chlorinated polyvinyl chloride. These polymers may be employed alone or in combination.

Generally, the amount of the stannic oxide hydrate added to the aqueous spinning dope is preferably about 1 to 3.5% by weight as tin, based on the total amount of the polymers employed.

Example 1: A metallic tin having a comparatively wide surface was added to a 33.5% by weight nitric acid maintained at 100°C and the resultant mixture was agitated at a temperature of not less than 100°C for 6 hours to oxidize the metallic tin. The obtained aqueous suspension was regulated to pH 3 with aqueous ammonia. The precipitate was washed with water by repeated decantation and the last supernatant was sufficiently removed to give a slurry of β-stannic acid having a solid concentration of 40% by weight. The β-stannic acid in the slurry had a free water content of 39% by weight and an impurity content of 0.9% by weight.

The obtained slurry was fed into a closed kettle provided with an agitator and a steam jacket and subjected to wet-heat treatments under various conditions. The humidity in the kettle was regulated by determining the amounts of steam and air exhausted from the kettle and regulating the amounts of heated steam and air to be fed into the kettle. The pH value, treatment temperature, treatment period and humidity employed in the wet-heat treatment are shown in the table below. Thus, stannic oxide hydrates having various bound water contents were prepared.

Experiment		1	2	3	4	5	6	7 (β-stannic acid)
Wet-heat-treatment	pH	3.0	3.2	2.8	3.0	3.2	2.8	—
	Temperature (° C.)	130	130	120	120	100	85	—
	Period (hr.)	16	10	10	6	4	3	—
	Humidity (kg. of steam/1 kg. of dry air)	2.0 to 10.0	2.0 to 10.0	2.0 to 10.0	2.0 to 10.0	0.7 to 1.5	0.5 to 1.0	—
Number of bound water		0.39	0.48	0.55	0.64	0.71	0.83	1.0
Averge particle size (μ)		0.81	0.48	0.45	0.42	0.43	0.41	0.40
Aggregating property	Average particle size of aggregated PVA (μ)	12	5.4	4.5	4.9	6.8	13	25
	Weight fraction ≥20μ of particles (% by weight)	29	15	8	11	17	33	65
Filtering property (KW value)		270	180	120	100	110	150	210

The obtained stannic oxide hydrates and β-stannic acid employed as a starting material were pulverized by means of a sand grinder for 8 hours. The pulverized substances were subjected to the determination of average particle sizes. Determinations of aggregating property and filtering property of the pulverized substances were made in an aqueous solution of polyvinyl alcohol. The results are shown in the above table.

The average particle size was determined with a light-transmission type device for determining particle size distribution by centrifugal sedimentation (SKN model, Seishin Kogyo KK). As a dispersing agent, a 2% by weight aqueous solution of sodium tripolyphosphate was employed.

The aggregating property was estimated as follows: 100 parts by weight of a 15% by weight aqueous dispersion of the stannic oxide hydrate (or β-stannic acid) and 15 parts by weight of 10% by weight aqueous solution of polyvinyl alcohol having a degree of polymerization of 1,700 and a degree of hydrolysis of 98.5% by mol were admixed and heated with agitation at 145 rpm at 80°C for 2 hours. The resultant solution was diluted with distilled water to give a sample solution.

The particle size distribution of the sample solution was determined by the abovementioned device for determining particle size distribution. From the obtained particle size distribution, there was calculated the average particle size of the particles in the sample solution (referred to as average particle size of aggregated PVA) and the weight fraction of particles having a particle size of not less than 20 μ among the whole particles in the sample solution (referred to as weight fraction of particles ⩾20 μ).

The filtering property was estimated as follows: 100 parts by weight of a 16% by weight aqueous solution of polyvinyl alcohol having a degree of polymerization of 1,700 and a degree of hydrolysis of 98.5% by mol, and 5 parts by weight of a 16% by weight aqueous dispersion of the stannic oxide hydrate were admixed and heated with agitation at 80°C for 3 hours to give a sample solution. The obtained sample solution was filtered according to a constant pressure filtering method employing a pressure of 2 kg/cm^2 and Polyfron Paper having an average pore size of 20 μ, and at the same time KW value was determined. Small KW value means that filtering property is good and hence operating efficiency is high.

As shown in the table on the preceding page, in case of the stannic oxide hydrate having a number of bound water less than about 0.5, its pulverizing property is poor, lowering its production efficiency; and the filtering property of the aqueous solution of polyvinyl alcohol admixed with the stannic oxide hydrate is poor; and average particle size of aggregated PVA is large. In the case of stannic oxide hydrate having a number of bound water of more than about 0.7, average particle size of aggregated PVA is large and the filtering property is poor. Accordingly, the condition that the number of bound water of stannic oxide hydrate is within the range of 0.5 to 0.7 is essential for obtaining a stannic oxide hydrate having a good dispersibility.

Example 2: Into an autoclave were supplied 32 parts by weight of vinyl chloride, 2 parts by weight of polyvinyl alcohol having a degree of polymerization of 500 and a degree of hydrolysis of 99.9% by mol, 0.07 part by weight of potassium persulfate, 0.9 part by weight of sodium lauryl sulfate, and 65 parts by weight of demineralized water. The mixture was agitated at 45°C for 6 hours to give an aqueous emulsion containing as main polymer ingredients a homopolymer of vinyl chloride and a graft polymer in which a part of vinyl chloride was grafted to polyvinyl alcohol (the concentration of the total polyvinyl chloride component including the polyvinyl chloride unit in the graft polymer in the emulsion: 30% by weight; the concentration of the vinyl chloride homopolymer in the emulsion: 26.9% by weight; the concentration of the graft polymer in the emulsion: 4.7% by weight; the ratio of the amount of the grafting polyvinyl chloride unit and the amount of the polyvinyl alcohol in the graft polymer: 66:34 by weight).

To 300 parts by weight of the aqueous emulsion were added 467 parts by weight of a 18% by weight aqueous solution of polyvinyl alcohol having a degree of polymerization of 1,650 and a degree of hydrolysis of 99.5% by mol and 30 parts by weight of a slurry of the stannic oxide hydrate prepared in Example 1 (the concentration of the stannic oxide hydrate: 12% by weight as tin) to give an aqueous spinning dope at 78°C. The stability of the spinning dope was good and the filtering property was also good (the specific KW value: 1.1).

The spinning dope was extruded into a solution of salt cake. The spinning pressure was 1.5 kg/cm^2 and breaking of single filaments occurred infrequently. The spun filaments were successively passed through a saturated solution of salt cake maintained at 95°C and washed with water. After drying, the filaments were subjected to heat stretching and heat treatment to give filaments consisting of polyvinyl chloride, polyvinyl alcohol and stannic oxide hydrate, 50:50:2, as tin, by weight.

The filaments were dipped in an acetalization bath consisting of 14 parts by weight of sulfuric acid, 11 parts by weight of salt cake, 6 parts by weight of formaldehyde, and 69 parts by weight of water at 70°C for 60 minutes and pressed to remove the bath liquid. The resultant was sufficiently washed with hot water of 40°C and neutralized and again washed with room-temperature water. The washed filaments were subjected to scouring treatments such as oiling and cut to give staple fibers of two deniers. The staple fibers were spun and the obtained yarns were woven into a fabric.

The fabric passed completely the flame resistance test provided in DOC FF 3-71. The fabric was subjected to the vertical flame resistance test provided in AATCC-34-69. In this test, a part of the fabric in contact with an igniting flame was decomposed and carbonized, but a flame did not remain after the igniting flame was removed and no glow was generated. The fabric was also subjected to determination of Oxygen Index according to the method provided in ASTM D-2863, to give an O.I. of 33.4%. This Oxygen Index shows a sufficient self-fire extinguishing property as a flame resistant fabric.

The fabric was subjected 50 times to washing tests provided in AATCC-135-70 and 10 times to dry cleaning tests provided in AATCC-85-63. The O.I. of the fabric after washing is 33.2% and 33.3% after dry cleaning tests.

MISCELLANEOUS PROCESSES

ASPHALT BASED ROOFING SHINGLES CONTAINING GRANULAR GLASS

S.A. Rashid, H.S. Bray and E.J. Boarini; U.S. Patent 4,082,885; April 4, 1978; assigned to United States Gypsum Company describe a fire resistant and weathering resistant shingle or surface covering material ecologically desirable in its utilization of disposable glass. The shingle has an asphalt coating with specified characteristics. The asphalt coating contains from 10 to 55% by weight granular glass and preferably a combination of granular glass and nonglass granular mineral filler a major portion of which is granular glass, with each having a specified particle size distribution.

Example 1: A conventional felt of the type normally used in asphalt shingles was saturated with asphalt of a standard type to produce the base for a conventional asphalt roofing shingle. A collection of disposable bottles was ground to give a particle size distribution (on a U.S. Standard Screen) as follows:

Cumulative	Weight Percent
On 60 mesh	19.7
On 80 mesh	28.05
On 100 mesh	34.25
On 140 mesh	46.00
On 200 mesh	46.10
On 325 mesh	78.30
Through 325 mesh	98.90

Asphalt having a penetration of about 30 mm at 77°F and a softening point of about 235°F was heated until it was soft and workable. The ground glass was thoroughly mixed with the asphalt to give a filled asphalt coating material. The asphalt coating material was then applied to the saturated felt base until the filled asphalt comprised about 40% by weight of the finished material. Upon cooling, the resulting asphalt shingle had good fire resistance properties. Weather-Ometer tests were conducted on the shingle according to the test of ASTM-D529 (1971). Using 26 or more pinholes on a standard test specimen of 2¾" x 5⅞" as a failure point, the shingle reached 166 average cycles before failure, an improvement of 67.8% over the same shingle not having a filler and an improvement of 13.15% over a shingle using only a carbonate as a filler.

Example 2: Example 1 was repeated with the exception that the disposable glass was replaced with a commercially produced glass not made into bottles but ground to have a particle size distribution as follows:

Cumulative	Weight Percent
On 60 mesh	4.50
On 80 mesh	12.80
On 100 mesh	20.65
On 140 mesh	37.45
On 200 mesh	48.30
On 325 mesh	78.10
Through 325 mesh	99.75

Again, the resulting shingle had good fire resistance and weathering properties, matching the Weather-Ometer tests in Example 1.

Example 3: A conventional felt of the type normally used in asphalt shingles was saturated with asphalt of a standard type to produce the base for a conventional asphalt roofing shingle. A collection of disposable bottles was ground to give a particle size distribution as in Example 1. A limestone carbonate was divided until it had a particle size distribution as follows:

Cumulative	Weight Percent
On 60 mesh	1.80
On 80 mesh	5.80
On 100 mesh	10.70
On 140 mesh	24.25
On 200 mesh	35.60
On 325 mesh	78.20
Through 325 mesh	98.25

Asphalt having a penetration of about 30 mm at 77°F and a softening point of about 235°F was heated until it was soft and workable. Six parts by weight of the ground glass and four parts by weight carbonate were mixed and added to the asphalt until it comprised about 50% by weight of the asphalt to give a filled asphalt coating material. The asphalt coating material was then applied to the saturated felt base until the filled asphalt comprised about 40% by weight of the finished material.

Upon cooling, the resulting asphalt shingle had good fire resistance. Weather-Ometer tests were conducted on the shingle according to the test of ASTM-D529. Using 26 or more pinholes as a failure point, the shingle reached 162 average cycles before failure, an improvement of 63.7% over the same shingle not having a filler and an improvement of 6.7% over a shingle using only the carbonate as a filler.

PLASTISOL MOLDING

J.W. Pedlow; U.S. Patent 4,018,983; April 19, 1977 describes electric arc and flame protective sheath or boot material for electrical, particularly for high voltage electrical, cables and complex electrical splices and junctures.

The sheath is precast preferably from a liquid polyvinyl chloride plastisol, preformed into a selected shape of a protective sheath or boot, reinforced and protected against destruction by premixing with a filler of heat resistant fiber containing flame and fire retardant plasticizer as well as fire retardant, preferably intumescent, and heat stabilizing substances. The precast sheath or boot is flexible and elastomeric and may be sized and shaped to be slipped upon the cable or joint protectively to further insulate and protect it against electrical arcing, heat or fire. The following example illustrates the preparation of the protective composition.

Example: Polyvinyl chloride plastisol in quantity of 72 parts mixed with 28 parts of polyvinyl extender resin, a coarser polyvinyl chloride of particle size in the range of about 50 to 100 mesh, is blended in a dough mixer with 92 parts of cresyl diphenyl phosphate, 23 parts of chlorinated paraffin wax (Chlorowax 500) and an extender of 24 parts of heavy aromatic oil, such as a mixture of lower alkyl benzenes (Eseoflex 150) and a heat stabilizer, a mixture of barium, cadmium and zinc thiolate with small quantities of lead naphthenate in quantity of 2 parts, all parts being by weight.

Into the mix is added 3 parts of antimony oxide, 60 parts of hydrous alumina having 38% of water of hydration and 40 parts cenospheres (fly ash). There is further added 2 parts of heat stable fibers which are fire resistant aramide having mixed lengths between ¼" and ⅜".

Finally, 7 parts of titanium dioxide pigment, 1 part of calcium oxide and 1 part of fungicide 10,10'-oxybisphenoxarsene (Vinyzene) are added. The components are mixed in a dough mixer at room temperature, first supplying all but about one-quarter of the plasticizer and extender liquids, and then in three small increments the polyvinyl chloride, first blending one to complete liquidity before the next is added, and thereafter the solids are stirred into the mixture with slow continuous mixing.

A small amount of a final quantity of plasticizer is added as a final thinner with the last added cenospheres. The final fluid mix has a viscosity of about 10,000 to 45,000 cp at 25° to 27°C measured with a number six spindle at about 10 revolutions per minute. A stainless steel mandrel which has a lower portion of a diameter of about 1½ to 3 inches and an upper flared portion with a diameter of about 3 to 6 inches respectively, flared smoothly from one diameter to the other, is preheated to 150° to 210°C and dip coated on its outer surface as it is slowly rotated with the composition as formed above.

A coating having a thickness of about 0.125 inch remains, and it is set by the heat transferred by the hot mandrel into a gelated film. Further heat may be applied to complete fusion by placing the coated part into an oven at 175° to 200°C or immersing in molten salt at 200°C.

On cooling, the film is slit longitudinally by a knife blade and peeled from the mandrel. The boot was mounted about an inner lead sleeve in close surface contact and the assembly was exposed by mounting it above the tip of a flame of a Fisher burner having a tip temperature of 1700°F (plus or minus 50°F). The heat did not melt or sag to expose the lead and therefore the lead did not melt during a 20 minute exposure to this temperature. Moreover, a thermocouple inserted within the lead sheath to measure the temperature during the test ranged from 75° to 195°F well below the melting point of the lead.

FLAME PROTECTIVE TAPE

In related work, *J.W. Pedlow; U.S. Patent 4,018,962; April 19, 1977* describes a flame resistant tape for use as a protective covering for electrical cables, splices, junctures and appurtenant electrical equipment, including high voltage, as well as critical circuits, to fireproof and protect them against electrical arcing, heat and flame damage such as from short circuiting, or from fire developed in neighboring cables and equipment.

The tape comprises a base sheet fabric which is coated with a low melting, easily formed thermoplastic resinous material, filled with a heat stable fiber to inhibit its fluidity before decomposition by heat, arcing and flame; and heat activated foaming, fireproofing and intumescing substances, to convert the coating to a fire and heat protective foam remaining protectively in place about a tape wrapped electrical element such as a cable, even after the base resin or fabric is decomposed by heat. The preparation of the coating is described in U.S. Patent 4,018,983.

To prepare the flame protective tape, the composition is coated by knife-coating the plastisol composition upon a woven fiber, such as a spun polyester terephthalate fiber. The wet-coated fabric base with the coating of about 0.055 inches thick, is passed through an oven at a temperature of 165°C in which it is slowly heated for a period of 20 minutes whereby the plastisol sets up to a strong gel. It is cooled in air, slit in appropriate widths and then wrapped about a cable as a flexible protective arc and fire protective wrapping.

AMIDE POLYPHOSPHATE CONDENSATE

K. Fukuba and Y. Ogushi; U.S. Patent 3,969,291; July 13, 1976; assigned to Sumitomo Chemical Company Limited, Japan describe an intumescent fire retardant coating composition containing, as a fire retardant ingredient and an intumescent solid, (A) a substantially water insoluble amide polyphosphate condensate, (B) a film-forming binder, and (C) a carbonaceous material, with (A), (B), and (C) comprising from 40 to 80% by weight, of the total composition.

The intumescent coating composition is particularly effective as a surface coating on insulated wires to prevent a reduction of insulation ability in electrical lines. The water-insoluble amide polyphosphate condensate used is a polyphosphate containing nitrogen in the form of an amide and can be prepared by heat-condensing together a phosphoric acid-containing material such as ammonium orthophosphate, orthophosphoric acid, polyphosphoric acid, phosphoric anhydride, urea phosphate and a mixture of these, and a nitrogen source such as a cyanamide derivative such as melamine, and dicyanamide in the presence of a condensing agent selected from the group consisting of urea, urea phosphate, etc.

Further, in view of the insolubility in water of the amide polyphosphate condensate produced, urea phosphate and melamine are preferably used to produce the present amide polyphosphate condensate.

The coating amount of the intumescent fire retardant coating composition can be varied depending on the kind and quality of materials to be coated and protected. More than about 300 g/m^2 is required in the case of three layer lauan plywood having a thickness of about 5.5 mm in order to achieve a fire retardancy of Grade 3 in the JIS A-1321 combustion test.

In the case of insulated wires, an amount of more than about 250 g/m^2 is required in order to obtain sufficient fire retardancy. When the wires are distributed through walls or boards, the composition can be packed into those spaces as a fire retardant sealing agent. Further, the composition has the ability to form an excellent intumescent heat-insulating layer and, accordingly excellent fire retardancy can be obtained when it is applied to iron beams or pillars.

Example 1: A substantially water-insoluble amide polyphosphate condensate (N, 31.48 wt %; P_2O_5, 47.27 wt %; percentage elution, 1.31%; pH 8.30 as a 1% solution) was prepared by heat-condensing a mixture of urea phosphate and melamine in a molar ratio of urea:phosphoric acid (as H_3PO_4):melamine of 1:1:0.5 for 3 hours at 260° to 280°C. Using the prepared amide polyphosphate condensate the following intumescent fire retardant coating composition was prepared.

Component	Parts by Weight
Amide polyphosphate condensate	40.0
Pentaerythritol	7.6
Ethylcellulose (1% aqueous solution)	7.6
Antifoam agent (Silicone KM72, The Shin-Etsu Chemical Industry Co., Ltd.)	0.2
Dispersing agent (polyethylene glycol alkylphenyl ether, Noigen EA-143, Daiichi Kogyo Seiyaku Co., Ltd.)	0.2
Ethylene-vinyl acetate copolymer emulsion (about 55% solids, Sumikaflex 510, Sumitomo Chemical Co., Ltd.)	10.4
Melamine resin (Sumimal M 50 w, Sumitomo Chemical Co., Ltd.)	9.0
Hardening agent (organic amine derivative, Sumitex Accelerator ACX, Sumitomo Chemical Co., Ltd.)	3.0
Water	22.0

The composition was applied to a panel of three-layer lauan plywood having a thickness of 5.5 mm at a coverage of 450 g/m^2, and the resulting panel was heat-dried for 90 minutes at 80°C to produce a fire retardant plywood panel. This fire retardant plywood panel was subjected to the test method for incombustibility of internal finish materials and procedures for buildings according to JIS A-1321 (Japanese Industrial Standard), and the results obtained are shown in Table 1.

TABLE 1

Exhaust Temperature, (°C)						Tdθ	Smoking	After-Flame	Cracks in	
1*	2	3	4	5	6	(°C x min)	Coefficient	(sec)	Back**	Judgement
85	95	105	192	245	275	0	11.7	0	not observed	passed the third grade of fire-retardancy

*Elapsed time in minutes
**According to JIS A-1321

Further, in the hot water resistance test prescribed in JAS (Japanese Agricultural Standard) Class 2 (immersion in hot water at 70°C for 2 hours), the condition of the coating surface was good and almost the same as that before immersion.

Example 2: Using the same amide polyphosphate condensate as described in Example 1, the following intumescent fire retardant coating composition was prepared.

Component	Parts by Weight
Amide polyphosphate condensate	45.0
Pentaerythritol	7.0
Ethylcellulose (1% aqueous solution)	5.8
Antifoam agent (same as Example 1)	0.2
Dispersing agent (same as Example 1)	0.2
Ethylene-vinyl acetate copolymer emulsion (about 50% solids, Sumikaflex 751, Sumitomo Chemical Co., Ltd.)	16.0
Melamine resin (same as Example 1)	6.0
Hardening agent (same as Example 1)	2.0
Water	17.8

The above composition was applied to a polyethylene sheath wire of a three-axial type having a size of 5.5 mm² at a coverage of 500 g/m², and the resulting wire was allowed to stand for 2 days at room temperature (i.e., 20° to 30°C). In the incombustibility test, a Meckel burner prescribed in JIS L-1091 was used, and the sample wire was held horizontal to the flame length of 65 mm, and the time required for the sample wire to ignite and other items were measured. Further, the sample wire was immersed in warm water for 2 days at 60°C and, after drying a similar test was conducted. The results obtained are shown in Table 2.

TABLE 2

Sample	Ignition Time	After Flame (sec)	After Glow (sec)
Allowed to Stand in a Room	No ignition for 20 Minutes	0	0
Immersed in Warm Water	"	0	0
Blank (not coated)	Ignited in 30 sec.	∞	—

As seen from the results in Table 2, both the sample which was allowed to stand in a room and the sample which was immersed in warm water exhibited excellent fire retardancy.

TREATMENT OF SOLID WASTE WITH CHLORINATED PARAFFIN AND ANTIMONY TRIOXIDE

G. Dreer; U.S. Patent 4,012,231; March 15, 1977 describes a process for the production of fillers from solid waste which is first ground, magnetically freed from iron particles, supplied with sewage sludge or water, fractionated into portions of different size and treated in a fermentation process for about 24 hours at 65°C while air is continuously fed. While the compost is being fermented, a solvent is added in a ratio of 1 to 2-4 parts by weight, compost to solvent, the treatment continuing for 10 to 16 hours, dried, and to the dried product, chlorinated paraffin and antimony trioxide are added.

Solvents which can be used are carbon tetrachloride, trichloroethylene, ketones (acetone), ether, styrene, xylene, hydrogen peroxide, aromatic hydrocarbons, and alcohols, either by themselves or mixed.

The fillers possess the following characteristics: they are noncombustible, water-repellent, resistant to mold formation, resistant to many acids, mineral oils, low-boiling petroleum distillates, they are stable at temperatures up to 1200°C, unaffected by abrupt temperature changes, resistant to cold, sound-absorbing, insulating with respect to transfer of heat and electricity, and they readily polymerize. In addition, they possess a high filling capacity, i.e., considerable quantities of such fillers can be incorporated into a given quantity of resin. These fillers can be incorporated into natural or synthetic resins for the manufacture of plastics without difficulties whereby the polymerization process is not impeded.

Resins which are filled with the fibrous or granular fillers can be processed to molded articles of all types. The following list cites some examples:

Walls and slabs for exterior and interior construction;
Floors;
Furniture;
Sound-absorbing masses;
Laminated construction elements;
Linoleum-like materials;
Coating compositions (when the fillers have small grain sizes); and
Adhesives (because of the protein content of the fillers).

Example: 3 parts by weight of compost and 5 parts by weight of styrene are introduced into the first heatable tank. As soon as the total batch has been introduced, the heating and stirring is switched off. The following time for ripening amounted to about 10 to 12 hours. As soon as a light-brown color of the mass has been reached, one admixes chlorinated paraffin and antimony trioxide while stirring and further heating. Thereby, the temperature is increased by 100°C within about 15 minutes. Admixed are 2% by weight of chlorinated paraffin and 3% by weight of antimony trioxide.

In order to use the obtained filler for manufacturing, for example, a noncombustible slab to be placed on the exterior surface of the front of a building, one mixes the filler with a synthetic resin in the ratio of 60:30 by weight and adds as well hardeners and accelerators, followed by casting or molding.

FOAMED MAGNESIUM OXYCHLORIDE CEMENT

H.C. Thompson; U.S. Patent 3,969,453; July 13, 1976; assigned to Thompson Chemicals, Inc. describes a method of making a foamed fireproof product of magnesium oxychloride cement. A porous substrate is impregnated with a foaming mixture of magnesium chloride, magnesium oxide, and frothing agent in water. The mixture hardens with small voids throughout the porous substrate, thus providing a fireproof product of low density. The product may be formed in a mold with fibrous elements introduced into the foaming mixture. The fireproof products of relatively low density are particularly valuable for building and construction purposes. Magnesium powder is the preferred frothing agent which, in combination with a surfactant, induces a large volume of small bubbles that remain in the composition as it is set.

Example 1: Simulated roofing shakes were prepared using molds having the configuration of wooden shakes in 2 courses of several shakes per course. 72.2 pounds of 20 mesh slate was blended with 24.8 pounds of magnesium oxide and 3 pounds of iron oxide pigment. After these ingredients are thoroughly mixed, 21.4 pounds of the blended mixture is added to one gallon of 22° Baume magnesium chloride liquid. The liquid and powder are thoroughly mixed and the mixture is sprayed onto plastic molds simultaneously with the application of chopped fiber glass strands at the rate of 0.8 ounces per square foot of mold area. The slate reacts with the cement ingredients to foam the cement in the same manner as magnesium powder does.

The fiber glass strands are cut into lengths between 1 and 2 inches and applied to the molds at the same time as the mixture is sprayed to give simultaneous application. The sprayed molds are stacked and held for 2 hours until they are hard enough to be removed. They are then trimmed and packaged for shipment. The molded product has the appearance of wooden shakes and can be easily applied as a roofing material with hammer and nails. The product is completely fireproof, as contrasted to wooden roofing shakes.

Example 2: In preparation of the froth, 19¼ pounds of magnesium chloride were added to 4¼ gallons of water. The specific gravity was measured and additional magnesium chloride was added to adjust the specific gravity to 22° Baume. The combined total makes about 5 gallons. 200 grams of lactic acid were added to the aqueous mixture. Magnesium oxide was then measured in an amount equal to 5.5 pounds magnesium oxide to each gallon of magnesium chloride in water. Then, 32 grams of powdered magnesium, which is about 0.1% of the dry weight of the magnesium oxychloride cement mixture and 100 grams of surfactant (Rohm & Haas X-100) were added to the magnesium oxide component. The surfactant constituted about 0.5% of the dry weight of the total mixture.

Chopped glass fibers, in an amount equal to 3% of the total were added to the dry component and mixed thoroughly. The dry magnesium oxide component was added to the aqueous magnesium chloride component and the mixture immediately began to froth. It was poured into the female part of a mold having a decorative configuration for a building ornament. The male part of the mold was closed over the female part and the cement permitted to set. After 2 hours the molded fireproof product was removed from the mold.

Example 3: Fireproof products may be molded in a desired configuration. Wooden roofing shakes were simulated by forming a mold by vacuum drawing synthetic resin over wooden shakes laid in courses as in roofing. The inside dimensions of the mold were 30 inches by 30 inches and resembled two courses of cedar shakes. Commercially available aqueous magnesium chloride was diluted with water to 22° Baume. For each gallon of this liquid, 15 pounds of magnesium oxide mixture having the composition on the following page was added.

	Percent
Magnesium oxide (50 min set)	37½
Limestone dust, 80 mesh	30
Limestone dust, 200 mesh	7
Silica flour	10
Asbestos fibers	2½
Diatomaceous silica	4
Pigment (brown and black)	1½
Waterproofed vermiculite	7½
Foaming agent	0.01

The frothing agent mixture was sprayed through an airless spray gun into the mold to completely cover the bottom of the mold. Simultaneously with the spraying of cement, chopped glass fibers were sprayed into the mold in an amount equal to one-half ounce fibers per square foot of mold. Since the bottom of the mold had an irregular surface to resemble the shakes, more frothing cement formed in the depressions than on the raised portion; the mold did have a complete ¼" cover. The rovings were ¾" to 1½" long. The fiber glass provides resiliency and strength, so care was taken to make certain that a uniform layer throughout the area of the mold is provided.

The mold was variable in depth from about ¼" to 1¼". It was filled with the fibers and cement and permitted to set for 2 hours. After the panel had thoroughly set, it was removed from the mold to give a strong, lightweight panel resembling wooden shakes but was completely fireproof.

The weight of the panel was approximately 15 pounds, and had an average thickness of ½ inch. It could be nailed onto conventional roofing surfaces and was completely incombustible upon heating to 1350°F. For molded products simulating wood, a density of between 30 and 40 pounds per cubic foot should be obtained. For most purposes, a density of 35 pounds per cubic foot is optimum for ease of manufacture and for resembling wood product density.

EEL GRASS SEAWEED

A.M. Primo; U.S. Patent 4,016,084; April 5, 1977 describes products having as an essential ingredient a seaweed known as eel grass. Such seaweed has been found to constitute a flame retardant, verminproof, soundproof, rotproof, weatherproof, and bugproof material. The chemical analysis of this seaweed is stated below.

	Percent
Total nitrogen	0.69
Total phosphorus	0.118
Potash (potassium carbonate)	16.29
Potassium	0.118

This material is processed as follows: The seaweed is collected and dried, reduced to form small particles, and mixed with a fluid carrier so that the resulting composition may be utilized by spraying. Such spray may be applied on fabrics, carpets or mixed with paint for paint spraying, in order to apply a coat of a flame retardant and soundproof material.

Other uses for eel grass seaweed include: flame retardant fabrics, incorporation into board-like materials, binder for bricks or blocks, and as a filler for cavities in building walls.

VISCOSITY INDEX MODIFIERS IN FLAME RETARDANT OIL EXTENDED ELASTOMERS

A.W. Morgan and R.G. Ribaudo; U.S. Patent 3,997,493; December 14, 1976; assigned to Monsanto Company describe a composition comprising (A) a rubber, (B) at least about 37.5 phr of an extender oil for the rubber, (C) a flame retardant and (D) from about 0.05 to 5.0 phr of a high viscosity liquid viscosity index modifier for the extender oil, the modifier having an aliphatic nonpolar backbone containing pendant carboxyl groups where the composition does not burn over a maximum of 90 seconds in the SBR Vertical Burn Test. The modifier increases the flash point of the extending oil and enhances the effects of the flame retardant thereby lowering the flammability of the rubber composition.

Representative synthetic rubbery polymers are the homopolymerization products of butadiene and its homologs and derivatives, as for example, methylbutadiene, dimethyl butadiene, pentadiene, and chloroprene (neoprene synthetic rubber), as well as copolymers such as those formed from butadiene, or its homologs or derivatives, with other unsaturated organic compounds.

Extender oils of particular utility are largely composed of naphthenic and alkylated naphthenic hydrocarbons and mixtures of these with various aromatic hydrocarbons. Such oils may be obtained from the high boiling fractions of the so-called naphthenic or mixed based crude oils. They may comprise distillate fractions boiling above about 200°C. Suitable fractions are those at least 90% of which boil above about 250°C.

The viscosity index modifiers which are used in this process, have the common characteristic of enhancing the effect of conventional flame retardants and of increasing the viscosity index of extender oils at high processing temperatures while having no effect on the viscosity of such extender oils or on the viscosity of the rubber composition at normal use temperatures. These materials are aliphatic acids, anhydrides and esters, including anhydrides and esters of polyacids.

Conventional flame retardant materials which are widely used include halogenated and nonhalogenated phosphorus acid esters, e.g., cresyl diphenyl phosphate, tris-β-chloroethyl phosphate, metal salts, oxides and hydrates, e.g., magnesium carbonate, antimony oxide, alumina hydrate, metal borates, ammonium salts and halogenated organic compounds of widely diverse nature such as halogenated aliphatic and aromatic hydrocarbons.

Example: The following master batch is prepared:

Ingredient	Parts by Weight
SBR rubber	100.0
Stearic Acid	5.3
Whiting	30.0
Clay	65.0
Sulfur	3.7

(continued)

Ingredient	Parts by Weight
Zinc oxide	5.0
TMTD*	2.0
MBT**	2.5
"Celogen"***	2.2
Sodium carbonate	4.0
Titanium dioxide	6.0
Alumina hydrate	125.0
"Aroclor 5460"†	12.0
Cresyl diphenyl phosphate	30.0
"Admex 710"††	4.0
Urea	4.0
Oil†††	34.0
	434.7

*Tetramethyl thiuram disulfide
**2-Mercaptobenzothiazole
***p,p'-Oxybis-(benzenesulfonyl hydrazide)
†Halogenated polyphenyl
††Epoxidized soybean oil
†††1200 sec^{-1} viscosity naphthenic processing oil

The master batch is prepared by banding the rubber on a two-roll mill for 2 minutes or until the stock reaches a temperature of 120°F. The solid and liquid additives are blended in a beaker and added to the branded stock in increments while keeping the temperature below 140°F. After the additives are thoroughly blended, the master batch is sheeted off at a stock thickness of about ⅛ to ¼".

Eleven samples of the master batch, each sample comprising six replicates, are prepared. To each of the replicates there is added 30 g of naphthenic extending oil and varying amounts of viscosity index modifiers. The replicate mixtures are Banburyed, sheeted at ⅜" on the mill and cured for 5 minutes at 385°F.

The eleven samples (of six replicates each) are exposed to the SBR Vertical Burn Test. In such test, a replicate sample measuring one inch wide by twelve inches long is vertically mounted in a cabinet. The bottom end of the replicate is exposed for twelve seconds to a 1½" flame from a ⅜" diameter barrel Bunsen burner supplied with natural gas. The sample should not burn over a maximum of 90 seconds.

In the table below, the "seconds to flame out" for each of the eleven samples represents the average "seconds to flame out" of the six replicates of each sample.

		Sample Number										
	Control	1	2	3	4	5	6	7	8	9	10	11
Ingredients, g												
Masterbatch	434.7	434.7	434.7	434.7	434.7	434.7	434.7	434.7	434.7	434.7	434.7	434.7
Acryloid 710*	–	0.64	–	–	–	–	1.28	0.32	–	–	–	–
Santolube 900**	–	–	0.64	–	–	–	–	–	1.28	0.32	–	–
PBSA***	–	–	–	0.64	–	–	–	–	–	–	1.28	0.32
PBSAE†	–	–	–	–	0.64	–	–	–	–	–	–	–
Modaflow††	–	–	–	–	–	0.64	–	–	–	–	–	–
Seconds to flame-out	†††	47	16	20	7	2	60	2	33	1	25	50

*Polyacrylic acid ester.
**Polybutenyl succinic anhydride ester.
***Polybutenyl succinic anhydride.
†Polybutenyl succinic anhydride ester.
††Liquid acrylic resin.
†††Did not self-extinguish.

The above data illustrate the increased flame retardant characteristics of oil-extended, flame retarded SBR rubber containing viscosity modifiers.

Comparable results are attained when using the viscosity index modifiers in the table throughout the range of from 0.05 to 5.0 parts modifier to 100 parts rubber. The effectiveness of the modifiers is greatest at low concentrations, decreases rapidly at middle and higher concentrations. Even at lowest effectiveness, however, samples pass the SBR Vertical Burn Test (i.e., the samples do not burn over a maximum of 90 seconds) in contrast to the Control which did not self-extinguish.

Similar results are attained when oil-extended and flame retarded SBR rubber is replaced by other oil-extended, flame retarded rubbers, such as butyl rubber, polymers of isoprene, butadiene and chloroprene, nitrile rubber, isoprene/acrylonitrile rubber, polyisoolefins such as polyisobutylene and natural rubber as well as natural and synthetic rubber latices.

THERMALLY PROTECTIVE MATERIALS

A process described by *F.A. Marion and H.J. McSpadden; U.S. Patent 4,001,126; January 4, 1977; assigned to Universal Propulsion Co.* relates to materials having properties of withstanding temperatures in the thousands of degrees for extended periods of time without gross decomposition.

The materials constituting this process include binders and fillers mixed with the binders. The materials form a char when subjected to the elevated temperatures. The surface of the char glows in the luminous to incandescent range and rejects incoming heat by thermal radiation. The fillers have the properties of decomposing through a number of successive steps, each step advanced in temperature from the previous step by a relatively small amount. The gaseous products of such successive decompositions flow through the char, which is porous so as to provide transpirational heating. The resultant gases produced at the surface of the material provide a gaseous barrier to absorb heat and to transport heat away from the surface of the material.

The binder may be formed from a mixture of a polysulfide and an epoxide. A preferred range is 70 to 30 parts by weight of polysulfide to 30 to 70 parts by weight of epoxide. Typically the two may be approximately equal in weight, or the epoxide may have a percentage by weight in the order of 40 to 45 parts per 100 parts of binder and the polysulfide may comprise the balance of the binder. An especially preferred amount is 57.5% by weight of polysulfide and 42.5% by weight of epoxide. The ratio is dependent upon the properties desired for the binder.

For example, the epoxide tends to be brittle and glass-like with high tensile strength and low elongation and is subject to shock. The polysulfide tends to provide a high elongation and low tensile strength. It is flexible and rubbery and neither glass-like nor plastic. Variations in the ratios of the polysulfide and the epoxide will affect the physical and structural properties of the cured binder matrix and the rate of cure. The polysulfides are mercapto-terminated polymers of the general formula $HS(R-SS)_nRSH$ where R is a polyvalent organic radical containing at least one methylene group and n is an integer of from 3 to 25. Preferred polysulfides include those in which R is $-C_2H_4OCH_2OC_2H_4-$.

The polysulfides may be prepared by condensation of an alkali metal polysulfide, e.g., sodium polysulfide, with an organic dihalide such as dichlorodiethyl formal, ethylene dichloride, or dichloroethyl ether, as described in *Industrial and Engineering Chemistry,* Volume 43, pp. 324-8 (1951). Small proportions, e.g., 0.5 to 2% of trichloropropane or other polyhalides are often included with the dihalide. The polysulfides range in property from mobile to viscous liquids to solids at room temperature, depending on molecular weight. Those of liquid form are ordinarily preferred.

The epoxides, also referred to as glycidyl polyether resins, are epoxy-terminated polymers of the general formula

where R' is the divalent organic radical of a dihydric alcohol or a dihydric phenol and n usually has a value of from about 1 to 20. Preferred epoxides include those in which R' is

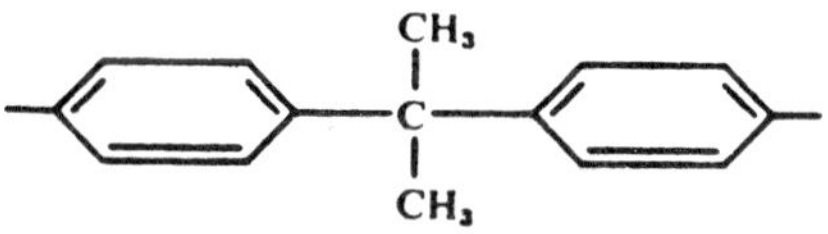

The epoxides may be prepared by condensation of epichlorhydrin in alkaline medium with a dihydric alcohol or a dihydric phenol such as bisphenol A. The epoxides range in property from viscous liquids to low melting solids, depending on molecular weight or degree of condensation. The degree of condensation is indicated by the epoxide equivalent, defined as the grams of resin per one gram equivalent of epoxy. Epoxides are prepared with epoxide equivalents ranging from about 140 to 4000, but those of about 185 to 300 are usually preferred since these are in liquid form at room temperature.

The filler may constitute a mixture of a phosphate and a borate, suitably in the form of ammonium and/or sodium salts. Preferably the filler may constitute a mixture of monobasic ammonium phosphate ($NH_4H_2PO_4$) or dibasic ammonium phosphate $(NH_4)_2HPO_4$ and anhydrous sodium borate ($Na_2B_4O_7$).

The heat-protective material of this process retains its insulating properties when subjected to severe aerodynamic heating and high aerodynamic shear loads. It provides insulation of vessels that must withstand high temperatures and high pressures, insulation of ducts and venturi sections that must contain pressurized gases flowing at high velocities and insulation of structures subjected to high levels of radiant heating and to high temperature combustion products.

FIREPROOF STAMPING MASS

A process described by *A.L. Renkey and H.-P. Engelhardt; U.S. Patent 4,017,316; April 12, 1977; assigned to Magnesital-Feuerfest GmbH, Germany* relates to a fireproof stamping mass for tap spouts and stack flues in blast furnaces consisting of one or several fireproof basic components having at least one means for inducing a bonding of the basic components.

The stamping mass comprises the following composition:

(1) 10 to 60% by weight pyrophyllite ($Al_2O_3 \cdot 4SiO_2 \cdot H_2O$),

(2) 5 to 30% by weight silicon carbide,

(3) 5 to 20% by weight plastic clay ($Al_2O_3 \cdot SiO_2 \cdot xH_2O$),

(4) 5 to 25% by weight quartzite,

(5) 5 to 15% by weight carbonaceous bonding agent having a fusion point of 110° to 200°C (without the components which become gaseous at operating temperature), and

(6) a quantity of water suitable for attaining optimal density.

The components of the mass preferably have the following grain size:

(1) pyrophyllite 0 to 8 mm, preferably 0 to 6 mm,

(2) silicon carbide 0 to 1 mm,

(3) plastic clay maximally 0.5 mm, preferably 0.1 mm,

(4) quartzite as coarse grain, preferably 3 to 8 mm, and

(5) carbonaceous bonding agent maximally 0.5 mm.

Operational experiments have shown that spouts and flues stamped from this mass last for several weeks up to more than a month.

FIRE SCREENING GLAZING PANEL

A process described by *F. Jacquemin, R. Terneu and J.-P. Voiturier; U.S. Patent 3,997,700; December 14, 1976; assigned to Glaverbel-Mecaniver SA, Belgium* relates to a fire screening glazing panel and a method for forming the panel. The panel is composed of at least two structural plies, at least one of which is a vitreous sheet, and a layer in between composed at least in part of a heat convertible barrier-forming material.

The material forms a solid porous or cellular thermally insulating body when subjected to sufficient heat, as during a fire. The barrier-forming material is an hydrated metal salt selected from the group consisting of aluminates, plumbates, stannates, alums, borates, and phosphates. The barrier-forming material does not significantly reduce the transparency of the structural sheets, if they are transparent, even over extended use during normal conditions.

Example 1: A fire screening panel as shown in Figure 11.1 was made comprising two sheets **1, 2** of soda-lime glass 4 mm thick and located between them a heat convertible layer of hydrated aluminum phosphate which, when subjected to heat formed a fire screening barrier. The heat convertible layer **3** had a thickness of 0.5 mm and was applied as follows:

An aqueous solution of 3.5 mols of hydrated aluminum phosphate was obtained by mixing hydrated aluminum chloride ($AlCl_3 \cdot 6H_2O$) and phosphoric acid (H_3PO_4)

in stoichiometric proportions. The solution prepared in this way was applied onto the upper face of the first glass sheet which had previously been laid substantially horizontal. The layer thus applied was then ventilated in a current of warm air in order to dry it. After the heat convertible layer **3** was dried the second glass sheet **2** was applied to the layer.

FIGURE 11.1: PANEL FIRE SCREENING

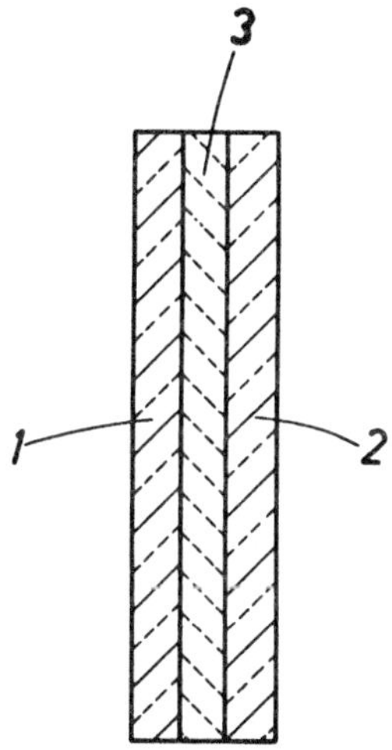

Source: U.S. Patent 3,997,700

It has been found that a fire screening glazing panel made in this way has several advantages. It is very simple to position the panel in order to close an opening and there is no necessity for any other auxiliary installation. Furthermore, the glazing panel constitutes an efficacious fire screening barrier. At the advent of fire the heat convertible layer **3** of hydrated aluminum phosphate becomes converted into a layer of anhydrous aluminum phosphate which is porous and opaque. This anhydrous layer is thicker than the heat convertible layer from which it derives and forms a screen against infrared radiation.

During the course of this conversion, the combined water of the hydrated aluminum phosphate is driven off and this contributes to a limitation in the rise in temperature of the side of the panel removed from the fire during the course of the conversion.

These phenomena allow the temperature of the face of the panel which is not directly subjected to the action of the fire to remain at an acceptable level. Furthermore, it has been found that on the advent of fire the layer of anhydrous aluminum phosphate thus formed constitutes a more refractory layer than does for example, sodium silicate, and also, the aluminum phosphate layer formed in this way binds itself more strongly to the adjacent sheets of glass.

It has been found that the transparency of this glazing panel is conserved satisfactorily during the course of time particularly when it is compared with an otherwise identical panel comprising a layer of hydrated sodium silicate in place of the hydrated aluminum phosphate. Indeed it has been found that fire screening glazing panels comprising an hydrated sodium silicate layer lose their transparency fairly rapidly during the course of time and that blebs appear therein within a few months of manufacture. The glazing panel maintains its transparency for a much longer period.

As a variation, a fire screening glazing panel was made comprising a heat convertible layer **3** of hydrated sodium phosphate, but otherwise identical with the fire screening panel previously described in this example. In another similar panel, the heat convertible layer **3** constituted hydrated potassium phosphate. It was found that these two modified panels had the advantages of being able to form an effective screening barrier to maintain their transparency for a considerable period before advent of fire.

Example 2: A fire screening panel, as shown in Figure 11.1, was made comprising sheets **1** and **2** of soda-lime glass each 4 mm thick and a heat convertible layer **3** of hydrated sodium borate 0.1 mm thick.

This hydrated sodium borate layer was obtained by laying the first sheet **1** of glass horizontal and applying to its upper face a solution of hydrated sodium borate. The solution thus poured on was allowed to spread out so that substantially the whole of the upper face of the sheet was covered with a layer of sodium borate solution of substantially uniform thickness and this was then dried by directing currents of warm air across it in order to drive off the unbound water, i.e., the water of solution.

After drying of the heat convertible layer, the panel was completed by placing the second sheet of glass **2** on there and it was found that a panel made in this way had the same advantages as the three panels described in Example 1. As a modification, the glass sheets **1** and **2** of this panel were chemically tempered before being assembled to form the panel. Such a chemical tempering treatment involves the exchange of potassium ions from a bath of potassium nitrate maintained at a temperature of 470°C in which the sheets were immersed.

The advantages of transparency-maintaining and fire resistance of this modified panel were the same as the previously mentioned panels, but this modification has the particular advantage of strong resistance to breakage due to thermal shock which may be encountered during the first few minutes of a fire.

In another modification for use in a situation where there is only a slight fire risk on one side of the partition, the sheet of glass to be directed to that side was replaced by a sheet of plastics material, and in a third modification the second sheet **2** of the panel was replaced by a laminated pane comprising two sheets of glass bonded together by an intervening layer of polyvinyl butyral. These two modifications were also able to maintain their transparency for a long period until the advent of fire when their heat convertible layers were converted to form an efficacious fire screening barrier.

U.S. PATENT NUMBER INDEX

INVENTOR INDEX

COMPANY INDEX

The company names listed below are given exactly as they appear in the patents, despite name changes, mergers and acquisitions which have, at times, resulted in the revision of a company name.

NOTICE

Nothing contained in this Review shall be construed to constitute a permission or recommendation to practice any invention covered by any patent without a license from the patent owners. Further, neither the author nor the publisher assumes any liability with respect to the use of, or for damages resulting from the use of, any information, apparatus, method or process described in this Review.

THE TEXTILE INDUSTRY
ENVIRONMENTAL CONTROL AND ENERGY CONSERVATION 1978

by S. G. Cooper

Pollution Technology Review No. 42
Energy Technology Review No. 28

Raw material and energy consumption by textile mills and related enterprises is on a steady, albeit slow increase somewhat proportional to the increase of the world's population.

In former years cotton was the principal raw material. However, in 1968 man-made fiber consumption in the U.S. and Europe surpassed cotton and has claimed a progressively larger share of the market. In the mid-seventies about 70% of the fibers used were synthetics. In the USA the processing of wool is down to about 1% of all the fibers utilized.

For these reasons the emphasis in this book is on environmental control and energy conservation that can be carried out in the processing of polyester, nylon, acrylic, olefin and cotton fibers as well as in the finishing of a great variety of fabrics and fabric blends into a multitude of end products.

Many approaches to saving energy and avoiding pollution during the manufacture and finishing of textiles are presented in this volume, which is based upon various technological studies and recent U.S. patents. The bibliography at the end of the volume lists the high important government reports as well as other sources used in the preparation of this book. A partial and condensed table of contents follows here. Chapter headings and important subtitles are given.

ISBN 0-8155-0702-X

385 pages

University of Lowell
Alumni/Lydon Library